Raymond L Pickholtz

Communication Systems Engineering Theory

Wiley Series on Systems Engineering and Analysis
HAROLD CHESTNUT, Editor

Chestnut
Systems Engineering Tools

Wilson and Wilson
Information, Computers, and System Design

Hahn and Shapiro
Statistical Models in Engineering

Chestnut
Systems Engineering Methods

Rudwick
Systems Analysis for Effective Planning: Principles and Cases

Sunde
Communication Systems Engineering Theory

Communication Systems
Engineering Theory

Erling D. Sunde
Bell Telephone Laboratories (Ret.)

John Wiley & Sons, Inc.
New York · London · Sydney · Toronto

Copyright © 1969 by John Wiley & Sons, Inc.

All Rights reserved. No part of this book may be reproduced by any means, nor transmitted, nor translated into a machine language without the written permission of the publisher.

10 9 8 7 6 5 4 3 2 1

Library of Congress Catalog Card Number: 78-78477

SBN 471 83560 9

Printed in the United States of America

Preface

In the last two decades significant advances in communication systems performance and economy have come about by virtue of technological progress in the form of new devices—transistors, solar batteries, masers, information storage and retrieval devices, and space vehicles for satellite repeaters. Modern communication systems incorporating these and other devices may include a number of links that employ different transmission media, such as underground or submarine cable, microwave radio relays, troposcatter paths and communication satellites, each with a different analog or digital modulation method and each with its particular or dominant type of transmission impairment. The sources of impairment may be random noise in the transmission medium, various kinds of amplifier nonlinearity, unavoidable attenuation and phase deviations over the transmission band, random time-varying fluctuations in channel transmittance, and finally, mutual interference between systems.

Design for optimum performance and economy in the face of these various limitations requires judicious systems engineering. To this end it is desirable to have a unified, comprehensive, and realistic transmission–performance theory, which is the objective of this book. An exposition is given of analytical methods by which various kinds of transmission impairment can be adequately assessed, together with tables and charts for finding representative numerical results.

In spite of the comprehensive literature relating to communication systems theory, no exposition of this particular kind has appeared. It differs in objective and certain other respects from present textbooks as discussed further in the Introduction that follows, and may serve as a supplement to these books, which are concerned principally with the theoretical exploration of performance in transmission over highly idealized channels with various methods of statistical and other detection of signals perturbed by Gaussian noise. Valuable as such an idealized mathematical communication theory may be in its own right and in certain applications, it has not had a significant impact on the design and performance of extensive analog and digital communication systems of the kind just mentioned. There are two principal reasons for this, each sufficient in itself. One is the importance of transmission impairments from various kinds of channel imperfections. The other is that elaborate encoding and decoding procedures afford only minor improvements in performance over simple conventional methods that have been long in use.

For conciseness and improved perspective much of the mathematical detail has been omitted in the derivations or covered by reference to pertinent literature. The exposition presumes prior acquaintance with conventional modulation principles and basic analytical methods. These are expounded in several modern textbooks, so that repeated detailed exposition seemed superfluous. However, for convenient reference and as an aid to the exposition the applicable analytical methods are reviewed briefly in the appendices.

As already noted, the primary objective of this book is to provide an integrated and realistic performance theory for engineering applications. As a supplement to present textbooks, it may also serve to acquaint students with some basic communication systems engineering philosophy and with various physical limitations on performance not considered in most textbooks. Hopefully it may also help to bridge certain communication gaps, discussed further in the introduction, that seem to exist between more abstract theorists concerned principally with probabilistic aspects of information and its transmission over highly idealized channels and those engaged in the application of more realistic theory to communication channels with a variety of imperfections already mentioned.

An exposition of this kind appears to be timely for reasons just mentioned, particularly in view of the rapid growth of communication networks within most countries and of intercontinental links by communication satellites and submarine cable. This expansion entails increasing difficulties in achieving and maintaining a high transmission quality as the networks approach global proportions. Hence the greater present need for an integrated exposition of the many factors that determine performance as an aid to the increasing number of participants in the engineering, development, manufacture and administration of communication facilities and in the teaching of communication systems theory.

This effort toward development and exposition of a communication systems engineering theory has been motivated and influenced by my work in this field at Bell Telephone Laboratories during the last two decades. In this work I was aided and influenced by the several basic papers of Dr. S. O. Rice and Dr. W. R. Bennett, who over a period of three decades have developed fundamental analytical methods, and achieved concrete results, with regard to Gaussian noise and other sources of transmission

Preface

impairments of importance in extensive communication systems. I also had the advantage of consultations with them and the benefit of discussing problems with many other associates concerned with communication systems engineering.

Springfield, New Jersey

E. D. SUNDE

Contents

Introduction   1

1 Basic Transmission Concepts and Relations   12

    1.0 General   12
    1.1 Signal Resolution into Elementary Components   13
    1.2 Signal Transmission over Baseband Channels   16
    1.3 Signal Transmission over Bandpass Channels   17
    1.4 Properties of Transmission-Frequency Characteristics   22
    1.5 Idealized Channels with Sharp Complete Cutoffs   29
    1.6 Idealized Channels with Gradual Complete Cutoffs   33
    1.7 Channels with Raised Cosine Amplitude Characteristics   37
    1.8 Channels with Gaussian Amplitude Characteristics   41
    1.9 Impulse Equivalent Baseband Transmittance   43
    1.10 Impulse Equivalent Bandpass Transmittance   44
    1.11 Ideal Frequency-Shift Transmission Characteristics   48
    1.12 Optimum Digital Transmission and Detection Methods   52
    1.13 Suboptimum Digital Transmission and Detection Methods   54
    1.14 Imperfections in Transmission Characteristics   56

2 Basic Modulation Concepts and Objectives   58

    2.0 General   58
    2.1 Characterization of Message Waves   61
    2.2 Statistical Properties of Speech Waves   63
    2.3 Volume Modification by Syllabic Compandors   67
    2.4 Message Modification by Instantaneous Compandors   70
    2.5 Message Modification by Filters   72
    2.6 Sampling and Quantization of Continuous Waves   73
    2.7 Coding of Quantized Samples   75
    2.8 Baseband Signal Waves   77
    2.9 Carrier Signal Waves   78

2.10 Received Carrier Waves — 82
2.11 Carrier Wave Demodulation — 84
2.12 Threshold Performance of PCM and FM Systems — 87
2.13 Frequency-Division Multiplexing — 91
2.14 Time-Division Multiplexing — 94
2.15 Cumulation of Distortion in Repeater Chains — 96
2.16 Choice of Modulation Method — 97
2.17 Basic Performance Criteria for Analog Systems — 99
2.18 Basic Performance Criteria for Digital Systems — 101
2.19 Statistical Coding and Detection — 103

3 Statistical Properties of Gaussian Random Variables — 107

3.0 General — 107
3.1 Analytical Representation of Noisy Sine Waves — 108
3.2 Power Spectra and Autocorrelation Functions — 109
3.3 Amplitude Variations of Gaussian Noise — 111
3.4 Amplitude Variations of Noisy Sine Wave — 112
3.5 Envelope Variations of Noisy Sine Wave — 114
3.6 Mean and Variance of Envelope — 116
3.7 Phase Variations of Noisy Sine Wave — 119
3.8 Variance of Phase Deviations — 119
3.9 Frequency Deviations of Noisy Sine Wave — 121
3.10 Variance of Frequency Deviations — 123
3.11 Power Spectra of Phase and Frequency Variations — 126
3.12 Probability Distribution of Envelope Time Derivative r' — 129
3.13 Probability Distribution of $\theta''(t)$ — 129

4 Transmission Systems Optimization against Gaussian Noise — 133

4.0 General — 133
4.1 Basic Transmission Parameters — 134
4.2 Optimum Combined Equalization without Noise — 139
4.3 Optimum Combined Equalization with Noise — 140
4.4 Baseband Transmission Systems — 142
4.5 Carrier Amplitude Modulation Systems — 143
4.6 Frequency and Phase Modulation Systems — 144
4.7 Analog Pulse Transmission Systems — 148
4.8 Carrier Pulse Transmission Systems — 152
4.9 Comparison with Matched Filter Reception — 157
4.10 Single-Band Digital Transmission Systems — 159
4.11 Multiband Digital Transmission Systems — 161
4.12 Limiting Channel Capacity with Optimum Encoding — 163

Contents

    4.13 Digital Baseband Transmission over Metallic Pairs   164
    4.14 Optimization against Impulse Noise   169

5 Digital Transmission and Modulation Methods   172

    5.0 General   172
    5.1 Transmission and Modulation Methods   174
    5.2 Carrier Demodulation Methods   177
    5.3 Power Spectra of Pulse Trains   181
    5.4 Methods of Timing Wave Extraction   184
    5.5 Timing Wave Deviations   187
    5.6 Derivation of Reference Carrier   189
    5.7 Transmission Limitation by Gaussian Noise   193
    5.8 Transmission Capacity of Baseband Systems   194
    5.9 Duobinary Baseband Transmission   197
    5.10 Symmetrical Sideband Amplitude Modulation   199
    5.11 Orthogonal Carrier and Vestigial Sideband Transmission   202
    5.12 Phase Modulation with Synchronous Detection   203
    5.13 Phase Modulation with Differential Phase Detection   204
    5.14 Binary FM with Dual Filter Detection   206
    5.15 Binary FM with Frequency Discriminator Detection   208
    5.16 Sub-binary Transmission by Reduced Pulse Rate   210
    5.17 Sub-binary Transmission by PPM   211
    5.18 Sub-binary Transmission by Orthogonal Codes   213
    5.19 Biorthogonal Codes with Correlation Detection   215
    5.20 Sub-binary Transmission by Error-Correcting Codes   216
    5.21 Binary Transmission with Feedback Error Correction   218
    5.22 Special Multiplexing Methods   220
    5.23 Impulse Noise   221

6 Attenuation and Phase Distortion in Analog Systems   224

    6.0 General   224
    6.1 Characterization of Transmittance Deviations   225
    6.2 Signal Distortion from Transmittance Deviations   229
    6.3 Sinusoidal Distortion and Echoes in Low-Pass Channels   232
    6.4 Sinusoidal Distortion and Echoes in Bandpass Channels   234
    6.5 Signal Distortion with Fourier Series Transmittance Deviations   236
    6.6 Signal Distortion with Power Series Transmittance Deviations   237
    6.7 Message Distortion in Baseband Systems   238
    6.8 Message Distortion in AM Carrier Systems   240

6.9	Message Distortion in FM and PM Systems	242
6.10	Linear Amplitude Deviations in FM	243
6.11	Quadratic Phase Deviations in FM	244
6.12	Single-Echo Distortion in FM	246
6.13	Fine-Structure Transmittance Deviations in FM	251
6.14	Distortion from Flat Fine-Structure Deviations in FM	252
6.15	Signal-Power Spectra in FM and PM	257
6.16	Distortion from Channel Bandwidth Limitation in FM	260
6.17	Distortion from AM-PM Conversion	262
6.18	Distortion in FM Systems with Power Series Transmittance Deviations	265

7 Attenuation and Phase Distortion in Digital Systems — 266

7.0	General	266
7.1	Carrier Pulse Trains	268
7.2	Slicing Levels and Noise Margins	269
7.3	Evaluation of Transmission Impairments	270
7.4	Synchronous AM and Two-Phase Modulation	272
7.5	Quadrature Carrier AM and Four-Phase Modulation	274
7.6	Vestigial Sideband AM with Synchronous Detection	276
7.7	Phase Modulation with Differential Phase Detection	277
7.8	Basic Relations for In-Phase and Quadrature Envelopes	281
7.9	Linear Amplitude versus Frequency Deviation	282
7.10	Quadratic Delay Distortion	283
7.11	Linear Delay Distortion	287
7.12	Fine-Structure Deviations	291
7.13	Digital Transmission by Pure PM	294
7.14	Distortion from Bandwidth Limitation with Pure PM	297
7.15	Effect of Linear Delay Distortion with Pure PM	299
7.16	Pulse Distortion from Low-Frequency Cutoff	301
7.17	Low-Frequency Cutoff Effect with Dicode Transmission	305

8 Intermodulation Distortion in Nonlinear Channels — 308

8.0	General	308
8.1	Analytical Representation of Nonlinear Transmittance	311
8.2	Steady-State Sine-Wave Transmittance	312
8.3	Output Signal and Distortion Powers	314
8.4	Autocorrelation Functions and Power Spectra	317
8.5	Solution by Transform Method	318
8.6	Square Law Nonlinearity	320
8.7	Cube-Law Nonlinearity—Multiple Sine-Wave Transmission	322
8.8	Cube-Law Nonlinearity—Gaussian Wideband Signals	325

Contents　　　　　　　　　　　　　　　　　　　　　　　　　　　xiii

8.9   Cube-Law Nonlinearity—Gaussian Narrowband Signals　　328
8.10 Power Series Nonlinearity　　330
8.11 Linear Limiter Distortion—Sine-Wave Transmission　　332
8.12 Linear Limiter Distortion—Gaussian Wideband Signals　　337
8.13 Linear Limiter Distortion—Gaussian Narrowband Signals　　339
8.14 AM-PM Conversion Factors　　340
8.15 Intermodulation Distortion by AM-PM Conversion　　342
8.16 Combined Intermodulation with AM-PM Conversion　　346
8.17 Intermodulation Distortion versus Number of Phase Modulated Carriers　　348
8.18 Combination of Intermodulation Distortion and Random Noise　　352
8.19 Output Distortion in Multicarrier FM Systems　　353
8.20 Intelligible Interference in Multicarrier FM Systems　　354
8.21 Nonlinear Distortion in Digital Systems　　356

9 Random Multipath and Troposcatter Transmittances　　358

9.0   General　　358
9.1   Transmittance Function　　360
9.2   Envelope and Phase Distribution　　362
9.3   Time Autocorrelation Functions　　364
9.4   Frequency Autocorrelation Function　　365
9.5   Distribution of Envelope Time Derivative $r'(t)$　　367
9.6   Distribution of Envelope Frequency Derivative $\dot{r}(u)$　　367
9.7   Distribution of Instantaneous Frequency $\varphi'(t)$　　369
9.8   Distribution of Envelope Delay $\dot{\varphi}(u)$　　369
9.9   Distribution of Time Derivative $\varphi''(t)$　　370
9.10 Distribution of Frequency Derivative $\ddot{\varphi}(u)$　　371
9.11 Troposcatter Channels　　372
9.12 Observed Transmission Loss Variation　　373
9.13 Observed Time Autocorrelation of Troposcatter Channels　　374
9.14 Differential Delay Δ of Troposcatter Channels　　374
9.15 Observed Frequency Autocorrelation of Troposcatter Channels　　376
9.16 Measured versus Predicted Intermodulation in FM Systems　　377
9.17 Diversity Transmission Objectives and Methods　　381
9.18 Maximal Ratio Combining　　383
9.19 Equal-Gain Combining　　386
9.20 Pre-detection versus Post-detection Combining　　387
9.21 Selection Diversity　　388
9.22 Intermodulation Distortion with Diversity Transmission　　389

10 Digital Transmission over Random Multipath Channels — 390

- 10.0 General — 390
- 10.1 Errors from Gaussian Noise in Binary FM and PM — 392
- 10.2 Binary AM versus Binary PM and FM — 394
- 10.3 Combined Rayleigh and Slow Log-Normal Fading — 396
- 10.4 Errors in PM from Phase Variations over Signal Interval — 397
- 10.5 Errors in FM from Frequency Variations over Signal Interval — 400
- 10.6 Errors from Frequency-Selective Fading — 401
- 10.7 Combined Error Probabilities — 405
- 10.8 Diversity Transmission for Error Reduction — 407
- 10.9 Error Probabilities with Optimum Diversity Methods — 408
- 10.10 Error Probabilities with Equal-Gain Diversity — 411
- 10.11 Error Probabilities with Selection Diversity — 412
- 10.12 Diversity Improvements with Selective Fading — 413
- 10.13 Error Reduction by Error-Correcting Codes — 415
- 10.14 Combination of Diversity Transmission and Error Correction — 417
- 10.15 Multiband Transmission — 418
- 10.16 Experimental Results — 420
- 10.17 High-Frequency Ionospheric Transmission — 420

11 Intersystem Interference — 422

- 11.0 General — 422
- 11.1 Basic Considerations — 425
- 11.2 Interference between Single-Sideband Systems — 428
- 11.3 Interference between AM Systems — 430
- 11.4 Interference between FM Systems — 433
- 11.5 Interference between FM Systems with Gaussian Modulation — 437
- 11.6 Interference between Baseband Pulse Systems — 441
- 11.7 Interference between Digital PM Systems — 444
- 11.8 Interference between Digital and Analog PM Systems — 448
- 11.9 Coupling Factors of Twisted Pairs — 451
- 11.10 Coupling Factors of Coaxial Lines — 454

Appendix 1: Basic Transform Concepts and Relations — 457

1. Fourier Integral Transforms — 457
2. Laplace Integral Transforms — 459
3. Hilbert Integral Transforms — 460
4. Convolution Integral Transforms — 461
5. Fourier Integral Energy Relations — 463

6. Resolution of Functions into Components	464
7. Series Representation of Band-Limited Functions	465

Appendix 2: Basic Statistical Concepts and Relations — 468

1. Random Variables and Probability Functions	468
2. Averages and Moments of Random Variables	470
3. Sums of Random Variables	472
4. Time Correlation Functions and Power Spectra	475
5. Functions of Random Variables	477
6. Gaussian Probability Functions	479
7. Rayleigh Probability Functions	481
8. The Log-Normal Probability Function	481

Appendix 3: Nonlinear Transformations of Power Spectra — 483

1. Basic Relations	484
2. Power Law Transformations	485
3. Cosine and Sine Transformations	486
4. Ideal Limiter Transformations	487

Appendix 4: Determination of Optimum Filter Characteristics — 489

1. General	489
2. Optimum Receiving Filter	490
3. Matched Filter Detection	491
4. Optimum Smoothing Filter	491
5. Channel Capacity with Multiband Transmission	492

References — 494

Index — 503

Introduction

1 Communication Systems Engineering Theory

Extensive modern communication systems or networks, such as the Bell System in the U.S.A., comprise a variety of transmission facilities. Among them are transcontinental coaxial cable and radio relay routes, intercontinental submarine cable, troposcatter links over inaccessible territory, and satellite links between the communication networks of different continents. Except for satellite and troposcatter links, each such component facility generally includes several hundred repeater sections in tandem, and may span several thousand miles. The channels of these systems are designed to carry various kinds of information, such as speech, telephoto, television, telemetry or digital data. Such information is ordinarily converted into various kinds of signals for transmission over the different facilities. These signals usually consist of amplitude modulated carrier waves for cable systems and frequency modulated waves for the other facilities mentioned earlier. More recently, digital systems have come into use and will no doubt provide many of the communication highways of the future.

In communication systems engineering, the broad objective is design for optimum performance, considering all the factors involved therein, such as economy, transmission quality, reliability, flexibility in providing various services, compatibility with existing systems, time required for development, and several others. Each of these particular considerations may require elaborate study. However, not all entail the use or even comprehension of communication systems theory, ordinarily understood as an application of mathematical and physical disciplines in the engineering and subsequent detailed design. Certain analytical methods to this end constitute communication systems* engineering theory, as defined here.

---

* Internationally, the designation "telecommunication systems" is used to distinguish them from other kinds of communication systems, such as highways, railroads, or even lines of command in organizations.

One of the several problems in communication systems engineering is to provide high-quality transmission. To this end, it is necessary to avoid excessive disturbances from noise inherent in the transmission medium and also noise or interference from other sources. Beside noise it is also necessary to consider disturbances that originate in a variety of unavoidable imperfections in the transfer characteristics of the channels. Among them are various kinds of nonlinearities in the hundreds of amplifiers, reflections due to irregularities in the transmission media, attenuation and phase distortion within the frequency bands of the channels inherent in practicable filters, random fluctuations in received signals in radio links, and interference between various systems.

These channel imperfections and the resultant transmission impairments are usually of greater concern than additive random noise in communication systems engineering. The reason for this is that in virtually all extensive systems the signals must of necessity be much stronger than the noise to facilitate efficient communication, even in binary systems. Transmission impairments in the form of various kinds of distortion due to channel imperfections are proportional to, or, in some other way, related to, the signal strength, and thus give rise to "multiplicative" noise that unlike "additive" noise, cannot be overcome by increasing the signal power. Determination and control of such impairments is a principal concern in systems engineering.

Communication systems theory as presented here is not concerned with the design and detailed analysis of the many kinds of active and passive circuit elements of communication systems—the subject matter of a large body of literature on different kinds of transmission media, electrical transducers and filters, amplifiers, oscillators, frequency converters, modulators, and detectors. The concern is rather with overall transmission performance as related to various system parameters that reflect the properties of such components, together with the various kinds of imperfections mentioned previously. A principal objective in the analysis of performance, as related to these various parameters, is the formulation and specification of various basic requirements that must be imposed in systems design and operation in order to insure satisfactory performance. Specification of such requirements is essential for reliable economical evaluations that precede any decision on development of new systems. Such specifications are also desirable for guidance in the detailed design and development of the systems as well as in their subsequent operation and maintenance.

The foregoing objectives differ from those in recent textbooks and most other literature on communication systems theory. Most of this literature is concerned with the exposition of various basic principles underlying communication systems, with certain probabilistic aspects of information, and with performance in transmission over highly idealized channels incapable of physical realization, in which the only source of transmission impairment is assumed to be additive random noise. The relation of the present exposition to that in most books on communication systems theory will be considered in further detail next.

2 Present Books on Communication Systems Theory

While early communication systems theorists recognized certain statistical properties of information, they were also concerned with certain transmission limitations inherent in the properties of physical transducers serving as communication channels. This complicated matters and precluded neat analytical relations for transmission performance. Shannon, by formulating a purely "Mathematical Theory of Communication," could thereby avoid questions of physical realizability, as regards both the channels and certain postulated encoding and detection methods. This resulted in a simple relation for the channel capacity, as limited by white Gaussian noise, that could in some sense be interpreted as an unattainable upper bound on the performance of physical channels. In his famous paper, Shannon was concerned mostly with the nature of information, with measures of information content and with certain other aspects of communication theory that need not be considered here. Of principal interest in the present context is that his assumption of an idealized mathematical channel has been adopted in most subsequent textbooks and other literature on communication systems theory.

Shannon's formulation of information theory started a flood of publications under such titles as "Information Theory," "Statistical Communication Theory," "Detection of Signals in Noise," and "Modern Communication Theory." In these the emphasis is on various probabilistic aspects of information, on statistical methods of encoding and detecting signals perturbed by additive Gaussian noise, and on ultimate performance that in principle can be realized in transmission over highly idealized mathematical channels.

The existing literature also includes several texts on "Modulation Theory" and "Principles of Communication." These books are concerned mainly with the basic principles underlying various conventional modulation and detection techniques, with their implementation and with optimum performance in transmission over the same overly idealized channels as assumed in the other books just mentioned.

These books are, in a sense, strictly mathematical; they are not concerned with the properties of the channels as electrical transducers, such a concern being the province of network and transmission theory. This separation is a distinct advantage from the standpoint of exposition in academic courses, since no knowledge need be presumed about electrical network theory. But it is also very misleading from the standpoint of application to physical channels that are subject to the laws of electrical network theory. These books thus fail to recognize and deal with the various kinds of channel imperfections mentioned previously.

3 Communication Gap between Communication Theorists

The aforementioned differences in emphasis or approach appear to be responsible for a certain communication gap between theorists at academic institutions and those engaged in communication systems engineering. This is particularly the case in dealing with extensive networks where the signals for acceptable quality must be much stronger than the noise, as even in binary systems. The situation is different in the application of mathematical communication theory to the detection of certain weak signals, often considered in the province of communication theory. One such application is in the detection of radar echoes. Another more recent application is in telemetry from space vehicles. In these cases the noise may even exceed the signals in strength and certain statistical encoding and detection procedures can be used to advantage. The various kinds of channel imperfections mentioned previously can then be disregarded for several reasons, all related to the circumstance that efficient communication is not the primary objective, as in extensive communication networks. In these the situation is radically different in that the channels must be designed to carry simultaneously a variety of different signals that are all amplified by hundreds of common amplifiers in tandem, so that excessive signal distortion or interference between signals can occur owing to the various kinds of channel imperfections mentioned before. Hence the preoccupation of systems engineers with appropriate channel design rather than with elaborate detection procedures that can hardly be used to advantage in extensive communication systems.

There are also differences in mathematical "models," or assumptions in dealing with such matters as error probability in digital systems, that can be traced to the different applications of theory mentioned earlier. In academic work it is ordinarily assumed that the received pulses in the absence of noise are rectangular in shape and are applied to a "matched filter receiver." The latter integrates the signal over the duration of a pulse, with the aid of a narrow band filter, and samples the output at the end of a pulse interval. This method can have advantages in mathematical analysis and seems to originate in certain arrangements for detection of repetitive weak signals. However, it is not the method ordinarily used in efficient communication systems. For one thing, rectangular received pulses would not be used because of excessive channel bandwidth requirements. For another, an at least equally effective method with band-limited pulses is to use sampling at the midpoint of each pulse. The latter method is readily implemented by electronic gates for pulse sampling, even at very high digital speeds.

As indicated earlier, in engineering theory the emphasis is on realistic idealization of channels or other system components. This may entail unruly mathematics and hence the need for approximations in the derivations rather than in the models. Some theoretical work that would seem to imply realism can in fact yield unrealistic and quite misleading results as applied to actual systems. A suitable example is the question of signal distortion due to channel bandwidth restriction, as in FM systems where the carrier sideband spectrum in theory has infinite bandwidth. In such systems, bandpass filters are inserted at the transmitter to confine the spectrum within a prescribed band, and at the receiver to limit noise. Such filters have two separate effects on signal distortion. The first and ordinarily more important effect is that they introduce phase distortion over the pass-band and resultant distortion in the demodulated message wave. It can be reduced by increasing the bandwidth of the filters and by application of phase equalization. The resultant signal distortion depends on residual phase deviations over the pass-band and is difficult to evaluate accurately. The second and ordinarily minor effect of the filters is the distortion that results from elimination of spectral components. Engineering theorists will focus their attention on the first and more important effect, and will go by the semi-empirical Carson bandwidth rule, of some thirty years' standing, for making the distortion due to the second effect negligible.

By contrast, theorists who postulate mathematical filters will concern themselves with the second minor effect, without recognition of the first. A realistic solution entails consideration of both effects and will involve several parameters. This example is representative of others relating to the effect of bandwidth limitation by filters, as encountered in the case of low-frequency cutoff or with such devices as instantaneous compandors that expand the signal spectrum beyond the available channel band.

In the literature dealing with the performance of analog or digital systems in the presence of random noise it is customary and convenient to postulate an idealized channel characteristic with linear phase. While this may be a legitimate approximation in some cases, it may lead to highly misleading results in others, where the performance is determined to a greater extent by small unavoidable departures from the assumed ideal channel characteristic than by additive random noise. The latter situation is encountered in analog video or picture transmission and also in multi-level digital transmission by amplitude, phase or frequency modulation.

As a striking example of illusory performance predictions that may result from ignoring certain channel imperfections, consider multilevel digital transmission by amplitude, phase, or frequency modulation over channels with a high signal-to-noise ratio, such as the conventional voice channels of commercial systems. The maximum practicable number of levels is not, in this case, determined by random noise but principally by distortion caused by irregular attenuation and phase deviations over the channel band that remain even after elaborate equalization. Although these and certain other channel imperfections do not significantly impair voice communication, they impose a severe limitation on the digital transmission capacity. In fact, the maximum capacity so far realized with refined digital transmission methods in conjunction with equalization is less than one-fifth of that predicted from the nominal channel bandwidth and signal-to-noise ratio.

As noted before, to insure realistic performance predictions it is necessary in communication systems engineering theory to include all principal sources of transmission disturbances. This complicates the analysis and virtually precludes exact solutions for the combined transmission distortion. However, for engineering purposes certain conventional approximations suffice, as outlined in the next section.

4 Basic Premises of Communication Systems Engineering Theory

The condition of physical realizability, as this term is used in network theory, is obtained by invoking the condition that the response of a transducer must of necessity be zero prior to the application of an electric force. This leads to the Laplace transform relation between the steady-state sine-wave transmittance as a function of frequency and the impulse response as a function of time. Under this condition, a Hilbert transform relation exists between the amplitude versus the frequency characteristic and the phase characteristic of the channel transmittance. Hence the introduction of this Hilbert relation suffices to insure that the theory applies to physical channels. Although Hilbert transforms are often introduced in textbooks in connection with signal analysis, the use of "analytic signals," they have not been used for the purpose considered here. In this text, Hilbert transforms are introduced in the first chapter to show the relation just mentioned between the amplitude and the phase characteristics. This injection of realism serves to show some of the basic problems and transmission limitations encountered in actual bandlimited channels, such as the filter problems mentioned earlier.

The condition for physical realizability, just mentioned, can be met by a multitude of idealized channel characteristics that may be taken as a point of departure in engineering theory: for example, a raised cosine amplitude versus frequency characteristic. However, it must be recognized that these idealizations cannot be duplicated in actual channels, because of various unavoidable imperfections in manufacture and installation. Hence it is necessary to derive analytical relations for transmission impairments resulting from various kinds of imperfections in order to facilitate specification of tolerable deviations from the prescribed idealized characteristics. The tolerable imperfections of each kind depends on the maximum allowable transmission impairment from all sources, which, in turn, is related to the particular system under consideration. The second required augmentation of mathematical communication theory is thus to provide relations for transmission impairment that are due to each kind of imperfection encountered in actual channels.

In efficient communication systems, the tolerable degradation in transmission quality from each source must be quite small. Moreover, hardly any correlation exists between transmission impairments from each of the several sources. Hence it is legitimate to deal separately with each source in analyses of transmission impairments, and to obtain the resultant signal distortion by random or power addition.

Thus a flexible engineering theory can be provided by augmenting conventional mathematical communication systems theory as just outlined. It should be recognized that such a theory is not concerned with the physical embodiment of any particular type of system, of which there can be an almost endless variety, but rather with certain basic requirements that must be met for a specified performance, regardless of the particular implementation.

5 Organization and Scope of Book

As just outlined, the objective is a comprehensive communication systems engineering theory that recognizes both noise and various other sources of transmission impairment. The present effort toward this end is based on certain realistic idealizations of channels and their imperfections in deriving various theoretical relations for engineering applications. The exposition presumes prior acquaintance with communication systems principles and basic theory, as presented in several modern textbooks and is intended to supplement these. The more important results in such books are reviewed, with emphasis on those aspects that are of prime importance in systems engineering. This serves as a point of departure for more detailed treatment of random noise and various other transmission limitations.

For a particular type of analog or digital communication system, these various sources of transmission impairments are all involved, each to a different degree, depending on the kind of modulation and detection employed. Rather than considering separately each kind of system that might come into consideration, the book is organized into chapters each dealing with a particular kind of channel imperfection and its effect on different systems. With this arrangement, greater flexibility is insured in application to various prospective systems other than the comparatively few now in use. The emphasis in the exposition is thus on basic analytical methods and concrete theoretical relations, rather than on specific illustrative applications in analog or digital systems engineering for voice, telephoto, television, telemetry or data transmission.

For conciseness and better exposition of basic relations and engineering considerations, and thus improved perspective, mathematical details are omitted in most derivations or are covered by reference the pertinent literature. A prominent source of theoretical literature relating to communication systems engineering is the *Bell System Technical Journal*. Other notable sources, particularly of more recent papers, are the *IEEE Proceedings* and the *IEEE Transactions* on "Communication Technology" and "Information Theory." It has been necessary to cite only those papers that are more directly related to the problems under consideration and give further detailed derivations or other additional information. This, by necessity, confines the references to but a small portion of a bibliography in this field, one that by now is so vast that some pertinent contributions may inadvertently have been overlooked. For reasons mentioned previously, much of the communication systems literature in recent years applies to overly idealized or "mathematical channels." Hence the results of many publications may not be directly applicable to systems engineering without scrutiny of the basic premises.

6 Review of Chapter Contents

The following brief review of chapter contents is intended to give further orientation to the reader. To facilitate continuity and ready reference in the exposition, certain basic transform and statistical methods used in communication systems theory are outlined in appendices that are not reviewed here.

Chapter 1 deals with certain basic transmission concepts and relations applicable to bandlimited functions representing signals and to idealized transducers simulating communication channels. A point of novelty is the emphasis given to the minimum phase shift relations between the amplitude and phase characteristics of physical transducers. This injection of realism at an early stage in the presentation is necessary to exhibit some of the basic problems and transmission limitations encountered with actual communication channels and also to resolve the paradox of "response before zero time." Various idealized impulse transmission characteristics are discussed, in order to arrive at the preferable types for actual pulse transmission systems. With baseband or carrier pulses of finite duration as actually used, the desired ideal impulse response can be obtained by simple filter modifications, which facilitate analysis and design for optimum performance.

Chapter 2 reviews various fundamental modulation concepts and is intended to impart a perspective of basic objectives and methods used in systems engineering to approach optimum performance. The basic statistical parameters of information or message waves are discussed, together with the modification of these that may be desirable prior to transmission. Conventional methods of analog and digital baseband and carrier modulation are reviewed, together with various important kinds of transmission impairment, the conventional performance criteria and certain primary considerations that dictate the choice of transmission and modulation method.

Chapter 3 reviews and collects various analytical relations for the statistical properties of Gaussian random variables, published by Rice and others. These relations are used in later chapters in determining the effect of additive random noise in communication channels, in dealing with signals that have properties approaching those of random noise, and in evaluating transmission distortion in random time-varying multipath channels, such as troposcatter paths.

Chapter 4 is devoted to derivations of optimum shapes of transmitting and receiving filters in the presence of additive Gaussian random noise. Analog and digital baseband and carrier modulation systems are considered. It turns out that these optimum shapes conform with the conventional optimum "matched filters" only when the noise has a flat power spectrum and the attenuation in the transmission medium is constant over the transmission band. Moreover, when the noise power or signal power spectra vary with frequency, the optimum shapes differ from those obtained from relations published by Shannon for these cases. The difference is due to a tacit assumption, frequently overlooked, in Shannon's derivation of channel capacity in the presence of other than white noise.

Chapter 5 deals with digital pulse transmission by various conventional baseband and carrier modulation and detection methods, and with certain basic considerations in systems engineering. Among them are the extraction from the received signals of a carrier demodulating wave and of a timing wave for sampling of the demodulated baseband pulse train. This extraction entails certain unavoidable timing deviations or timing jitter, which is a source of transmission impairment. Formulas and tables are given for probability of error due to additive Gaussian noise and comparisons are made of different methods of reducing the error probability in exchange for increased bandwidth. Certain basic questions relating to impulse noise are also discussed.

In summary, the first five chapters present a realistic transmission and modulation theory that permits determination of transmission impairments by random noise in analog and digital systems and optimum design to minimize these impairments. The remaining chapters are devoted to an exposition of analytical methods for determining the several other kinds of transmission impairment.

Chapter 6 presents methods for evaluation of transmission impairment in analog systems arising from various kinds of attenuation and phase deviations over the channel band. Baseband transmission is considered, together with amplitude, phase and frequency modulation. To facilitate determination of the resultant average intermodulation distortion in FM and PM systems, the signal is assumed to have the statistical properties of Gaussian noise. This assumption is ordinarily used as a common basis of comparison of various modulation methods. Charts are presented for evaluation of intermodulation distortion in FM and PM systems due to small signal echoes together with relations applicable to fine structure deviations in the attenuation and phase characteristics.

Chapter 7 deals with evaluation of transmission impairment in digital systems arising from attenuation and phase deviations over the channel band. Curves are presented from which can be determined the maximum degradation in signal-to-noise ratio in baseband and various carrier modulation systems, resulting from linear and quadratic delay distortion. In addition, fine structure deviations in the attenuation and phase characteristics are considered, together with the effects of a low-frequency cutoff.

Chapter 8 is devoted to methods for determining intermodulation distortion due to various kinds of nonlinear amplification. Relations are presented for evaluation of intermodulation distortion resulting from AM-PM conversion, which is encountered in microwave transmitters, and is an important consideration in multicarrier FM satellite systems.

Chapter 9 deals with certain important statistical properties of the transmittances of random time-varying multipath channels, such as troposcatter paths, which at present are the only practicable communication links across certain territories. Relations are also given for the probability distribution of signal-to-noise ratios with various methods of diversity transmission. In addition, a method is presented for approximate evaluation of intermodulation distortion in analog FM troposcatter systems, and comparisons are made with comprehensive experimental data.

Chapter 10 develops relations for determination of error probabilities in digital transmission at various rates over random multipath or troposcatter channels, considering both additive random noise and distortion from frequency-selective fading. Charts are presented from which can be determined the error probability in binary transmission by FM and PM as related to the transmission rate and certain basic parameters of troposcatter links. Relations are given for the error reduction afforded by various methods of diversity transmission and by certain error correcting codes in conjunction with diversity transmission.

Chapter 11 deals with intersystem interference caused by overlaps of the signal spectra of two systems of the same or different kinds, owing to electromagnetic coupling of the transmission paths. Evaluation and control of such interference is a problem of first importance in extensive communication systems employing different transmission media, such as open-wire lines, coaxial cable pairs, radio relays, and communication satellites. Various considerations relating to such interference are reviewed, and basic analytical relations are presented for the more important interference situations and modulation methods.

The several sources of transmission impairments mentioned above reside in the channel, including its repeaters. Apart from the channel, transmission impairments may arise from imperfections in terminal equipment required for encoding and decoding, carrier modulation and demodulation, timing wave provision and other necessary transmission functions. Although these are important they are usually determined more readily and reliably by laboratory tests than by theory. Hence they are not considered here.

1
Basic Transmission Concepts and Relations

1.0 General

A comprehensive communication network in general includes a number of transmission systems that employ different transmission media, such as open-wire lines, aerial, buried, or submarine cable, waveguides, radio waves, or radio beams. In each such system the original message waves undergo certain conversion or modulation processes before they appear as signals at the input to a channel. These modulation processes together with the inverse demodulation processes at the receiver are dealt with in Chapter 2. This chapter is concerned with various basic concepts and relations in the transmission of the signals over a communication channel, which in general is an electric transducer.

In analyzing transmission performance, or in synthesizing channels to comply with specified performance requirements, it is convenient to regard the signal as consisting of certain elementary components, and to determine the channel or transducer response to such components. The received wave can then be formulated in terms of the elementary components present in the signal wave and the transducer response to such components as will be discussed here.

The performance of a channel or transducer for any signal is ordinarily formulated in terms of two basic response functions. One of these is the "transmission-frequency" characteristic, which gives the steady state sine-wave response as a function of the sine-wave frequency. The other is the unit "impulse-transmission" characteristic which specifies the response to a current or electromotive force of very high intensity and infinitesimal duration such that the area under the impulse is unity. The time responses of transmission systems to these two basic time functions are interrelated so that one may be obtained when the other is known.

The impulse characteristic has certain inherent advantages in the formulation of communication theory. This comes about because discrete messages can be represented by samples or impulses and because continuous messages or related signals can be uniquely represented by samples or impulses taken at appropriate finite intervals. The instantaneous amplitude of a received wave can thus be formulated by an infinite series involving the amplitudes of transmitted signal samples and the impulse transmission characteristic of the channel. In essence, modern communication theory consists of an analysis of deviations or errors in the amplitudes of the received samples that are due to random noise and various channel imperfections which are reflected in the impulse characteristic.

Although formulation in terms of the system impulse characteristic has inherent advantages in communication theory, formulation in terms of the related transmission-frequency characteristic is preferable in system design and synthesis to meet specified performance requirements. For one thing, the transmission-frequency characteristics of various existing facilities are known and, for new facilities, can be determined more readily by calculations or measurements than the impulse characteristic. But the more fundamental reason is that the transmission-frequency characteristic of various system components connected in tandem or parallel can readily be combined to obtain the resultant overall transmission characteristic, while this is not the case for the impulse characteristic. It is thus possible to analyze complicated systems with the transmission-frequency characteristic as a basic parameter and to specify requirements that must be imposed on the transmission-frequency characteristic of the systems and its components for a given transmission performance.

The synthesis of signals from elementary components is reviewed here, together with certain basic properties of transmission-frequency and related-impulse characteristics. The transmittances of representative channels can be approximated by certain idealized band-limited transmission-frequency characteristics, which are considered here together with the corresponding impulse characteristics. This approach simplifies the theoretical analysis and facilitates a general formulation, by avoiding the more complex question of detailed network design to realize adequate approximations to the idealized characteristics. The effect of various kinds of departures from the idealized characteristics is dealt with in later chapters.

Signals as considered here at the input to channels are related to the original message waves by various modulation processes outlined in Chapter 2.

1.1 Signal Resolution into Elementary Components

Let $G(t)$ be a function of time designating a signal. It can be represented by the following Fourier integral in terms of the elementary cisoidal time function $e^{i\omega t}$ of complex amplitude distribution $S(i\omega)$, where $\omega = 2\pi f$

$$G(t) = \frac{1}{2\pi} \int_{-\infty}^{\infty} S(i\omega) e^{i\omega t}\, d\omega, \tag{1.1}$$

and, conversely,

$$S(i\omega) = \int_{-\infty}^{\infty} G(t) e^{-i\omega t}\, dt. \tag{1.2}$$

The first integral could be multiplied by a factor c, provided that the second integral is multiplied by c^{-1}, and in some expositions the choice $c = (2\pi)^{1/2}$ is made to give the integral pair a symmetrical appearance. This can also be accomplished by changing the variable of integration in (1.1) to f in place of ω, in which case the factor $1/2\pi$ is replaced by 1.

In the present application of Fourier integrals, $G(t)$ will be a real function, which must vanish prior to a certain time t_0 at which the signal originated, and which is ordinarily chosen as time origin.* The lower limit in (1.2) can then be taken as O, and it is evident that in that case $S(i\omega)$ will in general have both a real and an imaginary component. This complex function, $S(i\omega)$, is usually referred to as the spectrum of the real function $G(t)$, a designation that originated in the application to optics. When ω is taken in radians/second, the spectrum has the inverse dimension. The function $G(t)$ is then a numeric, indicating the amplitude of the voltage or current under consideration, as the case may be.

Let $g(t)$ be another time function and $s(i\omega)$ its spectrum, in which case,

$$g(t) = \frac{1}{2\pi} \int_{-\infty}^{\infty} s(i\omega) e^{i\omega t}\, d\omega, \tag{1.3}$$

$$s(i\omega) = \int_{-\infty}^{\infty} g(t) e^{-i\omega t}\, dt. \tag{1.4}$$

The following relations apply for a time function with the combined spectrum $s(i\omega)\, S(i\omega)$

$$G_0(t) = \frac{1}{2\pi} \int_{-\infty}^{\infty} s(i\omega)\, S(i\omega) e^{i\omega t}\, d\omega, \tag{1.5}$$

$$= \int_{-\infty}^{\infty} g(t - \tau)\, G(\tau)\, d\tau, \tag{1.6}$$

$$= \int_{-\infty}^{\infty} g(\tau)\, G(t - \tau)\, d\tau. \tag{1.7}$$

The two latter integrals are known as convolution integrals for the function $G_0(t)$. In these two relations, the lower limit can be replaced by

---

* This point is discussed further in Appendix 1.

Signal Resolution into Components

O and the upper limit by t, provided the time origin can be so chosen that g and G vanish for negative values of τ and $t - \tau$. In the latter form, with limits O and t, the integrals are basic to the analysis of the transient behaviors of linear networks. In such applications $g(t)$ is the unit impulse response when $G(t)$ is the applied voltage of current. Or $g(t)$ may be taken as the unit step response, if $G(t)$ is replaced by the time derivative $G'(t)$.

Communication systems transmission theory is basically concerned with the transient responses of a transducer, the channel, to message functions $G(t)$. When certain restrictions are imposed on $G(t)$, the function $G_0(t)$ can be represented by an infinite series of spectral components or elementary functions $g(t)$, as outlined below.

Let $G(t)$ have a spectrum confined to the frequency band $-\Omega < \omega < \Omega$, in which case (1.5) becomes

$$G_0(t) = \frac{1}{2\pi} \int_{-\Omega}^{\Omega} s(i\omega) \, S(i\omega) e^{i\omega t} \, d\omega. \tag{1.8}$$

Let, further, the spectrum of $g(t)$ be constant over the above frequency range, and zero outside the range, that is,

$$\begin{aligned} s(i\omega) &= 1 \quad \text{for} \quad -\Omega < \omega < \Omega \\ &= 0 \quad \text{for} \quad -\Omega > \omega > \Omega. \end{aligned} \tag{1.9}$$

In this case, (1.3) yields

$$g(t) = \frac{\Omega}{\pi} \frac{\sin \Omega t}{\Omega t}. \tag{1.10}$$

Relation (1.8) then becomes

$$G_0(t) = G(t) = \frac{1}{2\pi} \int_{-\Omega}^{\Omega} S(i\omega) e^{i\omega t} \, d\omega. \tag{1.11}$$

In view of (1.6),

$$G_0(t) = G(t) = \frac{\Omega}{\pi} \int_{-\infty}^{\infty} G(\tau) \frac{\sin \Omega(t - \tau)}{\Omega(t - \tau)} \, d\tau. \tag{1.12}$$

The function $G(t)$ is here represented by an integral involving the function itself in conjunction with the elementary time function $g(t)$ given by (1.10). This integral can be replaced by the following infinite series [2]

$$G(t) = \sum_{n=-\infty}^{\infty} G\left(\frac{n\pi}{\Omega}\right) \frac{\sin (\Omega t + n\pi)}{\Omega t + n\pi}. \tag{1.13}$$

Here $G(n\pi/\Omega)$ represents samples of $G(t)$ taken at intervals $T = \pi/\Omega = 1/2B$, known as the Nyquist interval [1]. The function $G(t)$ can thus be represented by an infinite sum of elementary composing functions or signals $\sin \Omega t / \Omega t$ centered on the sampling instants. This is known as the sampling theorem for band-limited functions [2].

1.2 Signal Transmission Over Baseband Channels

A channel can in general be regarded as a transducer with a transmission-frequency characteristic

$$T(i\omega) = A(\omega)e^{-i\psi(\omega)}, \quad (1.14)$$

where $A(\omega)$ is the amplitude characteristic and $\psi(\omega)$ is the phase characteristic.

When a signal $G(t)$ with spectrum $S(i\omega)$ is applied to the channel, the received signal is given by

$$H(t) = \frac{1}{2\pi} \int_{-\infty}^{\infty} T(i\omega) S(i\omega) e^{i\omega t} d\omega. \quad (1.15)$$

In the particular case of a rectangular pulse of amplitude a and duration from $t = -\delta/2$ to $t = \delta/2$, the spectrum obtained from (1.2) is

$$S(i\omega) = a\delta \frac{\sin \omega\delta/2}{\omega\delta/2}. \quad (1.16)$$

As $\omega\delta \to O$, the spectrum approaches $a\delta$. If $a \to \infty$, and $\delta \to O$, such that $a\delta = 1$ second, the function $H(t)$ represents the unit impulse-transmission characteristic of the channel. It is convenient to omit the dimensional factor a of one second required to make $H(t)$ a numeric. The impulse characteristic thus obtained with (1.16) in (1.15) is

$$P(t) = \frac{1}{2\pi} \int_{-\infty}^{\infty} T(i\omega) e^{i\omega t} d\omega, \quad (1.17)$$

$$= \frac{1}{\pi} \int_{O}^{\infty} A(\omega) \cos [\omega t - \psi(\omega)] d\omega. \quad (1.18)$$

For an input, $G(t)$, the received wave, $H(t)$, given by (1.15) can alternately be expressed by the convolution integral

$$H(t) = \int_{-\infty}^{\infty} P(t - \tau) G(\tau) d\tau, \quad (1.19)$$

$$= \int_{-\infty}^{\infty} P(\tau) G(t - \tau) d\tau. \quad (1.20)$$

The same remarks apply with respect to the lower and upper limits in the latter integrals as in connection with (1.7).

Signal Transmission Over Bandpass Channels

When the signal, $G(t)$, is sampled at intervals T, the amplitude $W(t)$ of the received wave at the time t from a reference sampling instant $t = O$ is

$$W(t) = \sum_{n=-\infty}^{\infty} G(nT) \, P(t + nT). \tag{1.21}$$

The amplitude $W(t)$ of the received wave will in general differ from that of the transmitted wave unless the frequency band of $G(t)$ is limited to Ω, and $P(t)$ is of the basic form (1.10). As will be seen in Section 1.5, this basic form is possible only with infinite transmission delay in the channel. However, for other functions, $P(t)$, it is possible for the received wave to conform with the transmitted wave at all sampling instants $t = O$, except for a fixed factor. At a sampling instant,

$$W(O) = \sum_{n=-\infty}^{\infty} G(O) \, P(nT), \tag{1.22}$$

and

$$W(O) = P(O) \, G(O), \tag{1.23}$$

provided that

$$P(nT) = O \quad \text{for} \quad n \neq O. \tag{1.24}$$

Relations (1.23) and (1.24) can, in principle, be satisfied or closely approached for a variety of transmission-frequency characteristics and corresponding impulse characteristics, provided that adequate transmission delay is allowed. The correct signal samples can in turn be applied to an ideal flat low-pass filter with linear phase characteristic and bandwidth Ω, to recover the original wave $G(t)$. This is the basis for analog pulse transmission of continuous message waves. In digital pulse transmission the received samples need not conform exactly with the transmitted samples, that is, it is not essential for relation (1.24) to be satisfied, as discussed in Chapter 2.

1.3 Signal Transmission Over Bandpass Channels

The relations in the preceding section apply for any type of communication channel. For bandpass channels, as shown in Fig. 1.1, however, it is convenient, from the standpoint of general analysis as well as numerical evaluations, to choose a reference frequency ω_0 within the transmission band. That is, to employ the translation $\omega = \omega_0 + u$, $d\omega = du$. With the notation

$$\begin{aligned} \mathscr{A}_0(u) &= A(\omega) = A(u + \omega_0), \\ \Psi_0(u) &= \psi(\omega) - \psi(\omega_0) = \psi(\omega) - \psi_0, \end{aligned} \tag{1.25}$$

(1.17) can be written:

$$P(t) = P_0(t) = R_0(t) \cos(\omega_0 t - \psi_0) - Q_0(t) \sin(\omega_0 t - \psi_0), \quad (1.26)$$
$$= \bar{P}_0(t) \cos[\omega_0 t - \psi_0 + \varphi_0(t)], \quad (1.27)$$

where

$$R_0(t) = \bar{P}_0(t) \cos \varphi_0(t) = \frac{1}{\pi} \int_{-\omega_0}^{\infty} \mathscr{A}_0(u) \cos[ut - \Psi_0(u)] \, du, \quad (1.28)$$

$$Q_0(t) = \bar{P}_0(t) \sin \varphi_0(t) = \frac{1}{\pi} \int_{-\omega_0}^{\infty} \mathscr{A}_0(u) \sin[ut - \Psi_0(u)] \, du, \quad (1.29)$$

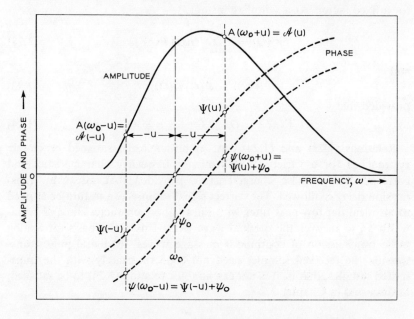

Figure 1.1 Transfer of reference frequency from $\omega = 0$ to $\omega = \omega_0$.

$$\bar{P}_0(t) = [R_0^2(t) + Q_0^2(t)]^{1/2}, \quad (1.30)$$

$$\varphi_0(t) = \tan^{-1}[Q_0(t)/R_0(t)]. \quad (1.31)$$

In application to most bandpass channels, $\mathscr{A}_0(u)$ will virtually vanish for $|u|_{\max} \ll \omega_0$ and the lower limit can then formally be replaced by $-\infty$. Under this condition, $R_0(t)$ and $Q_0(t)$ are Hilbert transforms, discussed further in Appendix 1. These functions are ordinarily called the in-phase and quadrature components of the impulse-transmission characteristic.

The transmission-frequency characteristic may correspondingly be regarded as made up of a component with even symmetry and another component with odd symmetry about ω_0, as indicated in Fig. 1.2. These two components, together with the in-phase and quadrature components, will depend on the choice of ω_0. However, $P_0(t)$, as given by (1.26), and the envelope, as given by (1.30), will remain the same, since a single-impulse characteristic is associated with a given transmission-frequency characteristic.

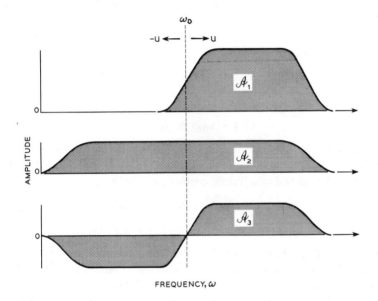

Figure 1.2 Decomposition of amplitude characteristic $\mathscr{A}_1$ asymmetrical with respect to ω_0 into a component $\mathscr{A}_2$ of even symmetry and a component $\mathscr{A}_3$ of odd symmetry about ω_0. When the phase shift is linear, $\mathscr{A}_1 = \mathscr{A}_2 + \mathscr{A}_3$.

With the customary pulse-transmission methods, the reference frequency ω_0 may be identified with a modulating or carrier frequency, which has a special significance when the envelope of a sequence of received pulses is considered. Although, for a single pulse, the envelope is always the same, for a sequence of pulses the resultant envelope of the received pulse train will depend on the in-phase and quadrature components. The reason for this is that one has an even and the other an odd symmetry about the peak amplitude of the envelope for a single pulse, when the phase characteristic is linear.

To compare the transmission performance as the reference or carrier frequency is changed, it is necessary to determine the in-phase and quadrature components for each carrier frequency under consideration. One method is to evaluate integrals (1.28) and (1.29) for each carrier frequency, which may be facilitated by resolving the transmission-frequency characteristic into symmetrical and antisymmetrical components as indicated in Fig. 1.2. This, however, is a rather elaborate procedure which can be avoided with the aid of a simple translation from one reference or carrier frequency to another, as shown below, provided the in-phase and quadrature components or the envelope has been determined for one reference frequency.

Let the reference frequency be changed from ω_0 to ω_1. Relation (1.27) can then be written

$$P(t) = P_1(t) = \bar{P}_0(t) \cdot \cos [\omega_1 t - \varphi_1(t)], \tag{1.32}$$

where

$$\varphi_1(t) = \varphi_0(t) - (\omega_1 - \omega_0)t + (\psi_1 - \psi_0)$$
$$= \varphi_0(t) - \omega_{10} t + \psi_{10}. \tag{1.33}$$

Thus, when the reference frequency is changed by $\omega_{10} = \omega_1 - \omega_0$ and its phase by $\psi_{10} = \psi_1 - \psi_0$, the envelope remains the same and the modified in-phase and quadrature components become

$$R_1(t) = \cos [\varphi_0(t) - \omega_{10} t + \psi_{10}]\bar{P}_0(t), \tag{1.34}$$

$$Q_1(t) = \sin [\varphi_0(t) - \omega_{10} t + \psi_{10}]\bar{P}_0(t). \tag{1.35}$$

To summarize, when the in-phase and quadrature components $R_0(t)$ and $Q_0(t)$ have been determined for any reference frequency ω_0 from (1.28) and (1.29), and in turn $\bar{P}(t)$ and $\varphi_0(t)$ from (1.30) and (1.31), the components $R_1(t)$ and $Q_1(t)$ for another reference or carrier frequency can readily be determined with the aid of (1.34) and (1.35). In the particular case where the amplitude characteristic has even symmetry and the phase characteristic has odd symmetry about ω_0, the quadrature component disappears and the relations simplify to

$$R_1(t) = \cos [-\omega_{10} t + \psi_{10}]\bar{P}_0(t), \tag{1.36}$$

$$Q_1(t) = \sin [-\omega_{10} t + \psi_{10}]\bar{P}_0(t). \tag{1.37}$$

Let the transmitted signal have the bandpass spectrum $S_0(u)e^{-i\varphi_0(u)}$. From the above relations it follows that relation (1.15) for baseband transmission is replaced by

$$H(t) = H_0(t) = X_0(t) \cos \omega_0 t - Y_0(t) \sin \omega_0 t$$
$$= \bar{Z}_0(t) \cos [\omega_0 t + \theta_0(t)], \tag{1.38}$$

where now (1.28) through (1.31) are replaced by

$$X_0(t) = \frac{1}{\pi} \int_{-\omega_0}^{\infty} S_0(u)\mathscr{A}_0(u) \cos [ut - \Psi_0(u) - \varphi_0(u)] \, du, \quad (1.39)$$

$$Y_0(t) = \frac{1}{\pi} \int_{-\omega_0}^{\infty} S_0(u)\mathscr{A}_0(u) \sin [ut - \Psi_0(u) - \varphi_0(u)] \, du, \quad (1.40)$$

$$\bar{Z}_0(t) = [X_0^2(t) + Y_0^2(t)]^{1/2}, \quad (1.41)$$

$$\theta_0(t) = \tan^{-1} Y_0(t)/X_0(t). \quad (1.42)$$

Relation (1.38) can also be expressed as the real part of the complex function,

$$Z_0(t) = [X_0(t) + iY_0(t)]e^{i\omega_0 t}, \quad (1.43)$$

where now

$$X_0(t) + iY_0(t) = \frac{1}{\pi} \int_{-\omega_0}^{\infty} S_0(u)\mathscr{A}_0(u) \exp i[ut - \Psi_0(u) - \varphi_0(u)] \, du. \quad (1.44)$$

The function $X_0(t)$, obtained from the latter relation, conforms with (1.39), and $iY_0(t)$ with (1.40). A signal of the form (1.43) is known as an analytic or complex signal [3] and the notation $Y_0(t) = \hat{X}_0(t)$ is often used to indicate that $Y_0(t)$ is a Hilbert transform of $X_0(t)$, as is the case in view of relations (1.39) and (1.40), provided that $S_0(u)\mathscr{A}_0(u)$ vanishes for $|u|_{\max} \ll \omega_0$. On this premise, the following alternate Hilbert transform relation applies:

$$Y_0(t) = \hat{X}_0(t) = \frac{1}{\pi} \int_{-\infty}^{\infty} \frac{X_0(\tau)}{t - \tau} \, d\tau, \quad (1.45)$$

where the principal value of the integral is to be used. (See Appendix 1 and Section 1.4.)

The use of complex signals (1.43) in place of real signals (1.38) may afford some advantage with respect to symbolism and manipulation, but not as regards numerical evaluations in actual applications.

If the shape $G_0(t)$ of a transmitted wave is specified, rather than its spectrum, as above, relations (1.19) through (1.21) apply, provided that expression (1.26) for $P_0(t)$ is used. Thus (1.20) is replaced by

$$H_0(t) = \cos (\omega_0 t - \psi_0) \int_{-\infty}^{\infty} R_0(\tau) G_0(t - \tau) \, d\tau$$

$$- \sin (\omega_0 t - \psi_0) \int_{-\infty}^{\infty} Q_0(\tau) G_0(t - \tau) \, d\tau. \quad (1.46)$$

Similarly, (1.21) is replaced by

$$W_0(t) = \cos(\omega_0 t + \psi_0) \sum_{n=-\infty}^{\infty} G_0(nT) R_0(t + nT)$$

$$- \sin(\omega_0 t - \psi_0) \sum_{n=-\infty}^{\infty} G_0(nT) Q_0(t + nT). \quad (1.47)$$

The various preceding relations for signal transmission over baseband and bandpass channels involve the transmission-frequency characteristic of the channels. Certain relations apply between the amplitude and phase characteristics of the transmission-frequency characteristic that are important in application to physical channels and will be considered next.

1.4 Properties of Transmission-Frequency Characteristics

As noted before, a basic parameter of transmission systems is the transmission-frequency characteristic

$$T(i\omega) = A(\omega)e^{-i\psi(\omega)}, \quad (1.48)$$

in which $\omega = 2\pi f$ is the radian frequency, $A(\omega)$ is the amplitude, and $\psi(\omega)$ the phase characteristic. The transmission-frequency characteristic may designate the ratio of received voltage to transmitted current of received current to transmitted voltage, of received to transmitted current, or of received to transmitted voltage. The two latter ratios are not the same, except for symmetrical networks with impedance matching at both ends. For symmetrical structures having appreciable attenuation, such as transmission lines between repeaters, the ratios are virtually the same with impedance matching at the receiving end. In the following, $T(i\omega)$ will designate any of the above ratios, as the case may be.

When a number of networks are connected in series, as is usually the case in transmission systems, the resultant transmission characteristic is

$$T(i\omega) = T_1(i\omega) T_2(i\omega) \cdots T_n(i\omega)$$
$$= (A_1 A_2 \cdots A_n) e^{-(\psi_1 + \psi_2 + \cdots + \psi_n)}, \quad (1.49)$$

where $T_1, T_2, \ldots, T_n$ are the transmission characteristics of the individual networks with the same impedance terminations as encountered in the series arrangement, that is, as measured in place or with equivalent terminations.

The phase characteristic ψ can in general be regarded as the sum of three components. The first is the minimum phase-shift component, ψ^0, which has a definite relation to the amplitude characteristic of the system, and is of particular interest in connection with phase distortion with different types of amplitude characteristics. The second is a linear component, $\omega \tau_d$, which represents a constant transmission delay, τ_d for all frequencies, as in the case of an ideal delay network. Ladder-type structures and transmission lines have phase characteristics which can be represented by the above two components. The third component can be represented by a lattice structure with constant amplitude characteristic but varying phase. Such a network component may be present in a transmission system or may be inserted intentionally for phase equalization, that is, to supplement the first component above so as to secure a linear phase characteristic without altering the amplitude characteristic of the system.

The following discussion is concerned with the relationship of the first component to the amplitude characteristic of the system, or conversely.

The natural logarithm of the transmission-frequency characteristic given by (1.48) is

$$\ln T(i\omega) = \ln A(\omega) - i\psi(\omega). \tag{1.50}$$

The component $\ln A(\omega)$ is referred to as the attenuation characteristic, and when expressed in decibels equals $20 \log_{10} A(\omega) = 8.69 \ln A(\omega)$.

The following relations, also Hilbert transforms, exist between the attenuation and phase characteristics of minimum phase shift systems or system components: [4], [5]

$$\ln A(\omega) = -\frac{1}{\pi} \int_{-\infty}^{\infty} \frac{\psi^0(u)}{\omega - u} du = \frac{2}{\pi} \int_0^\infty \frac{u\psi^0(u)}{u^2 - \omega^2} du, \tag{1.51}$$

and

$$\psi^0(\omega) = \frac{1}{\pi} \int_{-\infty}^{\infty} \frac{\ln A(u)}{\omega - u} du = -\frac{2}{\pi} \int_0^\infty \frac{\omega \ln A(u)}{u^2 - \omega^2} du. \tag{1.52}$$

In the evaluation of these integrals, the principal values are to be used, that is, results of the form $\ln(-u)$ are to be taken as $\ln|-u|$ rather than $\ln|u| + i\pi$.

As an example, consider an attenuation characteristic as shown in Fig. 1.3, with $A(\omega) = A_0$ between $\omega = 0$ and ω_c and $A(\omega) = A_1$ between $\omega = \omega_c$ and ∞. Equation (1.52) then becomes

$$\psi^0(\omega) = -\frac{2\omega}{\pi} \left(\ln A_0 \int_0^{\omega_c} \frac{du}{u^2 - \omega^2} + \ln A_1 \int_{\omega_c}^\infty \frac{du}{u^2 - \omega^2} \right)$$

$$= \frac{1}{\pi} \ln(A_0/A_1) \ln \left| \frac{\omega_c + \omega}{\omega_c - \omega} \right|. \tag{1.53}$$

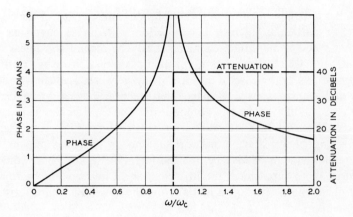

Figure 1.3 Low-pass transmission frequency characteristic with sharp cutoff.

In Fig. 1.3 is shown the phase characteristic for $A_0/A_1 = 100$, corresponding to a 40 db cutoff at $\omega = \omega_c$.

In Fig. 1.4 the attenuation and phase characteristics are shown as a function of ω/ω_c for $\omega < \omega_c$ and as a function of the inverse ratio ω_c/ω for $\omega > \omega_c$. It will be noticed that for the above case the phase characteristic is infinite for $\omega/\omega_c = 1$ and has even symmetry about this point, while the attenuation characteristic has odd symmetry with respect to the

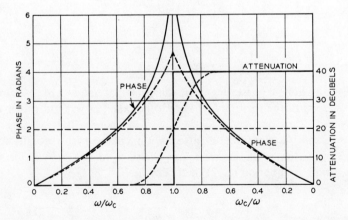

Figure 1.4 Solid curves same as in Fig. 1.3 but with inverse scale for $\omega/\omega_c > 1$. Dashed curves illustrate modification in phase characteristic with gradual cutoff in attenuation (not computed).

midpoint of the amplitude discontinuity. As noted previously, the phase characteristic can be made linear by phase equalization. When the phase characteristic becomes infinite at $\omega = \omega_c$, this will entail a phase characteristic with infinite slope and thus infinite transmission delay after equalization.

The phase characteristic may be modified by a gradual cutoff in the attenuation characteristic, as illustrated in the figure. It is possible to shape the attenuation characteristic to obtain a linear phase characteristic in the transmission band, that is, between $\omega/\omega_c = 0$ and 1. Since transmission systems with a linear phase characteristic in this range are of particular importance in pulse transmission, this case will be considered further.

It will be assumed that the phase characteristic has even symmetry when expressed in the scales of Fig. 1.4, in which case, the phase characteristic as shown by the solid lines in Fig. 1.5 is given by

$$\psi^0(\omega) = \omega\tau, \qquad \frac{\omega}{\omega_c} < 1,$$

$$= \frac{\omega_c^2 \tau}{\omega}, \qquad \frac{\omega}{\omega_c} > 1.$$

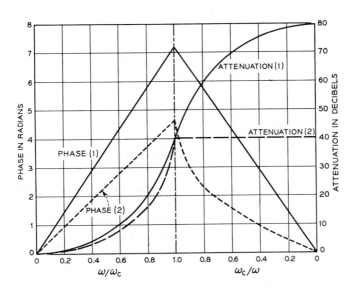

Figure 1.5 Low-pass transmission frequency characteristics with natural linear phase shift for $\omega/\omega_c < 1$.

With these expressions in (1.51) the attenuation characteristic becomes:

$$\ln A(\omega) = \frac{2\omega_c \tau}{\pi}\left[1 + \frac{1}{2}\left(\frac{\omega_c}{\omega} - \frac{\omega}{\omega_c}\right)\ln\frac{1 + \omega/\omega_c}{1 - \omega/\omega_c}\right]. \quad (1.54)$$

For $\omega = O$, the latter expression approaches the limit $\ln A(O) = 4\omega_c\tau/\pi$, so that

$$\ln \frac{A(\omega)}{A(O)} = -\frac{2\omega_c\tau}{\pi}\left[1 + \frac{1}{2}\left(\frac{\omega}{\omega_c} - \frac{\omega_c}{\omega}\right)\ln\frac{1 + \omega/\omega_c}{1 - \omega/\omega_c}\right], \quad (1.55)$$

which is the attenuation characteristic shown in Fig. 1.5.

Other attenuation characteristics with a linear phase characteristic between $\omega/\omega_c = O$ and 1 are possible with other types of variations in the attenuation or phase characteristic for $\omega/\omega_c > 1$ than assumed earlier. For example, the attenuation characteristics may be assumed constant for $\omega/\omega_c > 1$, in which case, the attenuation characteristic will be somewhat different for $\omega/\omega_c < 1$, and the phase characteristic different for $\omega/\omega_c > 1$, as illustrated in Fig. 1.5. (The solution for the latter case is given in Reference 4.) It will be noticed that there is a comparatively minor difference between the attenuation characteristics for $\omega/\omega_c < 1$ in the foregoing cases, so that the attenuation characteristic for $\omega/\omega_c > 1$ has a relatively minor effect, provided that there is no discontinuity near $\omega/\omega_c = 1$ The transmission loss characteristics shown in Fig. 1.5 represent a close approximation to the type of characteristic employed in pulse transmission systems, as will be shown later.

The above examples illustrate that with an appropriate gradual and incomplete cutoff, the phase characteristic can be made linear with finite slope over a portion $\omega < \omega_c$ of the total band. This portion of the band is ordinarily referred to as the "transmission band" of the channel. Although the remainder of the channel can, for practical purposes, be disregarded because of the high attenuation, the shape of this attenuation has a significant effect on the phase characteristic within the transmission band.

In the above examples, low-pass characteristics were assumed. For high-pass characteristics, the algebraic sign of the phase is reversed with respect to the amplitude characteristic as indicated in Fig. 1.6, which also illustrates relationships for bandpass characteristics. The bandpass characteristics are obtained by connecting low-pass and high-pass networks in tandem. The resultant attenuation and phase characteristics are obtained by adding the low- and high-pass attenuation and phase characteristics, as illustrated in the figure. In the second case, shown in the figure, the bandpass characteristic is assumed to have a linear phase characteristic in the transmission band, in which case the attenuation characteristic will not be symmetrical about the midband frequency, unless the latter is high

Transmission-Frequency Characteristics

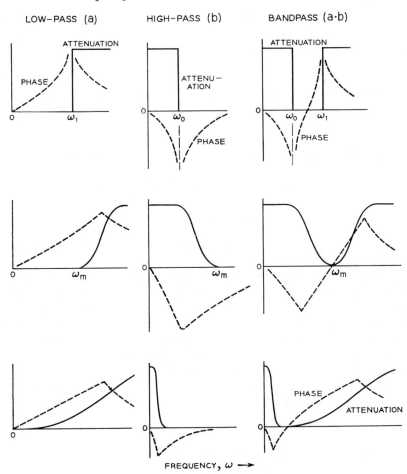

Figure 1.6 Attenuation and phase shift for various types of transmission frequency characteristics.

in relation to the bandwidth. The third case illustrates the type of bandpass characteristic encountered in wire systems with a low-frequency cutoff. There will then be phase distortion at the low end of the band, since it is not feasible with a fairly sharp low-frequency cutoff to obtain a linear phase characteristic in the transmission band. This phase distortion may be a more important source of transmission impairment than the elimination of the low-frequency components.

If the amplitude or attenuation characteristic of a transmission system is modified, it will be accompanied by a modification in the phase characteristic. Of basic importance are cosine modifications in the attenuation and amplitude characteristics. Let the modified amplitude characteristic be of the form

$$A(\omega) = A_0(\omega)e^{a \cos \omega \tau}, \tag{1.56}$$

where $A_0(\omega)$ is the original amplitude characteristic. The modified attenuation characteristic is then

$$\ln A(\omega) = \ln A_0(\omega) + a \cos \omega \tau. \tag{1.57}$$

In accordance with (1.52) the modified phase characteristic becomes

$$\psi^0(\omega) = \frac{1}{\pi}\int_{-\infty}^{\infty} \frac{\ln A_0(u)}{\omega - u}\,du + \frac{a}{\pi}\int_{-\infty}^{\infty} \frac{\cos u\tau}{\omega - \tau}\,du,$$

$$= \psi_0(\omega) + a \sin \omega \tau, \tag{1.58}$$

where $\psi_0(\omega)$ is the phase characteristic of the original amplitude characteristic $A_0(\omega)$.

Thus for any cosine modification in the attenuation characteristic, there is a corresponding sine modification in the phase characteristic; and for any sine modification in the phase characteristic, a corresponding cosine modification in the attenuation characteristic. In general any modification in the attenuation characteristic may be represented by a Fourier cosine series, in which case the modification in the phase characteristic will be the corresponding Fourier sine series.

With a cosine modification in the amplitude rather than in the attenuation characteristic,

$$A(\omega) = A_0(\omega)[1 + a \cos \omega \tau], \tag{1.59}$$

and the corresponding phase characteristic becomes

$$\psi^0(\omega) = \frac{1}{\pi}\int_{-\infty}^{\infty} \frac{A_0(u)[1 + a \cos u\tau]}{\omega - u}\,du,$$

$$= \psi_0(\omega) + 2 \tan^{-1} \frac{r \sin \omega \tau}{1 + r \cos \omega \tau},$$

$$= \psi_0(\omega) + 2\left[r \sin \omega \tau + \frac{r^2}{2} \sin 2\omega \tau + \frac{r^3}{3} \sin 3\omega \tau + \cdots\right] \tag{1.60}$$

where $r = (1/a)[1 \mp \sqrt{1 - a^2}]$, and the minus sign is to be used.

Thus a cosine modification in the amplitude characteristic is accompanied by an infinite series of sine deviations in the phase characteristic. For sufficiently small values of a, $r \simeq a/2$ and (1.60) reduces to (1.58).

From the various previous relations, it appears that any idealized amplitude characteristics with complete cutoffs will be accompanied by a phase characteristic with infinite slope. The transmission delay through such channels would thus be infinite and they are not physically realizable except as limiting cases. Nevertheless they afford convenient approximations to actual amplitude versus frequency characteristics and for this reason are ordinarily postulated in communication theory in determining the associated impulse characteristics. A number of these idealized transmission-frequency and associated impulse-transmission characteristics have been dealt with elsewhere by methods outlined previously [6]. The more important idealized characteristics will be considered in the sections that follow.

1.5 Idealized Channels With Sharp Complete Cutoffs

In pulse-transmission theory, particularly in dealing with transmission capacity of idealized transmission systems, an ideal low-pass transmission-frequency characteristic is ordinarily assumed, with constant amplitude and delay in the transmission band, together with an abrupt cutoff at the top frequency and zero amplitude beyond, as shown in Fig. 1.7. As is evident from Fig. 1.3, this type of characteristic is an abstraction which cannot be physically realized, since it will have phase distortion and infinite transmission delay. It can, however, be approached with sufficiently elaborate phase equalization.

For the above type of characteristic $A(\omega) = 1$ between $\omega = O$ and Ω, while $\psi(\omega) = \omega\tau_0$, where τ_0 is the transmission delay. With these values in (1.18):

$$P(t) = \frac{\Omega}{\pi} \frac{\sin \Omega t_0}{\Omega t_0}, \qquad (1.61)$$

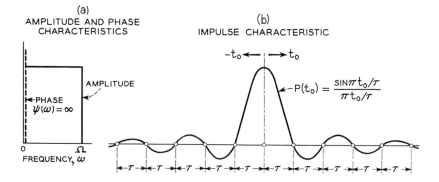

Figure 1.7 Idealized low-pass characteristic with sharp complete cutoff.

where $t_0 = t - \tau_0$ is the time referred to the peak amplitude of the received pulse.

The resultant impulse transmission characteristic is shown in Fig. 1.7, with the factor Ω/π omitted. The peak amplitude is attained after an infinite time, since the above type of characteristic can be realized only with $\tau_0 \to \infty$. The impulse characteristic is zero when $\Omega t_0 = \pm n\pi$, or $t_0 = \pm T, \pm 2T, \pm 3T, \ldots$, where

$$T = \frac{1}{2B}. \tag{1.62}$$

Impulses can thus be transmitted at the latter intervals without mutual interference between the peaks of the received pulses. This is a basic theorem underlying the determination of the transmission capacity of idealized channels.

It will be noted that $P(t)$ is of the same form as the elementary signal function $g(t)$ considered in Section 1.1. There the resolution of a signal into elementary components $g(t)$ was a purely mathematical operation that involved no physical considerations. By contrast, the determination of the impulse characteristic of a transducer or channel, as considered here, entails consideration of the phase characteristic and resultant transmission delay. The distinction is not ordinarily noted in purely mathematical treatises in communication or information theory, and this has been the source of some confusion in applications to physical channels.

For an idealized bandpass channel with cutoffs at ω_a and ω_b it follows from (1.26) through (1.29) with $\psi_0(u) = u\tau_0$ that the impulse characteristic with respect to the midband frequency ω_0 is

$$P_0(t) = 2\cos[\omega_0 t_0 - \psi_0]\bar{P}_0(t), \tag{1.63}$$

where $\bar{P}_0(t)$ is given by (1.30). Here $\psi_0 = \Psi_0 - \omega_0 \tau_0$ is the phase intercept at zero frequency. For the transmission characteristic to be ideal in the sense that the peak pulse amplitude occurs when $t_0 = t - \tau_0 = 0$, it is necessary that $\psi_0 = \pm n\pi$, where n is an integer. The phase intercept is of no consequence if the bandwidth is small in relation to the midband frequency. There will then be a large number of cycles of the modulating frequency within the envelope $\bar{P}_0(t)$, and the latter can be recovered by envelope detection regardless of the phase of the modulating frequency.

With $\psi_0 = \pm n\pi$,

$$P_0(t) = \frac{2\Omega}{\pi} \cos \omega_0 t_0 \frac{\sin \Omega t_0}{\Omega t_0}, \tag{1.64}$$

$$= \frac{\omega_b}{\pi} \frac{\sin \omega_b t_0}{\omega_b t_0} - \frac{\omega_a}{\pi} \frac{\sin \omega_a t_0}{\omega_a t_0}, \tag{1.65}$$

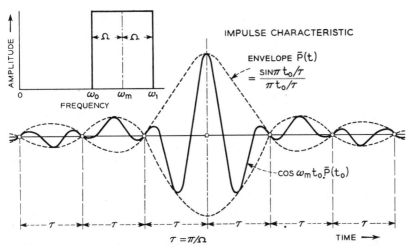
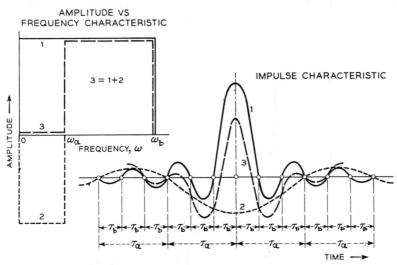

Figure 1.8 Idealized band-pass characteristics and corresponding impulse transmission characteristics.

where $\omega_0 = (\omega_a + \omega_b)/2$ is the midband frequency and $\Omega = (\omega_b - \omega_a)/2$ is the semi-bandwidth.

The shape of the impulse characteristic as given by (1.64) is illustrated in the upper half of Fig. 1.8. Alternatively the impulse characteristic may be regarded as made up of two components in accordance with (1.65). The first component corresponds to a low-pass characteristic of bandwidth ω_b, the second component to a negative low-pass characteristic of bandwidth ω_a, as indicated in the lower part of the Fig. 1.8.

The factor $\sin \Omega t_0 / \Omega t_0$ in (1.64) is zero at the same intervals as for a low-pass characteristic of bandwidth Ω, as shown in Fig. 1.8, so that pulses may be transmitted at the same rate without mutual interference between pulse peaks. The bandwidth in the present case, however, is $2\Omega = \omega_b - \omega_a$, so that for the same bandwidth the pulse transmission rate is half as great as for a low-pass characteristic.

An exception to this is the particular case when $\omega_b = 2\omega_a$, so that the total bandwidth is ω_0. The factor $\sin \omega_a t_0 / \omega_a t_0$ in (1.65) is then zero at intervals $1/2f_b = 1/4f_a$, while the factor $\sin \omega_b t_0 / \omega_b t_0$ is zero at intervals $1/2f_b = 1/4f_a$, as shown in Fig. 1.9. Pulses may accordingly, in principle, be transmitted without mutual interference at the same rate as for a low-pass characteristic of bandwidth ω_a, or at the same rate as with single sideband transmission over a bandpass system of bandwidth ω_a. More generally,

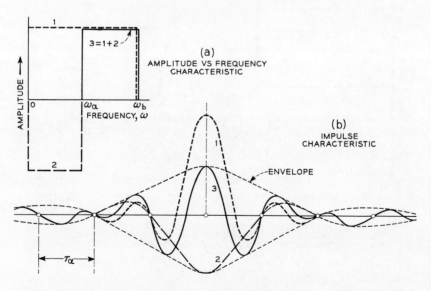

Figure 1.9 Special case of idealized band-pass characteristic in which $\omega_b = 2\omega_a$. Resultant impulse characteristic is zero at intervals $T_a = \pi/\omega_a$.

pulses can in principle be transmitted without mutual interference between pulse peaks at the same rate as for a low-pass characteristic of bandwidth $\omega_b - \omega_a = 2\Omega$ if ω_a is a multiple of $\omega_b - \omega_a$. It should be noted, however, that this pulse transmission rate cannot actually be realized since the phase characteristic will have infinite slope, so that the transmission delay will be infinite. In addition, the zero frequency phase intercept ψ_0 must be $\pm n\pi$, a condition which cannot be attained or remain stable in view of the infinite slope of the phase characteristic.

With the envelope given by the factor $\sin \Omega t_0 / \Omega t_0$ in (1.64), the in-phase and quadrature components for any reference frequency can be determined with the aid of (1.36) and (1.37). If the lower band edge is selected, that is, $\omega_0 = -\Omega$. With a linear phase characteristic $\psi_0 = \omega \tau_0$, so that $-\omega_{10} t + \psi_{10} = \Omega t_0$. The in-phase and quadrature components are accordingly obtained by multiplying the envelope by $\cos \Omega t_0$ and $\sin \Omega t_0$, respectively.

As an alternate method, the two components can be obtained from (1.26) through (1.29) with $-\omega_0 = O$. This yields

$$P_0(t) = \cos \omega_0 t_0 R_0(t) - \sin \omega_0 t_0 Q_0(t), \tag{1.66}$$

with

$$R_0(t) = \frac{1}{\pi} \int_0^{2\Omega} \cos ut_0 \, du$$

$$= \frac{2\Omega}{\pi} \frac{\sin 2\Omega t_0}{2\Omega t_0} = \frac{2\Omega}{\pi} \frac{\sin \Omega t_0}{\Omega t_0} \cos \Omega t_0, \tag{1.67}$$

$$Q_0(t) = \frac{1}{\pi} \int_0^{2\Omega} \sin ut_0 \, du$$

$$= \frac{2\Omega}{\pi} \frac{1 - \cos 2\Omega t_0}{2\Omega t_0} = \frac{2\Omega}{\pi} \frac{\sin \Omega t_0}{\Omega t_0} \sin \Omega t_0, \tag{1.68}$$

where 2Ω is the channel bandwidth. It will be noted that $R_0(t)$ and $Q_0(t)$ are obtained by multiplying the envelope by $\cos \Omega t_0$ and $\sin \Omega t_0$ in accordance with (1.36) and (1.37).

1.6 Idealized Channels With Gradual Complete Cutoffs

The idealized transmission characteristics discussed previously are of principal interest in that they indicate the physical limitations on pulse-transmission rates for a given bandwidth. Even if these impulse characteristics could be realized without undue difficulties from the standpoint of phase equalization, they would be impracticable in most applications. Their oscillatory nature would entail the use of discrete pulse positions and precise synchronized sampling at fixed intervals and would preclude certain methods of pulse modulation and detection. Moreover, because of pulse overlaps the peak amplitude of a pulse train will be much higher than the average amplitude, even in binary systems. Hence severe requirements may be imposed on channel linearity if excessive non-linear distortion is to be avoided.

The nonlinearity in the phase characteristic as well as the oscillations in the impulse characteristic and the peak pulse train amplitude can be reduced with a gradual rather than a sharp cutoff, as illustrated in Fig. 1.10. Following Nyquist [1], it is assumed that an ideal characteristic with a sharp cutoff is supplemented by an amplitude characteristic α_x which has odd symmetry about the cutoff frequency Ω, that is, $\mathscr{A}_x(-u) = -\mathscr{A}_x(u)$.

If the latter component alone is considered, and a linear phase characteristic assumed, it follows from (1.27) through (1.29) with $\Omega = \omega_0$ that

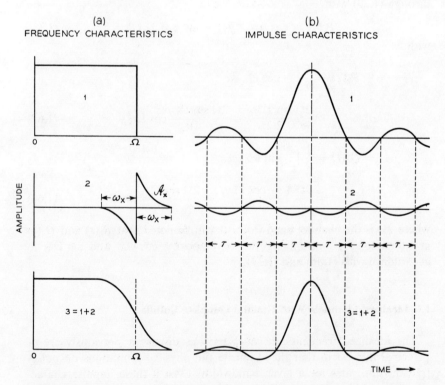

Figure 1.10 Idealized transmission characteristic with gradual cutoff, 3, obtained by superposition of characteristic with sharp cutoff, 1, and characteristic, 2, with odd symmetry about Ω. Linear phase shift assumed.

the effect of this component on the pulse-transmission characteristic is given by

$$P_1(t) = -Q_1(t) \sin \Omega t_0, \tag{1.69}$$

where $t_0 = t - \tau_0$ and

$$Q_1(t) = \frac{2}{\pi} \int_0^{\omega_x} \mathscr{A}_x(u) \sin u t_0 \, du. \tag{1.70}$$

The function $P_1(t)$ will be zero at the same points as the original pulse transmission characteristic with a sharp cutoff at Ω and under certain conditions also at other points. It will modify the original impulse characteristic by reducing the oscillatory tail, as illustrated in Fig. 1.10, but the zero points remain unchanged.

With the above modification, the resultant impulse characteristic obtained by superposition of (1.61) and (1.69) becomes

$$P(t) = \frac{1}{\pi} \sin \Omega t_0 \left(\frac{1}{t_0} - 2 \int_0^{\omega_x} \mathscr{A}_x(u) \sin u t_0 \, du \right)$$

$$= \frac{1}{\pi} \sin \Omega t_0 F(t), \tag{1.71}$$

where

$$F(t) = \left(\frac{1}{t_0} - 2 \int_0^{\omega_x} \mathscr{A}_x(u) \sin u t_0 \, du \right). \tag{1.72}$$

In the following, an expression for $F(t)$ is given for the case when the band edge is modified by the supplementary characteristic of the form

$$\mathscr{A}_x(u) = \tfrac{1}{2}\left(1 - \sin \frac{\pi u}{2\omega_x}\right), \quad u < \omega_x,$$
$$= 0, \quad u > \omega_x. \tag{1.73}$$

This form of $\mathscr{A}_x$ represents a close approximation to actual modifications of band edges by a gradual cutoff and also results in rather simple expressions for the modified impulse characteristic.

With (1.73) in (1.72),

$$F(t) = \frac{1}{t_0} - \int_0^{\omega_x} \left(1 - \sin \frac{\pi u}{2\omega_x}\right) \sin u t_0 \, du,$$

$$= \frac{1}{t_0} - \omega_x \left(\frac{1 - \cos \omega_x t_0}{\omega_x t_0} + \frac{\cos \omega_x t_0}{\pi + 2\omega_x t_0} - \frac{\cos \omega_x t_0}{\pi - 2\omega_x t_0} \right)$$

$$= \omega_x \cos \omega_x t_0 \left(\frac{1}{\omega_x t_0} + \frac{1}{\pi - 2\omega_x t_0} - \frac{1}{\pi + 2\omega_x t_0} \right),$$

$$= \frac{1}{t_0} \frac{\cos \omega_x t_0}{1 - (2\omega_x t_0/\pi)^2}. \tag{1.74}$$

The impulse characteristic obtained from (1.71) is

$$P(t) = \frac{\Omega}{\pi} \frac{\sin \Omega t_0}{\Omega t_0} \frac{\cos \omega_x t_0}{1 - (2\omega_x t_0/\pi)^2}. \qquad (1.75)$$

For the particular case shown in Fig. 1.11 the value of ω_x is taken to be $\Omega/2$.

For a symmetrical bandpass characteristic, as shown in Fig. 1.12,

$$P_0(t) = 2 \cos (\omega_0 t_0 - \psi_0) \bar{P}_0(t). \qquad (1.76)$$

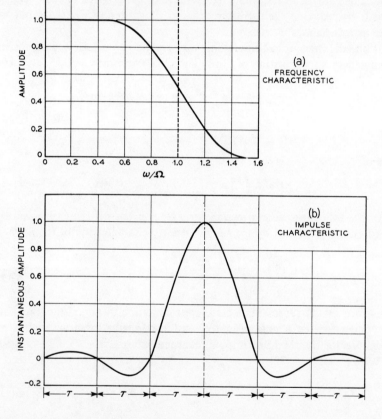

Figure 1.11 Low-pass characteristic with gradual cutoff and associated impulse characteristic. Linear phase characteristic assumed.

In accordance with (1.30), $\bar{P}_0(t) = R_0(t)$ and conforms with $P(t)$ as given by (1.75). Hence

$$\bar{P}_0(t) = \frac{\Omega}{\pi} \frac{\sin \Omega t_0}{\Omega t_0} \frac{\cos \omega_x t_0}{1 - (2\omega_x t_0/\pi)^2}. \qquad (1.77)$$

For the particular case shown in Fig. 1.12, the value of ω_x is taken to be $\Omega/2$.

The in-phase and quadrature components with respect to any frequency are obtained from (1.36) and (1.37) with $\psi_0 = \omega\tau_0$, and are shown in Fig. 1.12 for the particular case in which the reference frequency is displaced from the midband frequency by $\omega_{10} = \Omega$.

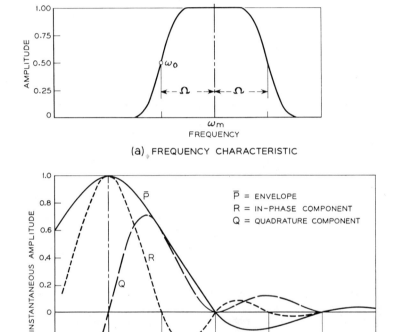

Figure 1.12 Symmetrical band-pass characteristic with gradual cutoff and associated impulse characteristic. In-phase and quadrature components shown with respect to $\omega_0 = \omega_m - \Omega$.

1.7 Channels With Raised Cosine Amplitude Characteristics

With the type of amplitude characteristics discussed earlier it is necessary to employ phase equalization to obtain a linear phase characteristic. Furthermore, oscillations of appreciable amplitude remain in the impulse

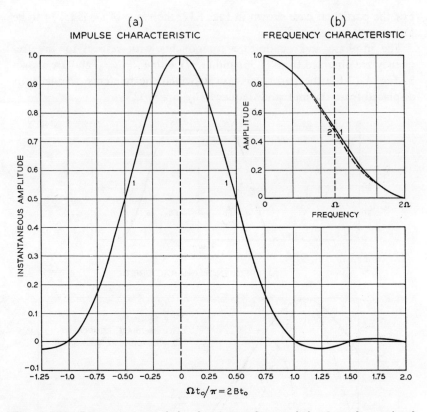

Figure 1.13 Low-pass transmission-frequency characteristic, 1, and associated impulse characteristic. Frequency characteristic, 2, is same as shown by solid lines in Fig. 1.5 and has a linear phase characteristic between $\omega = 0$ and 2Ω.

characteristic. A virtually linear phase characteristic together with a reduction of these oscillations can be attained by a further extension of the gradual cutoff in Fig. 1.10 such that $\omega_x = \Omega$. An amplitude characteristic of this type, together with the corresponding impulse characteristic, is shown in Fig. 1.13. The supplementary amplitude characteristic and the impulse characteristic are obtained by making $\omega_x = \Omega$ in (1.73) and (1.75).

The resultant amplitude characteristic between $\omega = 0$ and $\omega = 2\Omega$ in this case becomes a so-called "raised cosine."

$$A(\omega) = \frac{1}{2}\left(1 + \cos\frac{\pi\omega}{2\Omega}\right) = \cos^2\frac{\pi\omega}{4\Omega}. \qquad (1.78)$$

The corresponding impulse characteristic is

$$P(t) = \frac{\Omega}{\pi}\frac{\sin 2\Omega t_0}{2\Omega t_0[1 - (2\Omega t_0/\pi)^2]}. \qquad (1.79)$$

This impulse characteristic has the properties

$$P\left(\pm\frac{T}{2}\right) = \tfrac{1}{2}P(O), \tag{1.80}$$

$$P\left(\pm\frac{mT}{2}\right) = O, \quad m = 2, 3, 4, \ldots \tag{1.81}$$

A unique feature of an impulse characteristic with these properties is that it satisfies both the first and second Nyquist criteria for impulse characteristics that permit distortionless transmission with appropriate detection methods. The various impulse characteristics considered previously have the property that $P(t_0) = O$ for $t_0 = \pm nT$, $n = 1, 2, 3, \ldots$. This is the first Nyquist criterion [1]. In the particular case of binary pulse trains it is also possible to realize distortionless reception if the intervals between transition points in the received pulse train is preserved at a specified amplitude, say $\tfrac{1}{2}$ or O. To this end, it is necessary that the zero points in the impulse characteristic occur at $t_0 = \pm mT/2$ with $m = 3, 5, 7, \ldots$. This is the second Nyquist criterion [1]. Like the first, it can be satisfied with an infinity of amplitude characteristics with bandwidth exceeding Ω. The amplitude characteristic of minimum bandwidth Ω is given by $A(\omega) = \cos \pi\omega/2\Omega$ for $O < \omega < \Omega$ and $A(\omega) = O$ for $\omega > \Omega$. Duobinary transmission, discussed in Section 5.9, exemplifies a particular embodiment of the second Nyquist criterion or principle.

When the impulse characteristic has the above properties (1.80) and (1.81) and an "on-off" binary impulse train is transmitted, the received pulse train has the properties indicated in Fig. 1.14. The intervals between

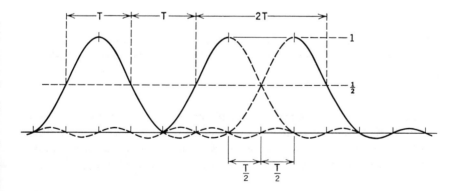

Figure 1.14 Properties of "on-off" binary pulse trains with pulses having raised cosine spectra. The amplitudes at sampling instants are 0 or 1. The intersections of the pulse train with a slicing level 1/2 occur at intervals that are multiples of T.

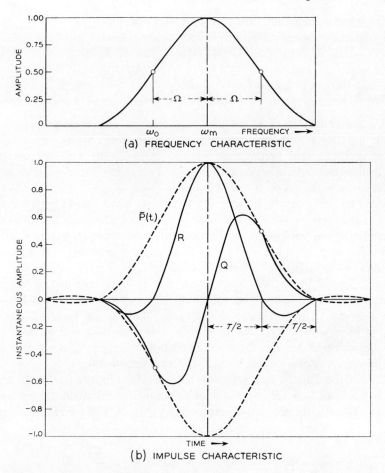

Figure 1.15 Symmetrical band-pass characteristic with linear phase shift and corresponding impulse characteristic. In-phase and quadrature components shown with respect to $\omega_0 = \omega_m - \Omega$.

intersections of the pulse train with a slicing level set at $\frac{1}{2}$ will be multiples of T. With bipolar transmission the interval between zero points will be multiples of T. Impulses triggered at these intersections or zero points afford a convenient means of deriving a timing wave for sampling of the pulse train.

Another desirable property of this impulse characteristic is that the oscillations in its tail are the minimum possible with zero points at intervals $T = 1/2B$ for channels with limited bandwidth. This permits the use of this impulse characteristic for pulse systems with discrete pulse positions with minimum intersymbol interference and considerable tolerance on synchronization. Since the oscillations in the impulse characteristic are inappreciable, it can also be used for pulse systems without discrete pulse positions and with other methods of detection than synchronized instantaneous sampling.

For a symmetrical bandpass characteristic, as shown in Fig. 1.15, the impulse characteristic is given by (1.76) and the envelope by (1.77) with $\omega_x = \Omega$, or

$$\bar{P}_0(t) = \frac{\Omega}{\pi} \frac{\sin 2\Omega t_0}{2\Omega t_0[1 - (2\Omega t_0/\pi)^2]}. \quad (1.82)$$

The in-phase and quadrature components shown in Fig. 1.15 with respect to a reference frequency at the midpoint of the band edge are obtained from (1.36) and (1.37) with $\omega_{10} = \Omega$. This gives $\bar{P}_0(t) \cos \Omega t_0$ for the in-phase and $\bar{P}_0(t) \sin \Omega t_0$ for the quadrature component.

1.8 Channels With Gaussian Amplitude Characteristics

Another type of amplitude characteristic resembling that shown in Fig. 1.13 and frequently considered in connection with pulse transmission is a Gaussian characteristic;

$$A(\omega) = e^{-\sigma \omega^2}. \quad (1.83)$$

The corresponding impulse characteristic is

$$P(t) = \frac{1}{2(\pi\sigma)^{1/2}} e^{-t_0^2/4\sigma}. \quad (1.84)$$

A striking feature of these characteristics is that the frequency and time variations are of the same form. Another intriguing property of these shapes is that the product of rms bandwidth and rms duration has the minimum possible value [3]. Moreover, the phase characteristic associated with $A(\omega)$ is linear and of infinite slope, that is, $\tau_0 = \infty$. The simple form of the characteristics can be an advantage in certain theoretical analyses. For these reasons, Gaussian shapes have received much attention in the literature. Nevertheless, they do not afford an advantage over raised cosine characteristics in applications to pulse transmission systems, as the following comparison will show.

Let the amplitude of $P(t)$ be reduced to 1 percent of the peak value after an interval $t_0 = \pi/\Omega$, corresponding to the first zero point of an ideal

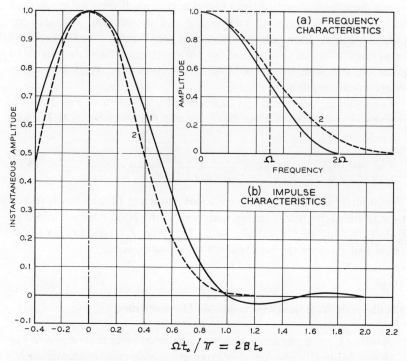

Figure 1.16 Comparison of two representative frequency and impulse transmission characteristics. Frequency characteristic 1:

$$T(\omega) = (1/2)[1 + \cos \pi\omega/2\Omega].$$

Frequency characteristic 2:

$$T(\omega) = \exp -0.54(\omega/\Omega)^2.$$

impulse characteristic, it is necessary that $t_0^2/4\sigma = 4.6$ or $\sigma = 0.54/\Omega^2$. The corresponding amplitude and impulse characteristics are

$$A(\omega) = e^{-0.54(\omega/\Omega)^2}, \tag{1.85}$$

$$P(t) = \frac{\Omega}{0.83\pi} e^{-0.46(\Omega t_0)^2} \tag{1.86}$$

In Fig. 1.16, comparison is made of the raised cosine and the Gaussian frequency characteristics, and of the corresponding impulse characteristics. The comparison shows that for the same pulse-transmission rate and with negligible intersymbol interference, a somewhat wider band must be provided with a Gaussian amplitude characteristic. This is a disadvantage, particularly when the band is restricted within prescribed limits by considerations of interference in adjacent transmission bands, as in radio systems.

1.9 Impulse Equivalent Baseband Transmittance

Let $P(t)$ be an ideal impulse characteristic with zero points at intervals T, as determined in the preceding sections. If a pulse $G(t)$ is applied in place of an impulse, it is possible to determine the shape $H(t)$ of the received pulse with the aid of (1.15). Unlike the ideal impulse characteristic, the pulse $H(t)$ would not have zero points at intervals T. It is possible, however, by appropriate modification of the transmittance $T(i\omega)$ to realize the same shape of a received pulse as for impulse transmission, as shown below.

Let $T(i\omega)$ be the transmittance resulting in a spectrum $S(i\omega)$ of the received pulses with impulse transmission. When pulses having a spectrum $S_1(i\omega)$ are applied, the spectrum of the received pulses will be the same as for impulse transmission provided the channel transmittance $T(i\omega)$ is modified into

$$T_1(i\omega) = T(i\omega) \frac{S(i\omega)}{S_1(i\omega)}. \tag{1.87}$$

By way of example, for rectangular pulses of duration δ, from $t = -\delta/2$ to $t = \delta/2$, the spectrum is in accordance with (1.16)

$$S_1(i\omega) = \delta \frac{\sin \omega\delta/2}{\omega\delta/2}. \tag{1.88}$$

With impulse transmission, $\delta \to O$ and the spectrum of a received pulse is

$$S(i\omega) = \delta T(i\omega). \tag{1.89}$$

For a pulse with spectrum (1.88), the modified transmittance is, in accordance with (1.87),

$$T_1(i\omega) = T(i\omega) \frac{\omega\delta/2}{\sin \omega\delta/2}. \tag{1.90}$$

In the above case, the spectrum $S_1(i\omega)$ does not have an imaginary component and it suffices to modify the amplitude characteristic $A(\omega)$ into

$$A_1(\omega) = A(\omega) \frac{\omega\delta/2}{\sin \omega\delta/2}, \tag{1.91}$$

$$= A(\omega) \quad \text{for} \quad \delta \to 0. \tag{1.92}$$

When the duration δ of the applied pulses is equal to the interval T between zero points in the impulse characteristic, the modified amplitude characteristic becomes

$$A_1(\omega) = A(\omega) \frac{\omega T/2}{\sin \omega T/2}. \tag{1.93}$$

In the particular case of a raised cosine amplitude characteristic (1.78), relation (1.93) yields

$$A_1(\omega) = \cos^2 \frac{\pi\omega}{4\Omega} \frac{\omega T/2}{\sin \omega T/2}$$

$$= \frac{\pi\omega/4\Omega}{\tan \pi\omega/4\Omega}, \quad (1.94)$$

where the last relation is obtained by noting that $T = \pi/\Omega$. The characteristic $A_1(\omega)$ is shown in Fig. 1.17.

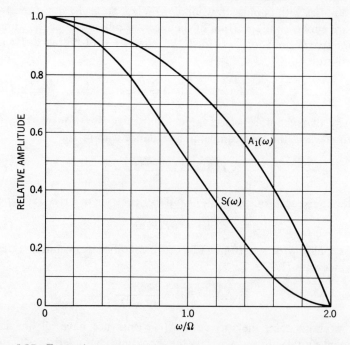

Figure 1.17 Transmittance $A_1(\omega)$ for ideal baseband pulse transmission with rectangular pulses of duration $T = \pi/\Omega$ and raised cosine pulse spectrum $S(\omega)$ at detector input.

1.10 Impulse Equivalent Bandpass Transmittance

In Section 1.6, it was shown that impulse transmission without intersymbol interference is possible with bandpass characteristics having the properties indicated in Fig. 1.18. With impulse transmission, these characteristics also conform in shape with the spectrum of the envelope of received pulses. Though impulse transmission is possible in principle, the more practicable method is by transmission sine-wave or "carrier pulses." In this case, the input to the bandpass channel is of the form

$$P_1(t) = \bar{P}_1(t) \cos(\omega_0 t + \varphi_0). \tag{1.95}$$

The frequency ω_0 can be identified with the reference frequency in previous sections, and may or may not coincide with the midband frequency of the bandpass channel.

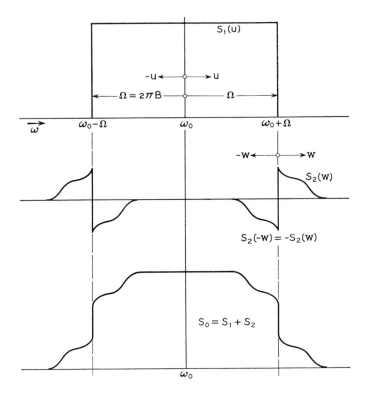

Figure 1.18 Properties of spectra at detector input such that, in double-sideband amplitude modulation, $P(0) = 1$, $P(mT) = 0$, $T = \pi/\Omega$, $m = 1, 2, 3,\ldots$

The spectrum $S_1(u)$ of the received pulse can be determined from (1.2) and becomes

$$S_1(u) = \int_{-\infty}^{\infty} \bar{P}_1(t) \cos(\omega_0 t + \varphi_0) e^{-i\omega t} dt. \tag{1.96}$$

The same shape of the envelope of the received pulses as with impulse transmission is obtained with the modified amplitude characteristic,

$$\mathscr{A}_1(u) = \frac{\mathscr{A}_0(u)\, S(u)}{S_1(u)}. \tag{1.97}$$

In the particular case of sine-wave pulses of duration δ, (1.96) becomes

$$S_1(u) = \int_{-\delta/2}^{\delta/2} \cos(\omega_0 t + \varphi_0) e^{-i\omega t}\, dt$$

$$= \frac{\delta}{2} e^{i\varphi_0} \frac{\sin u\delta/2}{u\delta/2} + \frac{\delta}{2} e^{-i\varphi_0} \frac{\sin(2\omega_0 + u)\delta/2}{(2\omega_0 + u)\delta/2}, \tag{1.98}$$

where $u = \omega - \omega_0$. The two components of the spectrum are indicated in Fig. 1.19.

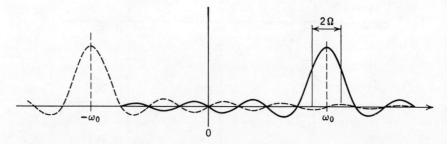

Figure 1.19 Spectrum of rectangular carrier pulse of frequency ω_0 and duration $T = \pi/\Omega$. In the transmission band 2Ω the component of the spectrum represented by the dashed curve can be neglected.

For narrow bandpass channels the last term in (1.98) can be disregarded, so that

$$S_1(u) = \frac{\delta}{2} e^{i\varphi_0} \frac{\sin u\delta/2}{u\delta/2}. \tag{1.99}$$

The factor $e^{i\varphi_0}$ introduces a phase change in the sine wave within the envelope, which can be disregarded here. (With envelope detection this factor is immaterial, and with synchronous detection it can be removed by an appropriate phase of the demodulating sine wave.)

For a sine-wave pulse of duration T the modified amplitude characteristic becomes

$$\mathscr{A}_1(u) = \mathscr{A}_0(u) \frac{uT/2}{\sin uT/2}. \tag{1.100}$$

Impulse Equivalent Bandpass Transmittance

For the particular case of a raised cosine amplitude characteristic $\mathscr{A}_0(u)$ that is symmetrical with respect to ω_0, the modified amplitude characteristic becomes

$$\mathscr{A}_1(u) = \cos^2 \pi u/2\Omega \frac{uT/2}{\sin uT/2}$$

$$= \frac{u\pi/4\Omega}{\tan u\pi/4\Omega}, \qquad (1.101)$$

where the relation $T = \pi/\Omega$ has been used.

The amplitude characteristic $\mathscr{A}_1(u)$ is shown in Fig. 1.20.

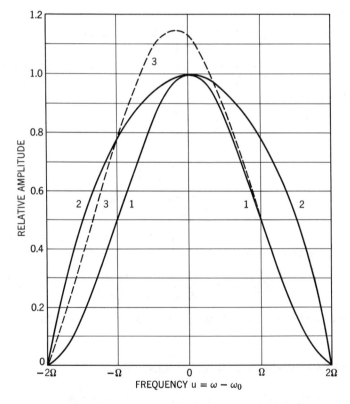

Figure 1.20 Ideal channel transmittances for carrier pulse transmission with raised cosine pulse spectrum at detector input. 1. Raised cosine spectrum $S(u)$ at detector input. 2. Transmittance $\mathscr{A}_1(u)$ for rectangular carrier pulses of frequency ω_0 and duration $T = \pi/\Omega$. 3. Transmittance $\mathscr{A}_1(u)$ for rectangular carrier pulses of frequency $\omega_0 + \Omega$ and duration $T/2 = \pi/2\Omega$.

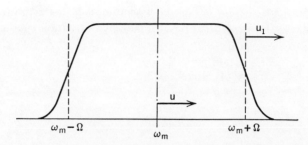

Figure 1.21 Carrier frequency in vestigial sideband transmission at $\omega_0 = \omega_m + \Omega$ or at $\omega_0 = \omega_m - \Omega$. In frequency shift pulse transmission the carrier frequency would be at $\omega_m - \Omega$ for a "space" and at $\omega_m + \Omega$ for a mark.

In the case of vestigial sideband transmission the sine-wave frequency would be $\omega_0 = \omega_m + \Omega$ or $\omega_m - \Omega$, as indicated in Fig. 1.21. As shown previously, pulses can in this case be transmitted without intersymbol interference at intervals $T/2 = \pi/2\Omega$. The spectrum of a rectangular pulse of this duration is

$$S_1(u) = \frac{T}{4} \frac{\sin u_1 T/4}{u_1 T/4}$$

$$= \frac{T}{4} \frac{\sin (u - \Omega)T/4}{(u - \Omega)T/4}$$

$$= \frac{\pi}{4\Omega} \frac{\sin \pi(u - \Omega)/4\Omega}{\pi(u - \Omega)/4\Omega}, \quad (1.102)$$

where $u_1 = \omega - \omega_0$ and $u = \omega - \omega_m$ as in Fig. 1.21. For a raised cosine pulse spectrum the modified amplitude characteristic becomes

$$\mathscr{A}_1(u) = \cos^2 \frac{\pi u}{4\Omega} \frac{\pi(u - \Omega)/4\Omega}{\sin \pi(u - \Omega)/4\Omega}. \quad (1.103)$$

This modified amplitude characteristic is shown in Fig. 1.20.

1.11 Ideal Frequency-Shift Transmission Characteristics

With one method of binary transmission, "on" and "off," or "marks" and "spaces," are represented by carriers or sine waves of frequencies $\omega_m + \Omega$ and $\omega_m - \Omega$, or conversely, as indicated in Fig. 1.21. At the receiving end, the amplitude of the signal is applied to a detector with a response proportional to the time derivative of the phase of the received

Ideal Frequency-Shift Transmittances

wave, that is, to the instantaneous frequency deviation. It can be shown [7] that with this method of frequency-shift transmission the resultant demodulated pulses can be made to have zero points at intervals T, with a channel bandwidth no greater than that required when "marks" and "spaces" are represented by "on" and "off" transmission of a carrier of frequency ω_0. To this end, it is necessary to have appropriate shaping of the amplitude characteristic $\mathscr{A}_1(u)$, which differs from that required with on-off transmission of a sine wave of frequency ω_0.

With the above method, let a "mark" of frequency $\omega_m + \Omega$ and duration $T = \pi/\Omega$ be preceded and followed by a continuing space of frequency $\omega_m - \Omega$. The resultant demodulated pulse $P(t_0)$ is then given by [7]

$$P(t_0) = \frac{\mu}{2} \frac{\mu p^2 - p \cos \Omega t_0 - (p'/\Omega) \sin \Omega t_0}{\sin^2 \Omega t_0 + (\cos \Omega t_0 + \mu p)^2}, \qquad (1.104)$$

where $p = p(t_0)$ and $p' = dp(t_0)/dt_0$ and

$$\mu = 1/A(-\Omega) = 1/A(\Omega) = 2. \qquad (1.105)$$

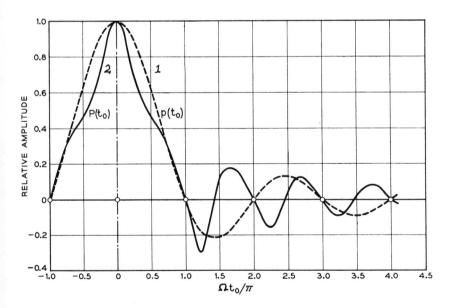

Figure 1.22 Received demodulated pulses with "on-off" transmission of midband carrier, 1, and with frequency-shift transmission, 2, for transmitted rectangular pulses of duration $T = \pi/\Omega$ and for flat spectrum of bandwidth 2Ω at output of bandpass channel, i.e., detector input.

The function $p(t_0)$ represents the shape of the envelope of a carrier pulse of frequency ω_0 and envelope spectrum $S(u)$, and is given by

$$p(t_0) = \frac{1}{\pi} \int_{-\omega_0}^{\infty} S(u) \cos ut_0 \, du. \tag{1.106}$$

The above relation (1.104) yields zero points in $P(t_0)$ at intervals $T = \pi/\Omega$ provided the channel has an amplitude characteristic [7]

$$\mathscr{A}_1(u) = \frac{S(u)}{S_1(u)}, \tag{1.107}$$

where

$$S_1(u) = S_1(-u) = \frac{T}{2} \frac{4}{\pi} \frac{\cos(\pi u/2\Omega)}{1 - (u/\Omega)^2}. \tag{1.108}$$

A further requirement is that the bandpass channel be narrow, that is, $\Omega \ll \omega_0$.

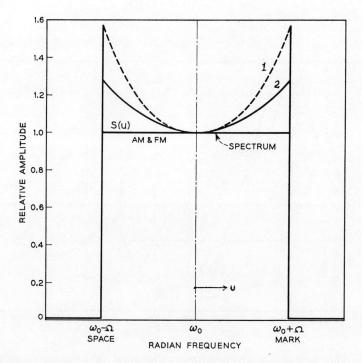

Figure 1.23 Ideal channel transmittance $\mathscr{A}_1(u)$ for "on-off" transmission of midband carrier, 1, and for frequency shift transmission, 2, for flat spectrum of total bandwidth 2Ω, at output of bandpass channel.

Ideal Frequency-Shift Transmittances

By way of illustration, in Fig. 1.22, $P(t_0)$ is compared with $p(t_0)$ for a flat spectrum of bandwidth 2Ω of the pulse $p(t_0)$ at the channel output, that is, detector input. The required amplitude characteristic in this case is shown in Fig. 1.23 and is

$$\mathscr{A}_1(u) = \frac{\pi}{4} \frac{1 - (u/\Omega)^2}{\cos(\pi u/2\Omega)}. \tag{1.109}$$

For purposes of comparison with AM the amplitude characteristic is normalized to unity at $u = 0$.

As another example, in Fig. 1.24 are shown the pulse shapes $P(t_0)$ and $p(t_0)$ for a raised cosine spectrum of bandwidth 4Ω at the channel output, that is, the detector input. In this case, the requisite amplitude characteristic for binary FM is [7]

$$\mathscr{A}_1(u) = \frac{\pi}{4} \frac{1 + \cos(\pi u/2\Omega)}{2\cos(\pi u/2\Omega)} \left[1 - \left(\frac{u}{\Omega}\right)^2\right]. \tag{1.110}$$

This amplitude characteristic is shown in Fig. 1.25, together with that for amplitude modulation of a carrier of frequency ω_0, as given by (1.100), with both normalized to unity at ω_0.

With pulse spectra and corresponding channel transmittances determined as above, the received pulses with frequency shift transmission

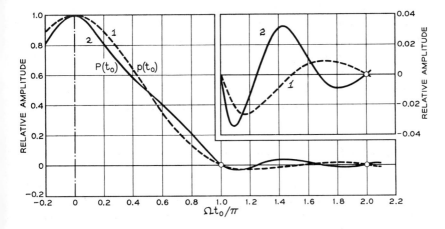

Figure 1.24 Received demodulated pulses with "on-off" transmission of midband carrier, 1, and with frequency-shift transmission, 2, for transmitted rectangular pulses of duration $T = \pi/\Omega$ and for raised cosine spectrum of total bandwidth 4Ω at output of bandpass channel, i.e., detector input.

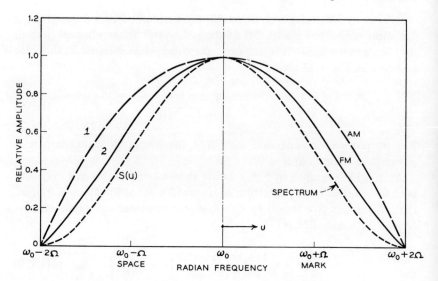

Figure 1.25 Ideal channel transmittance $\mathscr{A}_1(u)$ or "on-off" transmission of midband carrier, 1, and for frequency-shift transmission, 2, for raised cosine spectrum of total bandwidth 4Ω, at output of bandpass channel.

satisfy the first Nyquist criterion, mentioned in Section 1.7. As can be shown [8], it is also possible to determine pulse spectra and channel transmittances such that the received pulses satisfy the second Nyquist criterion. Moreover, both criteria can be satisfied with a particular spectrum and channel transmittance with frequency-shift keying, as in the case of baseband and single-frequency carrier transmission. As in the latter cases, the bandwidth of the spectrum of received pulses that satisfies both requirements is twice the minimum bandwidth of spectra of received pulses that satisfy either the first or the second Nyquist requirement [8].

1.12 Optimum Digital Transmission and Detection Methods

In digital systems it is possible to employ a variety of channel transmittances and detection methods with different properties. As shown previously, intersymbol interference can be avoided with instantaneous sampling of pulse trains at the midpoint of a signal interval of duration T. To this end, the pulse spectra must have the attributes indicated in Figs. 1.10 and 1.18 for low-pass and bandpass channels. It is possible to have an infinity of spectra for each specified spectrum bandwidth $\Omega_o > \Omega$ and thus an infinity of impulse characteristics that afford distortionless transmission with pulse intervals $T = \pi/\Omega$.

A common feature of digital systems employing transmittances with the above properties is that for Gaussian noise with a flat power spectrum the error probability is the same for a given ratio ρ of average signal power to average noise power in the band Ω. This applies for instantaneous sampling provided the spectrum shaping is divided equally between transmitting and receiving filters and the attenuation in the transmission medium does not vary with frequency over the channel band [7]. On the above premises the error probability with bipolar transmission and r equally spaced amplitudes is given by (5.64) or

$$\epsilon = \left(1 - \frac{1}{r}\right) \text{erfc} \left(\frac{3\rho/2}{r^2 - 1}\right)^{1/2}. \tag{1.111}$$

This relation applies for baseband transmission and for carrier amplitude modulation with synchronous demodulation.

The above detection method yields optimum performance as regards error probability with digit-by-digit detection. It is readily implemented by electronic gates that sample the pulse train over a small fraction of the interval T, even at very high transmission rates.

Since the error probabilities are equal, it is necessary to examine other properties, in order to determine the preferable transmittance, considering both its spectrum and the associated impulse characteristic. From the standpoint of filter design, a raised cosine spectrum appears desirable, as regards both the shaping and the approximate linearity of the associated phase characteristic. As noted in Section 1.7, the impulse characteristic has certain desirable properties as regards the required sampling accuracy and timing-wave extraction for such sampling. With this impulse characteristic the maximum amplitude of a random pulse train is greater than the peak amplitude of a single pulse by a factor 1.06. This compares with a minimum factor 1 for an impulse characteristic with triangular spectrum of the same bandwidth 2Ω as a raised cosine spectrum [2].

Another feature of the raised cosine spectrum is that the impulse characteristic has virtually the same shape as the spectrum, if the small oscillations in the tail are disregarded. The energy in this tail is but a small fraction $2 \cdot 10^{-4}$ of the total pulse energy. Hence it is a legitimate approximation to assume a raised cosine pulse shape. Raised cosine pulses can be generated by gating a biased sine wave, which is an advantage from the standpoints of implementation, timing, and synchronization.

Because of the above virtues, a raised cosine pulse spectrum is considered a desirable objective in most pulse systems and is assumed in certain numerical evaluations of performance in later chapters. Other pulse spectra with a partial raised cosine roll-off may be preferable in certain situations, as in some applications to existing transmission facilities of restricted bandwidth not specifically designed for digital transmission.

1.13 Suboptimum Digital Transmission and Detection Methods

Although instantaneous sampling, as considered above, affords optimum performance and is readily implemented, methods that employ integration over a signal interval T are considered in the literature. As applied to band-limited signals, these methods are suboptimum, in that error probability is greater and implementation more difficult than with instantaneous sampling.

Straight integration may be employed, in which case the integral of the pulse trains (1.21) over the interval $-T/2$ to $T/2$ becomes

$$I = \sum_{n=-\infty}^{\infty} G(nT)\lambda_n, \qquad (1.112)$$

where

$$\lambda_n = \int_{-T/2}^{T/2} P(t + nT)\, dt. \qquad (1.113)$$

An alternate method is correlation integration or detection, in which (1.113) is replaced by

$$\mu_n = \int_{-T/2}^{T/2} P(t)\, P(t + nT)\, dt. \qquad (1.114)$$

Both integration methods yield the same error probability (1.111), as with instantaneous sampling for the particular case of rectangular received pulses of duration T. Although such pulses entail infinite spectrum and channel bandwidths, they are often assumed or implied in probabilistic work on error probability that ignores transmission aspects. For pulses with limited bandwidth, the error probability will be increased, for the reason that all of the pulse energy is not then contained in the interval T. To optimize performance with integration it is necessary to determine the maximum possible energy in the interval T with bandwidth constraint as considered in some literature [9,10,11,12].

Integration can be accomplished with the aid of a filter of narrow bandwidth $\Omega_x \ll \Omega$, which must be deactivated after each integration during an interval much smaller than T to erase memory of previous integrations. To this end, it is necessary after each integration to apply a deactivating pulse of short duration and appropriate amplitude in proportion to the integral I.

Suboptimum Digital Detection Methods

As shown by Nyquist [1], it is possible to have pulse shapes $P(t)$ such that λ_n vanishes except for $n = O$, so that intersymbol interference is avoided. This is his third criterion for distortionless transmission and, like the first and second, can be realized with an infinity of pulse spectra with bandwidth in excess of $\Omega = \pi/T$. Under the latter condition, it is also possible to have pulse shapes such that μ_n vanishes except for $n = O$. Thus both integration methods in principle afford distortionless transmission at the same rate as with instantaneous sampling. However, the error probability is greater than with instantaneous sampling, both with correlation detection [9] and the less efficient method of straight integration.

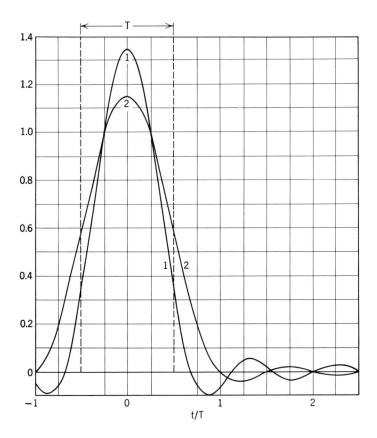

Figure 1.26 Comparison of two pulses of spectral bandwidth $1/T$ cycles/sec. normalized to unit total energy. 1. Pulse with maximum energy in interval T. 2. Pulse with raised cosine spectrum.

By sacrificing the requirement on zero intersymbol interference, it is possible to increase the energy in the interval T for a given spectrum bandwidth [10,11]. In this case, μ_0 as obtained from (1.114) is increased and μ_n does not vanish. When these pulse shapes are used in conjunction with correlation detection over the interval T, the error probability is greater than with instantaneous sampling [9], partly because of intersymbol interference and partly because all of the pulse energy is not contained in the interval T. For a bandwidth 2Ω, the maximum energy in the interval T is about 98 percent of the total energy and about 5.3 percent greater than for a pulse with a raised cosine spectrum. The pulse shapes for these cases are compared in Fig. 1.26. Curve 1 was obtained by computerized numerical integration of relations in [10].

In summary, the optimum pulse shape for digital systems depends on several factors set forth in Section 1.12 and is not obtained by maximizing the energy in a signal interval T for a specified bandwidth. Nor does it suffice to maximize this energy, subject to additional constraints, such as a bounded peak amplitude for a random pulse train [9] or a minimum ratio or peak to rms amplitude for a single pulse [12].

1.14 Imperfections in Transmission Characteristics

In this chapter, certain idealized channel characteristics were assumed that cannot be fully realized in actual systems. For one thing, a low-frequency cutoff will be encountered in actual systems where transformers are required for increased transmission efficiency or for other reasons. For another, smooth amplitude and phase characteristics, as assumed here, cannot be fully attained, partly because lumped networks will be required for equalization and partly because of unavoidable reflection points in actual transmission media. Moreover, a linear phase characteristic if it could be realized, would entail either incomplete channel cutoffs or infinite transmission delay. Furthermore, in most channels, amplifiers are required, which introduces some transmittance nonlinearity and resultant transmission distortion. Finally, the channel transmittances are not time-invariant as assumed here, but exhibit slow variations in metallic circuits owing to temperature changes and more rapid fluctuations or "fading" in radio circuits, caused by various kinds of turbulence in the transmission medium.

The effect of these various departures from the idealized characteristics postulated here depends on the modulation and detection methods and will be dealt with in later chapters. Consideration is also given in later chapters to various transmission impairments not related to imperfections in the channel transmittance. Among them are imperfect synchronization and sampling in the detection of signals resulting from unavoidable timing-wave deviations with actual implementations, inteference between systems owing to electromagnetic coupling, and additive random noise inherent in the transmission medium.

2

Basic Modulation Concepts and Objectives

2.0 General

Chapter 1 dealt with certain basic concepts in signal transmission over communication channels. The signals considered there are generally related to the original message waves by various conversion or modulation processes intended to facilitate transmission or improve performance; the latter will be outlined here.

In general, a comprehensive communication network includes a number of transmission systems employing different transmission media. As indicated in Fig. 2.1, each such transmission system consists of an information source, a transmitter, a channel, a receiver, and a destination. The channel ordinarily comprises a number of repeater sections in tandem, sometimes several hundred. In general, a two-way system is provided, with separate channels for opposite directions of transmission, though in some cases a two-way channel with bidirectional repeaters is used.

Information sources produce a variety of message types, such as numbers or letters in telegraph, teletype and data systems, variable acoustic waves in speech, and light beams of variable intensity in television. As a first modulation step, these messages are converted from their original

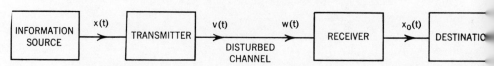

Figure 2.1 Basic transmission system arrangement. $x(t)$ = message wave at transmitter input; $v(t)$ = signal wave at transmitter output; $w(t)$ = disturbed signal wave at receiver input; $x_0(t)$ = disturbed message wave at receiver output.

states into equivalent electrical message waves. In the transmitter, these may undergo further processing before they appear as signals at the input to the channel or transmission medium. At the receiver the inverse of the transmitter operation is performed, so that in the absence of various kinds of imperfections the original message would be recovered without distortion or error.

In the channel the signal suffers distortion, owing to various kinds of imperfection. Among them can be mentioned random noise, channel bandwidth limitation, unwanted variations in attenuation and phase characteristics over the channel band, nonlinearity of the channel transfer characteristic, unpredictable time variations in the characteristic, or interference from external sources.

The wave, $x(t)$, from the information source is converted in the transmitter into a wave, $v(t)$, at the channel input. Because of transmission distortion and noise, the wave at the channel output is $w(t)$ and at the receiver output is $x_0(t)$. The objective in transmission systems engineering is generally to meet, in the most economical manner, certain requirements on error probability in the received wave $x_0(t)$, or mean squared error in continuous wave transmission. For a channel of specified transmission and noise properties, certain limitations are imposed on the properties of the signal waves, $v(t)$, such as the bandwidth and midband frequency, the admissible average or peak signal power from the standpoint of overloading channel amplifiers, or as regards interference in other communication channels. Nevertheless, the requirements on errors or noise in the signal, $x_0(t)$, can ordinarily be met with a number of different modulation methods, of different complexity and cost. When channels of different transmission and noise properties are allowed, rather than a specific channel, the possible transmission and modulation methods are of course much greater. It then becomes necessary to consider various system possibilities and their compatibility with existing facilities.

Translation from a message wave, $x(t)$, to a signal wave, $v(t)$, in the transmitter can be accomplished by various methods that in general involve a number of processes, as indicated in Fig. 2.2, for one direction of transmission. Some of the steps indicated in the figure are not required with some modulation methods and can be removed. For example, the sampling process is not involved in direct continuous wave transmission, but can in principle be inserted without affecting performance. Quantization would not be used in direct continuous wave transmission, nor if the message is of digital nature. Coding is not always used, nor modulation from a baseband to a carrier signal wave. The diagram in Fig. 2.2 can be considered to represent the various processes that may be involved, though some may be omitted with the simpler modulation methods.

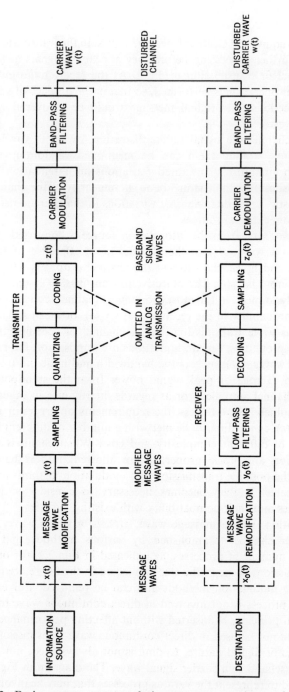

Figure 2.2 Basic message wave translation steps at transmitter and receiver.

Characterization of Message Waves

Various basic modulation concepts, methods and objectives are reviewed here, together with the conventional analytical characterization of the properties of message waves and some of the principal problems and considerations in systems design. More detailed treatment of various modulation methods and their implementation is given elsewhere [1], with emphasis on certain other aspects than dealt with here.

2.1 Characterization of Message Waves

A continuous message wave, $x(t)$, can be represented by a Fourier integral as

$$x(t) = \frac{1}{\pi} \int_0^\infty c_x(\omega) \cos\left[\omega t - \psi_x(\omega)\right] d\omega, \tag{2.1}$$

where $c_x(\omega)$ and $\psi_x(\omega)$ are the amplitude and phase functions for the particular message wave. Thus $x(t)$ may designate a single pulse, a sequence of pulses, or a continuous wave representing a particular word or sentence in speech.

The message waves by their very nature will vary continually and in evaluating the performance of transmission systems it is convenient to use statistical analysis and to characterize various kinds of messages ensembles, such as speech waves, by certain statistical parameters. These include the probability density and distribution, the mean amplitude, the mean squared amplitude and the variance, together with the spectral density and autocorrelation function. These various statistical parameters are discussed in some further detail in Appendix 2.

If $p(x)$ is the probability density for an amplitude, x, of the message wave, the mean amplitude is

$$m_1 = E(x) = \int_{-\infty}^\infty x\, p(x)\, dx. \tag{2.2}$$

The mean squared amplitude is referred to as the average power P_x, and is given by:

$$m_2 = E(x^2) = P_x = \int_{-\infty}^\infty x^2\, p(x)\, dx. \tag{2.3}$$

The central moment about the mean, or the variance of $x(t)$, is given by

$$\sigma_x^2 = m_2 - m_1^2 \tag{2.4}$$

and represents the average power of the variable component of $x(t)$. No special name is accorded this component in modulation theory, but it represents the average power of the information component and might be referred to as average information power P_i as contrasted with average message power $m_2 = P_x$.

The power spectra of message waves and related signals find various applications in systems design, as in (1) channel optimization against random noise, (2) determination of average distortion owing to various channel imperfections, (3) evaluation of intersystem interference and (4) determination of certain properties of timing waves extracted from signals in digital systems. Such applications are exemplified in later chapters and may also involve the autocorrelation function, $R(\tau)$, of the message or signal waves, which are related to their power spectra, $W(\omega)$, by:

$$R_x(\tau) = \int_0^\infty W_x(\omega) \cos \omega\tau \, d\omega, \tag{2.5}$$

$$R_x(O) = m_2 = \int_0^\infty W_x(\omega) \, d\omega = P_x, \tag{2.6}$$

$$W_x(\omega) = \frac{2}{\pi} \int_0^\infty R_x(\tau) \cos \omega\tau \, d\tau. \tag{2.7}$$

For random processes that are stationary in the wide sense,* the average power obtained from (2.6) must conform with (2.3). For a given $W_x(\omega)$, this is possible with an infinite number of functions, $p(x)$, and conversely.

The message produced by an information source may be a continuous wave, as just assumed, or it may consist of a sequence of symbols or states, such as numbers or letters. These can be represented by a sequence of impulses of n different amplitudes occurring at different times. Ensembles of such sequences can be characterized by the probability, p_i, of encountering amplitude, i, together with the mean amplitude, m_1, the mean squared amplitude, m_2, and variance, σ_x^2, obtained by replacing the integrals in (2.2) and (2.3) by summation signs.

A property of n possible symbols or messages is the average information content per symbol or message, defined by Shannon as [2]

$$H(x) = -\sum_{i=1}^n p_i \log_2 p_i, \tag{2.8}$$

where p_i is the probability of a particular symbol or message $x_i(t)$. The logarithm is to the base 2, with $\log_2 x \simeq 3.33 \log_{10} x$. The unit of information content is a bit, which is an acronym for binary digit.

* See Appendix 2.

A maximum value of H is obtained when all states are equally probable, in which case $p_i = 1/n$ and

$$H_{\max} = \log_2 n. \tag{2.9}$$

For a continuous probability distribution, the average information content is defined by Shannon as

$$H(x) = -\int_{-\infty}^{\infty} p(x) \log_2 p(x)\, dx. \tag{2.10}$$

With sequences of m symbols, each with n different states, it is possible to form n^m combinations or messages. The corresponding average information content per message obtained from (2.8) is then proportional to m. Ordinarily, however, the actual average information content per message is less, for the reason that certain constraints may exist on permissible combinations. The resultant reduction in information content compared to the maximum obtained with all possible combinations is known as "redundancy." As an example, only certain combinations of letters are used to form pronounceable words, and grammatical rules place limitations on sentence structures. Knowledge of such constraints at the destination constitutes *a priori* information, and facilitates correct interpretation of messages that may be garbled in transmission. In some instances, it may be expedient to inject redundancy at the source to facilitate error detection or error correction, as in error-correcting codes. This entails increased channel bandwidth for a given information rate. In other situations it may be desirable and feasible to reduce channel bandwidth requirements by encoding and decoding methods that reduce redundancy at the source and restore it at the destination. This is exemplified by the use of voice-coding devices (vocoders), for speech transmission, as discussed in the next section. Such encoding and decoding procedures are difficult to implement and entail some message distortion in exchange for improved bandwidth utilization. However, the increasing need for improved radio frequency spectrum utilization may dictate the use of various methods of redundancy reduction [3]. This applies particularly to picture transmission, where significant bandwidth compression is often feasible and highly desirable, as in telephoto transmission from space vehicles [3]. Because of the rather special applications, redundancy reduction methods are not considered in this book.

2.2 Statistical Properties of Speech Waves

Speech is by far the most common type of message wave in communication systems. Hence its various statistical properties are of prime importance in systems engineering and have been determined by comprehensive measurements. Among these properties are the distribution of average speech power among individuals, the instantaneous amplitude probability distribution of speech waves, and the power spectrum, which will be reviewed briefly here.

It has been found that the probability distribution of average speech power among talkers closely follows the log-normal law [4], discussed further in Appendix 2. Let S be the average speech power of a particular talker, and S_r some arbitrary reference power; and let

$$s = \frac{S}{S_r}, \qquad (2.12)$$

$$v = \ln s. \qquad (2.13)$$

The probability distribution is then given by

$$P(v < v_1) = \frac{1}{2}\left(1 + \operatorname{erf} \frac{v_1 - v_0}{\sqrt{2}\sigma_v}\right), \qquad (2.14)$$

or

$$P(s < s_1) = \frac{1}{2}\left(1 + \operatorname{erf} \frac{\ln s_1/s_0}{\sqrt{2}\sigma_v}\right). \qquad (2.15)$$

As shown in Appendix 2, the following relation applies for the average value $\bar{s}$ of s and the corresponding value $\bar{v}$ of v.

$$\bar{v} = \ln \bar{s} = v_0 + \frac{\sigma_v^2}{2}. \qquad (2.16)$$

When the ratio s is expressed in decibels, the following relations apply

$$V = 10 \log_{10} s, \qquad (2.17)$$

$$\bar{V} = 10 \log_{10} \bar{s} = 4.34\left(v_0 + \frac{\sigma_v^2}{2}\right)$$

$$= V_0 + 0.115\sigma_V^2, \qquad (2.18)$$

where

$$V_0 = 4.34v_0 \quad \text{and} \quad \sigma_V = 4.34\sigma_v$$

and

$$P(V < V_1) = \frac{1}{2}\left(1 + \operatorname{erf} \frac{V_1 - V_0}{\sqrt{2}\sigma_V}\right). \qquad (2.19)$$

The quantity, V, is referred to as the speech volume. Thus V_0 is the average volume and S_0 is the average power corresponding to V_0, while $\bar{V}$ is the volume corresponding to the average speech power, $\bar{S}$. When the

Statistical Properties of Speech Waves

reference power S_r is 1 milliwatt, the volumes are in dB relative to 1 milliwatt, abbreviated dBm. A reference power $S_r = 0.7$ mW is also used, in which case the volumes are in volume units, designated VU, with 1 VU = -1.4 dBm.

The average volume V_0 and the standard deviation σ_V depends on the properties of the transmission path between the talker and the point of measurement. At the input to a toll-transmitting board, the mean volume at zero dB transmission level is about $V_0 = -15$ VU, corresponding to -16.4 dBm and $\sigma_V \simeq 5.4$ dB. These values apply for U.S.A. and differ somewhat from those observed elsewhere [5]. A 15 dB higher volume or a 15 dB lower volume would be encountered with a probability of about 0.01.

The instantaneous amplitude distribution of speech is virtually independent of the average speech power, but depends on the type of microphone and the point of measurement in the telephone circuit. It also depends on whether or not the measurement relates to natural telephone conversation. A distribution, as shown in Fig. 2.3, has been obtained for

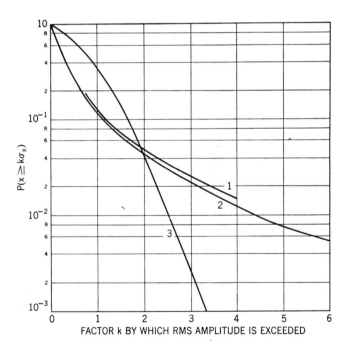

Figure 2.3 Instantaneous amplitude probability distribution of speech waves and random noise. 1. Measured probability distribution. 2. $P(x \geq k\sigma_x) = \exp -(ka)^{1/2}$; $a = (4!)^{1/2}$. 3. $P(x \geq k\sigma_x) = \operatorname{erfc}(k/2^{1/2})$.

telephone conversation using a commercial telephone set in conjunction with a representative line between the subscriber and central office [4]. In the same figure is shown for comparison a Gaussian amplitude probability distribution applying for random noise.

For some analytical purposes it is convenient to represent the instantaneous amplitude distribution by an approximate expression. As indicated in Fig. 2.3, the following relation approximates the distribution observed for $k \leq 4$:

$$P(X \geq k\sigma_x) = e^{-(ka)^{1/2}}, \qquad (2.20)$$

where σ_x is the rms amplitude and $a = (4!)^{1/2} \simeq 4.9$.

Other distributions have been observed under different conditions [5].

The normalized acoustic power spectrum of speech is shown in Fig. 2.4, and is almost independent of average speech power. It will be noted that most of the power is confined to the frequency band below 1 kc, whereas the nominal bandwidth of voice channels is about 3 kc.

In the case of speech, it is necessary to distinguish between the "literal" and the "subjective" information contents. The former depends on the rate at which words are uttered and on the number of words or letters in the alphabet. Assuming 120 words per minute and an average of 6 letters per word, the number of letters per second is 12. With an alphabet of 26 letters the corresponding literal information rate is $12 \log_2 26 \simeq 55$

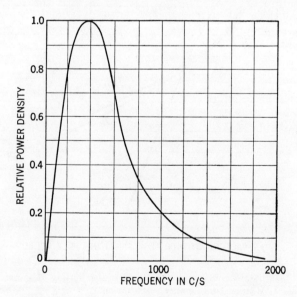

Figure 2.4 Approximate relative acoustic power spectrum of speech.

bits/sec. This information rate could be realized by binary transmission over a channel of about 30 c/s bandwidth.

The subjective information content, which permits recognition of a speaker by his voice, is much greater than the literal information content of spoken utterances. Speech waves can be adequately represented by instantaneous sampling with about 100 different quantized amplitudes. For a speech spectrum of 3000 c/s bandwidth, these samples must be transmitted at a rate of 6000 per second. The corresponding subjective information rate is $6000 \log_2 100 \simeq 40{,}000$ bits/sec. This assumes that speech is a random process without correlation between the above instantaneous samples. On this premise, a channel of the same bandwidth as the acoustic power spectrum (about 3000 c/s) would be required for direct transmission of speech with imperceptible impairment of quality. In addition, a high ratio of average speech power to average noise power would be required, about 50 dB.

However, spectrographic analysis reveals speech has a certain systematic structure and is made up of sequences of elementary sounds known as phonemes. About 40 phonemes are required for the words of the English language. Because of these properties, correlation exists between speech-wave samples so that the subjective information content is less than indicated above for a random wave without such correlation (40,000 bits/sec). Consequently, it is possible in principle by appropriate encoding and decoding procedures to transmit speech over a channel of reduced bandwidth without significant reduction quality, or subjective information content. With the aid of voice-coding devices (vocoders) about a 10-fold reduction in channel bandwidth has been realized, with a speech quality that is acceptable for some applications. It is possible to essentially maintain speech quality with more refined vocoders that permit about a 4-fold reduction in channel bandwidth [6].

Vocoders may find applications in special situations where bandwidth reduction is sufficiently important to warrant the extra complication and cost of terminal equipment.

2.3 Volume Modification by Syllabic Compandors

A communication channel may be used in turn for messages from different sources. The average message power may then vary widely, as in the case of speech from various individuals. This must be considered in system design by providing sufficient signal power so that adequate performance is realized in the presence of random noise or other interference even for messages of low power.

If to the above end the signal power is made proportional to the message power, the required maximum transmitter power may be excessive, or the repeater spacing unduly short, resulting in costly design. Moreover, a high signal power may cause excessive interference into other systems. However, a more fundamental limitation on the above procedure may be imposed in the case of crosstalk interference between systems of the same kind, such as those on different pairs in the same cable. In this case, a message of high power in one system may cause excessive interference with a message of low power on another system. This situation is not improved by making the signal powers proportional to the message powers, since the ratio of message power to interference power will remain the same.

Such difficulties may be overcome by a compression of the message power range at the transmitter, with a complementary expansion at the receiver, a process known as companding. In this process, the power of the weaker messages is increased in relation to that of the stronger, without an increase in the maximum message or signal powers. The ratio of message power to interference power is thereby increased for the weaker messages, in exchange for a reduction in this ratio for the strongest messages, as will be illustrated shortly.

The compression just mentioned can be automatically controlled by the average power of each message, and will then occur at sufficiently slow rates so that the effect of transients can be disregarded. That is to say, the bandwidth of the transmitted signals will not be significantly greater than that of the message waves. As applied to voice transmission, this process is known as syllabic companding, since the insertion of compression and complementary expansion is at a syllabic rate. The improvements in performance that can be realized is exemplified below for a particular kind of compandor.

Let $x = \sigma_x/\hat{\sigma}_x$ and $y = \sigma_y/\hat{\sigma}_x$, where σ_x is the rms amplitude of a particular message wave, $\hat{\sigma}_x$ that of the maximum message wave, and σ_y that of the compressed wave. With change in compression at an adequately slow rate, as in syllabic companding, the following general nonlinear relation applies for the compressor characteristic, $g(x)$, and the expandor characteristic, $h(y)$

$$y = g(x), \tag{2.21}$$

$$x = h(y), \tag{2.22}$$

where $h(y)$ is obtained by solution of (2.21) for x.

Nearly optimum performance can be realized with logarithmic types of compressor characteristics that are also conveniently implemented. A characteristic of this type that is also convenient from the standpoint of analysis is represented by

$$y(x) = \frac{\sinh^{-1} \mu x}{\sinh^{-1} \mu} \qquad (2.23)$$

$$= \pm \frac{\ln [\pm \mu x + \sqrt{\mu^2 x^2 + 1}]}{\ln (\mu + \sqrt{\mu^2 + 1})}, \qquad (2.24)$$

where the negative signs apply for negative values of x. The inverse expandor characteristic is then

$$h(y) = \frac{1}{\mu} \sinh (y \sinh^{-1} \mu). \qquad (2.25)$$

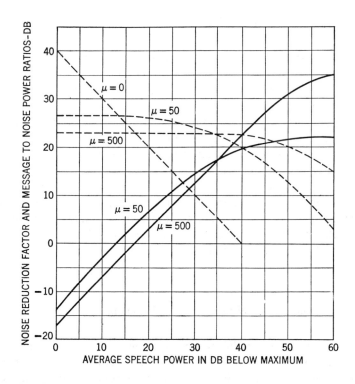

Figure 2.5 Noise reduction factors and message-to-noise power ratios for a syllabic compandor with compressor characteristic $g(x) = \sinh^{-1} \mu x/\sin h^{-1} \mu$. ——— Noise reduction factor $-10 \log_{10} \eta$. - - - - Ratio of average message power to average noise power.

At the expandor input, let the rms noise amplitude be σ_n, let $n = \sigma_n/\hat{\sigma}_x$ and at the output let $n_0 = \sigma_n^{(O)}/\hat{\sigma}_x$. The following relation then applies

$$x + n_0 = \frac{1}{\mu} \sinh\left[(y + n)\sinh^{-1}\mu\right], \tag{2.26}$$

from which can be determined n_0 given n, μ, and x. For $n\sinh^{-1}\mu \ll 1$, as would ordinarily be the case, the following approximation applies

$$x + n_0 \cong x + n\frac{\sinh^{-1}\mu}{\mu}\cosh(\sinh^{-1}\mu x)$$

$$= 1 + n(1 + \mu^2 x^2)^{\frac{1}{2}}\frac{\sinh^{-1}\mu}{\mu}. \tag{2.27}$$

The noise reduction factor is thus

$$\eta \cong \frac{\sinh^{-1}\mu}{\mu}(1 + \mu^2 x^2)^{\frac{1}{2}}. \tag{2.28}$$

In Fig. 2.5 is shown the above noise reduction factor in dB as a function of the ratio x in dB, for $\mu = 50$ and $\mu = 500$. In the same figure are shown the ratios of average message power to average noise power on the premise that this ratio is 40 dB for $x = O$ and $\mu = O$. It will be noted that with compandors it is possible to obtain a nearly constant ratio over about a 40 dB range of variation in speech volumes.

In the case of crosstalk interference between similar systems, evaluation of the interference reduction afforded by compandors is somewhat more complex and is discussed further elsewhere [7]. Nearly the same reduction as for random noise is obtained under the worst condition, in which the interfering talker has a high volume and the disturbed talker a low volume. With interference from a multiplicity of talkers with different volumes, the reduction is less than for random noise.

2.4 Message Modification by Instantaneous Compandors

Syllabic compandors modify the average power but have a minor effect on the instantaneous amplitude probability distribution and other message parameters. By contrast, with an instantaneous compressor, the message $x(t)$ would be modified into $y(t)$ with different parameters, $p(y)$, P_y, and $W_y(\omega)$. In the previous relations $g(x)$ and $h(y)$, $x = x(t)$, and $y = y(t)$ are now rapidly varying functions.

If $x(t)$ has a probability density, $p(x)$, then that of $y(t)$ becomes for monotonic functions $g(x)$ and thus $h(y)$.

$$p(y) = p[h(y)]h'(y), \qquad (2.29)$$

where $x = h(y)$ is the expandor characteristic (2.22).

If $P_x(x \leq x_1)$ is the probability distribution of $x(t)$, then that of $y(t)$ becomes

$$P_y(y \leq y_1) = \int_{-\infty}^{y_1} p(y)\, dy = P_x[x \leq h(y_1)]. \qquad (2.30)$$

For example, if $P(x \leq x_1) = 1 - e^{-\alpha x_1^{1/2}}$ and $y = x^{1/2}$, then $x = h(y) = y^2$ and $P_y(y < y_1) = P_x(x < y_1^2) = 1 - e^{-\alpha y_1}$.

The average power at the compressor output becomes

$$\mathsf{P}_y = \int_{-\infty}^{\infty} y^2 p(y)\, dy \qquad (2.31)$$

$$= \int_0^{\infty} W_y(\omega)\, d\omega. \qquad (2.32)$$

The power spectrum can in principle be determined by methods outlined in Appendix 3, but actual determination for representative compressor characteristics becomes very difficult. The nonlinear instantaneous compression would be accompanied by an expanded and, in theory, infinite bandwidth of the signal power spectrum. Hence direct application of instantaneous compandors to analog messages entails increased channel bandwidth if excessive distortion is to be avoided. To the same end, a nearly linear phase characteristic is required, a condition not imposed with syllabic compandors. Because of the above disadvantages, instantaneous compandors are not used in direct analog transmission. The principal advantage of an instantaneous compandor is that implementation is simpler than with syllabic compandors. Moreover, certain transmission impairments can be avoided that may be encountered when channels provided with syllabic compandors are used for other than voice communication.

An increase in channel bandwidth is not inherent in instantaneous compandors, and can be avoided with the aid of signal sampling, as discussed later, and by other more complicated methods of decompression, rather than with simple complementary expansion, as considered here. Because of the advantages over syllabic compandors just noted, instantaneous compandors may be used in time-division systems. In this application there is the further advantage that a single instantaneous compandor can be used in common for all channels, whereas with frequency-division multiplexing and direct analog transmission, a syllabic compandor would be required in each channel.

2.5 Message Modification by Filters

To improve performance, the statistical parameters of message waves may be modified by appropriate linear filters. These are intended to modify the power spectrum $W_x(\omega)$ into $W_y(\omega)$, but this will ordinarily be accompanied by modification of other statistical parameters.

With a linear transducer of transmittance $T(i\omega)$ the power spectrum is modified into

$$W_y(\omega) = |T(i\omega)|^2 W_x(\omega). \tag{2.33}$$

The average transmitted power becomes

$$R_y(O) = P_y = \int_O^\infty W_y(\omega)\, d\omega. \tag{2.34}$$

If the channel between transmitter and receiver has a transmittance $L(i\omega)$, then the receiving filter must have a transmittance $R(i\omega) = [L(i\omega)T(i\omega)]^{-1}$ if distortion is to be avoided. With a noise power spectrum, $W_n(\omega)$, at the input to the receiving filter, the average noise power at the output becomes

$$N_0 = \int_O^\infty \frac{W_n(\omega)\, d\omega}{|L(i\omega)|^2 |T(i\omega)|^2}. \tag{2.35}$$

As shown in Chapter 4, by appropriate choice of $T(i\omega)$ it is possible to minimize N_0 for a given transmitter power P_y. The improvements that can be realized by this means are ordinarily rather small, a few dB.

The phase characteristic of the modifying transducer does not influence the power spectrum, $W_y(\omega)$, or the average power, P_y, but will affect the probability density, $p(y)$, of the transmitted wave unless the phases of the spectral components of $W_x(\omega)$ are statistically independent, as in the case of random noise. The process of spectrum modification just mentioned ordinarily is referred to as frequency pre-emphasis, since the frequency spectrum is modified in a specific way, though this may be accompanied by a modification in probability distribution.

Idealized impulse noise is characterized by a linear variation in the phase of the spectral components with frequency. The peak amplitude of such impulse noise can be reduced by introducing phase distortion at the receiver, with complementary phase correction at the transmitter to avoid transmission distortion. This method may come into consideration in digital pulse systems, where impulse noise can be an important source of error in certain situations.

2.6 Sampling and Quantization of Continuous Waves

As shown in Chapter 1, a continuous wave of bandwidth B can be represented uniquely by narrow samples taken at intervals $T = 1/2B$. Such sampling introduces certain new possibilities in systems design besides direct transmission. Thus samples from various messages can be interleaved in time, as in time-division multiplexing, to be discussed later. This permits a single instantaneous compandor to be used in common for all message waves, without the need for increased channel bandwidth. As applied to speech transmission, the noise reduction afforded by such compandors is much the same as discussed for syllabic compandors in Section 2.3.

The samples may modulate a signal wave, or may be quantized prior to such modulation. Moreover, the quantized samples instead of modulating a signal wave can be coded into a sequence of samples or pulses prior to such modulation. This permits a reduction in signal power in exchange for increased channel bandwidth. Quantization and coding permit the use of regenerative repeaters so that cumulation of the effects of noise or other transmission distortion along a repeater chain can be reduced.

In exchange for the above various advantages, it is necessary to introduce a certain quantizing error at the transmitting end in continuous wave transmission, and to accept the resultant quantizing-error noise. With uniform quantizing steps, s, the maximum error in the quantization of an infinite narrow signal sample is $\pm s/2$, and the minimum error is O, as indicated in Fig. 2.6. If all signal amplitudes between O and $\pm s/2$ are equally probable, then the rms quantizing error is $s/(2\sqrt{3})$. The resultant quantizing noise incurred at the transmitting end is equivalent to noise in the demodulated output at the receiving end. The average quantizing noise power becomes

$$N_0 = \frac{s^2}{12}. \qquad (2.36)$$

If a symmetrical wave of amplitude range, $\pm \hat{A}$, is represented by L quantizing levels, then each step is $s = 2\hat{A}/L$ and $s^2 = 4\hat{P}_0/L^2$, where $\hat{P}_0 = \hat{A}^2$ is the peak power. The ratio of average quantizing noise power

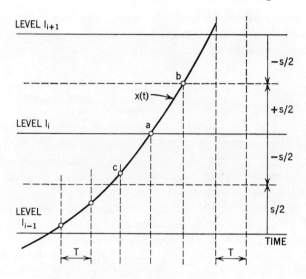

Figure 2.6 Quantizing errors at encoder. At three sampling points a, b and c, the continuous message wave $x(t)$ would be represented by the same quantized amplitude l_i, with a quantizing error between $-s/2$ and $+s/2$.

to peak power is thus

$$\frac{N_0}{P_0} = \frac{1}{3L^2}. \tag{2.37}$$

In addition to quantizing noise it is also necessary to consider noise from errors in the transmitted digits, as discussed in Section 2.12.

As mentioned before, instantaneous compandors can be applied to signal samples that are not quantized, as a means of reducing the effect of channel noise. When the samples are quantized, such compandors reduce quantizing noise power, particularly for messages of low power [8,9]. In application to speech transmission, the reduction in quantizing noise is about the same as the reduction in channel noise afforded by syllabic compandors.

Although a certain quantizing noise is incurred, it is possible, in principle, to avoid the effect of noise or other distortion in the channel, provided that the peak distortion is less than $s/2$. In principle, quantization permits the use of regenerative repeaters, so that cumulation of the effect of noise along a repeater chain can be avoided. However, since the quantizing steps must of necessity be small, only minor distortion by noise or other imperfections can be tolerated. For this reason, quantization and regeneration cannot be used to full advantage, unless the quantizing steps in the transmitted signal are increased by coding each quantized message sample into a sequence of pulses, as discussed in the following section.

2.7 Coding of Quantized Samples

A particular discrete amplitude can be represented by the following sequence of m digits

$$A = x_{m-1}r^{m-1} + x_{m-2}r^{m-2} + \cdots + x_0 r^0, \tag{2.38}$$

where r is the radix and $x_{m-1}, x_{m-2}, \ldots, x_0$ are the digital values. In a decimal system $r = 10$ and the digits $x_{m-1}, x_{m-2}, \ldots, x_0$ can each have the values $0, 1, 2, \ldots, 9$. In a binary system $r = 2$ and the digits can have the values 0 or 1. The number of digital values including zero is equal to r.

The number of levels that can be represented by m digits and a radix or base r is

$$L_r = r^m. \tag{2.39}$$

In communication theory, a binary system is used as reference, in which case

$$L_2 = 2^m, \tag{2.40}$$

and

$$\log_2 L_2 = m. \tag{2.41}$$

Each binary digit represents one unit of information, referred to as a bit. The number of information units represented by (2.39) is

$$I_r = \log_2 r^m = m \log_2 r. \tag{2.42}$$

As shown in Chapter 1, pulses or digits can be transmitted without intersymbol interference at a rate

$$s = 2B \tag{2.43}$$

over a flat channel of bandwidth B c/s and linear phase characteristic. The information transmitted by m digits during m seconds is thus $2BI_r$ and the information rate per second when the message is encoded in accordance with (2.38) is

$$R_r = 2B \log_2 r = B \log_2 r^2 \tag{2.44}$$

Such a method of coding message waves is known as pulse-code modulation (PCM), and the particular case of $r = 2$ as binary PCM. The latter method may entail a large increase in channel bandwidth ($m = 5$ to 9).

The advantage that may be derived from this increased channel bandwidth is that the effect of channel noise and other channel imperfections can be reduced, to the extent that the principal transmission distortion in the received message is quantizing noise. Pulse-code modulation thus affords a means of overcoming moderate channel imperfections other than by improvements in the channel transmission characteristics by such means as rather accurate gain and phase equalization.

PCM is the most efficient and practicable method of digital transmission of analog messages, as regards channel bandwidth requirements, but entails more complicated encoding and decoding equipment than certain other methods that require greater bandwidths. Among them can be mentioned pulse-duration modulation (PDM) and pulse-position modulation (PPM), which can be used for direct analog transmission or for transmission of quantized samples. In PDM a quantized level l would be represented by a pulse having the duration of l sampling intervals, which is equivalent to a sequence of l pulses of equal amplitudes and the duration of a sampling interval. In PPM the level l is represented by a displacement of l sampling intervals of a transmitted pulse from a time origin. With L quantizing levels the channel bandwidth with these methods must be greater than the message bandwidth by a factor L, as compared to a factor $m = \log_2 L$ in binary PCM. (For example, with $L = 64$, $m = 6$, so that over a 10-fold increase in channel bandwidth would be required.)

Instead of transmitting the quantized message amplitudes at each sampling point, as just assumed, it is possible with any one of the above pulse modulation methods (PCM, PDM, PPM) to transmit differences in amplitudes between one sampling instant and the next. This is known as differential or delta transmission, or as differential PCM, PDM, or PPM. If the channel bandwidth is reduced with differential transmission, all the slopes of the message waves cannot be adequately represented, so that in addition to quantizing noise, "slope overload noise" is also encountered. With certain kinds of message waves where the probability of encountering excessive slopes is adequately small, the combined distortion may be acceptable, even with substantially smaller bandwidth than required with direct transmission of quantized samples [10]. Thus delta PDM, or delta modulation, can be used for transmission of speech of acceptable quality over a channel of about the same bandwidth as required with PCM. As another example, differential PCM, is advantageous in picture transmission [3]. In both of these cases, the message waves contain considerable redundancy. The aforementioned differential methods would afford no advantage in applications to message waves with the statistical properties of flat Gaussian noise.

2.8 Baseband Signal Waves

In analog systems, the baseband signal waves, $z(t)$, are ordinarily the same as the message waves, $x(t)$, or the modified message waves, $y(t)$. In pulse systems, the wave samples, $y(t)$, may be transmitted as pulses of amplitude proportional to $y(t)$. In other systems, the samples, $y(t)$, may be coded into a sequence of pulses, as in PCM, or they may be represented by the time displacements of pulses, as in PPM. In all cases, the baseband input is ordinarily in the form of narrow rectangular pulses. These are then applied to a low-pass filter, and a sequence of such pulses gives rise to a baseband modulating wave. In general this wave is continuous, owing to the shaping of pulses by the low-pass filter, and a particular wave can be represented by

$$z(t) = \frac{1}{\pi} \int_0^\infty c_z(\omega) \cos\left[\omega t - \psi_z(\omega)\right] d\omega. \tag{2.45}$$

In the case of direct analog transmission, the functions $c_z(\omega)$ and $\psi_z(\omega)$ are the same as $c_y(\omega)$ and $\psi_y(\omega)$. With coded digital transmission, the functions for each wave, $z(t)$, depend on the coding method and on the

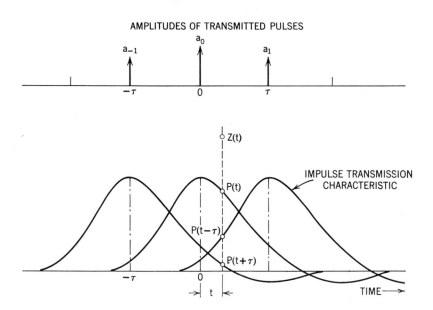

Figure 2.7 Representation of pulse train amplitude $z(t)$ in terms of impulse transmission characteristic $P(t)$. $z(t) = \sum_{n=-\infty}^{\infty} a(n) P(t - nT)$. In the above illustration $a(n) = 0$ for $1 < n < -1$.

characteristics of the low-pass filter noted above. It is, in this case, ordinarily preferable to relate $z(t)$ to the coded pulse sequence representing the wave $y(t)$, which can be done with the aid of the pulse-transmission characteristic, $P(t)$, of the low-pass filter as discussed in Chapter 1. In Fig. 2.7 the pulses are assumed to occur at intervals, T, and the amplitude of the pulse at time nT is denoted $a(n)$. The wave, $z(t)$, can then be represented by

$$z(t) = \sum_{n=-\infty}^{\infty} a(n) P(t - nT). \qquad (2.46)$$

Representation (2.46) has definite advantages over (2.45) in analyzing the performance of pulse systems.

2.9 Carrier Signal Waves

In most cases, transmission is over a bandpass channel, rather than over a baseband channel. It is then convenient to use a reference frequency ω_0 in the pass band, which ordinarily would coincide with the frequency of a sine-wave carrier, as discussed in Chapter 1 and indicated in Fig. 1.1. The transmitted signal wave is then of the general form

$$v(t) = \bar{v}(t) \cos [\omega_0 t - \varphi_0 + \varphi_v(t)] \qquad (2.47)$$

$$= X_v(t) \cos (\omega_0 t - \varphi_0) - Y_v(t) \sin (\omega_0 t - \varphi_0), \qquad (2.48)$$

where

$$X_v(t) = \bar{v}(t) \cos \varphi_v(t) = \frac{1}{\pi} \int_{-\omega_0}^{\infty} c_v(u) \cos [ut - \psi_v(u)] \, du, \qquad (2.49)$$

$$Y_v(t) = \bar{v}(t) \sin \varphi_v(t) = \frac{1}{\pi} \int_{-\omega_0}^{\infty} c_v(u) \sin [ut - \psi_v(u)] \, du, \qquad (2.50)$$

$$\bar{v}(t) = [X_v^2(t) + Y_v^2(t)]^{1/2} \quad \text{and} \quad \varphi_v(t) = \tan^{-1} \left[\frac{Y_v(t)}{X_v(t)} \right]. \qquad (2.51)$$

In the same manner as discussed in Section 1.3 for the impulse characteristic of narrow bandpass channels, a signal of the form (2.48) can be represented as the real part of the analytic signal

$$[X_v(t) + iY_v(t)]e^{i(\omega_0 t - \varphi_0)} = [X_v(t) + i\hat{X}_v(t)]e^{i(\omega t - \varphi_0)}, \qquad (2.52)$$

where $\hat{X}_v(t)$ is a Hilbert transform of $X_v(t)$. This facilitates a general exposition of various possibilities of carrier modulation and detection [11,12]. The present discussion will be confined to conventional methods.

Carrier Signal Waves

In the case of pure amplitude modulation of the carrier wave, it is necessary that $\varphi_v(t) = 0$ in (2.47) and hence that $Y_v(t) = 0$. This, in accordance with (2.50), entails that $c_v(-u) = c_v(u)$ and that $\psi_v(-u) = -\psi_v(u)$. In this case, $\bar{v}(t) = X_v(t)$ and of the following form

$$\bar{v}(t) = X_v(t) = [k_0 + kz(t)], \tag{2.53}$$

where $z(t)$ represents the baseband wave, while k_0 and k are constants. The carrier wave in this case is

$$v(t) = [k_0 + kz(t)] \cos(\omega_0 t - \varphi_0) \tag{2.54}$$

By appropriate demodulation processes, to be discussed later, the wave, $X_v(t)$, as given by (2.53), can be recovered, and in turn the baseband wave, $z(t)$, together with the message wave, $x(t)$.

The power spectrum of the wave ensembles represented by (2.54) is indicated in Fig. 2.8 and is symmetrical about ω_0. The spectrum, $W_v(u)$, to each side of the carrier frequency is referred to as the sideband spectrum and is equal to the spectrum of the baseband signal $[k_0 + kz(t)]$.

It is possible to eliminate the carrier and the spectrum to one side of the carrier frequency, as in single sideband transmission, with the aid of transmitting filters that remove one of the two sideband spectra present in the original symmetrical sideband wave.

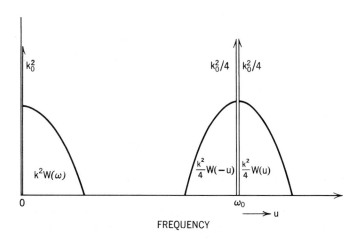

Figure 2.8 Power spectra of baseband modulating wave and amplitude modulated carrier wave. Note: The spectral components at $-u$ and u have equal phases so that they combine in the demodulator output by direct voltage addition, rather than by power addition with no correlation between phases.

If the spectrum is eliminated for $u < 0$, then the functions $X_v(t)$ and $Y_v(t)$ become

$$X_v(t) = \frac{1}{\pi} \int_0^\infty c_v(u) \cos[ut - \psi_v(u)] \, du, \qquad (2.55)$$

$$Y_v(t) = \frac{1}{\pi} \int_0^\infty c_v(u) \sin[ut - \psi_v(u)] \, du \qquad (2.56)$$

The signal wave is thus the general form (2.47) and contains both amplitude and phase modulation, since $\varphi_v(t) \neq 0$. The quadrature component of (2.48) can be eliminated by appropriate demodulation methods and the modulating wave $z(t)$ can thus be recovered together with the original message wave $x(t)$. Single sideband transmission affords a twofold reduction in channel bandwidth as compared to double sideband transmission. Of even greater importance in applications to message waves with a high ratio of peak to average power, such as a speech, is a radical reduction in transmitter power that comes about from elimination of the carrier.

For improved transmission performance in certain applications, it may be preferable to increase the bandwidth of the signal wave. One means to this end is to quantize and code the message wave, as discussed previously for PCM. Another conventional method is to modulate a sine-wave carrier in phase or in frequency. In (2.47) the envelope $\bar{v}(t)$ is then constant while $\varphi_v(t)$ is related to the baseband modulating wave $z(t)$. The transmitted wave is thus of the general form

$$v(t) = \bar{v} \cos\{\omega_0 t - \varphi_0 + \varphi[z(t)]\}. \qquad (2.57)$$

In pure PM

$$\varphi[z(t)] = kz(t), \qquad (2.58)$$

where k is a constant with appropriate dimension.

In pure FM

$$\varphi[z(t)] = k \int_0^t z(t) \, dt. \qquad (2.59)$$

The phase derivative φ' represents a frequency deviation from ω_0 ordinarily referred to as the instantaneous frequency deviation. With PM, the frequency deviation is proportional to $z'(t)$, and with FM, proportional to $z(t)$. With the aid of the inverse process at the receiver, called frequency discriminator detection, the modulating wave $z(t)$ can be derived and in turn the message wave $x(t)$, as discussed in Section 2.11.

It will be noted that in general (2.57) represents a nonlinear operation on the function $z(t)$. For this reason, in FM and PM the bandwidth of the

Carrier Signal Waves

power spectrum of the signal ensembles $v(t)$ is wider than in AM where a linear operation is involved in accordance with (2.54). In theory, the bandwidth of the sideband spectrum is infinite. However, for practical purposes the spectrum is ordinarily considered to be confined within a band $2(\varphi'_{max}(t) + \Omega_z)$, where Ω_z is the bandwidth of the spectrum of the modulating waves and φ'_{max} is the maximum frequency deviation, or the deviation exceeded with an adequately small probability. This is the Carson rule for the bandwidth of the bandpass channel, required in order that distortion from elimination of spectral components can be ignored in FM systems. As shown in Section 6.16, this bandwidth is adequate when the message or modulating wave has the properties of Gaussian noise with a flat power spectrum of bandwidth Ω_z, and the maximum frequency deviation is taken as 4 times the rms value, a common assumption.

It is important to note that with actual channel bandpass filters, pronounced phase distortion will be encountered in the pass-band, and that the resultant distortion in the demodulated wave is of greater importance than distortion from elimination of spectral components when the Carson rule is satisfied. Hence elaborate phase equalization is required. Distortion remaining after such equalization depends on residual attenuation and phase deviations, as discussed further in Chapter 6.

In the case of digital transmission by PM, $z(t)$ in (2.58) is of the general form (2.46), so that (2.57) becomes

$$v(t) = \bar{v} \cos \left[\omega_0 t - \varphi_0 + k \sum_{n=-\infty}^{\infty} a(n) P(t - nT)\right]. \quad (2.60)$$

With FM, $P(t - nT)$ is replaced by the integral of this function, in accordance with (2.59).

In the case of digital FM or PM systems, a significantly smaller bandwidth than indicated by the Carson rule may suffice without excessive distortion at sampling instants, particularly with binary FM or PM. There may then be appreciable envelope fluctuations in the received wave, which is permissible in systems with nearly linear repeaters. In fact, as discussed in Section 1.11, it is possible to avoid intersymbol interference at sampling instants in binary FM with a bandwidth no greater than required with double sideband AM, provided envelope fluctuations are allowed.

It may be noted that single sideband transmission is possible both with synchronous product demodulation, as discussed previously, and with frequency discriminator detection [11]. (U.S. Patent 3054073 by K. H. Powers.) Both of these methods entail pronounced envelope fluctuations, so that only minor repeater nonlinearities can be tolerated if excessive distortion is to be avoided. A principal advantage of FM systems is that highly nonlinear repeaters can be employed, since envelope fluctuations are absent. This advantage would be sacrificed with single sideband FM or PM. Moreover, generation of single sideband FM entails a complicated modulation process, since more is involved than elimination of one sideband if conventional frequency-discriminator detection is to be used. Finally, a twofold reduction in bandwidth is not realized except for small frequency deviation ratios.

2.10 Received Carrier Waves

The wave at the receiving end of the channel will differ from the transmitted signal wave, owing to imperfections in the channel transmission characteristic and channel noise $n(t)$. The received carrier wave can thus be written

$$w(t) = v_0(t) + n(t), \tag{2.61}$$

where $v_0(t)$ is related to the signal wave $v(t)$ through the channel transmittance. A particular wave $v_0(t)$ can be expressed in the same way as $v(t)$ by

$$v_0(t) = \bar{v}_0 \cos [\omega_0 t - \psi_0 + \varphi_{v_0}(t)]$$
$$= X_{v_0}(t) \cos [\omega_0 t - \psi_0] - Y_{v_0}(t) \sin (\omega_0 t - \psi_0), \tag{2.62}$$

where

$$X_{v_0}(t) = \bar{v}_0 \cos \varphi_{v_0}(t) = \frac{1}{\pi} \int_{-\omega_0}^{\infty} c_{v_0}(u) \cos [ut - \psi_{v_0}(u)] \, du, \tag{2.63}$$

$$Y_{v_0}(t) = \bar{v}_0 \sin \varphi_{v_0}(t) = \frac{1}{\pi} \int_{-\omega_0}^{\infty} c_{v_0}(u) \sin [ut - \psi_{v_0}(t)] \, du. \tag{2.64}$$

If the channel has a transfer characteristic

$$T(u) = A(u) e^{-iB(u)}, \tag{2.65}$$

then the functions X_{v_0} and Y_{v_0} are given by

$$X_{v_0}(t) = \frac{1}{\pi} \int_{-\omega_0}^{\infty} c_v(u) \, A(u) \cos [ut - \psi_v(u) - B(u)] \, du \tag{2.66}$$

$$Y_{v_0}(t) = \frac{1}{\pi} \int_{-\omega_0}^{\infty} c_v(u) \, A(u) \sin [ut - \psi_v(u) - B(u)] \, du. \tag{2.67}$$

In determining distortion from the channel imperfections, differences between $v_0(t)$ and $v(t)$ in the form of a constant factor can be disregarded, together with a fixed transmission delay τ. Hence $A(u)$ and $B(u)$ can be normalized to

$$A(u) = 1 + a(u), \qquad (2.68)$$

$$B(u) = u\tau + b(u), \qquad (2.69)$$

where $a(u)$ represents the departure from a mean constant amplitude characteristic and $b(u)$ the departure from a mean linear phase characteristic. For small values of $a(u)$ and $b(u)$ it follows with the aid of the Fourier integral energy theorem that the energy difference between transmitted and received waves is

$$\int_{-\infty}^{\infty} [v_0(t - \tau) - v(t)]^2 \, dt = \frac{1}{\pi} \int_{-\omega_0}^{\infty} c_v^2(u)[a^2(u) + b^2(u)] \, du. \qquad (2.70)$$

When the transmitted signal has a power spectrum $W_z(u)$, the variance in the received wave $v_0(t)$ relative to the transmitted waves $v(t)$ is

$$\sigma^2_{(v_0 - v)} = \int_{-\omega_0}^{\infty} W_z(u)[a^2(u) + b^2(u)] \, du. \qquad (2.71)$$

The variance $\sigma^2_{(v_0 - v)}$ is often referred to as characteristic distortion, since it is related to the transmission characteristic of the channel. It is also referred to as multiplicative noise, since it is proportional to the signal power, as distinguished from additive channel noise.

In general, the transmission frequency characteristic, as represented by (2.65), will not be linear, but will contain a nonlinear component. This nonlinear component will give rise to transmission distortion often referred to as nonlinear distortion or intermodulation noise, which is also related to signal power, though not linearly as in the case of characteristic distortion. Intermodulation distortion is also encountered with a linear characteristic having deviations $a(u)$ and $b(u)$, provided that the modulation method is nonlinear, such as FM.

When the transmission frequency characteristic of the channel varies with time, as in radio systems, there will in general be a variable flat attenuation over the channel band, together with time variations in $a(u)$ and $b(u)$. In this case, characteristic distortion is not always the same for a given signal wave, and cannot in principle be corrected by equalization if the time variation is unpredictable.

The noise wave $n(t)$ appearing in (2.61) can be represented in a similar manner as the signal wave by

$$n(t) = \bar{n}(t) \cos [\omega_0 t - \psi_0 + \varphi_n(t)]$$

$$= X_n(t) \cos (\omega_0 t - \varphi_0) - Y_n(t) \sin (\omega_0 t - \varphi_0), \qquad (2.72)$$

where $\bar{n}(t)$ is the noise envelope and

$$X_n(t) = \frac{1}{\pi} \int_{-\omega_0}^{\infty} c_n(u) \cos\left[ut - \psi_n(u)\right] du, \quad (2.73)$$

$$Y_n(t) = \frac{1}{\pi} \int_{-\omega_0}^{\infty} c_n(u) \sin\left[ut - \psi_n(u)\right] du. \quad (2.74)$$

In the case of random noise, $c_n(u)$ is independent of $\psi_n(u)$, and there is thus no correlation between $c_n(u)$ and $\psi_n(u)$ for different values of u.

In view of (2.62) and (2.72), the received wave $w(t)$ can be represented by

$$w(t) = X_w(t) \cos(\omega_0 t - \psi_0) - Y_w(t) \sin(\omega t - \psi_0) \quad (2.75)$$

$$= \bar{w}(t) \cos\left[\omega_0 t - \psi_0 + \varphi_w(t)\right], \quad (2.76)$$

where

$$X_w(t) = X_{v_0}(t) + X_n(t), \quad (2.77)$$

$$Y_w(t) = Y_{v_0}(t) + Y_n(t), \quad (2.78)$$

$$\bar{w}(t) = [X_w^2(t) + Y_w^2(t)]^{1/2}, \quad (2.79)$$

$$\varphi_w(t) = \tan^{-1}\left[\frac{Y_w(t)}{X_w(t)}\right], \quad (2.80)$$

$$\varphi_w'(t) = \frac{Y_w'(t) X_w(t) - Y_w(t) X_w'(t)}{X_w^2(t) + Y_w^2(t)}. \quad (2.81)$$

2.11 Carrier Wave Demodulation

The first step in the demodulation process is to derive from the carrier signal wave $w(t)$ a baseband wave $z_0(t)$ which will differ from the transmitted baseband wave $z(t)$ because of characteristic distortion and channel noise. The next step is to derive from the modified baseband wave $z_0(t)$ a message wave $x_0(t)$.

In amplitude modulation systems where $\varphi(t) = 0$, the baseband wave $z_0(t)$ can be derived with the aid of product demodulation. In this process, the wave $w(t)$ is applied to one pair of terminals of a demodulator and a steady-state sine wave $a(\cos \omega_0 t - \psi_0)$ to another pair. The demodulator output becomes

$$aw(t)(\cos \omega_0 t - \psi_0) = \frac{a}{2} X_w(t) + \frac{a}{2} X_w(t) \cos(2\omega_0 t - 2\psi_0) \quad (2.82)$$

With actual implementation there will be some phase error in the demodulating sine wave, which is disregarded here.

Carrier Wave Demodulation

The last term is eliminated with the aid of a low-pass filter with a cutoff above the maximum frequency component of the baseband wave $z(t)$. The demodulated baseband wave thus becomes

$$z_0(t) = \frac{a}{2} X_w(t)$$

$$= \frac{a}{2} [X_{v_0}(t) + X_n(t)]. \tag{2.83}$$

In view of (2.53), this can be written, with $a/2$ normalized to 1

$$z_0(t) = k_0 + kz(t) + X_d(t) + X_n(t), \tag{2.84}$$

where

$$X_d(t) = X_{v_0}(t) - X_v(t). \tag{2.85}$$

The component $X_d(t)$ arises from transmission distortion, owing to imperfections in the channel characteristic, as represented by (2.68) and (2.69). This component is related to the baseband wave and represents so-called multiplicative noise. With time invariant channels, multiplicative noise can in principle be compensated for, as distinguished from the component $X_n(t)$ representing additive noise. With a nonlinear transfer characteristic, there will be an additional component in (2.84) that is not linearly related to the baseband wave.

With envelope detection, the demodulated baseband signal is proportional to $\bar{w}(t)$, as given by (2.79). Since $\varphi_v = O$, $Y_v(t) = O$ and $Y_{v_0}(t) \approx O$. With $Y_n(t) \ll X_{v_0}(t)$, the following approximation applies for practicable systems:

$$\bar{w}(t) \cong X_{v_0}(t) + X_n(t). \tag{2.86}$$

Hence the demodulated baseband wave, which is proportional to $\bar{w}(t)$, is of the same form as (2.83).

In single sideband systems the transmitted wave is of the general form (2.48), with $X_v(t)$ and $Y_v(t)$ given by (2.55) and (2.56). By product demodulation with a sine wave $\cos(\omega_0 t - \psi_0)$ the component $Y_w(t)$ of the received wave (2.75) can be eliminated in the demodulator output and the component $X_w(t)$ recovered in the same manner as for double sideband AM. However, since a carrier frequency component is absent in the received wave it is not possible to derive a sine wave of frequency ω_0 for synchronous demodulation. A departure from the carrier frequency will give rise to a corresponding frequency shift in the spectrum of the demodulated wave, together with some distortion, owing to the presence of the quadrature component $Y_w(t)$. This turns out to be of no concern in application to speech transmission, provided the frequency error is limited to about 10 c/s, which is possible with present stable oscillators at the transmitter and receiver. In other applications where the wave shape must be better preserved, it is necessary to transmit a pilot carrier component for automatic frequency control to facilitate synchronous demodulation. If a sufficiently strong carrier component is transmitted, it is also possible to employ envelope detection to recover the component $X_w(t)$, since the quadrature component $Y_w(t)$ will then have a negligible effect on the envelope fluctuations. Other methods of envelope detection are possible that do not require a very strong carrier frequency component, provided advantage is taken of the relationship between envelope and phase variations in single sideband transmission [12].

In phase modulation the received wave $w(t)$ is applied to a detector with an output proportional to $\varphi_w(t)$, as given by (2.80), while in frequency modulation the detector output is proportional to $\varphi'_w(t)$ as given by (2.81). Since the output is not linearly related to $X_w(t)$ or $Y_w(t)$ or time derivatives of these waves, determination of the effect of distortion and noise becomes more complicated than for AM where a linear relation exists.

Ordinarily, frequency discriminator detection is used regardless of the functional relation $\varphi[z(t)]$ in the modulating process indicated by (2.57). It is then necessary to employ the inverse process at the output of the discriminator to recover the modulating wave $z(t)$. Thus, with pure PM it is necessary to integrate the output of the frequency discriminator. The combination of a frequency discriminator and a post-detection integrator thus constitutes a phase detector.

In technical literature it is customary to use the designation FM for those methods that employ frequency discriminator detection as a first step in the carrier demodulation process, regardless of the modulating procedure. Thus in single sideband FM the modulation process entails a combination of amplitude and frequency modulation, and this also applies to certain digital FM methods to be considered in Chapter 5. This convention will be followed here. In mathematical work the general designation "angle modulation" is often used, which precludes convenient abbreviation, lest it be confused with AM.

After the baseband wave $z_0(t)$ has been derived from the carrier wave, the message waves $y_0(t)$ and $x_0(t)$ are in turn obtained with the aid of demodulation processes that are the inverse of the modulation processes used in the transmitter. Thus if $z(t)$ is a coded input wave as in PCM, the wave $z_0(t)$ must be decoded into a message $y_0(t)$ consisting of quantized message samples at fixed intervals. These samples are in turn applied to an appropriate low-pass filter in order to obtain a continuous message wave $x_0(t)$ of the same form as the input wave $x(t)$ except for distortion owing to quantizing noise and digital errors in transmission and imperfections in the final low-pass filter. In analog systems where coding is not employed, the baseband wave $z_0(t)$ is modified into $y_0(t)$ and in turn into $x_0(t)$ by appropriate filtering and modification of the amplitude distribution, to compensate for similar modulation steps at the transmitting end.

2.12 Threshold Performance of PCM and FM Systems

In transmission of analog message waves by digital methods, it is necessary to consider not only noise from quantizing errors in the encoder, but also noise due to errors in the transmitted pulses representing the digits, caused by random noise in the channel and other imperfections mentioned previously. Errors in digits will cause related errors in the decoded samples representing the analog wave. Noise from such errors is equivalent to that caused by pulses of the same duration as the samples and amplitudes equal to that of the committed error. As the signal-to-noise ratio at the receiver input is reduced below a certain threshold value, digital error noise begins to become noticeable, and will ultimately predominate over quantizing noise, as will be discussed presently for PCM systems.

With equal probability of errors in all digits, it can be shown that the squared rms error in a demodulated sample is $L^2/3m$, where $L = r^m$ is the number of quantizing levels with r digit amplitudes and m digits per sample. (See Section 2.7.) Such a relation applies regardless of the amplitude probability distribution of the message wave. The mean squared amplitude of the message wave is L^2/k^2, where k is the ratio of peak to rms amplitude. The ratio of average error power per sample to average message power is thus $k^2/3m$. If the probability of error in a digit is ϵ, then that in a sample is $m\epsilon$, and the ratio of average digital error noise power to average message power becomes $\epsilon k^2/3$.

In accordance with (2.37) the ratio of average quantizing noise power to average message power becomes $k^2/3L^2$. The following relation is thus obtained for the ratio of average message power to combined average output noise power:

$$\rho_0 = \frac{3}{k^2} \frac{L^2}{1 + \epsilon L^2}. \qquad (2.87)$$

By way of numerical illustration, for sine-wave modulation $k^2 = 2$. If additive Gaussian noise is the only source of errors, the probability ϵ is obtained from (5.64), both for bipolar baseband transmission and for bipolar carrier amplitude modulation with synchronous demodulation.

For purposes of later comparison with FM, binary bipolar AM or two-phase modulation will be assumed, in which case $\epsilon = \frac{1}{2} \text{erfc } \rho_i^{1/2}$, where $\rho_i = \rho/2$ is the ratio of average carrier power to average noise power in the band $1/T$ of the predetection filter. With $m = 6$ digits, the following relation is then obtained for binary PCM: $\rho_0 = 6140/(1 + 2048 \text{ erfc } \rho_i^{1/2})$.

The latter relation is shown in Fig. 2.9. With actual implementation a certain increase in ρ_i would be required for a given ρ_0. Thus, if digital transmission by binary FM were assumed the increase would be about 3 dB. Ideal PCM with synchronous demodulation, as assumed above, would be expected to represent optimum performance as regards the threshold relation between ρ_i and ρ_0. With other modulation methods that permit a noise reduction in exchange for increased bandwidth, a less favorable relation would be expected, as illustrated below for FM.

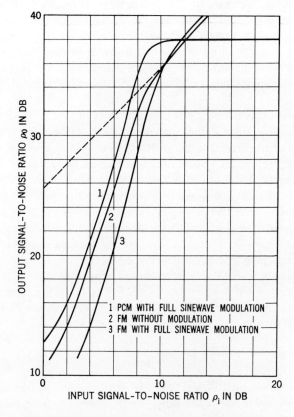

Figure 2.9 Relation between input and output signal-to-noise ratios in PCM and FM systems for sinewave modulation with $m = 6$ in PCM and $\hat{D} = 5$ in FM.

Threshold Performance in PCM and FM

A property of FM that follows upon further examination of (2.81) is that the output message-to-noise ratio diminishes rapidly as the input signal-to-noise ratio decreases below a certain threshold value, about 10 dB. This comes about because of the greatly increased probability that noise will make the denominator of (2.81) small, which gives rise to impulses or "clicks" in the output $z_0(t)$.

The relation between output message-to-noise ratio and input signal-to-noise ratio for pure FM is given by (4.59) as a function of the rms frequency deviation ratio $\underline{D}$, for input signal-to-noise ratios above about 10 dB. For lower signal-to-noise ratios it is necessary to introduce the modification given by relation (3.99). Assuming the Carson rule bandwidth for the predetection filter and sine-wave modulation with peak frequency deviation ratio $\hat{D}$, the following relation is obtained by combining (4.59) and (3.99)

$$\rho_0 = \rho_i \frac{3(1 + \hat{D})\underline{D}^2}{1 + (12/3^{1/2})(1 + \hat{D})^2 \rho_i \, \text{erfc} \, \rho_i^{1/2}}, \qquad (2.88)$$

where $\hat{D} = 2^{1/2}\underline{D}$ for sine-wave modulation, ρ_0 is the output signal-to-noise ratio and ρ_i the ratio of average carrier power to average noise power in the band of the predetection filter. This relation applies in the absence of modulation, the output power with sine-wave modulation being used merely as a reference in specifying ρ_0. With modulation, a more complicated relation applies, as determined by Rice in Reference 8 of Chapter 3.

In Fig. 2.9 relation (2.88) is illustrated for the case $\hat{D} = 5$, such that the channel bandwidth is the same as for the PCM system. The relation between ρ_0 and ρ_i in the presence of sine-wave modulation as determined by Rice in the above reference is also shown. The latter relation is of primary interest as regards performance comparison. It will be noted that the difference is insignificant for ρ_i above 10 dB.

For a modulating wave with the properties of flat Gaussian noise it is customary to neglect peaks in excess of 4 times the rms value. Hence $k^2 = 16$ in place of 2 for sine-wave modulation. In Fig. 2.9 the values of ρ^0 are then reduced by $10 \log_{10} 8 = 9$ dB. The rms frequency deviation ratio is in this case $\underline{D} = \hat{D}/4$ in place of $\hat{D}/2^{1/2}$. Because of the smaller $\underline{D}$ the effect of modulation is reduced so that the curve for full modulation will approach that for no modulation.

It should also be noted that with noise modulation the ratio ρ_0 will not increase indefinitely with ρ_i as indicated in Fig. 2.9 for sine wave modulation, but will reach an upper bound comparable to that with PCM. This comes about because of intermodulation distortion in FM, owing to attenuation and phase deviations over the channel band in conjunction with channel bandwidth limitation, as considered further in Chapter 6.

From Fig. 2.9 it appears that FM entails a 2 to 3 dB increase in ρ_i compared to PCM. The threshold performance of FM can, in principle, be improved by automatic adjustment of the bandwidth of the predetection filter to the instantaneous frequency deviation. An alternate means to the same end is to use a predetection filter of fixed reduced bandwidth, and reduce the instantaneous frequency deviation ahead of this filter by appropriate control of the frequency of the oscillator producing the IF frequency. Such control can be accomplished, at least to some degree, by negative feedback, as discussed in further detail elsewhere [1,13]. Complete control by this means is not possible because of transmission delay around the feedback loop. Moreover, the noise reduction is partly offset by intermodulation distortion produced in the feedback path, owing principally to phase nonlinearity, which cannot be reduced without increasing the transmission delay. Such intermodulation distortion depends on the properties of the message wave and on the degree of modulation. For a given FM receiver, the effectiveness of negative feedback thus depends on the nature of the message wave and the degree of modulation or the system load. Furthermore, when negative feedback is employed, it is necessary to employ amplifying devices in the FM receiver, which provide an essentially flat gain over a much wider frequency band than without feedback, in order to insure stable operation. (See Reference 4 of Chapter 1.) Hence there is a certain device limitation on the maximum channel bandwidth for which negative feedback will be effective in reducing the threshold signal-to-noise ratio.

Theoretical evaluation of the threshold improvement has been made for the case of an unmodulated carrier [13]. However, for the more important case of full carrier modulation it is necessary to rely on experimental data obtained with various messages waves, because of the several factors mentioned above. Comprehensive experimental data are available for an FM receiver designed for a system with 2 mc bandwidth of the baseband channel, a peak deviation ratio $\hat{D} = 5$ and a predetection filter of 25 mc bandwidth [13]. For sine-wave modulation at frequencies below 1 mc and peak frequency deviations of ± 10 mc, a 3 to 4 db improvement in the threshold carrier-to-noise ratio ρ_i was realized for output signal-to-noise ratios below 30 db. About the same improvement was obtained with modulation by flat Gaussian noise for bandwidths up to 1 mc. However, for a bandwidth of 2 mc no improvement was realized, owing to increased intermodulation distortion in the negative feedback circuit.

With message waves containing much redundancy, such as TV waves, the subjective effect of intermodulation distortion is reduced. Subjective tests indicated about 4 db improvement with TV waves of about 3 mc bandwidth and a peak frequency deviation of ± 7 mc.

Frequency Division Multiplexing

It is possible that somewhat greater improvements in the threshold performance can be realized by other methods of threshold extension than negative feedback as used in the above tests [1]. With any method the improvements will diminish as the frequency deviation ratio is reduced. Hence threshold extension methods would hardly come into consideration for small frequency deviation ratios, as used in digital systems discussed in Chapter 5 and in typical microwave radio relay systems.

2.13 Frequency Division Multiplexing

The baseband wave $z(t)$ in the previous discussion may consist of a number of individual baseband waves combined in frequency division. In this process, the individual baseband waves modulate various subcarriers, in amplitude, phase or frequency. The resultant baseband modulating wave is the sum of these individual waves, and has a frequency spectrum, as indicated in Fig. 2.10. When the subcarriers are modulated in amplitude, the individual subcarriers may or may not be suppressed, and one or both sidebands on the carriers may be transmitted. Regardless of the method of

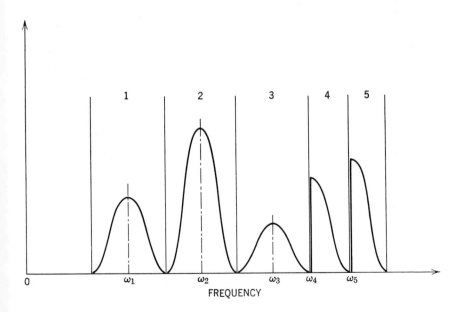

Figure 2.10 Signal spectrum with frequency division multiplexing of baseband channels. The individual carriers may be modulated in amplitude or frequency, or single sideband transmission may be employed, as indicated for channels 4 and 5.

subcarrier modulation, the combined baseband wave $z(t)$ may be transmitted by a number of modulation methods, such as AM, FM or PCM.

When a number of message waves or individual channels are combined as above, the various channels will not in general be active simultaneously. Let $\lambda(n,\epsilon_\lambda)$ designate the fraction of the total number of channels n that are active with an expectancy ϵ_λ of being exceeded. Thus $\lambda(100,0.01) = 0.4$ indicates that 40 or more channels are active during 1 percent of the total time. The average power equalled or exceeded for a fraction ϵ_λ of the total time is then for n message waves of equal average P_1

$$P(n,\epsilon_\lambda) = \lambda(n,\epsilon_\lambda)nP_1 \qquad (2.89)$$

Let $\hat{P}(n,\epsilon)$ be a power peak exceeded with an expectancy ϵ, when the average power is $P(n,\epsilon_\lambda)$

$$\hat{P}(n,\epsilon) = \mu(n,\epsilon_\mu) P(n,\epsilon_\lambda)$$
$$= \mu(n,\epsilon_\mu) \lambda(n,\epsilon_\lambda) nP_1, \qquad (2.90)$$

where $\epsilon \simeq \epsilon_\lambda \epsilon_\mu$. The factor $\mu(n,\epsilon_\mu)$ can be equal to or greater than $\mu(1,\epsilon_\mu)$. For example, if each wave is a sine wave, then the peak factor $\mu(n,\epsilon_\mu)$ is greater than $\mu(1,\epsilon_\mu)$. However, for speech waves the peak factor $\mu(n,\epsilon_\mu)$ for n waves in combination is smaller than $\mu(1,\epsilon_\mu)$ for a single wave.

In the following the expectancies ϵ, ϵ_λ and ϵ_μ are omitted for convenience in notation.

With a single message wave $\lambda(1) \simeq 1$ for $\epsilon_\lambda \leq 0.01$ so that $\hat{P}_1 = \mu(1)P_1$. With n message waves in frequency-division multiplex, the peak power that is exceeded with a given expectancy is greater than for a single message wave by the factor

$$\frac{\hat{P}_n}{\hat{P}_1} = \frac{\mu(n)\lambda(n)n}{\mu(1)}. \qquad (2.91)$$

With transmission of n message waves over individual channels, the combined peak transmitter power would be $n\hat{P}_1$. Thus the required peak transmitter power with frequency-division multiplexing differs from that with individual transmission by the factor

$$\gamma(n) = \frac{\hat{P}_n}{n\hat{P}_1} = \frac{\mu(n)\lambda(n)}{\mu(1)}, \qquad (2.92)$$

with a probability $\epsilon = \epsilon_\mu \epsilon_\lambda$ that this factor is exceeded. This factor can be smaller or greater than 1. For example, for a sine wave $\mu(1) = 2$, while for a very large number of sine waves a Gaussian probability distribution is approached, with a probability $\epsilon_\mu = 10^{-3}$ that a factor $\mu(n) \simeq 11$ is exceeded. Thus if $\lambda(n) = \frac{1}{4}$ for $\epsilon_\lambda = 0.01$, $\gamma(n) \simeq 1.4$ for $n \to \infty$. With an instantaneous amplitude distribution of speech waves as shown in Fig. 2.3, a factor $\mu(1) = 90$ is exceeded with a probability of 10^{-3}, while $\mu(n) = 11$ as for random noise. Thus in this case $\gamma(n) \simeq 0.03$ nor $n \to \infty$. This corresponds to about 15 dB reduction in peak power.

In systems design, the above factor $\gamma(n)$ is important from the standpoint of peak load rating of common channel amplifiers in baseband and carrier amplitude modulation and SSB systems. In FM systems it determines the peak frequency deviation, which, in conjunction with the bandwidth of the message wave, dictates channel bandwidth requirements.

It is possible to formulate the amplitude probability distribution of message waves in frequency division multiplex, as required to determine the factors $\mu(n)$ given $\mu(1)$. However, numerical evaluation is difficult, particularly in the case of speech waves, where the distribution of average speech power among individuals and the instantaneous amplitude distributions are as discussed in Section 2.2. An approximate semi-empirical evaluation of peak power requirements has been made by Holbrook and Dixon [4]. Based on their results, the factor $\gamma(n)$ for $\epsilon = 10^{-5}$ is about as shown in Fig. 2.10, for a factor $\lambda(n)$ that varies between 1 for $n = 1$ and $\frac{1}{4}$ for large values of n. This is the activity approached during the busiest hours in actual systems with a probability $\epsilon_\lambda = 0.01$ of being exceeded. More recently the multichannel load factor has been determined with greater accuracy by a computerized Monte Carlo method for various expectancies of being exceeded [14].

By way of comparison, in Fig. 2.11 is also shown the approximate factor $\gamma(n)$ for sine waves based on curves of the probability distribution published by Bennett [15] and the same factor $\lambda(n)$ as for speech. Curve 2 would also approximate the factor $\gamma(n)$ for a number of binary channels in frequency division multiplex.

Regardless of whether or not frequency division multiplexing affords any advantage from the standpoint of peak power, it has certain disadvantages. Thus transmission distortion, owing to a nonlinear transfer characteristic of the common channel, will give rise to interference among the individual message waves, ordinarily referred to as intermodulation distortion [16]. This places a limitation on the allowable nonlinearity of common amplifiers when the combined wave is transmitted directly or by amplitude modulation of a common carrier, as discussed in Chapter 8. When the combined wave is transmitted by phase or frequency modulation of a common carrier, attenuation and phase deviations over the channel band are sources of intermodulation distortion, considered further in Chapter 6. If the combined wave is transmitted by pulse modulation methods, such as PAM or PCM, timing deviations in the demodulated samples of the wave will give rise to interchannel interference [17,18]. Thus frequency division multiplexing of a number of waves imposes

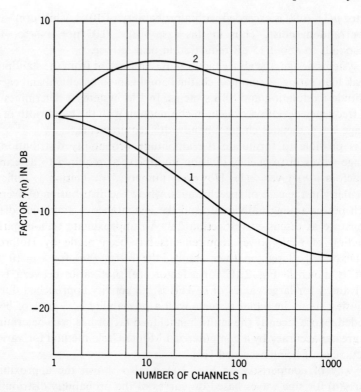

Figure 2.11 Approximate factors $\gamma(n)$ exceeded with an expectancy $\epsilon = 10^{-5}$. 1. Speech waves with equal average power. 2. Sine waves of equal average power.

requirements of one kind or another on the transmission properties of the common channel that ordinarily are not imposed when the waves are transmitted over individual channels.

2.14 Time-Division Multiplexing

In systems where the signals are sampled, it is possible to transmit samples from different channels in time division, or to transmit in time division coded sequences of pulses representing the samples, as in PCM. The duration of each pulse must then be shorter in inverse proportion to the number of individual channels, so that the bandwidth of the common channel is proportional to the number of channels. For an average power P_1 per wave and a nonsimultaneous load factor $\lambda(n)$ the combined average power with n channels in time division is given by (2.89), as in the case of frequency division. The peak power in this case is that which is in an individual channel, and is therefore greater than the average (2.89) by the factor $\mu(1)$ regardless of the number of channels and the activity $\lambda(n)$. Hence the peak power that is exceeded with a given expectancy is

$$\hat{P}_n = \mu(1)nP_1. \tag{2.93}$$

This is the same as the combined peak power required with transmission over n individual channels in frequency division. Hence in this case (2.92) is replaced by $\gamma(n) = 1$. Since it is possible that $\gamma(n) \gtrless 1$ with frequency division, there may or may not be a disadvantage in time-division multiplexing as regards peak power requirements. With a large number of speech waves, there is a disadvantage in this respect. This may, however, be offset with the aid of an instantaneous compandor, as will be discussed presently.

A distinct advantage of time-division multiplexing is that an unfavorable amplitude probability distribution of the combined message wave can be advantageously modified with the aid of a common instantaneous compandor for all channels [8,9]. Moreover, in time-division systems the use of an instantaneous compandor does not entail increased bandwidth. In systems where the samples are transmitted directly, as in pulse amplitude modulation, the reduced peak amplitude with compandors is reflected in reduced peak power. When the samples are coded as in PCM, a reduced peak amplitude of the message wave results in less quantizing noise for a given number of digits.

Each of the n component signals in time division may consist of a number of message waves in frequency division. In the case of speech, this may significantly reduce the peak factor. As indicated in Fig. 2.11, this reduction may be as much as 15 dB, yielding a corresponding reduction in quantizing noise in each message channel. It can be shown that with a Gaussian amplitude distribution that is approached with a large number of channels in frequency division, a further reduction of about 3 dB can be attained with a common instantaneous compandor. The combined reduction is about the same as afforded by a common instantaneous compandor with time-division multiplexing of individual voice channels.

With time-division multiplexing, a principal consideration is intersymbol interference, owing to overlap of adjacent pulses, as discussed in Chapter 7. In direct transmission of signal samples without coding, this will give rise to interchannel interference. In the case of speech transmission such interference, like noise, is reduced with the aid of a common instantaneous compandor. In coded pulse transmission, such as PCM, intersymbol interference reduces the tolerance to noise in the channel so that increased signal power is required for a specified error probability.

Another problem in time-division systems is the provision of timing waves for sampling of pulse trains. Moreover, in extensive time-division communication networks, synchronization of various network branches is desirable for efficient operation and is a difficulty not encountered with frequency-division multiplexing.

2.15 Cumulation of Distortion in Repeater Chains

A channel, as indicated in Fig. 2.1, generally includes a number of repeater sections in tandem, often several hundred. In analog systems the combined power of additive noise is proportional to the number of repeater sections. This also applies to characteristic distortion of multiplicative noise, provided that the source of such noise varies in an erratic manner between one repeater section and the next. This will be the case for small fine structure deviations in the attenuation and phase characteristics of the repeater sections. The above cumulation law may not then apply for each channel, but represents the expectation for a large number of channels with repeaters in tandem. By contrast, in the case of repeater nonlinearity, the source of distortion will be virtually the same in each repeater section. The resultant nonlinear distortion will then contain certain components that can cumulate by direct amplitude addition, provided that each repeater section after equalization has a linear phase characteristic [16]. Distortion power from these components will then be proportional to the square of the number of repeater sections. This condition may prevail for most third-order products resulting from amplifier nonlinearity in AM systems, as discussed in Chapter 8.

In the case of digital transmission, the channel may have a number of regenerative repeaters, with or without intermediate analog repeaters. Cumulation of additive noise and characteristic distortion between regenerative repeaters will then follow the same laws as for analog systems. With complete regeneration the error probability will be virtually proportional to the number of regenerative repeaters in tandem. However, owing to unavoidable imperfections in the timing waves derived from the pulse trains, random timing deviations will be encountered in regenerated pulses and will tend to increase with the number of repeaters [19]. Hence the error probability will increase at a faster rate than with complete regeneration.

With a number of transmission systems or links in tandem, additive random noise in analog systems and quantizing noise in digital systems will combine by power addition. This also applies for characteristic distortion of various kinds, with the possible exception of intermodulation distortion, owing to repeater nonlinearity. In the latter case the combination may be an amplitude addition basis, for third order modulation products, provided the phase characteristic is made linear in each repeater section by equalization and also that the various links are of the same kind and carry the same signal. Cumulation of third order products on an amplitude basis is avoided in extensive systems by signal "frogging" at appropriate intervals.

2.16 Choice of Modulation Method

The various modulation methods have unique properties that must be considered together with those of the transmission medium in the choice of modulation scheme. By way of example, double sideband AM is used widely in radio transmission, since it permits simple radio receivers. Because of the many receivers, this is a more important economic consideration than the greater transmitter power and bandwidth required with double- than with single-sideband transmission. For much the same reason double-sideband transmission is used on certain medium length carrier systems, since it permits simple terminal equipment [20]. On the other hand, single-sideband transmission is the preferred method for long-distance transmission over metallic circuits because the reduced signal power and bandwidth permit longer repeater spacing, which is here a more important economic consideration than simple terminal equipment [21].

Frequency modulation has a number of unique features that are used to advantage in various radio systems. For one thing, the absence of envelope fluctuations permits the use of highly nonlinear amplifiers and power transmitters, which is of first importance from the standpoint of maximum power output at radio frequencies. Another advantage of FM is that it facilitates fast gain adjustments in the presence of rapid fading, by predetection limiting in conjunction with automatic gain control. With AM the pronounced fluctuations in signal amplitudes caused by modulation makes it difficult to provide adequate automatic gain control against rapid fades. Finally, FM affords a reduction of the effects of both additive and multiplicative noise in exchange for increased channel bandwidth, by use of an adequately large frequency deviation ratio.

As an example, wideband FM is used for high-fidelity radio transmission over rather limited areas, since it permits a significant reduction in transmitter power. If FM with a smaller deviation ratio and reduced bandwidth were used, the fidelity would suffer, but a wider area could be covered with a given transmitter power and a specified rate of reception outages, owing to fading.

As another example, FM with about unity peak deviation ratio is used for microwave radio relay systems with many links in tandem. This deviation ratio affords better bandwidth utilization than a much larger ratio, together with a better compromise between outages owing to deep fades and transmission performance in the absence of fades. In this application the first two properties of FM mentioned above facilitate nonlinear repeaters and rapid gain control, which represent distinct advantages over AM or single-sideband transmission.

As a third example, FM with a large frequency-deviation ratio can be used advantageously in satellite systems to reduce satellite transmitter power requirements. Moreover, in this application, wide deviation FM facilitates the use of a common satellite repeater for several independently modulated carriers as a means of providing multiple access satellite systems, as discussed further in Chapter 8.

Finally, FM with a large frequency-deviation ratio is used in troposcatter systems, partly to reduce the effect of additive random noise and partly to limit intermodulation distortion caused by frequency selective fading, as discussed further in Chapter 9.

Binary PCM has desirable properties similar to those mentioned above for wide deviation FM. In addition it facilitates time-division multiplexing which can be a significant advantage in some applications. A most important additional attribute of binary PCM is that it facilitates the use of simple regenerative repeaters. The latter feature is a distinct advantage in systems involving a large number of stable repeater sections in tandem, as afforded by metallic circuits [22] or wave guides [17]. With regeneration, greater unwanted deviations in the attenuation and phase versus frequency characteristics can be tolerated than with single-sideband transmission or FM. The use of time-division multiplexing permits the use of a single instantaneous compandor. This is an advantage both from the standpoint of performance and cost in certain applications where syllabic compandors would be required in each voice channel with conventional single- or double-sideband, frequency-division multiplex systems.

However, in applications to radio relay systems, binary PCM would offer no advantage over FM, since it entails a wider band than FM with a small deviation ratio and more frequent outages, owing to fading. Moreover, regenerative repeaters would not be of much advantage here, for the reason that outages ordinarily come about because of severe fading in a single-repeater section, so that regeneration to avoid accumulation of noise in the remaining sections would not be of much aid. Nor would the regenerative feature of PCM be important in single-link satellite or troposcatter systems, although the possibility of employing time-division multiplexing may offer significant advantages.

2.17 Basic Performance Criteria for Analog Systems

The preferable modulation method in a given situation is generally dictated by certain basic considerations, such as those outlined in the preceding section. With any modulation method, systems engineering for efficient performance involves a variety of considerations, some of which are outlined in this section.

In transmission of analog messages by analog or digital modulation methods, the received wave generally includes the following components:

(a) The message wave of average power P, which, in general, will vary between zero and a full-load value.
(b) Additive random noise or quantizing noise of average power N.
(c) Linear characteristic distortion or multiplicative noise of average power $C_1 P$.
(d) Nonlinear or intermodulation distortion of average power $C_2 P^2 + C_3 P^3 + \ldots$.
(e) Interference from various external sources, including intersystem interference and impulse noise of average power X_e.

Generally, the combined distortion from all sources is thus

$$D(P) = N + C_1 P + C_2 P^2 + C_3 P^3 + \cdots + X_e. \tag{2.94}$$

The ratio of average distortion power to average message power becomes

$$\frac{D(P)}{P} = \frac{N}{P} + C_1 + C_2 P + C_3 P^2 + \cdots + \frac{X_e}{P}. \tag{2.95}$$

In the above general relations, the average noise power N and the various constants will depend on the modulation method and various system parameters.

In baseband, carrier amplitude modulation and single-sideband systems, N depends on the noise-power spectrum, together with the transmission-loss variation in the channel and the shapes of transmitting and receiving filters. The constant C_1 in this case depends on attenuation and phase deviations over the channel band, while C_2, C_3, etc., depend on amplifier nonlinearity. The transmitter or signal power is, in this case, proportional to the message power P, unless compandors are used.

In FM systems P determines the average frequency deviation, while N depends on the received carrier power, the noise power spectrum, the shapes of transmitting and receiving filters, and the frequency-deviation ratio. The constants $C_1, C_2, C_3, \ldots$, depend in this case on attenuation and phase deviations over the channel band, while amplifier nonlinearity can ordinarily be disregarded.

The relations apply for analog transmission with frequency-division as well as time-division multiplexing. In the latter case the constant C_1 depends on attenuation and phase deviations over the channel band, which gives rise to intersymbol interference of average power $C_1 P$. If the pulses are transmitted by amplitude modulation of a carrier, then the constants C_2, C_3 depend on amplifier nonlinearity, while with frequency modulation these constants depend on attenuation and phase deviations over the channel band.

The nature of the interference from external sources and its average power X_e depends on a variety of factors. It comprises interference between like and unlike systems, owing to electromagnetic coupling between them, as when they employ different metallic pairs in close proximity. In addition it may include impulse noise caused by atmospheric discharges or by switching transients in other systems.

With any modulation method, the ratio $D(P)/P$, or its equivalent, is ordinarily specified for a given system length and for the full-load condition resulting in the maximum value of $D(P)$. The specified value generally depends on the type of message under consideration. In the case of speech, considerable phase deviations among spectral components can be tolerated without significant impairment of intelligibility. Hence phase distortion over the transmission band of voice channels is of secondary importance to random noise and other interference, such as intermodulation noise. However, in picture or television transmission it is also necessary to place rather severe limitations on the permissible attenuation and phase distortion. The performance objectives in the foregoing various respects are of basic importance as regards transmission quality, particularly for extensive networks. Hence these objectives are subjects for international studies and recommendations by the International Telegraph and Telephone Consultative Committee (CCIT) and the International Radio Consultative Committee (CCIR).

Systems engineering for efficient performance with any modulation method involves the above and many other considerations. Among them are the cost of reducing the factors C_1, C_2, C_2, ..., by gain and phase equalization or by transmitter modifications to reduce nonlinearity and the cost of adequate suppression of interference from various external sources. Suffice it to note here that in extensive communication systems the evaluation and control of these factors and resultant distortion is ordinarily more of a problem than determination and control of transmission impairments from additive random noise.

In systems design it is ordinarily permissible to deal separately with the various kinds of transmission impairments noted earlier as is done in various chapters that follow, for the reason that any interaction between the different sources of transmission impairments is usually sufficiently weak to be disregarded.

In comparing systems intended to carry a variety of message waves, and in comparing the performance of different modulation methods, it is customary to assume a message wave with the properties of flat Gaussian noise. This assumption facilitates analytical evaluation of various kinds of distortion as well as experimental determination of performance. Certain modulation methods mentioned previously may afford advantages over others for certain message waves, such as speech or television. But such advantages usually disappear for message waves as encountered in multiplex systems and also for an assumed message wave with the properties of flat Gaussian noise. Hence the latter assumption insures a reasonably realistic and convenient basis of comparison.

2.18 Basic Performance Criteria for Digital Systems

Relations (2.94) and (2.95) also apply for digital transmission of analog message waves. In this case, N represents quantizing noise, which is related to the number of digits and the compandor characteristics. The constant C_1 depends on attenuation and phase distortion in terminal filters to which the message samples are applied, while $C_2, C_3, \ldots$, depend on nonlinearity in the coder-decoder combination and on random timing deviations in the decoded samples of the message wave. In addition there may be some noise owing to digital errors in transmission, usually small.

In analog systems the transmission impairments caused by interference from external sources depend principally on the average interference power. In digital systems where error probability is a basic criterion the probability distribution must also be considered and other statistical properties may also be important in the case of intermittent interference, such as impulse noise. Another complicating factor in digital systems is an unavoidable jitter in the timing wave derived at the receiver for sampling of the pulse train, as well as unavoidable deviations in reference carriers.

When the information to be transmitted is in the form of distinct symbols, such as numbers or letters in the alphabet, errors will be encountered, owing to additive random noise in conjunction with various other imperfections mentioned before. In this case the average distortion power at the demodulator output is given by a relation of the same form as (2.94). The error probability with a given number of digital levels depends on the ratio $D(P)/P$ as well as on the probability distribution of $D(P)$. Ordinarily, intersymbol interference, owing to attenuation and phase deviations, places a more important actual limitation on the practicable number of digital levels than does additive random noise in the transmission medium [23]. Nevertheless, it is customary to compare various digital modulation methods on the basis of their performance in the presence of random noise, on the premise that other impairments can in principle be eliminated, and for other reasons discussed in Section 5.7. The basic criterion in this case is error probability for a given information rate, bandwidth and transmitter power. In the particular case of transmission media with a flat loss over the channel band, the signal-to-noise ratio at the receiver input can serve as a basis for comparison, in place of the transmitter power.

With digital transmission, the permissible number of amplitudes, r, depends on the allowable error probability and on the ratio P/N of average signal power to average noise power in a band $B = 1/2T$ at the detector

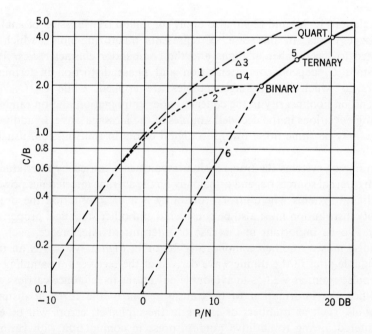

Figure 2.12 Channel capacities with various encoding methods. 1. $C/B = \log_2(1 + P/N)$; $H = \infty$; $\epsilon = 0$. 2. $C/B = 2\log_2 \dfrac{2}{1 + e^{-P/N}}$; $H = \infty$; $\epsilon = 0$. 3. C/B for $P/N = 10$ dB; $H = 1000$ bits; $\epsilon = 10^{-5}$. 4. C/B for $P/N = 10$ dB; $H = 100$ bits; $\epsilon = 10^{-5}$. 5. C/B with PCM; $H = 1$ bit; $\epsilon = 10^{-5}$. 6. C/B for sub-binary PCM; $H = 1$ bit, $\epsilon = 10^{-5}$.

input. The following relation, derived in Chapter 4, applies for r amplitudes of equal probability [24]

$$r = (1 + aP/N)^{1/2}. \qquad (2.96)$$

The factor a depends on the error probability and is about $a \simeq 0.12$ for a probability 10^{-6} of an error in a digit.

With PCM the corresponding channel capacity obtained from (2.44) is [24]

$$C = B \log_2 (1 + aP/N). \qquad (2.97)$$

This channel capacity is shown in Fig. 2.12 as a function of P/N for $r = 2$, 3, and 4, and an error probability, 10^{-5}. It is realized with digit-by-digit detection, using instantaneous sampling of the received pulse train at the midpoint of each signal interval of duration $T = 1/2B$.

The preceding conventional detection method entails a certain error probability for any finite ratio P/N. Other more complicated methods are possible that permit smaller ratios P/N for a given channel capacity and error probability. With sufficiently complicated coding methods it is even possible in principle to realize error-free communication in transmission over noisy channels for finite ratios, P/N. These methods are dealt with in literature on statistical communication and coding theory. Some of the principal results will be reviewed here, together with their prospective impacts on system performance.

2.19 Statistical Coding and Detection

The maximum information rate or channel capacity that in principle can be attained in the presence of Gaussian noise has been determined by Shannon from considerations of the average information content of the random received wave $x_0(t) = x(t) + n(t)$ and of the noise $n(t)$ alone. If the signal and the noise are independent random variables, then the maximum rate of information transfer that in principle can be attained is in accordance with Shannon [2].

$$R_{\max} = H'(x + n) - H'(n) \qquad (2.98)$$

$$= 2B[H(x + n) - H(n)], \qquad (2.99)$$

where the primes denote entropy or average information content per second, while $H(x + n)$ and $H(n)$ represent the entropy of the signal plus noise and of the noise alone, defined in accordance with (2.10). Relation (2.99) follows from (2.98) since $2B$ signal samples can be transmitted per second over an idealized low-pass channel of bandwidth B in c/s.

For a specified average power, N, the entropy of a random disturbance has its maximum possible value for a Gaussian probability density, and is

$$H(n) = \log_2 (2\pi e N)^{1/2}. \tag{2.100}$$

Similarly, the entropy of $x(t) + n(t)$ will be a maximum when both have a Gaussian probability density; and becomes

$$H(x + n) = \log_2 [2\pi e(P + N)]^{1/2}, \tag{2.101}$$

where P is the average signal power.

With (2.100) and (2.101) in (2.99)

$$C = R_{\max} = B \log_2 \left(1 + \frac{P}{N}\right). \tag{2.102}$$

To realize this maximum channel capacity it would be necessary to resort to complicated methods of encoding and decoding information. Thus for Gaussian noise, the message must be encoded into signal waves with the statistical properties of such noise. For example, if the message were represented by sequence of m binary digits, it would not suffice to transmit the corresponding binary pulse train over the channel and to use instantaneous digit-by-digit detection as in PCM. Rather, each sequence of m digits would be encoded into a wave with average power, P, and with the statistical properties of Gaussian noise. There would be a total of 2^m such signal waves, all of the same average power P but of different shapes. At the receiver these 2^m signal wave samples would be stored, together with the corresponding messages. Each signal-wave receiver would be compared with all the stored signals to determine the rms difference. In the absence of channel noise, such comparison would yield zero rms difference for the correct signal. In the presence of channel noise, the minimum rms difference would be expected for the correct transmitted signal, with a certain probability ϵ of error. This probability can be reduced by increasing m, or the duration of each encoded wave. The above maximum capacity can in principle be attained without errors for encoded waves or signals of infinite duration and thus infinite information content H.

In some applications, it may be desirable to obtain a lower signal-to-noise ratio than is possible with binary coding and digit-by-digit detection. This can be accomplished by various methods of subbinary transmission, discussed in Chapter 5, the simplest but least efficient of which would be repeated binary transmission. The maximum channel capacity that in principle can be realized when the signal consists of properly encoded binary sequences perturbed by Gaussian noise is given by [25]

$$C = 2B \log_2 \frac{2}{1 + e^{-P/N}}, \tag{2.103}$$

where P/N is defined as before. The latter capacity conforms with (2.102) for $P/N \ll 1$. To realize this channel capacity, it is necessary to use the same elaborate detection procedure as outlined in conjunction with (2.102), except that the information is now encoded into long binary sequences stored at the receiver, instead of signals having the properties of Gaussian noise.

In Fig. 2.12 a comparison is made of the foregoing limiting channel capacities with those obtained with PCM and digit-by-digit detection. For lower channel capacities than obtained with binary PCM, the curve shown in the figure applies for the least efficient method of repeated binary transmission and certain equivalent methods discussed in Chapter 5.

When the encoded message waves, and thus the stored signals, have finite information content, it is necessary to accept a certain error probability and reduction in channel capacity for a given P/N. Relations have been derived by Rice [26] for the channel capacity as a function of the error probability, the information content, and P/N. In Fig. 2.12 are shown two channel capacities obtained from these relations for $P/N = 10$ (10 dB) when the probability of an error in a digit is 10^{-5}, as for PCM. In one case the information content H of the encoded message is 1000 bits, in the second case 100 bits. It will be noted that in the latter case the reduction in signal-to-noise ratio compared to PCM is about 4 dB, instead of a maximum possible reduction of about 8 dB with encoded messages of infinite information content. When the information content of each message is 100 bits, the number of signals to be stored in the receiver is $2^{100} \simeq 10^{30}$. If 1000 signal waves could be stored in a cubic centimeter, then the receiver would have about the size of the entire earth. There would also be an incredible delay in the decoding of each signal. If a particular received noisy signal could be compared with 10^{12} stored noiseless signals per second, it would take about $3 \cdot 10^{10}$ years to determine the correct signal among the 10^{30} signals stored in the receiver. Evidently such a detection scheme is an abstraction of purely academic interest.

Even if implementation were not a basic obstacle, the above 4 dB reduction in signal-to-noise ratio would be of minor importance in most communication systems, as the following illustration will show. In both analog and digital systems a transmission loss of about 60 dB or more is feasible in a repeater section. The loss in metallic pairs and wave guides is proportional to length, so that the above 4 dB reduction in signal-to-noise ratio would permit at most 7% increase in repeater spacing.

The more promising application of coding would appear to be for systems operating at subbinary transmission rates and thus low signal-to-noise ratios. In such systems, additive random noise is a principal limitation on performance, while various kinds of distortion, mentioned previously, can ordinarily be disregarded. For such applications, certain binary error-correcting codes appear feasible and may afford about a 3 dB reduction in signal-to-noise ratio over no coding (or simple repetition), compared to a maximum possible reduction of about 10 dB, indicated in Fig. 2.12 for encoded binary sequences of infinite length. Such error-correcting codes may prove advantageous in special situations where a hard limitation is imposed on available transmitter power, but not on extra bandwidth, and where in addition the transmission loss does not increase with bandwidth, the transmission speed is sufficiently slow for decoder operation and cost is not a prime consideration. These conditions might prevail in telemetry from space vehicles, where transmitter power is derived from solar batteries.

A comprehensive exposition of coding is presented in some publications that also contain pertinent bibliography [25,27]. This book is concerned principally with methods of evaluating several kinds of transmission limitations in conventional extensive communication systems, where coding can hardly be used to advantage as a means of reducing transmitter power. For this reason, statistical detection and coding are not topics in the chapters that follow, except for a brief discussion in Chapters 5 and 10 of the performance of some efficient error-correcting codes.

3
Statistical Properties of Gaussian Random Variables

3.0 General

As discussed in Chapter 2, since messages and related signals, as well as noise, are by their nature random variables, it is necessary to use statistical concepts and analyses in assessing the performance of communication systems. In such analyses, it is ordinarily a great convenience and often a legitimate approximation to resort to the mathematical abstraction of Gaussian random variables. This is such a close approximation in the case of thermal noise that a distinction is hardly made, and it would also apply for other physical processes involving the random motion of large numbers of charged particles. It is also a legitimate engineering approximation for certain important message waves and related signals, such as those encountered with frequency-division multiplexing of a sufficiently large number of digital or analog messages, regardless of the statistical properties of the component waves. The statistical properties of Gaussian random variables are also approached as regards variations with time and with frequency in the transmittance of random multipath channels, such as troposcatter links.

In most problems concerning transmission impairment by random noise, it is a legitimate and convenient approximation to assume that the received wave is a steady-state sine wave perturbed by noise. In digital transmission of pulses by discrete amplitude, phase, or frequency modulation of a carrier, the signal over a brief sampling interval can be regarded as a sine wave perturbed by noise, and the error probability can be determined accordingly. In continuous wave amplitude and frequency modulation systems, the transmitted wave in the absence of modulation is a steady state sine wave. Noise at the detector output under this condition is virtually the same as with full modulation, and this also applies to single-sideband systems. Moreover, the statistical properties of a sine wave plus a Gaussian variable component is of interest in application to time-variant channels in which the transmittance contains a fixed component together with a random multipath component.

Various statistical properties of Gaussian random variables are dealt with comprehensively in the papers by Rice on random noise [1] and sine wave plus random noise [2], and in a number of other publications [3, 11]. The basic analytical methods are recited here, together with specific results that will be applied in later chapters. In this exposition, no distinction is made between a Gaussian random variable and random noise, designated $n(t)$.

3.1 Analytical Representation of Noisy Sine Waves

Returning to Section 2.10, we can write the received wave in a communication channel as

$$w(t) = v_0(t) + n(t). \tag{3.1}$$

In the present case $v_0(t)$ is a steady-state sine wave

$$v_0(t) = a \cos \omega_0 t \tag{3.2}$$

and

$$n(t) = X_n(t) \cos \omega_0 t - Y_n(t) \sin \omega_0 t, \tag{3.3}$$

where

$$X_n(t) = \frac{1}{\pi} \int_{-\omega_0}^{\infty} c_n(u) \cos [ut - \psi_n(u)] \, du, \tag{3.4}$$

$$Y_n(t) = \frac{1}{\pi} \int_{-\omega_0}^{\infty} c_n(u) \sin [ut - \psi_n(u)] \, du. \tag{3.5}$$

Relations (2.75) to (2.81) now become

$$w(t) = [a + X_n(t)] \cos \omega_0 t - Y_n(t) \sin \omega_0 t \tag{3.6}$$

$$= r(t)[\cos \omega_0 t + \theta(t)], \tag{3.7}$$

$$r(t) = \{[a + X_n(t)]^2 + Y_n^2(t)\}^{1/2}, \tag{3.8}$$

$$\theta(t) = \tan^{-1} \frac{Y_n(t)}{a + X_n(t)}, \tag{3.9}$$

$$\theta'(t) = \frac{[a + X_n(t)] Y_n'(t) - Y_n(t) X_n'(t)}{[a + X_n(t)]^2 + Y_n^2(t)}. \tag{3.10}$$

Power Spectra and Autocorrelation Functions

In these relations $\theta(t)$ is the phase variation caused by a particular noise wave $n(t)$, and $\theta'(t)$ the corresponding instantaneous frequency deviation.

The noise waves vary continually, and in determining the effect of noise it is convenient to characterize the wave ensembles by certain statistical parameters. These include average power, the probability densities, the power spectrum and autocorrelation function, as discussed in Appendix 2. These quantities depend on the type of noise and are generally quite complex, except for Gaussian noise, as considered here.

3.2 Power Spectra and Autocorrelation Functions

The power spectrum $W(\omega)$ and autocorrelation functions $R(\tau)$ of random variable $n(t)$ are related by

$$R_n(\tau) = \int_0^\infty W_n(\omega) \cos \omega\tau \, d\omega, \tag{3.11}$$

$$W_n(\omega) = \frac{2}{\pi} \int_0^\infty R_n(\tau) \cos \omega\tau \, d\tau. \tag{3.12}$$

In these relations $W_n(\omega)$ is the one-sided power spectrum, as discussed further in Appendix 2.

The average noise power is given by

$$\sigma_n^2 = R_n(O) = \int_0^\infty W_n(\omega) \, d\omega. \tag{3.13}$$

The corresponding quantities for the functions $X_n(t)$ and $Y_n(t)$ given by (3.4) and (3.5) are

$$R_x(\tau) = R_y(\tau) = \int_{-\omega_0}^\infty W_n(u) \cos u\tau \, du, \tag{3.14}$$

$$\sigma_n^2 = R_x(O) = R_y(O) = \int_{-\omega_0}^\infty W_n(u) \, du = R_n(O), \tag{3.15}$$

where $W_n(u)$ is the power spectrum in relation to the reference frequency ω_0, as indicated in Fig. 3.1, and is the same for both components (3.4) and (3.5) when $\varphi_n(t)$ is a random variable with equal probability of all values between O and $\pm\pi$.

In view of (3.15), the mean squared value of the envelope of random noise becomes

$$\overline{w^2(t)} = \overline{(X_n^2 + Y_n^2)} = R_x(O) + R_y(O) = 2\sigma_n^2. \tag{3.16}$$

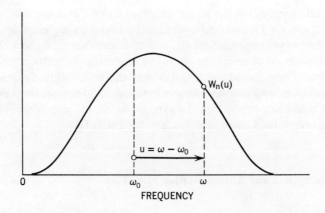

Figure 3.1 Power spectrum $W_n(u)$ in relation to carrier frequency ω_0.

Since the average value of the factor $\cos^2 \omega_0 t$ multiplying the envelope is $\frac{1}{2}$, the average power within the envelope is σ_n^2.

For the time derivative $n'(t)$ of random noise, the autocorrelation function and average power is given by

$$R_{n'}(\tau) = \int_O^\infty \omega^2 W_n(\omega) \cos \omega\tau \, d\omega, \tag{3.17}$$

$$R_{n'}(O) = \sigma_{n'}^2 = \int_O^\infty \omega^2 W_n(\omega) \, d\omega. \tag{3.18}$$

The autocorrelation function for the quantities $X_n'(t)$ and $Y_n'(t)$ are correspondingly

$$R_{x'}(\tau) = R_{y'}(\tau) = \int_{-\omega_0}^\infty u^2 W_n(u) \cos u\tau \, du, \tag{3.19}$$

$$R_{x'}(O) = R_{y'}(O) = \int_{-\omega_0}^\infty u^2 W_n(u) \, du. \tag{3.20}$$

The autocorrelation function $R_n(\tau)$, as given by (3.11), can alternately be written

$$R_n(\tau) = \bar{R}(\tau) \cos [\omega_0 t + \Phi(\tau)], \tag{3.21}$$

where

$$\bar{R}(\tau) = [R_{\cos}^2(\tau) + R_{\sin}^2(\tau)]^{\frac{1}{2}}, \tag{3.22}$$

$$\Phi(\tau) = \tan^{-1} \frac{R_{\sin}(\tau)}{R_{\cos}(\tau)}; \tag{3.23}$$

and

$$R_{\cos}(\tau) = \int_{-\omega_0}^{\infty} W_n(u) \cos u\tau \, du \qquad (3.24)$$

$$R_{\sin}(\tau) = \int_{-\omega_0}^{\infty} W_n(u) \sin u\tau \, du. \qquad (3.25)$$

The average noise power becomes

$$R_n(0) = [R_{\cos}(0) + R_{\sin}(0)] = R_{\cos}(0). \qquad (3.26)$$

For a spectrum with even symmetry about ω_0, $R_{\sin}(\tau) = 0$, $\Phi(\tau) = 0$ and $\bar{R}(\tau) = R_{\cos}(\tau)$. A spectrum with odd symmetry is not possible since $W_n(u) > 0$, but it is possible to have a component with odd symmetry such that $W_n(u) \geq 0$.

In the particular case of noise with a flat power spectrum $W_n(\omega) = W_n$ between $\omega_0 - \Omega$ and $\omega_0 + \Omega$ (3.14) yields

$$R_x(\tau) = R_y(\tau) = W_n \int_{-\Omega}^{\Omega} \cos u\tau \, du = \sigma_n^2 \frac{\sin \Omega\tau}{\Omega\tau}, \qquad (3.27)$$

where $\sigma_n^2 = 2\Omega W_n$ is the average noise power in the band 2Ω.

Expression (3.20) yields for the variance of the derivative of flat random noise

$$R_{x'}(0) = R_{y'}(0) = \sigma_{n'}^2 = \frac{\sigma_n^2 \Omega^2}{3}. \qquad (3.28)$$

$\bar{R}(\tau)$ as given by (3.22) is the envelope of the autocorrelation functions $R_{\cos}$ and $R_{\sin}$. This is to be distinguished from the autocorrelation function $R_r(\tau)$ of the envelope $r_n(t)$ for which the following relation applies [5]

$$R_r(\tau) = \bar{R}(0)\{2E[\kappa(\tau)] - [1 - \kappa^2(\tau)]K[\kappa(\tau)]\}, \qquad (3.29)$$

where $\kappa(\tau) = \bar{R}(\tau)/\bar{R}(0)$, $\bar{R}(\tau)$ is given by (3.22), and

E = complete elliptic integral of second kind,
K = complete elliptic integral of first kind.

3.3 Amplitude Variations of Gaussian Noise

Since $n(t)$ in accordance with (3.3) is the sum of a very large number of random variables with random phases, the probability density is Gaussian and given by

$$p(n) = \frac{1}{\sigma_n \sqrt{2\pi}} e^{-n^2/2\sigma_n^2}. \qquad (3.30)$$

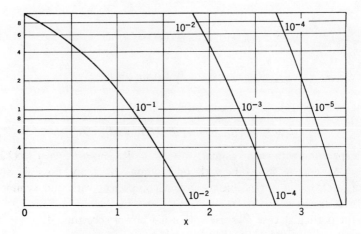

Figure 3.2 Error function complement erfc(x).

The same probability density applies for the components X_n and Y_n, with $\sigma_x = \sigma_y = \sigma_n$, and also for the time derivatives X'_n and Y'_n of random noise, with σ_n replaced by $\sigma_{n'}$, as obtained from (3.20).

In accordance with relations in Appendix 2, the probability that $n \leq k\sigma_n$ becomes

$$P(n \leq k\sigma_n) = \frac{1}{2}\left[1 + \text{erf}\left(\frac{k}{2^{1/2}}\right)\right]. \tag{3.31}$$

The complementary probability that the instantaneous noise amplitude exceeds the rms noise amplitude by a factor k, becomes

$$P(n \geq k\sigma_n) = \frac{1}{2}\left[1 - \text{erf}\left(\frac{k}{2^{1/2}}\right)\right] = \frac{1}{2}\text{erfc}\left(\frac{k}{2^{1/2}}\right). \tag{3.32}$$

The probability that $|n| \geq k\sigma_n$ is twice that given by (3.32) or

$$P(|n| \geq k\sigma_n) = \text{erfc}\left(\frac{k}{2^{1/2}}\right). \tag{3.33}$$

The above relations (3.31) through (3.33) also apply to X_n and Y_n, with $\sigma_x = \sigma_y = \sigma_n$, and to X'_n and Y'_n with σ_n replaced by $\sigma_{n'}$. The function erfc (x) is shown in Fig. 3.2.

3.4 Amplitude Variations of Noisy Sine Waves

Let the noisy sine wave be represented by

$$w(t) = a\cos(\omega_0 t + \psi_0) + n(t). \tag{3.34}$$

Amplitude Variations of Noisy Sine Waves 113

The probability density function of $w(t)$ can be determined from the characteristic function $C_w(u)$, which in turn is the product of the characteristic function $C_n(u)$ of $n(t)$ and $C_s(u)$ of a cos $\omega_0 t$.

In accordance with relations in Appendix 2

$$C_n(u) = e^{-\sigma_n^2 u^2/2}. \tag{3.35}$$

In cos $(\omega_0 t + \psi_0) = \cos\theta(t)$, $\theta(t)$ can have any value between 0 and 2π with equal probability. The characteristic function is in accordance with relations in Appendix 2

$$C_s(u) = E[e^{iau\cos\theta}] = \frac{1}{2\pi}\int_0^{2\pi} e^{iua\cos\theta}\, d\theta = I_0(au). \tag{3.36}$$

where I_0 is a Bessel function in the usual notation.

Hence, the characteristic function of $w(t)$ is

$$C_w(u) = e^{-\sigma_n^2 u^2/2} I_0(au). \tag{3.37}$$

The corresponding probability density of $w(t)$ is [3]

$$p(w) = \frac{1}{2\pi}\int_{-\infty}^{\infty} e^{-\sigma_n^2 w^2/2} I_0(au) e^{iuw}\, du$$

$$= \frac{1}{\sigma_n\sqrt{2\pi}}\frac{1}{\pi}\int_{-\pi/2}^{\pi/2} e^{-\rho[w/a - \sin\theta]^2}\, d\theta, \tag{3.38}$$

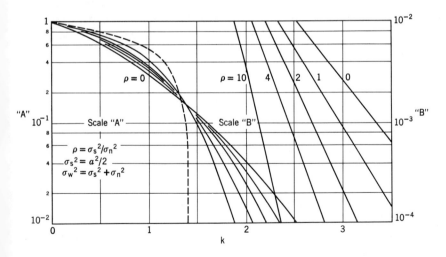

Figure 3.3 Probability $P(w \geq k\sigma_w)$ that the instantaneous amplitude w of a noisy sine wave exceeds the rms amplitude σ_w by a factor k, for various values of ρ.

where ρ is the ratio of average sine-wave power to average noise power

$$\rho = \frac{a^2}{2\sigma_n^2}. \tag{3.39}$$

The variance of $w(t)$ is $\sigma_w^2 = \sigma_s^2 + \sigma_n^2$, where $\sigma_s^2 = a^2/2$. The probability that $w < k\sigma_w$ is

$$P(w \leq k\sigma_w) = \int_{-\infty}^{k\sigma_w} p(w)\, dw. \tag{3.40}$$

This probability has been determined by numerical integration [6]. The complementary probability $P(w \geq k\sigma_w)$ is shown in Fig. 3.3 as a function of the parameter k for various values of ρ.

The above distribution of instantaneous amplitudes applies regardless of the bandwidth of the channels. In the case of narrow-band channels the probability density and distribution of the envelope of a noisy sine wave is of greater interest and is discussed in the following section.

3.5 Envelope Variations of Noisy Sine Wave

The probability density and distribution of the envelope of a sine wave plus random noise is of interest in connection with error probability in pulse transmission over narrow-band channels with envelope detection.

In (3.6) the noise voltages $X_n(t)$ and $Y_n(t)$ are independent random variables with zero mean, each having Gaussian probability density as given by (3.30) with $X_n = x$ and $Y_n = y$. The joint probability density is therefore

$$p(x,y) = p(x) \cdot p(y) = \frac{1}{2\pi\sigma_n^2} e^{-(x^2+y^2)/2\sigma_n^2}. \tag{3.41}$$

From (3.9) it follows that

$$x = r\cos\theta - a$$
$$y = r\sin\theta. \tag{3.42}$$

Hence $x^2 + y^2 = r^2 + a^2 - 2ar\cos\theta$. With the latter relation in (3.41) and with transformation into polar coordinates, so that $dx\, dy = a\, d\theta$, the following relation for the probability density of the envelope is obtained

$$p(r,\theta) = \frac{r}{2\pi\sigma_n^2} e^{-(r^2+a^2-2ar\cos\theta)/2\sigma_n^2}. \tag{3.43}$$

Envelope Variations of Noisy Sine Wave

The probability density of the envelope, obtained by integration with respect to θ between O and 2π is [2], [3], [4]

$$p(r) = \int_0^{2\pi} p(r,\theta)\, d\theta = \frac{r}{\sigma_n^2} I_0\left(\frac{ar}{\sigma_n^2}\right) e^{-(r^2+a^2)/2\sigma_n^2}, \qquad (3.44)$$

where I_0 is a Bessel function in the usual notation.

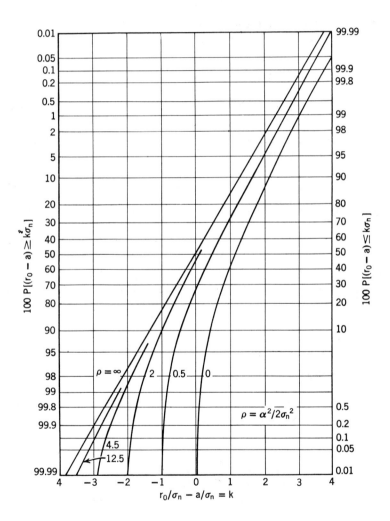

Figure 3.4 Probability $P[(r_0 - a) \geq k\sigma_n]$ that envelope variation $r_0 - a$ exceeds the rms noise amplitude σ_n by a factor k, and the complementary probability $P[(r_0 - a) \geq k\sigma_n]$.

The probability that $r \leq r_0$ becomes

$$P(r \leq r_0) = \int_0^{r_0} p(r)\, dr = e^{-(a^2 + r_0^2)/2\sigma_n^2} \sum_{i=1}^{\infty} \left(\frac{r_0}{a}\right)^i I_i\left(\frac{a r_0}{\sigma_n^2}\right), \qquad (3.45)$$

where I_i is a Bessel function in the usual notation.

For $a = 0$, a Rayleigh distribution is obtained

$$P(r \leq r_0) = 1 - e^{-r_0^2/2\sigma_n^2}. \qquad (3.46)$$

This type of distribution is approached for $a r_0/\sigma_n^2 \ll 1$.

For $a r_0/\sigma_n^2 \gg 1$, the following approximation is obtained [1]

$$P(r \leq r_0) \simeq \tfrac{1}{2} + \tfrac{1}{2}\operatorname{erfc}\left(\frac{r_0 - a}{\sqrt{2}\,\sigma_n}\right)$$

$$+ \frac{\sigma_n}{a\sqrt{8\pi}}\left(1 - \frac{r_0 - a}{4a} + \frac{\sigma_n^2 - (r_0 - a)^2}{8a^2}\right)e^{-(r_0 - a)^2/2\sigma_n^2}. \qquad (3.47)$$

As $a r_0/\sigma_n^2 \to \infty$, the distribution of $r_0 - a$ approaches the distribution of random noise as given by (3.31) or

$$P[(r_0 - a) \leq k\sigma_n] = \frac{1}{2}\left[1 + \operatorname{erf}\left(\frac{k}{2^{1/2}}\right)\right]. \qquad (3.48)$$

The complementary probability distribution becomes

$$P[(r_0 - a) \geq k\sigma_n] = \tfrac{1}{2}\operatorname{erfc}\left(\frac{k}{2^{1/2}}\right). \qquad (3.49)$$

Curves of the probability $P(r_0 \leq v\sigma_n)$ are given by Rice [2], with $v = r_0/\sigma_n$. The same curves also apply for the probability $P[(r_0 - a) \leq k\sigma_n]$ shown in Fig. 3.4 as a function of $(r_0 - a)/\sigma_n = k$, for various values of $\rho = a^2/2\sigma_n^2$. The curve for $a = 0$, and thus $\rho = 0$, is given by (3.46) and that for $\rho = \infty$, by (3.48).

3.6 Mean and Variance of Envelope

In connection with envelope detection of narrow band continuous-wave signals, the mean value and the variance of the envelope of a noisy sine wave is of interest. The mean value or d-c current with linear half-wave rectification is smaller than the mean value of the envelope, as will be considered presently, by a factor $1/\pi$, and in the case of a full-wave rectifier by a factor $2/\pi$. These factors come about because of the linear response of these devices to the sine-wave variations within the envelope.

The mean value of the envelope is

$$\bar{r} = \int_0^{\infty} r p(r)\, dr, \qquad (3.50)$$

Mean and Variance of Envelope

where $p(r)$ is given by (3.44). Evaluation of (3.50) gives [1,2]

$$\bar{r} = \left(\frac{\pi\sigma_n^2}{2}\right)^{1/2} e^{-\rho/2} \left[(1 + \rho)I_0\left(\frac{\rho}{2}\right) + \rho I_1\left(\frac{\rho}{2}\right)\right], \tag{3.51}$$

where I_0 and I_1 are Bessel functions in the usual notation. Except for a constant, the same expression can be derived [1] by integrating $wp(w)$ between O and ∞, when $p(w)$ is given by (3.38).

The following approximations apply:
For $x \ll 1$:

$$I_0(x) = 1 + \frac{x^2}{2} + \frac{x^4}{2^2 \cdot 4^2} + \frac{x^6}{2^2 \cdot 4^2 \cdot 6^2} + \cdots, \tag{3.52}$$

$$I_1(x) = \frac{x}{2}\left[1 + \frac{x^2}{2^2 \cdot 2} + \frac{x^4}{2 \cdot 2^4 \cdot 2 \cdot 3} + \cdots\right]. \tag{3.53}$$

For $x \gg 1$:

$$I_0(x) = \frac{e^x}{\sqrt{2\pi x}}\left(1 + \frac{1}{8x} + \cdots\right), \tag{3.54}$$

$$I_1(x) = \frac{e^x}{\sqrt{2\pi x}}\left(1 - \frac{3}{8x} + \cdots\right). \tag{3.55}$$

With these approximations in (3.51)

$$\bar{r} \simeq \left(\frac{\pi\sigma_n^2}{2}\right)^{1/2} e^{-\rho/2}\left(1 + \frac{\rho}{2}\right)^2; \quad \text{for} \quad \rho \ll 1 \tag{3.56}$$

$$= \left(\frac{\pi\sigma_n^2}{2}\right)^{1/2}; \quad \text{for} \quad \rho = 0 \tag{3.57}$$

$$\simeq a; \quad \text{for} \quad \rho \gg 1. \tag{3.58}$$

As noted above, with half-wave linear rectification the mean values are smaller by a factor $1/\pi$.

The mean squared value of the envelope becomes [2,3]

$$\overline{r^2} = \int_0^\infty r^2 p(r) \, dr = 2\sigma_n^2(1 + \rho). \tag{3.59}$$

With half-wave linear rectification the value is smaller by $(1/\pi)$.

The variance of the envelope about $\bar{r}$ is $\sigma_r^2 = \overline{r^2} - (\bar{r})^2$ or

$$\sigma_r^2 = 2\sigma_n^2\left\{(1 + \rho) - \frac{\pi}{4}e^{-\rho}\left[(1 + \rho)I_0\left(\frac{\rho}{2}\right) + \rho I_1\left(\frac{\rho}{2}\right)\right]^2\right\}. \tag{3.60}$$

The variance with linear rectification is smaller than this value by $1/\pi^2$ and with full-wave rectification by the factor $(2/\pi)^2$.

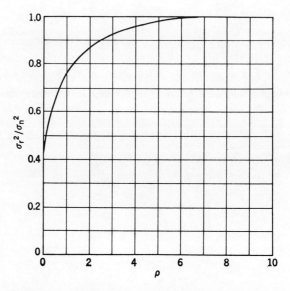

Figure 3.5 Ratio σ_r^2/σ_n^2 of envelope variance σ_r^2 about the mean value, to average noise power σ_n^2, as a function of the ratio ρ of average sine wave power to average noise power.

When $\rho \ll 1$, application of (3.52) and (3.53) gives

$$\sigma_r^2 = 2\sigma_n^2 \left\{ (1+\rho) - \frac{\pi}{4} e^{-\rho} \left(1 + \rho + \frac{3}{8}\rho^2 \right)^2 \right\} \tag{3.61}$$

$$= \frac{\sigma_n^2(4-\pi)}{2} \quad \text{for} \quad \rho = 0. \tag{3.62}$$

For $\rho \gg 1$, application of (3.54) and (3.55) gives

$$\sigma_r^2 \simeq 2\sigma_n^2 \left\{ (1+\rho) - \frac{1}{4\rho}\left(2\rho + \frac{1}{2} + \frac{1}{4\rho}\right)^2 + \cdots \right\} \simeq \sigma_n^2\left(1 - \frac{1}{8\rho}\right) \simeq \sigma_n^2. \tag{3.63}$$

Comparison of (3.63) with (3.62) shows that the noise at the output of an AM receiver is greater in the presence of an unmodulated carrier than when the carrier is absent by a factor $2/(4-\pi) = 2.33$, or about 3.7 dB.

The ratio σ_r^2/σ_n^2 is shown in Fig. 3.5 as a function of ρ. It will be noted that approximation (3.63) is quite accurate for $\rho \geq 6$.

For $\rho \gg 1$, the ratio of σ_r to the mean value $\bar{r}$ is

$$\frac{\sigma_r}{\bar{r}} = \frac{\sigma_n}{a}, \tag{3.64}$$

which also applies with linear half-wave or full-wave rectification.

3.7 Phase Variations of Noisy Sine Wave

The probability density of the phase $\theta(t)$ in (3.43) is of interest in connection with the probability of errors in digital pulse transmission by phase modulation, and is obtained by integration of $p(r,\theta)$ with respect to r between O and a. With $p(r,\theta)$ as given by (3.43), the following expression is obtained [3]

$$p(\theta) = \frac{1}{2\pi} e^{-\rho}[1 + \pi^{1/2}\alpha(1 + \mathrm{erf}\,\alpha)e^{\alpha^2}], \tag{3.65}$$

where ρ is given by (3.39) and $\alpha = \rho^{1/2} \cos \theta$.

Since $p(-\theta) = p(\theta)$, the probability that $|\theta| < \hat{\theta}$ is

$$P(-\hat{\theta} < \theta < \hat{\theta}) = 2\int_0^{\hat{\theta}} p(\theta)\,d\theta. \tag{3.66}$$

The complementary probability that $|\theta| \geq \hat{\theta}$ becomes

$$P(|\theta| \geq \hat{\theta}) = 1 - 2\int_0^{\hat{\theta}} p(\theta)\,d\theta. \tag{3.67}$$

The latter integral can be solved for $\hat{\theta} = \pi/2$ and $\hat{\theta} = \pi/4$, to yield the relations

$$P(|\theta| \geq \pi/2) = \tfrac{1}{2}\,\mathrm{erfc}\,\rho^{1/2}, \tag{3.68}$$

$$P(|\theta| \geq \pi/4) = \mathrm{erfc}\,(\rho/2)^{1/2}. \tag{3.69}$$

For large values of ρ and $\hat{\theta} < \pi/4$, so that $\alpha \gg 1$, it is permissible to make the approximation $\mathrm{erf}\,\alpha = 1$, in which case (3.67) becomes

$$P(|\theta| \geq \hat{\theta}) = 1 - \frac{1}{\pi}\int_0^{\hat{\theta}} [e^{-\rho} + 2\pi^{1/2}\rho^{1/2}\cos\theta\, e^{-\rho \sin^2 \theta}]\,d\theta. \tag{3.70}$$

With $\rho^{1/2} \sin \theta = u$; $du = \rho^{1/2} \cos \theta\,d\theta$, the following relation is obtained

$$P(|\theta| \geq \hat{\theta}) \cong \mathrm{erfc}\,(\rho^{1/2} \sin \hat{\theta}) - \frac{\hat{\theta}}{\pi} e^{-\rho} \tag{3.71}$$

$$\cong \mathrm{erfc}\,(\rho^{1/2} \sin \hat{\theta}). \tag{3.72}$$

The latter relation conforms with (3.69) for $\hat{\theta} = \pi/4$.

3.8 Variance of Phase Deviations

The variance of the phase deviation, owing to noise, is of interest in connection with noise in continuous wave transmission by phase modulation, and is obtained from

$$\sigma_\theta^2 = \int_{-\pi}^{\pi} \theta^2 p(\theta)\,d\theta, \tag{3.73}$$

where $p(\theta)$ is given by (3.65).

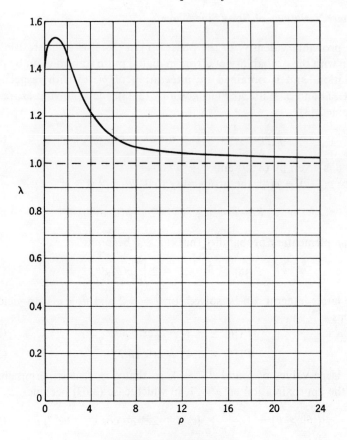

Figure 3.6 Function $\lambda(\rho)$.

The latter integral can be written in the form

$$\sigma_\theta^2 = \frac{1}{2\rho}\lambda(\rho), \qquad (3.74)$$

where

$$\lambda(\rho) = \frac{2\rho}{\pi}\int_0^\pi \theta^2 e^{-\rho}[1 + \pi^{1/2}\alpha(1 + \operatorname{erf}\alpha)e^{\alpha^2}]\,d\theta, \qquad (3.75)$$

where $\alpha = \rho^{1/2}\cos\theta$.

For $\rho \to O$, (3.75) yields $\lambda(\rho) = 2\rho\pi^2/3$. For $\rho \to \infty$, it is difficult to determine the limiting value of λ from (3.75). However, from (3.9) and (3.15) it follows that for $a \to \infty$, and thus $\rho \to \infty$ that

$$\sigma_\theta^2 = \frac{\sigma_n^2}{a^2} = \frac{1}{2\rho}. \qquad (3.76)$$

Hence for $\rho \to \infty$, (3.75) approaches the limit

$$\lambda(\infty) = 1. \tag{3.77}$$

In the table below are given values of $\lambda(\rho)$ obtained by numerical integration of (3.75)

Values of $\lambda(\rho)$

$\rho =$	0	1	2	4	8	10	32	∞
$\lambda(\rho) =$	0	1.521	1.474	1.223	1.079	1.034	1.016	1

The function $\lambda(\rho)$ is shown in Fig. 3.6. It will be noted that (3.77) and thus (3.76) is a close approximation for $\rho \geq 20$, corresponding to 13 dB ratio of average sine-wave power to average noise power.

3.9 Frequency Deviations of Noisy Sine Wave

In accordance with (3.10), the instantaneous radian frequency deviation is

$$\theta'(t) = \frac{aY_n' + Y_n'X_n - X_n'Y_n}{(a + X_n)^2 + Y_n^2}, \tag{3.78}$$

where Y_n, X_n, X_n', and Y_n' are functions of time.

The probability density of the instantaneous frequency deviation is of interest in connection with digital frequency shift systems, in which the noise power spectrum is not generally symmetrical about the sine-wave frequencies representing the digits. The probability density $p(\theta')$ applicable to digital FM can be obtained from general relations published by Rice [2]. A general formulation for the probability distribution of θ' has been given by Salz and Stein [7], applicable to an asymmetrical noise power spectrum and a sine wave with variable rather than constant envelope.

For the simpler case of a symmetrical noise band the following relation applies

$$p(\theta') = \frac{1}{2}\left(\frac{\sigma_n}{\sigma_{n'}}\right)^{1/2}(1 + z^2)^{-3/2}e^{-\rho + y/2}\left[(1 + y)I_0\left(\frac{y}{2}\right) + yI_1\left(\frac{y}{2}\right)\right] \tag{3.79}$$

where I_0 and I_1 are Bessel functions

$$z^2 = \frac{\sigma_n}{\sigma_{n'}}\theta'^2; \quad y = \frac{\rho}{1 + z^2}. \tag{3.80}$$

In these relations, σ_n^2 is the variance of the envelope functions $X_n(t)$ and $Y_n(t)$ or of the noise voltage $n(t)$, as given by (3.15); $\sigma_{n'}^2$ is the variance of the time derivative of $X(t)$ and $Y(t)$ as given by (3.20). Thus, with $\omega_0 \to \infty$:

$$\sigma_n^2 = \int_{-\infty}^{\infty} W_n(u)\, du, \qquad (3.81)$$

$$\sigma_{n'}^2 = \int_{-\infty}^{\infty} W_n(u) u^2\, du, \qquad (3.82)$$

where $W_n(u)$ can be taken as the power spectrum of $X_n(t)$ or $Y_n(t)$ or of $n(t)$.

The probability that $\theta'(t)$ exceeds a specified value θ_1' is

$$P(\theta' \geq \theta_1') = \tfrac{1}{2} P(|\theta'| \geq \theta_1') = \int_{\theta_1}^{\infty} p(\theta')\, d\theta' = \frac{\sigma_{n'}}{\sigma_n} \int_{z_1}^{\infty} p(\theta')\, dz, \qquad (3.83)$$

where

$$z_1 = \left(\frac{\sigma_n}{\sigma_{n'}}\right)\theta_1'$$

With (3.79) in (3.83)

$$P(\theta' \geq \theta_1') = \frac{e^{-\rho}}{2} \int_{z_1}^{\infty} \frac{e^{-y/2}}{(1+z^2)^{3/2}} \left[(1+y) I_0\!\left(\frac{y}{2}\right) + y I_1\!\left(\frac{y}{2}\right) \right] dz \qquad (3.84)$$

$$= \frac{1}{2} - \frac{z_1}{2}(1 + z_1^2)^{-1/2} \qquad \text{for} \qquad \rho = 0 \qquad (3.85)$$

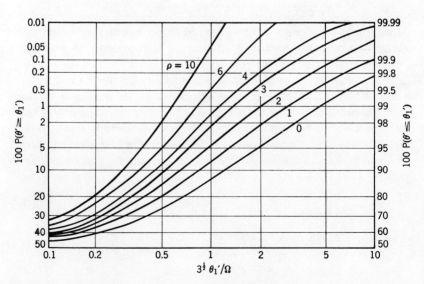

Figure 3.7 Probability $P(\theta' \geq \theta_1')$ that the frequency deviation θ' exceeds the frequency deviation θ_1', and complementary probability $P(\theta' \leq \theta_1')$, for narrow band noise with flat power spectrum between $\omega_0 - \Omega$ and $\omega_0 + \Omega$.

Variance of Frequency Deviations

The probability distribution $P(\theta' \geq \theta'_1)$ has been determined by Rice [2] by numerical integration for the particular case of a flat power spectrum $W_n(u) = W_n$ of bandwidth 2Ω. In this case $\sigma_n^2 = 2W_n\Omega$ and $\sigma_{n'}^2 = 2W_n\Omega^3/3$ so that $z_1 = 3^{1/2}\theta'_1/\Omega$. The function $P(\theta' \geq \theta'_1)$ is shown in Fig. 3.7 for various values of the ratio ρ of average sine-wave power to average noise power in the band 2Ω.

With the aid of the foregoing probability density $p(\theta')$ it is possible to determine the mean frequency deviation, which turns out to be [1]

$$\theta'_m = 2\int_0^\infty \theta' p(\theta')\, d\theta'$$

$$= \frac{\sigma_{n'}}{\sigma_n} e^{-\rho/2} I_0\left(\frac{\rho}{2}\right) \qquad (3.86)$$

$$= \frac{\sigma_{n'}}{\sigma_n} \quad \text{for} \quad \rho = 0. \qquad (3.87)$$

Determination of the variance of the frequency deviation presents a more difficult problem, as discussed in the next section.

3.10 Variance of Frequency Deviations

From the standpoint of noise in analog FM systems the rms frequency deviation caused by noise in the absence of a modulating wave is of interest. For Gaussian noise the mean value of the term $Y'_n X_n - X'_n Y_n$ in (3.78) vanishes. By neglecting second-order terms in the denominator of (3.78), the following approximation applies for $\rho > 10$:

$$\theta'(t) \simeq \frac{Y'_n(t)}{a}. \qquad (3.88)$$

In the particular case of a carrier at the center of the channel band, the mean squared value of $Y'_n(t)$ is $\sigma_n^2\Omega^2/3$ in accordance with (3.28), for a channel of bandwidth 2Ω and noise of average power σ_n^2 and a flat power spectrum. The variance of $\theta'(t)$ as given by (3.88) is accordingly for $\rho > 10$:

$$\sigma_{\theta'}^2 = \frac{\sigma_n^2 \Omega^2}{3a^2} = \frac{\Omega^2}{6\rho}. \qquad (3.89)$$

The above conventional expression for the variance is not entirely consistent with that obtained by a more rigorous analysis. Examination of (3.78) shows that it is possible for $X_n(t)$ and $Y_n(t)$ to have such values that the envelope and thus the denominator vanishes, in which case $\theta'(t)$

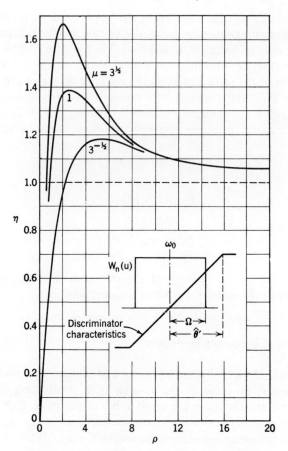

Figure 3.8 Function $\eta(\rho, \mu)$ for three values of μ.

becomes infinite over a vanishing interval. As shown by Rice [2], the variance of $\theta'(t)$ turns out to be infinite for the reason with the probability density of $\theta'(t)$ vanishes rather slowly, as θ^{-3}. However, in frequency modulation systems the output of the frequency discriminator is proportional to $\theta'(t)$ only over a limited range, and becomes constant or diminishes if $\theta'(t)$ exceeds a certain maximum value. Hence, with actual detectors the variance will be finite and the above approximation (3.89) applies for adequately large values of ρ. The minimum value of ρ for which this approximation applies depends on the characteristic of the frequency discriminator, as will be discussed shortly for certain idealized characteristics.

Variance of Frequency Deviations

In Fig. 3.8 is indicated a detector with linear response up to $\hat{\theta}'$ and a constant response equal to $\hat{\theta}'$ for $\theta' > \hat{\theta}'$. The variance in this case becomes

$$\sigma_{\dot{\theta}'}^2 = 2\int_0^{\hat{\theta}'} (\theta')^2 \, p(\theta') \, d\theta' + 2(\hat{\theta}')^2 \int_{\hat{\theta}'}^{\infty} p(\theta') \, d\theta' \qquad (3.90)$$

This relation can be written

$$\sigma_{\dot{\theta}'}^2 = \frac{\Omega_s^2}{6\rho} \eta(\rho,\mu), \qquad (3.91)$$

where

$$\mu = \frac{\hat{\theta}'}{\Omega} \qquad (3.92)$$

$$\eta(\rho,\mu) = 2\rho e^{-\rho} \int_0^{3^{1/2}\mu} \frac{z^2 e^{y/2}}{(1+z^2)^{3/2}} \left[(1+y) I_0\!\left(\frac{y}{2}\right) + y I_1\!\left(\frac{y}{2}\right) \right] dz$$

$$+ 6\rho\mu^2 e^{-\rho} \int_0^{3^{1/2}\mu} \frac{e^{y/2}}{(1+z^2)^{3/2}} \left[(1+y) I_0\!\left(\frac{y}{2}\right) + y I_1\!\left(\frac{y}{2}\right) \right] dz, \qquad (3.93)$$

where, as before, $y = \rho(1 + z^2)^{-1}$ and $z = 3^{1/2} \theta'/\Omega$.

By numerical integration of (3.93) with the aid of a digital computer the values in the following table were obtained.

Values of $\eta(\rho,\mu)$

$\rho =$	$\mu = 1/\sqrt{3}$	1	$\sqrt{3}$
0.5	0.3898	—	0.9688
1.0	0.6576	1.0361	1.4359
1.5	0.8406	—	1.6278
2.0	0.9646	1.3401	1.6748
2.5	1.0475	—	1.6503
3.0	1.1017	—	1.5947
4.0	1.1563	1.3467	1.4651
6.0	1.1685	1.2437	1.2755
8.0	1.1449	1.1722	1.1799
10.0	1.1196	1.1292	1.1309
20.0	1.0559	1.0560	1.0560
∞	1	1	1

For $\mu = \infty$ the factor η would approach infinity except when $\rho = \infty$. The factor $\eta(\rho,\mu)$ is shown in Fig. 3.8 as a function of ρ for the above three values of μ. It will be noted that η is only slightly greater than 1 for $\rho \geq 20$, corresponding to carrier-to-noise ratios in excess of $10 \log_{10} 20 = 13$ dB.

This is usually taken as the approximate threshold above which relation (3.89) is a valid approximation.

3.11 Power Spectra of Phase and Frequency Variations

When a noisy sine wave is applied to a detector that responds linearly to variations in either the envelope, the phase, or frequency of the sine wave, the power spectrum of the noise at the detector output will be related to these variations. The noise power can be reduced with the aid of a filter at the detector output, and the resultant reduction in noise power depends on the filter characteristic and on the power spectrum of the noise. The power spectra of the envelope, phase, and frequency variations can be determined from the autocorrelation function of these variables. Determination of these power spectra entails more elaborate analysis than determination of the probability density and the variances of these quantities. Exact analytical determinations have been made [1,2] but are quite difficult, even for certain simple cases. However, adequate approximate formulas can readily be derived for high ratios ρ as required for satisfactory transmission performance in the presence of random noise. These cases are considered here.

For adequately high ratios, $\rho = a^2/2\rho_n^2$, variations in the envelope are caused principally by the inphase components X_n of the noise voltage. Hence the power spectrum of the envelope variation is essentially that of the noise components X_n or

$$W_r(u) \simeq W_x(u). \tag{3.94}$$

The total average power of this component is in accordance with (3.15) equal to σ_n^2, which conforms with (3.63) for $\rho \gg 1$. From Fig. 3.6 it appears that (3.63) is closely approximated when $\rho \geq 6$, and this will also be the case for (3.94).

For sufficiently high ratios, ρ, the phase variations are caused principally by the quadrature components Y_n of the noise voltage, and the variance is given by (3.76). The corresponding power spectrum of the phase variation is

$$W_\theta(u) \simeq \frac{W_y(u)}{a^2}. \tag{3.95}$$

From Fig. 3.6 it appears that (3.76) and thus (3.95) are closely approximated for $\rho \geq 10$.

For adequately high ratios, ρ, the frequency variation is in accordance with (3.88), caused by the time derivative of the component $Y_n(t)$. In accordance with (3.18), the power spectrum of this time derivative is $u^2 W_n(u)$. The power spectrum of the frequency variation is thus on basis of (3.88) for $\rho \gg 1$,

$$W_{\theta'}(u) = \frac{u^2}{a^2} W_n(u). \tag{3.96}$$

The variance of $Y'_n(t)$ as given by (3.28) for a flat band of noise of total bandwidth 2Ω is $\sigma_n^2 \Omega^2/3$. The corresponding variance of θ' obtained by integration of (3.96) conforms with (3.89). From Fig. 3.8 it appears that (3.89) is a satisfactory approximation for $\rho \geq 10$, and this also applies to (3.96).

The following more accurate relation has been derived by Rice [8] for noise with a power spectrum $W_n(u)$

$$W_{\theta'}(u) = \frac{u^2}{a^2} W_n + \frac{\sigma_{n'}}{\sigma_n} \operatorname{erfc} \rho^{1/2}, \tag{3.97}$$

where σ_n, $\sigma_{n'}$, and ρ are given by (3.81), (3.82), and (3.39). This is the double-sided power spectrum, and applies for an ideal frequency discriminator without peak limitation, that is, $\mu = \infty$. The foregoing relation yields a satisfactory approximation for $\rho > 2$.

In FM a post-detection low-pass filter of bandwidth $\Omega_0 < \Omega$ is ordinarily used. With $W_n = \sigma_n^2/2\Omega$, the average noise power at the output of the post-detection filter becomes

$$
\begin{aligned}
N_0 &= c \int_{-\Omega_0}^{\Omega_0} W_{\theta'}(u) \, du \\
&= c \int_{-\Omega_0}^{\Omega_0} \left(\frac{u^2}{4\rho\Omega} + \frac{\Omega}{3^{1/2}} \operatorname{erfc} \rho^{1/2} \right) du \\
&= c \frac{\Omega_0^3}{6\Omega\rho} \left[1 + \left(\frac{\Omega}{\Omega_0} \right)^2 \frac{12}{3^{1/2}} \rho \operatorname{erfc} \rho^{1/2} \right],
\end{aligned}
\tag{3.98}
$$

where c is a detector constant.

The second bracket term vanishes as $\rho \to \infty$. The average noise power at the output of the post detection filter is thus greater than that obtained with a power spectrum (3.96) by the factor

$$\eta_0 = 1 + \left(\frac{\Omega}{\Omega_0} \right)^2 \frac{12}{3^{1/2}} \rho \operatorname{erfc} \rho^{1/2}. \tag{3.99}$$

For $\Omega = \Omega_0$ the values of η_0 are as follows:

η_0 vs. ρ for $\Omega/\Omega_0 = 1$

$\rho =$	2	4	6	8
$\eta_0 =$	1.56	1.1	1.02	1.003

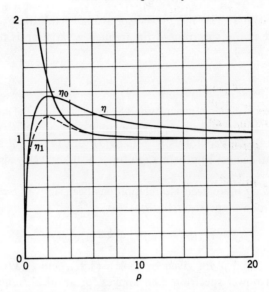

Figure 3.9 Factor η_0 for $\Omega_0 = \Omega$ and factor η for $\mu = 1$. The approximate combined factor is η_1.

This factor applies with a frequency discriminator having unlimited peak response to all noise amplitudes. In Fig. 3.9 is shown the above factor η_0, together with the factor η previously shown in Fig. 3.8 for $\mu = 1$. The factor η_1 in Fig. 3.9 represents the approximate combined effect of a postdetection filter and peak limitation in the frequency discriminator, for $\Omega_0 = \Omega$ so that the relations for η apply.

When the bandwidth of the post-detection filter is small in relation to the semibandwidth of the bandpass channel and last term in (3.99) becomes more important. For $\Omega_0 = \Omega/5$ the values of η_0 are as follows:

η_0 vs. ρ for $\Omega/\Omega_0 = 5$

$\rho =$	2	4	6	8	10
$\eta_0 =$	15	3.5	1.48	1.08	1.01

Thus the conventional relations based on a power spectrum (3.96) apply also in this case for $\rho \geq 10$. For smaller values of ρ the increase in η_0 and thus in the output noise power is much greater than for $\Omega/\Omega_0 = 1$.

3.12 Probability Distribution of Envelope Time Derivative r'

The probability density $p(r,r',\theta)$ has been derived by Rice [2] for a sine wave plus random noise. For a symmetrical noise power spectrum, the following relation applies

$$p(r,r',\theta) = \frac{r}{(2\pi)^{3/2}\sigma_n^2\sigma_{n'}} \exp - \frac{r^2 - 2r\cos\theta + a^2 + (\sigma_n/\sigma_{n'})^2 r'^2}{2\sigma_n^2}, \quad (3.100)$$

where σ_n and $\sigma_{n'}$ are given by (3.81) and (3.82) and a is the amplitude of the sine wave.

Integration with respect to θ between $-\pi$ and π yields the following expression

$$p(r,r') = g(a,r) \frac{(2\pi)^{-1/2}}{\sigma_{n'}} e^{-r'^2/2\sigma_{n'}^2}, \quad (3.101)$$

where

$$g(a,r) = \frac{r}{\sigma_n^2} e^{-(a^2+r^2)/2\sigma_n^2} I_0\left(\frac{ra}{\sigma_n^2}\right). \quad (3.102)$$

The probability that r' exceeds $\sigma_{n'}$ by a factor k becomes

$$P(r' \geq k\sigma_{n'}) = \tfrac{1}{2} e^{-a^2/2\sigma_n^2} G(a) \,\text{erfc}\left(\frac{k}{2^{1/2}}\right), \quad (3.103)$$

where

$$G(a) = \int_0^\infty \frac{r}{\sigma_n^2} e^{-r^2/2\sigma_n^2} I_0\left(\frac{ra}{\sigma_n^2}\right) dr \quad (3.104)$$

$$= 1 \quad \text{for} \quad a = 0. \quad (3.105)$$

The probability distribution of r' for Gaussian noise ($a = 0$) is thus

$$P(r' \geq k\sigma_{n'}) = \tfrac{1}{2} P(|r'| \geq k\sigma_{n'}) = \tfrac{1}{2}\,\text{erfc}\left(\frac{k}{2^{1/2}}\right). \quad (3.106)$$

3.13 Probability Distribution of $\theta''(t)$

The instantaneous frequency deviation is $\theta'(t)$, so that $\theta''(t)$ represents the time derivative of the instantaneous frequency deviation. The probability density $p(\theta'')$ can be obtained by integration of the following expression given by Rice [2] for the probability density $p(r,\theta,\theta',\theta'')$ for a sine wave plus Gaussian noise with a symmetrical power spectrum

$$p(r,\theta,\theta',\theta'') = \frac{r^3}{(2\pi)^2 \sigma_n \sigma_{n'} D^{1/2}(1+v^2)^{1/2}} \exp - N/2\sigma_n^2, \quad (3.107)$$

where

$$D = \sigma_n^2 \sigma_{n''}^2 - \sigma_{n'}^4; \qquad v = \frac{4\sigma_n^2 \sigma_{n'}^2 \theta'^2}{D}$$

$$\sigma_{n''}^2 = \int_{-\omega_0}^{\infty} W_n(u) u^4 \, du, \tag{3.108}$$

and

$$N = r^2 - 2ra \cos \theta + a^2 + \frac{\sigma_n^2 r^2 \theta'^2}{\sigma_{n'}^2} + \frac{(a\sigma_{n'}^2 + \sigma_n^2 r \theta'')^2}{(1+v)D}. \tag{3.109}$$

In the case of Gaussian noise alone, $a = O$, and the last relation yields

$$N = N_0 = r^2 + \frac{\sigma_n^2 r^2 \theta'^2}{\sigma_{n'}^2} + \frac{\sigma_n^4}{(1+v)D} r^2 \theta''^2. \tag{3.110}$$

The probability density $p(\theta'')$ is obtained by integration of (3.107) with respect to θ between O and 2π, with respect to θ' between $-\infty$ and ∞ and with respect to r between O and ∞. The probability distribution becomes

$$P(|\theta''| \geq \theta_1'') = 2 \int_{\theta_1''}^{\infty} p(\theta'') \, d\theta''. \tag{3.111}$$

For Gaussian noise alone ($a = O$) the following relation is obtained after considerable simplification

$$P(|\theta''| \geq \theta_1'') = 1 - \frac{2k}{\pi} \int_0^{\infty} \frac{dx}{[g(x) + k^2]g(x)} - \frac{2}{\pi} \int_0^{\infty} \frac{\tan^{-1}[k/g^{1/2}(x)]}{(1+x^2)^{3/2}} dx \tag{3.112}$$

where

$$k = \frac{\sigma_n^2 \theta_1''}{\sigma_{n'}^2} \tag{3.113}$$

$$g(x) = \left(\frac{\sigma_n^2 \sigma_{n''}^2}{\sigma_{n'}^4} - 1 + 4x^2\right)(1 + x^2). \tag{3.114}$$

In view of (3.113) the left hand side of (3.112) can alternately be written $P(|\theta''| \geq k\sigma_{n'}^2/\sigma_n^2)$. For adequately large values of k the following approximation applies

$$P\left(|\theta''| \geq \frac{k\sigma_{n'}^2}{\sigma_n^2}\right) \simeq \frac{2}{\pi k}\left[1 + \ln\left(\frac{k}{2} + 1\right)\right]. \tag{3.115}$$

For a flat power spectrum of bandwidth 2Ω

$$W(u) = \frac{\sigma_n^2}{2\Omega}, \qquad \sigma_{n'}^2 = \frac{\sigma_n^2 \Omega^2}{3} \quad \text{and} \quad \sigma_{n''}^2 = \frac{\sigma_n^2 \Omega^4}{5},$$

so that $\sigma_{n'}^2/\sigma_n^2 = \Omega^2/3$ and $\sigma_n^2 \sigma_{n''}^2/\sigma_{n'}^4 = 9/5$.

Probability Distribution of $\theta''(t)$

The probability distribution (3.112) as obtained by numerical integration is given in the following table, together with approximation (3.115) in brackets for $k = 100$

$P(|\theta''| \geq k\Omega^2/3)$ for Flat Power Spectrum

$k = 0$	1	2	3	4	5	10	20	50	100
1	0.538	0.381	0.321	0.269	0.238	0.158	0.100	0.051	0.031(0.03)

This probability distribution is shown in Fig. 3.10 as a function of k.

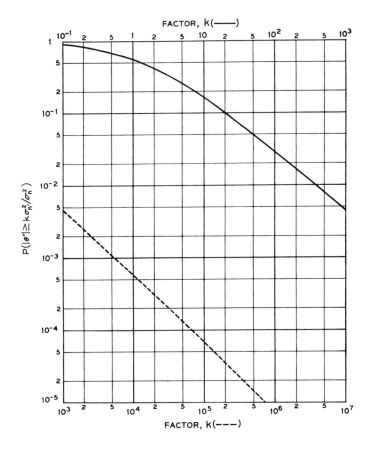

Figure 3.10 Probability $P(\theta'' \geq \theta_1'')$ that the time derivative $\theta''(t)$ of the instantaneous frequency deviation $\theta''(t)$ exceeds $\theta_1'' = k\sigma_n^2/\sigma_n^2$, for Gaussian narrow band noise with flat power spectrum.

For a Gaussian power spectrum

$$W(u) = \frac{\sigma_n^2}{(2\pi\Omega_0^2)^{1/2}} e^{-u^2/2\Omega_0^2}, \qquad (3.116)$$

$$\sigma_{n'}^2 = \sigma_n^2 \Omega_0^2 \quad \text{and} \quad \sigma_{n''}^2 = 3\sigma_n^2 \Omega_0^4,$$

so that $\frac{\sigma_n^2 \sigma_{n''}^2}{\sigma_{n'}^4} = 3.$

Numerical integration of (3.112) in this case yields the result given in the following table

$P(|\theta''| \geq k\Omega_0^2)$ **For Gaussian Power Spectrum**

$k =$	0	1	2	3	4	5	10	20	50	100
	1	0.595	0.447	0.369	0.317	0.280	0.182	0.113	0.057	0.033(0.03)

4

Transmission Systems Optimization against Gaussian Noise

4.0 General

As discussed in Sections 2.16 and 2.17, in communication systems design it is necessary to consider transmission impairments from a multiplicity of sources, including random noise inherent in the transmission medium, amplifiers, and detectors [1,2,3,4]. Such random noise results in part from thermal agitation of charged particles (thermal noise) and in part from their irregular motion in streams in response to electric forces (shot noise). In detectors and amplifiers operating at very low temperatures and very high frequencies, such as masers and lasers, another source of additive interference is quantum noise [4]. Additive random noise from these various sources have virtually the same statistical properties as Gaussian random variables, and impose an ultimate limitation on transmission performance for a given received signal power. For this reason, Gaussian random noise has been written about more extensively as a source of transmission impairment than other kinds of interference. In dealing with actual systems with several kinds of transmission distortion it is a legitimate procedure to consider each kind of distortion by itself, and to combine the distortion powers by direct addition, as discussed in Sections 2.16 and 2.17. Hence in this chapter, additive Gaussian noise alone in time-invariant channels will be considered, ignoring transmission impairments from other sources, to be dealt with in later chapters.

With any modulation method, transmitting and receiving filters are generally required for signal-spectrum shaping and limitation of receiver noise, as well as for equalization to avoid excessive transmission distortion. Optimum performance depends on appropriate shaping of these filters and, as far as additive random noise is concerned, on the optimum division of shaping between transmitting and receiving filters. This optimum division of shaping, commonly referred to as "pre-emphasis" and "de-emphasis," depends on the modulation method, on the power spectrum of channel noise, and on the attenuation characteristic of the transmission medium.

In the optimization of both analog and digital systems, it is necessary to recognize two different transmission methods. One is the conventional method of single-band transmission, with a single modulator or encoder and a single demodulator or decoder. The other is multiband transmission, in which the available channel band is divided into a number of narrower bands for transmission of signal components, with a separate modulator or encoder and a separate demodulator or decoder for each narrow band. The latter method can be optimized to provide better performance than the first, provided two conditions are met. One is that the spectral density of the noise, or the attenuation over the channel band, or both, varies with frequency. The other proviso is that modulation methods are used that afford a reduction in signal power in exchange for increased bandwidth, such as FM or PCM. Only single-band transmission will be considered here, except for a comparison with multiband digital transmission over baseband channels. The latter method is implied by Shannon in his determination of channel capacity as limited by Gaussian noise of variable spectral density [5]. Multiband transmission is to be distinguished from diversity transmission, in which the entire signal is transmitted over different channels, as dealt with in Chapters 9 and 10 for random multipath channels.

Both analog and digital systems are considered here, employing baseband transmission and certain conventional methods of carrier modulation and demodulation. In addition, certain related filter optimization problems are discussed briefly, such as "matched filters" for optimum peak detection of signals of known shape (such as radar pulses) in the presence of noise and "smoothing filters" for improved analog reception of random signals of known power spectra perturbed by random noise. A simple unified analytical procedure is used in the solution of these various optimization problems.

Digital transmission methods for other than the idealized conditions assumed here are dealt with in more detail in Chapter 5.

4.1 Basic Transmission Parameters and Criteria

In Fig. 4.1 is indicated a transmission system link comprising three basic elements: a transmitter, a channel or transmission medium, and receiver.

Basic Transmission Parameters and Criteria 135

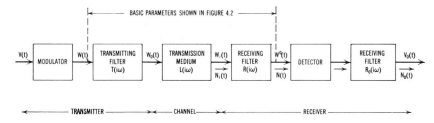

Figure 4.1 Transmission system arrangement.

At the transmitter input the messages may undergo various kinds of processing before they appear as waves $W(t)$ at the input to a transmitting filter with transmittance $T(i\omega)$. This preliminary processing may involve modification of the spectrum or the amplitude probability distribution of the original waves or gating processes for sampling of the waves with the aid of a premodulation network as outlined in Chapter 2. In the present analysis only the signal power P_0 at the output of the transmitter is of interest, as affected by the transmittance $T(i\omega)$ and the spectrum of the message waves.

At the receiver, the message wave $W(t)$ is restored with the aid of a receiving filter $R(i\omega)$, except for the addition of channel noise. An additional receiving filter of transmittance $R_0(i\omega)$ may be required for optimum performance, at the detector input or output, depending on the particular modulation and detection method. At the detector output the original message will appear, except for distortion or errors due to channel noise and other kinds of imperfections in the transmission system.

The transmission-frequency characteristics of the transmitter, the channel and the receiving filter will be designated as follows

$$T(i\omega) = \bar{T}(\omega)e^{-i\varphi_1(\omega)}, \qquad (4.1)$$

$$L(i\omega) = \underline{L}(\omega)e^{-i\varphi_2(\omega)}, \qquad (4.2)$$

$$R(i\omega) = \bar{R}(\omega)e^{-i\varphi_3(\omega)}. \qquad (4.3)$$

The corresponding power-transfer characteristics are designated

$$t(\omega) = |T(i\omega)|^2 = \bar{T}^2(\omega), \qquad (4.4)$$

$$l(\omega) = |L(i\omega)|^2 = \underline{L}^2(\omega), \qquad (4.5)$$

$$r(\omega) = |R(i\omega)|^2 = \bar{R}^2(\omega). \qquad (4.6)$$

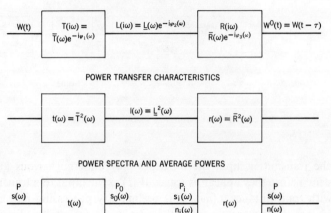

Figure 4.2 Basic transmission diagram and parameters.

The modulating waves $W(t)$ at the transmitter input are assumed to be random variables that can be characterized by a power spectrum $s(\omega)$, as indicated in Fig. 4.2.

The corresponding average modulating power at the transmitter input is

$$P = \int_0^\infty s(\omega)\, d\omega. \tag{4.7}$$

With carrier modulation, the spectrum $s(\omega)$ is confined to a pass-band extending from ω_1 to ω_2. The integration limits O and ∞ in the preceding and later integrals can then be replaced by ω_1 and ω_2.

The average modulating power at the transmitter output is

$$P_0 = \int_0^\infty s_0(\omega)\, d\omega = \int_0^\infty s(\omega) t(\omega)\, d\omega. \tag{4.8}$$

At the receiving filter input the received average power is

$$P_i = \int_0^\infty s_i(\omega)\, d\omega = \int_0^\infty s(\omega) l(\omega) t(\omega)\, d\omega. \tag{4.9}$$

If the power spectrum at the transmitter input is $s_1(\omega)$ in place of $s(\omega)$, the average transmitted power remains the same, provided that $t(\omega)$ is modified into $t_1(\omega)$, such that

$$s_1(\omega) t_1(\omega) = s(\omega) t(\omega). \tag{4.10}$$

Basic Transmission Parameters and Criteria

In pulse systems, the power spectrum $s_1(\omega)$ of pulse trains at the transmitter input differs in general from that at the receiver output. The modified power transfer characteristic $t_1(\omega)$ can then readily be determined from (4.10).

In general, the transmitted signal power is greater than the foregoing variable modulation component P_0 and given by

$$S_0 = P_0 + P_1. \tag{4.11}$$

The component P_1 is a steady-state component that does not carry information, but may be desirable or essential to facilitate the detection process, as when a carrier is transmitted in AM systems, or for other purposes.

In channels where $l(\omega) = l$, the received signal power differs from the transmitted power by a constant, so that

$$S_i = lS_0 = l(P_0 + P_1). \tag{4.12}$$

In this case, performance can be related to average received signal power, rather than average transmitted signal power S_0.

At the output of the first receiving filter, the noise power spectrum is

$$n(\omega) = r(\omega)n_i(\omega), \tag{4.13}$$

and the average noise power is

$$N = \int_0^\infty r(\omega)n_i(\omega)\,d\omega. \tag{4.14}$$

A quantity that enters into most transmission systems performance criteria is the signal-to-noise ratio at the receiving filter input. The ratio of average signal power to average noise power in a narrow-band $d\omega$ at the receiving filter input is

$$\rho_i(\omega) = \frac{s_i(\omega)}{n_i(\omega)}. \tag{4.15}$$

At the receiving filter output, or detector input

$$\rho(\omega) = \frac{s(\omega)}{n(\omega)} = \frac{s_i(\omega)r(\omega)}{n_i(\omega)r(\omega)} = \rho_i(\omega). \tag{4.16}$$

The ratio of average signal power in the total band Ω_i to average noise power in the total band at the receiving filter input becomes

$$\rho_i = \frac{\int_0^{\Omega_i} s_i(\omega)\,d\omega}{\int_0^{\Omega_i} n_i(\omega)\,d\omega}. \tag{4.17}$$

The corresponding ratio at the receiving filter output, or detector input, is

$$\rho = \frac{\int_0^\infty s(\omega)\,d\omega}{\int_0^\infty n(\omega)\,d\omega} = \frac{\int_0^\infty s_i(\omega)r(\omega)\,d\omega}{\int_0^\infty n_i(\omega)r(\omega)\,d\omega}. \tag{4.18}$$

The lower and upper integration limits can formally be taken as zero and infinity. The actual integration limits will depend on the bandwidth of the receiving filter.

The above relations between input and output signal-to-noise ratios apply for a passive transducer at the receiver of power transmittance $r(\omega)$, that is not itself a source of noise. In actual systems it is necessary to have not only an adequate output signal-to-noise ratio but also sufficient signal power for aural, visual, or other reception. To the latter end it is ordinarily necessary to introduce amplification, which entails a degradation in the output signal-to-noise ratio owing to intrinsic amplifier noise. This is ordinarily taken into account by introducing an appropriate multiplyer or "noise figure" that in general varies with frequency over the transmission band. Relations for the spectral densities of thermal and other random noise are given in various publications, together with definitions of "noise figures" and related "noise temperatures." (For example References 2 and 3.) In the analysis that follows it is assumed that the input power spectrum $n_i(\omega)$ has been determined from such relations and includes intrinsic noise contributed by amplifiers and other active circuit components, such as frequency converters. Thus $n_i(\omega)$ represents an equivalent input power spectrum to be used in conjunction with an assumed ideal noiseless transducer of power transmittance $r(\omega)$ obtained by an appropriate combination of an ideal amplifier and an attenuator.

In analog transmission, there is a certain ratio ρ_0 of average message or modulating power to average noise power in the detector output. This ratio depends on the modulation method, on the detector input signal-to-noise ratio ρ on the ratio β of channel bandwidth to message bandwidth and on the type of detector. In digital transmission the error probability ϵ_0 in the decoded message also depends on the above factors, as well as on several others. Thus the output signal-to-noise ratios and error probabilities may be formally designated

$$\rho_0 = F(\rho, \beta, \ldots), \tag{4.19}$$

$$\epsilon_0 = G(\rho, \beta, \ldots), \tag{4.20}$$

where F and G designate functional relations between basic parameters ρ and β together with certain others.

If ρ_0 and ϵ_0 are specified, together with the statistical properties of the modulating or message wave and of channel noise, it is possible by appropriate choice of modulation method, together with appropriate implementation, to realize minimum values of ρ, or of the average transmitter power.

4.2 Optimum Combined Equalization without Noise

In transmission by analog modulation methods, the ratio ρ_0 is independent of the phase characteristics appearing in (4.1), (4.2), and (4.3). In some applications, such as direct speech transmission, impairment of intelligibility, owing to phase distortion, can ordinarily be disregarded. In other instances, such as television transmission, phase distortion is a most important consideration, and distortionless transmission in the absence of noise is a desirable objective. In modulation theory, a common idealization in direct analog transmission is that in the absence of noise all output waves $W_0(t)$ be linearly related to the input waves $W(t)$, except for a transmission delay τ. That is, for time-invariant channels as assumed here

$$W_0(t) = cW(t - \tau). \tag{4.21}$$

The maximum allowable value of the constant c depends on considerations of singing, owing to feedback from output to input, and there may also be certain limitations on the minimum value of c. In the present analysis, the results are independent of c, and for convenience it is assumed that $c = 1$.

Condition (4.21) with $c = 1$ entails that the following relations be satisfied over the frequency band of the signals

$$\overline{T}(\omega)\,\underline{L}(\omega)\,\overline{R}(\omega) = 1, \tag{4.22}$$

and

$$\varphi_1(\omega) + \varphi_2(\omega) + \varphi_3(\omega) = \omega\tau. \tag{4.23}$$

In bandlimited channels, it is not possible to realize a linear phase characteristic $\omega\tau$, except in the limit $\tau \to \infty$. Even with infinite delay, it is not possible to realize (4.22) and (4.23) by gain and phase equalization without certain departures from a constant amplitude and linear phase characteristic, usually in the form of ripples over the channel band. Transmission impairments from such fine-structure deviations in analog and digital systems are dealt with in Chapters 6 and 7, and may be of even greater importance than random noise. Nevertheless, it is legitimate to ignore such fine-structure deviations in dealing with the effect of random noise.

4.3 Optimum Combined Equalization with Noise

Although (4.22) and (4.23) represent the optimum combined equalization in the absence of noise, it does not follow that they are also the optimum conditions in the presence of random noise. It will be assumed that at the output of the first receiving filter an additional filter with characteristic $R_0(i\omega)$ is inserted, as indicated in Fig. 4.1. To minimize distortion, it would be necessary for the phase characteristic to be linear, which entails elaborate phase equalization and infinite transmission delay. With an amplitude characteristic $\bar{R}_0(\omega) \neq 1$, some distortion will be introduced. Nevertheless, it is possible, with the aid of such an "infinite lag smoothing filter," to minimize the total average interference power in the output, in relation to the average undistorted output power [6,7], as will be discussed presently.

With an amplitude characteristic $\bar{R}_0(\omega) = 1$, the signal- and noise-power spectra would be $s(\omega)$ and $n(\omega)$, and there would be no distortion. With characteristic $\bar{R}_0(\omega) \neq 1$ the average output interference power is changed from N into

$$N' = \int_0^\infty \{[1 - \bar{R}_0(\omega)]^2 s(\omega) + \bar{R}_0^2(\omega) n(\omega)\}\, d\omega. \tag{4.24}$$

The average undistorted output power is changed from P into

$$P' = \int_0^\infty \{\bar{R}_0^2(\omega) - [1 - \bar{R}_0(\omega)]^2\} s(\omega)\, d\omega$$

$$= \int_0^\infty [2\bar{R}_0(\omega) - 1] s(\omega)\, d\omega \tag{4.25}$$

The ratio

$$\mu_0 = \frac{N'}{P'} \tag{4.26}$$

is to be minimized by appropriate choice of $\bar{R}_0(\omega)$.

As shown in Appendix 4, this optimum $\bar{R}_0(\omega)$ is given by

$$\bar{R}_0^0(\omega) = (1 + \mu_0) \frac{s(\omega)}{s(\omega) + n(\omega)}, \tag{4.27}$$

where μ_0 is given by (4.26) and is a constant.

The corresponding minimum ratio μ_0 with (4.27) in (4.26) becomes

$$\mu_0 = \frac{\int_0^\infty \frac{s(\omega)}{s(\omega) + n(\omega)} [\mu_0^2 s(\omega) + n(\omega)]\, d\omega}{\int_0^\infty \frac{s(\omega)}{s(\omega) + n(\omega)} [(1 + 2\mu_0)s(\omega) - n(\omega)]\, d\omega}. \quad (4.28)$$

Solution of the latter second-degree equation in μ_0 yields the relation

$$\mu_0 = \int_0^\infty \frac{s(\omega)n(\omega)\, d\omega}{s(\omega) + n(\omega)} \Big/ \int_0^\infty \frac{s^2(\omega)\, d\omega}{s(\omega) + n(\omega)}. \quad (4.29)$$

For $n(\omega) = \mu s(\omega)$, (4.29) gives $\mu_0 = \mu = N/P$, the ratio of average noise power to average signal power. When $n(\omega) \neq \mu s(\omega)$ the approximation $\mu_0 \simeq \mu$ applies provided $n(\omega) \ll s(\omega)$.

By way of illustration, let $n(\omega) = a\omega$ and $s(\omega) = s$ for $0 < \omega < \Omega$, and let $n(\omega)$ and $s(\omega)$ be zero for $\omega > \Omega$. In this case, solution of (4.29) yields with $\mu = N/P = a\Omega/2s$

$$\mu_0 = \frac{2\mu - \ln(1 + 2\mu)}{\ln(1 + 2\mu)}$$

$$\simeq \mu\left(1 - \frac{\mu}{3} - \frac{2\mu^2}{3} - \cdots\right). \quad (4.30)$$

In analog systems $\mu < 0.01$ so that $\mu_0 \simeq \mu$. With $\mu = 0.1$, corresponding to 10 dB signal-to-noise ratio, $\mu_0 = 0.97\mu$.

For continuous signal and noise-power spectra, as assumed in the foregoing illustration, the advantage of a smoothing filter is negligible in the range of signal-to-noise ratios encountered in communication systems. However, a significant advantage is afforded in the case of strong single-frequency interference. In this case, they serve as narrow-band suppression filters that in principle can eliminate interference without causing appreciable message distortion.

As noted at the end of Section 4.2, transmission impairments owing to small unavoidable fine-structure deviations from (4.22) and (4.23) may be of even greater importance than random noise. In extensive systems, such impairments will exceed by far the minor benefits to be derived from the small intentional modifications of (4.22) considered here, that may be smaller in amplitude than the unavoidable fine structure deviations and could hardly be implemented in systems design. For these reasons, it is assumed in the exposition that follows that the channel is equalized in accordance with (4.22) and (4.23), except for certain modifications that may be desirable for pulse systems, where it is not essential that the spectrum of the pulses at the transmitter input be the same as at the receiver output (as discussed further in Section 4.7).

4.4 Baseband Transmission Systems

In baseband transmission the channel bandwidth Ω_i is ordinarily related to the maximum message bandwidth Ω_m and the receiving filter bandwidth Ω_0 by

$$\Omega_i = \Omega_m = \Omega_0. \tag{4.31}$$

The average output signal-to-noise ratio (4.18) becomes

$$\rho = \rho_0 = \frac{\int_0^{\Omega_0} s(\omega)\,d\omega}{\int_0^{\Omega_0} n_i(\omega)r(\omega)\,d\omega}. \tag{4.32}$$

This ratio is to be maximized for a given average transmitter power P_0. Alternately, P_0 is to be minimized for a fixed ratio ρ_0. In either case the optimum condition is attained for the maximum value of the ratio

$$\frac{\rho_0}{P_0} = \frac{P}{\int_0^{\Omega_0} s(\omega)t(\omega)\,d\omega \int_0^{\Omega_0} n_i(\omega)r(\omega)\,d\omega} \tag{4.33}$$

$$= \frac{P}{\int_0^{\Omega_0} \frac{s(\omega)\,d\omega}{l(\omega)r(\omega)} \int_0^{\Omega_0} n_i(\omega)r(\omega)\,d\omega}. \tag{4.34}$$

As shown in Appendix 4 the maximum value of the latter ratio is obtained when the two integrals are equal, which is the case for the following optimum value of $r(\omega)$

$$r^0(\omega) = [\bar{R}^0(\omega)]^2 = c\left(\frac{s(\omega)}{n_i(\omega)l(\omega)}\right)^{1/2} \tag{4.35}$$

$$= c^2 \frac{s_i(\omega)}{n_i(\omega)l(\omega)} \tag{4.36}$$

$$= c^2 \frac{s_0(\omega)}{n_i(\omega)}, \tag{4.37}$$

where c is a constant. The second equation is obtained by inserting in (4.35) the relation $s(\omega) = s_i(\omega)r(\omega)$, and the third by observing that $s_i(\omega) = s_0(\omega)l(\omega)$.

The corresponding optimum transmitting filter characteristic is obtained from the relation $l(\omega)t(\omega)r(\omega) = 1$, and is

$$t^0(\omega) = [\bar{T}^0(\omega)]^2 = \frac{1}{l(\omega)r^0(\omega)} = \frac{1}{c}\left(\frac{n_i(\omega)}{s(\omega)l(\omega)}\right)^{1/2}, \tag{4.38}$$

where c is the same constant as in (4.35).

Carrier Amplitude Modulation Systems

The optimum ratio ρ_0 as obtained with (4.35) in (4.34) is

$$\rho^0 = \rho_0^0 = \frac{PP_0}{\left(\int_0^{\Omega_0} \gamma(\omega)\, d\omega\right)^2}, \qquad (4.39)$$

where

$$\gamma(\omega) = \left(\frac{s(\omega)n_i(\omega)}{l(\omega)}\right)^{1/2}. \qquad (4.40)$$

In the special case of $l(\omega) = l$, and $n_i(\omega) = n = N_i/\Omega_0$, (4.39) yields

$$\rho_0^0 = \frac{P_i}{N_i}, \qquad (4.41)$$

where P_i is the average modulating power at the receiving filter input and N_i the average noise power in the band $\Omega_0 = \Omega_i$.

4.5 Carrier Amplitude Modulation Systems

In single-sideband transmission, let the carrier frequency be ω_0. With ideal synchronous detection (4.39) is then replaced by

$$\rho_0^0 = \frac{PP_0}{\left(\int_0^{\Omega_0} \gamma(u)\, du\right)^2}, \qquad (4.42)$$

where $u = \omega - \omega_0$ and $\gamma(u)$ is obtained by replacing ω by u in (4.40).

In $\gamma(u)$ the power spectrum $s(u)$ is now the baseband spectrum translated to the channel band extending from $\omega = \omega_0$ to $\omega = \omega_0 + \Omega_0$, as indicated in Fig. 4.3. If the lower sideband is transmitted, the integration limits in (4.42) are from $-\Omega_0$ to 0 and $s(u)$ is the baseband spectrum translated to the channel band from $\omega_0 - \Omega_0$ to ω_0.

In double-sideband transmission the channel band extends from $\omega_0 - \Omega_0$ to $\omega_0 + \Omega_0$ and the optimum signal-to-noise ratio with ideal synchronous detection becomes

$$\rho_0^0 = \frac{2PP_0}{\left(\int_{-\Omega_0}^{\Omega_0} \gamma(u)\, du\right)^2}. \qquad (4.43)$$

The factor 2 comes about since the modulating waves in the two sidebands combine directly, whereas the noise waves add on a root-sum-square basis. In the particular case $l(u) = l$ and $s(u) = P/2\Omega_0$, (4.43) becomes

$$\rho_0^0 = \frac{2P_i}{N_i}, \qquad (4.44)$$

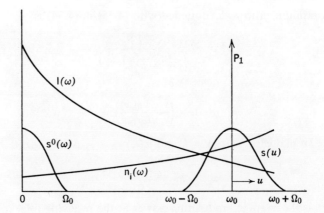

Figure 4.3 Carrier amplitude modulation. Power spectrum $s^0(\omega)$ of modulating wave translated to power spectrum $s(u)$ symmetrical about ω_0. The spike of area P_1 at ω_0 represents average power of steady state component. In single sideband transmission P_1 and spectrum to one side of ω_0 is eliminated by filter.

where N_i is the average noise power in an input band $2\Omega_0$. The noise power in a band Ω_0 is in this case $N_i' = N_i/2$ and $\rho_0^0 = P_i/N_i'$ as for baseband and single-sideband transmission, assuming that the carrier in double-sideband transmission is suppressed.

In (4.43) P_0 is the average transmitted sideband or modulating power, regardless of whether the carrier is transmitted or not. When a carrier of average power P_c is transmitted, the total average transmitter power is

$$S_0 = P_0 + P_c. \tag{4.45}$$

When the carrier is transmitted and is 100% modulated, $P_c = \hat{P}_0$ where $\hat{P}_0$ is the peak signal power. With speech and certain other types of waves $\hat{P}_0 \gg P_0$, so that $S_0 \gg P_0$. Hence, aside from the reduced bandwidth, a significant advantage of single-sideband transmission as applied to highly peaked modulating waves is that transmitter power requirements are drastically reduced.

4.6 Frequency and Phase Modulation Systems

In FM and PM systems the channel band is, ordinarily, sufficiently narrow relative to the carrier frequency so that both $n_i(\omega)$ and $l(\omega)$ can be assumed constant over the channel band.

The received wave at the detector input is of the form

$$e(t) = E \cos [\omega_0 t + \Phi(t)]. \tag{4.46}$$

Frequency and Phase Modulation Systems

The phase modulation $\Phi(t)$ is related to the modulating wave $V(t)$ by

$$\Phi(t) = \Delta_1 \int_0^t V(t)\, tu, \qquad (4.47)$$

where Δ_1 is the frequency deviation per unit amplitude of the modulating wave $V(t)$.

The instantaneous frequency deviation is therefore

$$\Delta(t) = \Phi'(t) = \Delta_1 V(t). \qquad (4.48)$$

If $\overline{V}$ is the rms value of $V(t)$, the corresponding rms value of $\Delta(t)$ is

$$\underline{\Delta} = \Delta_1 \overline{V}. \qquad (4.49)$$

If the message wave has a power spectrum $s(\omega)$, and a transmitting filter with a power transfer characteristic $t(\omega)$ is used, then the modulating voltage has a mean-squared value

$$\overline{V}^2 = \int_0^\infty s(\omega) t(\omega)\, d\omega. \qquad (4.50)$$

The squared rms frequency deviation is therefore given by

$$\underline{\Delta}^2 = \Delta_1^2 \int_0^\infty s(\omega) t(\omega)\, d\omega. \qquad (4.51)$$

The average message power in the demodulator output is

$$kP = k \int_0^\infty s(\omega)\, d\omega \qquad (4.52)$$

where k is a detector constant.

The power spectrum of the frequency variation resulting from random noise with a power spectrum $W_n(u) = n_i$ is in accordance with (3.96) for a sine wave of amplitude $a = E_i$ given by $u^2 n_i / E_i^2$. The corresponding noise-power density at the detector input is in accordance with (4.49) $u^2 n_i / E_i^2 \Delta_1^2$. With a post-detection low-pass filter of power transmittance $1/t(\omega)$, and with a detector constant k, as in (4.52), the average noise power at the output of a post-detection low-pass filter of bandwidth $\Omega_0 = 2\pi B_0$ becomes

$$N_0 = \frac{k}{E_i^2 \Delta_1^2} \left(n_i \int_{-\Omega_0}^{\Omega_0} \frac{u^2\, du}{t(u)} \right). \qquad (4.53)$$

This relation applies for adequately high signal-to-noise ratios at the detector input. With a modulated, rather than unmodulated carrier as assumed above, the analysis becomes more complicated, but (4.53) represents a close approximation also of this case, as shown in Section 2.12.

From (4.51), (4.52), and (4.53) the following relation is obtained for the ratio of average message power to average noise power at the output of the post-detection filter

$$\rho_0 = \rho_1 \underline{D}^2 \frac{\int_0^{\Omega_0} s(\omega)\, d\omega}{\frac{1}{\Omega_0}\int_0^{\Omega_0} \left(\frac{\omega}{\Omega_0}\right)^2 \frac{d\omega}{t(\omega)} \int_0^{\Omega_0} s(\omega) t(\omega)\, d\omega} \qquad (4.54)$$

where

$$\underline{D} = \underline{\Delta}/\Omega_0 = \text{rms deviation ratio},$$
$$\underline{\Delta} = \text{rms frequency deviation},$$

and

$$\rho_1 = \frac{P_i}{\Omega_0 n_i} = \frac{P_i}{N_1}. \qquad (4.55)$$

In the latter relation $P_i = E_i^2/2$ is the average received carrier power and N_1 is the noise power in a band Ω_0 at the detector input. The ratio of average received carrier power to noise power in the total input band is

$$\rho_i = \frac{P_i}{\Omega_i n_i}. \qquad (4.56)$$

Relation (4.54) is a satisfactory approximation of $\rho_i \geq 10$. The bandwidth Ω_i must be adequate so that transmission distortion can be ignored in comparison with channel noise, considering both elimination of spectral components by the bandpass channel filters as well as phase distortion owing to the filter cutoffs. The minimum permissible channel bandwidth Ω_i with a linear phase is ordinarily taken in accordance with the Carson rule as

$$\Omega_i = 2(\Omega_m + \hat{\Delta}), \qquad (4.57)$$

where $\hat{\Delta}$ is the peak frequency deviation and Ω_m the maximum bandwidth of the message waves. The latter ordinarily conforms with the bandwidth Ω_0 of the post-detection baseband filter, in which case

$$\Omega_i = 2\Omega_0(1 + \hat{D}), \qquad (4.58)$$

where $\hat{D} = \hat{\Delta}/\Omega_0$ is the peak deviation ratio.

Relation (4.58) is a legitimate approximation for determining ρ_i from (4.56). More exact determination of the minimum bandwidth may be required in evaluation of distortion owing to bandwidth restriction, as discussed in Section 6.16 for message waves with the properties of Gaussian random variables.

In the case of pure FM, $t(\omega) = t =$ constant and (4.54) gives

$$\rho_0 = \rho_1 3 \underline{D}^2. \qquad (4.59)$$

With pure PM, $t(\omega) = \omega^2$ and (4.54) yields

$$\rho_0 = \rho_1 D^2 \Omega_0^2 \frac{\int_0^{\Omega_0} s(\omega)\, d\omega}{\int_0^{\Omega_0} \omega^2 s(\omega)\, d\omega}. \qquad (4.60)$$

With a flat message power spectrum $s(\omega) = s$ of bandwidth Ω_m, (4.60) becomes

$$\rho_0 = \rho_1 3 \underline{D}^2 \left(\frac{\Omega_0}{\Omega_m}\right)^2. \qquad (4.61)$$

This is the same ratio as for pure FM for the particular case of $\Omega_m = \Omega_0$, that is, a flat message-power spectrum of the same bandwidth as the post-detection low-pass filter. However, when the baseband message power spectrum is concentrated near the lower end of the band, as, for example, in the case of speech waves, there is a significant advantage in PM over FM.

The maximum value of ρ_0 is obtained when the two integrals in the denominator of (4.54) are equal. (See Appendix 4.) This is the case for the following optimum transmitting filter-power transfer characteristic

$$t^0(\omega) = \frac{\omega}{[s(\omega)]^{1/2}}. \qquad (4.62)$$

The corresponding optimum ratio ρ_0 is

$$\rho_0^0 = \rho_1 D^2 \Omega_0^3 \frac{\int_0^{\Omega_0} s(\omega)\, d\omega}{\left(\int_0^{\Omega_0} \omega[s(\omega)]^{1/2}\, d\omega\right)^2}. \qquad (4.63)$$

With a flat power spectrum $s(\omega) = s$ of bandwidth Ω_m, (4.63) becomes

$$\rho_0^0 = \rho_1 4 \underline{D}^2 \left(\frac{\Omega_0}{\Omega_m}\right)^3. \qquad (4.64)$$

When $\Omega_0 = \Omega_m$ and the spectrum $s(\omega)$ is flat the optimum ratio is greater than for FM or PM by a factor 4/3, or about 1.3 dB.

In the more important present application of FM, the message wave consists of a large number of voice channels multiplexed in frequency division. Performance would then be considered optimum when the output signal-to-noise ratio is the same for all voice channels. This condition is realized with pure PM, which affords a three-fold increase in output message-to-noise ratio in the top channel over pure FM.

All of the above relations for ρ_0 apply for $\rho_i \geq 10$. For smaller values the analysis becomes very complicated, even for pure FM or PM, and this case is not considered herein. (See Sections 2.12 and 3.11.)

4.7 Analog Pulse Transmission Systems

In pulse systems the message or modulating wave is sampled at appropriate intervals, and the samples are transmitted as pulses.

Although pulses are deformed in transmission, owing to channel bandwidth limitation, it is possible to realize distortionless transmission in the absence of noise by appropriate amplitude and phase characteristics of the transmitting and receiving filters. Distortionless transmission can in this case be achieved with an asymmetrical pulse shape such as that illustrated in Fig. 4.4, in which intersymbol interference is absent at sampling instants. The above condition of distortionless pulse transmission with asymmetrical pulse shapes can be realized with an appropriate amplitude characteristic in conjunction with an appropriate nonlinear phase characteristic that in general could be realized with special phase modification networks. However, for optimum performance in the presence of noise it is essential that the pulse amplitude be a maximum at sampling instants. This entails a symmetrical pulse shape, such as that illustrated in Fig. 4.4 which is possible only with a linear phase characteristic. This class of optimum pulse shapes $P(t_0)$ has the general properties

$$P(-t_0) = P(t_0) \tag{4.65}$$

$$P(mT) = O; \quad m = \pm 1, \pm 2, \text{etc.} \tag{4.66}$$

$$P(O) = \hat{P} \tag{4.67}$$

where T is the sampling interval and $\hat{P}$ the peak pulse amplitude.

With the foregoing pulse shapes, it follows from the sampling theorem discussed in Chapter 1 that distortionless analog transmission is achieved when $T = 1/2B$ where B is the maximum frequency of the baseband modu-

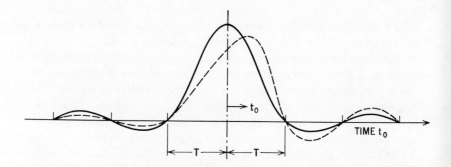

Figure 4.4 Symmetrical and asymmetrical pulses with zero points at intervals T.

Analog Pulse Transmission Systems

lating wave. The spectra of such pulses have the general properties indicated in Fig. 4.5. When impulses are applied at the transmitting end, the combined spectrum shown in Fig. 4.5 conforms in shape with the amplitude versus frequency characteristic of the channel.

While all spectra of the type illustrated in Fig. 4.5 permit distortionless transmission, the optimum type of spectrum or pulse shape depends on various transmission considerations. In actual systems it may even be preferable from the standpoint of simplified implementation to employ pulse shapes that entail some intersymbol interference, such as Gaussian pulses. In the analysis that follows it is assumed that the shape of the received pulses is specified, and thus the combined transmittance of transmitting and receiving filters.

As shown in Section 4.3, optimum performance in the presence of additive random noise may entail somewhat different combined equalization

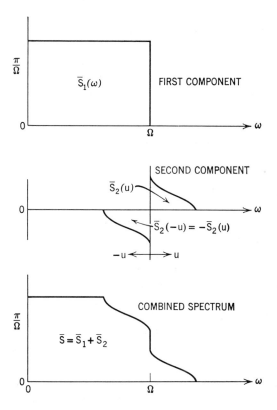

Figure 4.5 General properties of spectra of symmetrical pulses of unit peak amplitude and zero points at intervals $T = \pi/\Omega = 1/2B$.

than for distortionless transmission in the absence of noise. For representative signal-to-noise ratios the improvements that can be realized by departing from distortionless transmission are insignificant and will not be considered here.

A further assumption in the present analysis is that a single transmission band is used, rather than a number of separate and, in general, spaced transmission bands of the same combined bandwidth. Although the latter method would hardly be practicable, improved performance can in principle be realized with certain signal- and noise-power spectra, as shown in Section 4.11 and elsewhere [8]. An exact optimization has been formulated for baseband AM systems allowing optimum intersymbol interference and using an rms error criterion [9], as used in the analysis that follows.

In the case of digital transmission, error probability is a performance criterion and the determination of optimum equalization is much more complicated. A comprehensive analysis has been made for the particular case of binary pulse transmission on the premise that the received pulses have a Gaussian shape so chosen that intersymbol interference would be appreciable at sampling instants [10] The analysis indicates that optimum performance in this case is obtained by supplementing a matched receiving filter with tapped delay line equalization, such that intersymbol interference at sampling instants is virtually cancelled.

It should be recognized that intentional intersymbol interference for the purpose of optimizing performance is of the coarse-structure type discussed in Chapters 6 and 7, as contrasted with the fine-structure type encountered even after elaborate equalization, owing to small unavoidable residual irregularities in the amplitude and phase characteristics. Transmission impairments from such fine-structure interference is considered in Chapter 7 and will ordinarily exceed by far the small advantage to be derived from any intentional intersymbol interference as considered in the foregoing references. Consequently it is an entirely legitimate approximation to determine the joint optimization of transmitting and receiving filters on the premise of distortionless transmission in the absence of noise. On this premise, the optimization problem was formulated and resolved by the method used here, in an early investigation of baseband pulse transmission over wire circuits on behalf of the Signal Corps. ("PCM Transmission over Wire Circuits," *Technical Report No. 20*, May 16, 1951).

Let pulses of various amplitudes A be applied at the input to the transmitter, let the mean value be A_1 and the rms deviation from the mean $\underline{A}$. The mean and the rms deviation from the mean depends on the probability distribution of the amplitudes. If a message were absent from the

Analog Pulse Transmission Systems

input, there would be no fluctuation from the mean value, and the transmitter power corresponding to pulses of mean amplitude A_1 would be P_1. In the presence of a message, the average transmitter power is

$$S_0 = P_0 + P_1, \tag{4.68}$$

where P_0 is the average transmitter power variation, owing to the modulating pulses of rms amplitude $\underline{A}$.

Let $P(t)$ designate a symmetrical pulse of unit peak amplitude and $\bar{S}(\omega)$ the corresponding amplitude spectrum. When pulses of various amplitudes A are transmitted at intervals T, the average pulse train power variation is given by

$$P_0 = \left(\frac{\underline{A}^2}{T}\right) \int_{-\infty}^{\infty} P^2(t)\, dt = \left(\frac{\underline{A}^2}{\pi T}\right) \int_0^{\infty} \bar{S}^2(\omega)\, d\omega. \tag{4.69}$$

The power spectrum of the pulse train can accordingly be taken as

$$s(\omega) = \underline{a}^2 \bar{S}^2(\omega) \tag{4.70}$$

where

$$\underline{a}^2 = \frac{\underline{A}^2}{\pi T}. \tag{4.71}$$

With (4.70) in (4.35) the optimum receiving filter characteristic becomes

$$r^0(\omega) = [\bar{R}^0(\omega)]^2 = c_1 \bar{S}(\omega)[n_i(\omega)l(\omega)]^{-\frac{1}{2}}, \tag{4.72}$$

or

$$\bar{R}^0(\omega) = c_1 \bar{S}_i(\omega)[n_i(\omega)l(\omega)]^{-\frac{1}{2}}, \tag{4.73}$$

where the last relation is obtained with $\bar{S}(\omega) = \bar{S}_i(\omega)\bar{R}(\omega)$.

The associated optimum transmitting filter characteristic is in accordance with (4.38)

$$t^0(\omega) = [\bar{T}^0(\omega)]^2 = \frac{1}{c_1} \frac{1}{\bar{S}(\omega)} \left(\frac{n_i(\omega)}{l(\omega)}\right)^{\frac{1}{2}}. \tag{4.74}$$

In pulse systems, the pulses at the transmitting filter input ordinarily have a different shape from the received pulses. If the input pulses have an amplitude spectrum $\bar{S}_1(\omega)$ in place of $\bar{S}(\omega)$, the optimum transmitting filter characteristic is in accordance with (4.10)

$$t_1^0(\omega) = \frac{t^0(\omega)s(\omega)}{s_1(\omega)}, \tag{4.75}$$

or

$$\bar{T}_1^2(\omega) = \frac{1}{c_1} \frac{\bar{S}(\omega)}{\bar{S}_1^2(\omega)} \left(\frac{n_i(\omega)}{l(\omega)}\right)^{\frac{1}{2}}. \tag{4.76}$$

With impulses at the transmitter input, $\overline{S}_1(\omega) = $ constant. In this case, and with $n_i(\omega) = n_i = $ constant, $l(\omega) = l = $ constant, (4.76) gives

$$\overline{T}_1(\omega) = \overline{R}(\omega) = \frac{1}{c_2}\left(\frac{n_i}{l}\right)^{1/4}[\overline{S}(\omega)]^{1/2}. \tag{4.77}$$

In the case of pulse transmission, it is possible to consider two signal-to-noise ratios. One is the ratio of average signal power to average noise power, which will be designated $\bar{\rho}$. The other is the ratio of average instantaneous signal power at sampling instants to average noise power, which will be designated ρ.

The optimum value of the former of these ratios is in accordance with (4.39) and (4.70)

$$\bar{\rho}_0^0 = \bar{\rho}^0 = \frac{PP_0}{a^2\left\{\int_0^{\Omega_0}\overline{S}(\omega)\left(\frac{n_i(\omega)}{l(\omega)}\right)^{1/2}d\omega\right\}^2}. \tag{4.78}$$

Let $\hat{P}$ be the average power at sampling instants, which differs from the average power P. The optimum ratio $\hat{P}/N$ at the sampling instants is then

$$\rho_0^0 = \rho^0 = \bar{\rho}^0\hat{P}/P = \frac{\hat{P}P_0}{a^2\left\{\int_0^{\Omega_0}\overline{S}(\omega)\left(\frac{n_i(\omega)}{l(\omega)}\right)^{1/2}d\omega\right\}^2}. \tag{4.79}$$

When the combined phase characteristic is linear,

$$\hat{P} = \underline{A}^2\left(\frac{1}{\pi}\int_0^{\Omega_0}\overline{S}(\omega)\,d\omega\right)^2. \tag{4.80}$$

With (4.80) in (4.79) and $\Omega T = \pi$:

$$\rho_0^0 = \rho^0 = P_0\left[\frac{\int_0^{\Omega_0}\overline{S}(\omega)\,d\omega}{\int_0^{\Omega_0}\overline{S}(\omega)\left(\frac{\Omega n_i(\omega)}{l(\omega)}\right)^{1/2}d\omega}\right]^2. \tag{4.81}$$

Relation (4.81) is predicated on a linear phase characteristic so that (4.80) applies. Moreover, the pulse spectra must have the properties indicated in Fig. 4.5 so that intersymbol interference is absent at sampling instants. Thus (4.81) represents a modification of (4.39) applicable to idealized pulse systems.

4.8 Carrier Pulse Transmission Systems

The optimum signal-to-noise ratios with idealized double- and vestigial-sideband transmission is obtained by appropriate modification of the integration limits in (4.81).

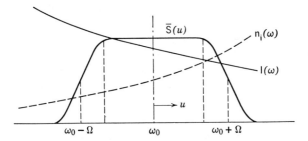

Figure 4.6 Sideband pulse spectrum $\bar{S}(u)$ with carrier amplitude and phase modulation. In double sideband transmission the carrier is at ω_0 and in vestigial sideband transmission at $\omega_0 + \Omega$ or $\omega_0 - \Omega$. In quadrature double sideband transmission two carriers of frequency ω_0 with quadrature phase relations are modulated independently in amplitude.

In double-sideband transmission with a carrier frequency at ω_0 the pulse spectrum extends from $\omega_0 - \Omega_0$ to $\omega_0 + \Omega_0$, as indicated in Fig. 4.6. The following relation then applies, in place of (4.81),

$$\rho_0^0 = \rho^0 = 2P_0 \left[\frac{\int_{-\Omega_0}^{\Omega_0} \bar{S}(u)\, du}{\int_{-\Omega_0}^{\Omega_0} \bar{S}(u)\left(\frac{\Omega n_i(u)}{l(u)}\right)^{1/2} du} \right]^2, \qquad (4.82)$$

where $u = \omega - \omega_0$ as indicated in Fig. 4.6. The factor 2 comes about because the two sideband pulse spectra combine directly after demodulation, whereas the noise spectra combine on a root-sum-square basis.

With double sideband transmission, it is possible in principle to transmit and demodulate two carriers displaced 90° in phase without mutual interference. With equal power in each of the two carriers and with a combined carrier power P_0, the signal-to-noise ratio then becomes, in place of (4.82),

$$\rho_0^0 = \rho^0 = P_0 \left[\frac{\int_{-\Omega_0}^{\Omega_0} \bar{S}(u)\, du}{\int_{-\Omega_0}^{\Omega_0} \bar{S}(u)\left(\frac{\Omega n_i(u)}{l(u)}\right)^{1/2} du} \right]^2. \qquad (4.83)$$

This is the same signal-to-noise ratio as in baseband transmission over a channel of bandwidth $2\Omega_0$. This method is thus equivalent to baseband transmission from the standpoint of channel bandwidth requirements and signal-to-noise ratios, for equal noise-power densities and transmission losses over the channel band. In actual implementation, a greater transmitter power will be required, so that the actual transmitter power is S_0 in place of P_0 in (4.83), as discussed in connection with (4.11).

An alternate method that in principle affords the same performance as baseband transmission is vestigial sideband carrier modulation, in which one sideband of the pulse spectrum is suppressed except for a vestige, as indicated in Fig. 4.6. With this method the shape of demodulated pulses with synchronous carrier demodulation will not be the same as in baseband or in double-sideband transmission, except in the limiting case of single-sideband modulation. The latter method is impracticable as applied to pulse transmission partly because of the pronounced phase distortion near the carrier frequency, and partly because some carrier must be transmitted to facilitate synchronous detection. In pulse transmission by the vestigial sideband method, it is desirable to have a gradual cutoff near the carrier frequency, so that the pulse spectrum at the detector input would be virtually the same as in double sideband transmission. This is illustrated in Fig. 4.7 for the particular case of a raised cosine spectrum. With double sideband transmission the shape of a demodulated pulse would in this case be that of $\bar{P}(t)$ in Fig. 1.15, while in vestigial sideband systems the pulse shape would be $R(t)$.

The optimum shapes of transmitting and receiving filters are determined in (4.86) and (4.87) for the particular case of a raised cosine pulse spectrum at the detector input. This spectrum has various advantages cited in Section 1.12 and is given by

$$\bar{S}(u) = \bar{S}(-u) = \frac{T}{2} \cos^2 \frac{\pi u}{4\Omega}; \quad u < 2\Omega \tag{4.84}$$

$$= O; \quad u > 2\Omega. \tag{4.85}$$

With this type of spectrum, pulses can be transmitted at intervals $T = \pi/\Omega = 1/2B$ without intersymbol interference. The shape of the pulses is indicated in Fig. 1.15.

Let the transmission loss and the noise power density be constant over the channel band, so that $l(u) = l$ and $n_i(u) = n_i$. The optimum receiving filter characteristic is then in accordance with (4.73) with u in place of ω.

$$\bar{R}^0(u) = \left(\frac{c_1 T}{2}\right)^{1/2} \left(\frac{1}{ln_i}\right)^{1/4} \cos \frac{\pi u}{4\Omega}. \tag{4.86}$$

The corresponding transmitting filter characteristic is, in accordance with (4.77),

$$\bar{T}_1^0(u) = \left(\frac{T}{2c_1}\right)^{1/2} \left(\frac{n_i}{l}\right)^{1/4} \frac{\cos \pi u/4\Omega}{\bar{S}_1(u)}. \tag{4.87}$$

With impulse transmission $\bar{S}_1(u) = T/2$ and

$$\bar{T}_1^0(u) = \left(\frac{2}{Tc_1}\right)^{1/2} \left(\frac{n_i}{l}\right)^{1/4} \cos \frac{\pi u}{4\Omega}. \tag{4.88}$$

Carrier Pulse Transmission Systems

The combined amplitude characteristic of the channel is then

$$A(u) = \overline{T}_1^0(u)\,\overline{R}(u) = \frac{1}{l^{1/2}} \cos^2 \frac{\pi u}{4\Omega}. \tag{4.89}$$

The factor $l^{-1/2}$ is the voltage amplification required in order that the received pulse has the peak amplitude $\hat{P}(O) = 1$, corresponding to the spectrum (4.84).

With rectangular pulses of duration T the spectrum at the channel input is

$$\overline{S}_1(u) = \frac{T}{2} \frac{\sin uT/2}{uT/2} \tag{4.90}$$

$$= \frac{T}{2} \frac{\sin \pi u/2\Omega}{\pi u/2\Omega}. \tag{4.91}$$

With (4.91) in (4.87)

$$\overline{T}_1^0(u) = \left(\frac{2}{Tc_1}\right)^{1/2} \left(\frac{n_i}{l}\right)^{1/4} \frac{\pi u/4\Omega}{\sin \pi u/4\Omega}. \tag{4.92}$$

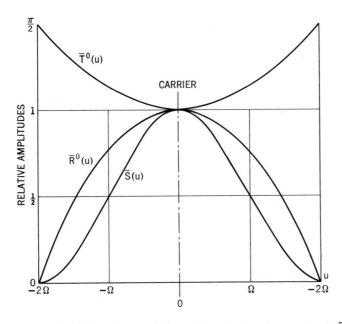

Figure 4.7 Double sideband transmission with raised cosine spectrum $\overline{S}(u)$ of received pulses. $\overline{R}^0(u)$ = optimum receiving filter characteristics. $\overline{T}^0(u)$ = optimum transmitting filter characteristics with rectangular transmitted pulses of duration $T = \pi/\Omega$.

The combined channel amplitude characteristic that will result in received pulses with the spectrum (4.84) and peak amplitude $\hat{P}(O) = 1$ is, in this case,

$$A(u) = \frac{1}{l^{\frac{1}{2}}} \frac{\pi u/4\Omega}{\tan \pi u/4\Omega}. \tag{4.93}$$

The above various characteristics are shown in Fig. 4.7.

With vestigial sideband transmission, the carrier frequency would be at $\omega_0 - \Omega$ or $\omega_0 + \Omega$. As shown in Chapter 1, Fig. 1.15, pulses can in this case be transmitted at twice the double sideband rate, that is, at intervals $T/2$. The spectrum of the received wave is now

$$\bar{S}_1(u) = \frac{T}{4} \frac{\sin \pi(u \pm \Omega)/4\Omega}{\pi(u \pm \Omega)/4\Omega}. \tag{4.94}$$

The positive signs apply for a carrier at $\omega_0 - \Omega$ and the negative sign when it is at $\omega_0 + \Omega$. If the carrier frequency is at $\omega_0 + \Omega$, the optimum transmitting filter characteristic obtained with (4.94) in (4.87) is

$$\bar{T}_1^0(u) = \left(\frac{4}{Tc_1}\right)^{\frac{1}{2}} \left(\frac{n_i}{l}\right)^{\frac{1}{4}} \frac{\pi(u - \Omega)/4\Omega}{\sin \pi(u - \Omega)/4\Omega} \cos \frac{\pi u}{4\Omega}. \tag{4.95}$$

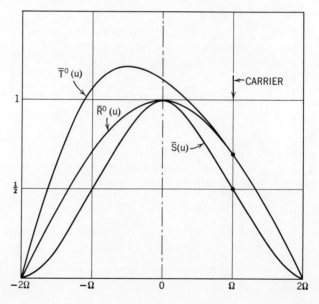

Figure 4.8 Vestigial sideband transmission with raised cosine spectrum of received pulses. $\bar{R}^0(u)$ = optimum receiving filter characteristic. $\bar{T}^0(u)$ = optimum transmitting filter characteristic with transmitted rectangular pulses of duration $T/2 = \pi/2\Omega$.

Comparison with Matched Filter Reception

The optimum receiving filter characteristic is the same as for double sideband transmission, and the combined amplitude characteristic of the channel becomes

$$A(u) = \left(\frac{2}{l}\right)^{1/2} \frac{\pi(u-\Omega)/4\Omega}{\sin \pi(u-\Omega)/4\Omega} \cos^2 \frac{\pi u}{4\Omega}. \qquad (4.96)$$

The above characteristics are shown in Fig. 4.8.

4.9 Comparison with Matched Filter Reception

A somewhat different optimization problem than considered herein is encountered in matched filter theory [11], which is concerned with detection of signal waves of known shapes in the presence of random noise, as in radar detection. Here spectrum $S_i(i\omega)$ of the received wave is known and the ratio of peak output power to average noise power is to be maximized by appropriate choice of $R(i\omega)$, much in the same way as in pulse transmission. However, intersymbol interference from pulse distortion need not be considered, since the interval between signals is adequately long. Moreover, there is no question of optimum shaping of the spectrum $S_i(i\omega)$ at the receiving filter output. For this reason the optimization problem is not the same as in transmission-system design and the optimum receiving-filter characteristics are not in general the same, as shown next.

Let the spectrum of the input voltage $U_i(t)$ be

$$S_i(i\omega) = \bar{S}_i(\omega)e^{-i\varphi(\omega)}. \qquad (4.97)$$

The signal voltage at the detector input is then

$$U(t) = \frac{1}{2\pi} \int_{-\infty}^{\infty} S_i(i\omega) R(i\omega) e^{i\omega t} d\omega, \qquad (4.98)$$

where $R(i\omega)$ is the receiving-filter characteristic.

The average noise power at the detector input is

$$N = \frac{1}{2} \int_{-\infty}^{\infty} n_i(\omega) \bar{R}^2(\omega) d\omega. \qquad (4.99)$$

The factor $\frac{1}{2}$ comes about since $n_i(\omega)$, as used previously, is the one-sided noise-power spectrum.

The power ratio $U^2(t)/N$ is

$$p_0(t) = p(t) = \frac{\left(\dfrac{1}{2\pi}\displaystyle\int_{-\infty}^{\infty} S_i(i\omega) R(i\omega) e^{i\omega t} d\omega\right)^2}{\dfrac{1}{2}\displaystyle\int_{-\infty}^{\infty} n_i(\omega) \bar{R}^2(\omega) d\omega}. \qquad (4.100)$$

As shown in Appendix 4, the maximum value of the foregoing ratio is obtained for a receiving filter characteristic

$$R^0(i\omega) = \bar{R}^0(\omega)e^{-i[\omega t - \varphi(\omega)]}, \qquad (4.101)$$

where

$$\bar{R}^0(\omega) = c\,\frac{\bar{S}_i(\omega)}{n_i(\omega)}, \qquad (4.102)$$

in which c is an arbitrary constant.

From (4.101) it is seen that the receiving filter must have a phase characteristic $+\varphi(\omega)$ as compared with $-\varphi(\omega)$ in (4.97). In general, this entails phase equalization with a transmission delay τ, and a symmetrical pulse shape in which the peak is reached at the time $t = \tau$. The optimum ratio (4.100) is obtained for $t = \tau$, such that the pulse is sampled at the peak. In this case (4.101) in (4.100) yields

$$\rho_0^0 = \left(\frac{1}{2\pi^2}\right)\int_{-\infty}^{\infty}\frac{\bar{S}_i^2(\omega)}{n_i(\omega)}\,d\omega. \qquad (4.103)$$

In the particular case of a flat noise-power spectrum, $n_i(\omega) = n_i$ (4.103) becomes

$$\rho_0^0 = \frac{1}{2\pi^2 n_i}\int_{-\infty}^{\infty}\bar{S}_i^2(\omega)\,d\omega$$

$$= \frac{2W_i}{2\pi n_i} = \frac{2W_i}{n_i'}, \qquad (4.104)$$

where W_i is the energy per pulse as given by

$$W_i = \frac{1}{2\pi}\int_{-\infty}^{\infty}\bar{S}_i^2(\omega)\,d\omega. \qquad (4.105)$$

In (4.104) n_i is the noise-power density per radian cycle per second and $n_i' = 2\pi n_i$ is the noise power density per c/s.

By way of comparison, let $l(\omega) = l$ and $n_i(\omega) = n_i$ in (4.81), in which case

$$\rho_0^0 = \frac{P_0 l}{\Omega n_i}. \qquad (4.106)$$

When pulses are transmitted at intervals T given by $\Omega T = \pi$, relation (4.106) becomes

$$\rho_0^0 = \frac{2P_0 lT}{2\pi n_i} = \frac{2W_i}{2\pi n_i}. \qquad (4.107)$$

Thus matched filter theory based on single pulses yields the appropriate result when applied to pulse trains in the foregoing instance. However, this is not generally the case when the attenuation or the noise-power density varies with frequency. The optimum receiving filter characteristic as obtained from (4.73) then becomes

$$\bar{R}^0(\omega) = c_1 \frac{\bar{S}_i(\omega)}{[n_i(\omega)l(\omega)]^{1/2}}. \quad (4.108)$$

Comparison of (4.108) with (4.102) shows that the latter relation does not apply to optimum pulse transmission except for the special case in which $n_i(\omega)l(\omega) = $ constant. This condition is not realized in transmission over metallic circuits when the transmission band occupies several octaves, since the transmission loss then increases appreciably with frequency, as illustrated in Section 4.13.

4.10 Single-Band Digital Transmission Systems

It will be assumed that the information is encoded into digital sequences as discussed in Section 2.4, with r possible amplitudes of each digit. Furthermore, that the digits are transmitted over a baseband channel without division of the available band into a number of narrower channels. If r is the maximum number of pulse amplitudes that can be transmitted with a certain error probability ϵ, the channel capacity with this error probability is in accordance with (2.44)

$$C = B \log_2 r^2. \quad (4.109)$$

The maximum value of r for a given error probability will be determined here for the case in which errors are caused by Gaussian noise and instantaneous digit-by-digit detection is used.

It will be assumed that all amplitude steps are equal. The maximum tolerable deviation $\hat{e}$ in the peak amplitudes before an error occurs is then, for r pulse amplitude levels,

$$\hat{e} = \frac{A_{\max} - A_{\min}}{2(r-1)}, \quad (4.110)$$

where $A_{\max}$ and $A_{\min}$ are the maximum and minimum peak amplitudes.

If all pulse amplitudes are assumed equally probable, then the mean pulse amplitude is

$$A_1 = \tfrac{1}{2}(A_{\max} + A_{\min}), \quad (4.111)$$

and the maximum deviation from the mean is

$$\pm \hat{A} = \tfrac{1}{2}(A_{\max} - A_{\min}). \quad (4.112)$$

The rms deviation from the mean is found to be [12]

$$\underline{A} = \hat{A}\left(\frac{r+1}{3(r-1)}\right)^{1/2}$$

$$= \hat{e}\left(\frac{r^2-1}{3}\right)^{1/2}. \tag{4.113}$$

The tolerable peak interference $\hat{e}$ before an error occurs is greater than the tolerable rms interference $\underline{e}$ by a factor $\kappa > 1$ that depends on the error probability. Thus with $\hat{e} = \kappa\underline{e}$ and with $\hat{A}^2/\underline{e}^2 = \hat{P}/N = \rho$ (4.113) can be written

$$\rho = \frac{\kappa^2}{3}(r^2 - 1), \tag{4.114}$$

or

$$r^2 = 1 + \frac{3\rho}{\kappa^2}, \tag{4.115}$$

or

$$\kappa = \left(\frac{3\rho}{r^2 - 1}\right)^{1/2}. \tag{4.116}$$

With random noise the probability of exceeding the rms amplitude by a factor κ is $(\frac{1}{2})$ erfc $(\kappa/2^{1/2})$, when the polarity of the noise amplitude is specified. The probability of an erroneous pulse amplitude is, in this case,

$$\epsilon = \left(1 - \frac{1}{r}\right) \text{erfc}\left(\frac{\kappa}{2^{1/2}}\right) \tag{4.117}$$

$$= \left(1 - \frac{1}{r}\right) \text{erfc}\left(\frac{3\rho/2}{r^2-1}\right)^{1/2} \tag{4.118}$$

$$= \tfrac{1}{2} \text{erfc}\left(\frac{\rho}{2}\right)^{1/2} \quad \text{for} \quad r = 2 \tag{4.119}$$

$$= \tfrac{3}{4} \text{erfc}\left(\frac{\rho}{10}\right)^{1/2} \quad \text{for} \quad r = 4. \tag{4.120}$$

The channel capacity obtained from (4.109) for various values of r will be associated with a certain error probability ϵ, which is obtained from (4.118) for various values of ρ. Alternately, for a specified r and ϵ, the requisite ρ can be determined from (4.118).

When the attenuation and noise-power density varies with frequency, the optimum value of ρ for a specified transmitter power P_0 and pulse spectrum $\bar{S}(\omega)$ is obtained from (4.81). Alternately, when r is specified and ρ is obtained from (4.118), the minimum average transmitter power is obtained from (4.81) with $\rho = \rho^0$.

In view of (4.115) the channel capacity can be written [12]

$$C = B \log_2 (1 + a\rho), \qquad (4.121)$$

where

$$a = \frac{3}{\kappa^2}. \qquad (4.122)$$

When the optimum ρ obtained from (4.81) is inserted in (4.121), the channel capacity becomes

$$C = B \log_2 \left\{ 1 + aP_0 \left[\frac{\int_0^{\Omega_i} \bar{S}(\omega)\, d\omega}{\int_0^{\Omega_i} \bar{S}(\omega) \left(\frac{\Omega n_i(\omega)}{l(\omega)} \right)^{1/2} d\omega} \right]^2 \right\}. \qquad (4.123)$$

Here $\Omega_i = 2\pi B_i$ depends on the pulse spectrum $\bar{S}(\omega)$. For a flat spectrum $B_i = B$ while for a raised cosine spectrum $B_i = 2B$.

For a channel with flat loss $l(\omega) = l$ and a flat noise-power spectrum $n_i(\omega) = n_i$, (4.123) yields the following expression for all pulse spectra with the properties indicated in Fig. 4.5

$$C = B \log_2 \left(1 + \frac{aP_i}{N_i} \right), \qquad (4.124)$$

where $P_i = P_0 l$ is the average power at the input to the receiving filter and $N_i = n_i \Omega$ is the average noise power in the band B at the input to the receiving filter. This expression is of the same form as that for the upper bound on channel capacity as given by Shannon [5], except that $a = 1$. For an error probability $\epsilon = 10^{-5}$ in a digit, $\kappa \simeq 4.8$ and $a \simeq 0.125$. Hence about an eight-fold increase in signal-to-noise ratio is required, as compared to the minimum possible with sufficiently complicated methods of encoding and decoding [Section 2.12].

4.11 Multiband Digital Transmission Systems

In the previous section it was assumed that the signal is transmitted over a single channel band. An alternate method would be to divide the total channel band into a number of narrower bands of bandwidth $d\Omega$. With ideal single sideband carrier modulation in each narrow band, the combined channel capacity would in this case be

$$C = \frac{1}{2\pi} \int_0^{\Omega_i} \log_2 [1 + a\rho(\omega)]\, d\omega, \qquad (4.125)$$

where $\rho(\omega) = \rho_i(\omega)$, as given by (4.15), or

$$\rho(\omega) = \frac{s_0(\omega)l(\omega)}{n_i(\omega)}. \tag{4.126}$$

As shown in Appendix 4, the maximum channel capacity for a given average transmitter power is obtained when the following relation given by Shannon [5] is satisfied

$$s_0(\omega) + \frac{n_i(\omega)}{al(\omega)} = \lambda = \text{constant}. \tag{4.127}$$

Integration of (4.127) over the band Ω_i gives the following relation for determination of λ

$$P_0 + \int_0^{\Omega_i} \frac{n_i(\omega)}{al(\omega)} d\omega = \lambda \Omega_i. \tag{4.128}$$

With (4.128) in (4.127), and with $s_i(\omega) = l(\omega)s_0(\omega)$, the following relation is obtained for the optimum ratio, $\rho(\omega) = s_i(\omega)/n_i(\omega)$

$$1 + a\rho(\omega) = \frac{aP_0 + \int_0^{\Omega_i} \frac{n_i(\omega)}{l(\omega)} d\omega}{\Omega_i n_i(\omega)/l(\omega)}. \tag{4.129}$$

The maximum channel capacity obtained with (4.129) in (4.125) is

$$C = \frac{1}{2\pi} \int_0^{\Omega_i} \log_2 \left(\frac{aP_0 + \int_0^{\Omega_i} n_i(\omega) \, d\omega/l(\omega)}{\Omega_i n_i(\omega)/l(\omega)} \right). \tag{4.130}$$

Returning to (4.127), a necessary requirement for $s_0(\omega)$ to be positive at all frequencies is that

$$\lambda - \frac{1}{a} \frac{n_i(\omega)}{l(\omega)} \geq 0. \tag{4.131}$$

This entails

$$\lambda = \frac{1}{a} \left(\frac{n_i}{l} \right)_{\max} + \frac{P_1}{\Omega_i}, \tag{4.132}$$

where $(n_i/l)_{\max}$ is the maximum value of $n_i(\omega)/l(\omega)$ and P_1/Ω_i represents an extra uniform power spectrum, as indicated in Fig. 4.9.

In view of (4.132), the average transmitter power (4.128) can be expressed as

$$P_0 = \frac{\Omega_i}{a} \left(\frac{n_i}{l} \right)_{\max} - \frac{1}{a} \int_0^{\Omega_i} \frac{n_i(\omega)}{l(\omega)} d\omega + P_1. \tag{4.133}$$

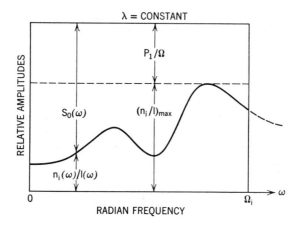

Figure 4.9 Optimum spectrum of transmitted power $s_0(\omega)$ in multiband digital transmission over a channel of bandwidth Ω_i with noise power spectrum $n_i(\omega)$ and power transmittance $l(\omega)$.

and the maximum channel capacity (4.130) can alternately be written

$$C = \frac{1}{2\pi} \int_0^{\Omega_i} \log_2 \left(\frac{aP_1 + \Omega_i(n_i/l)_{\max}}{\Omega_i n_i(\omega)/l(\omega)} \right) d\omega. \qquad (4.134)$$

When the noise-power spectrum is flat so that $n_i(\omega) = N_i/\Omega_i$, (4.134) simplifies to

$$C = \frac{1}{2\pi} \int_0^{\Omega_i} \log_2 \left(\frac{l(\omega)}{l_{\min}} + \frac{aP_1 l(\omega)}{N_i} \right) d\omega. \qquad (4.135)$$

In the particular case of a channel with flat power-transfer characteristic, $l(\omega) = l = l_{\min}$ over a band $\Omega_i = \Omega$ (4.135) yields

$$C = B \log_2 \left(1 + a\frac{P}{N} \right), \qquad (4.136)$$

where $B = \Omega/2\pi$, $P = P_i = P_1 l$ and $N_i = N = \Omega n_i$.

The foregoing relations only apply provided that the number of digital amplitudes can be chosen to conform with the signal-to-noise ratio $\rho(\omega)$ in each narrow band. This cannot be realized except for some special variation in the ratio $n_i(\omega)/l(\omega)$ appearing in the relations.

4.12 Limiting Channel Capacity with Optimum Encoding

The maximum channel capacity is obtained when both the signal and the noise have a flat power spectrum and a Gaussian amplitude distribution, as in white noise. With idealized channel characteristics and sufficiently complicated methods of encoding and decoding the information, as outlined in Section 2.18, the upper bound or channel capacity is

$$C = B \log_2 \left(1 + \frac{P}{N}\right). \tag{4.137}$$

When both the signal power and noise power varies over the channel band, the upper bound on channel capacity is obtained by dividing the total band into a number of sufficiently narrow bands, such that the signal and noise power spectra can be regarded as constant over each narrow band. This upper bound is given by

$$C = \frac{1}{2\pi} \int_0^{\Omega_i} \log_2 \left[1 + \rho^0(\omega)\right] d\omega. \tag{4.138}$$

The optimum value of $1 + \rho^0(\omega)$ is obtained from (4.129) with $a = 1$ and the channel capacity (4.138) from (4.134) with $a = 1$.

The channel capacity with single-band transmission and optimum methods of encoding and decoding cannot readily be determined. However, it will be less than that obtained with $a = 1$ in the expressions for channel capacity with single-band digital transmission. The reason for this is that the signal- and noise-power spectra are not flat over the channel band, as would be required to realize the channel capacity obtained with $a = 1$.

4.13 Digital Baseband Transmission over Metallic Pairs

The preferable modulation method for transmission over metallic pairs or coaxial units depends on several factors. Among them are the requirements on gain and phase equalization, interference between systems on different pairs in the same cable and repeater spacings as determined by additive random noise. Single-sideband analog modulation is presently used for long-distance broadband transmission over heavily shielded coaxial units. However, binary or multilevel PCM offers advantages in certain respects noted in Section 2.16 and may be preferable for a cable containing a large number of lightly shielded balanced pairs or coaxial units.

Fom the standpoints of equalization and intersystem interference, binary is preferable to multilevel PCM. However, if additive random noise were the only consideration, multilevel transmission would be preferable in that it would permit longer repeated spacings, as shown in the following illustrative application of relations derived previously.

In the case of coaxials, and metallic pairs in general, the following approximation applies for the attenuation factor $\underline{L}(\omega)$:

$$\underline{L}(\omega) = e^{-\alpha(\omega/\Omega)^{1/2}}, \tag{4.139}$$

in which case

$$l(\omega) = e^{-2\alpha(\omega/\Omega)^{1/2}}. \tag{4.140}$$

The pulse spectrum will be assumed to have a partial raised cosine cutoff, as given by the following relations.

For $(1 - \kappa)\Omega < \omega < (1 + \kappa)\Omega$:

$$\bar{S}(\omega) = \frac{\pi}{\Omega} \cos^2 \frac{\pi}{4} \frac{\omega + (1 + \kappa)\Omega}{\kappa\Omega}. \tag{4.141}$$

For $\omega > (1 + \kappa)$: $\bar{S}(\omega) = 0$ and for $\omega < (1 - \kappa)\Omega$: $\bar{S}(\omega) = \pi/\Omega$.

These equations are obtained from relations given in section 1.6.

The corresponding pulse shape obtained with the aid of relations in Section 1.6 is

$$P(t) = \frac{\sin \Omega t_0}{\Omega t_0} \frac{\cos \kappa \Omega t_0}{1 - (2\kappa\Omega t_0/\pi)^2}. \tag{4.142}$$

The noise-power spectrum will be taken

$$n_i(\omega) = n_i = \frac{N_i}{\Omega} \tag{4.143}$$

With the foregoing assumptions, the optimum signal-to-noise ratio (4.81) can be reduced to the following form:

$$\rho^0 = \frac{P_0 e^{-2\alpha}}{N_i} \frac{1}{L^2(\alpha, \kappa)}, \tag{4.144}$$

where

$$e^\alpha L(\alpha,\kappa) = \frac{2}{\alpha^2} \left\{ e^{\alpha(1-\kappa)^{1/2}} [\alpha(1-\kappa)^{1/2} - 1] + 1 \right\}$$

$$+ \int_{1-\kappa}^{1+\kappa} e^{\alpha x^{1/2}} \cos^2 \left[\frac{\pi}{4} \frac{x - (1 - \kappa)}{\kappa} \right] dx \quad (4.145)$$

The function L^2 obtained by numerical integration is given in Table 1 for some values of α and κ.

Table 1. Function $L^2(\alpha, \kappa)$

$\alpha =$	0	4	6	8	10 nepers
$\kappa = 0$	1	0.142	0.077	0.048	0.032
$\kappa = \frac{1}{2}$	1	0.181	0.124	0.104	0.101
$\kappa = 1$	1	0.335	0.380	0.571	1.052

In this table α is the attenuation in nepers at the frequency $\Omega = 2\pi B$, the attenuation in dB being 8.69α. With the type of pulse spectra just assumed, pulses can be transmitted without intersymbol interference at intervals $T = 1/2B$.

The function L^2 is shown in Fig. 4.10. It represents the factor by which the repeater power differs from that required with a channel of bandwidth Ω and constant attenuation α over the channel band, equal to that at the frequency $\Omega = 2\pi B$.

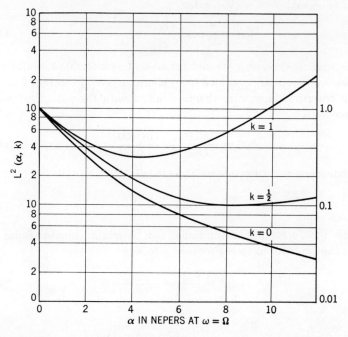

Figure 4.10 Function $L^2(\alpha, k)$ for transmission medium with power loss $l(\omega) = \exp - 2\alpha(\omega/\Omega)^{1/2}$.

With a flat noise-power spectrum, and the pulse spectrum assumed here, the following optimum receiving filter characteristic is obtained from (4.73). For $0 < \omega < \Omega(1 - k)$:

$$\bar{R}(\omega) = \frac{c}{N_i^{\frac{1}{2}}} e^{\frac{\alpha}{2}(\omega/\Omega)^{1/2}}. \qquad (4.146)$$

For $\Omega(1 - \kappa) < \omega < \Omega(1 + \kappa)$:

$$\bar{R}(\omega) = \frac{c}{N_i^{\frac{1}{2}}} e^{\frac{\alpha}{2}(\omega/\Omega)^{1/2}} \cos \frac{\pi}{4} \frac{\omega - (1 - \kappa)\Omega}{\kappa\Omega}. \qquad (4.147)$$

For $\omega > \Omega(1 + \kappa)$, $\bar{R}(\omega) = 0$.

Digital Baseband Transmission over Metallic Pairs

With impulse transmission, $\bar{S}_1(\omega) = 1$ in (4.76), and the optimum transmitting-filter characteristic differs from $\bar{R}(\omega)$ by the factor N_i. Their product is thus

$$\bar{T}(\omega)\,\bar{R}(\omega) = e^{\alpha(\omega/\Omega)^{1/2}}\bar{S}_1(\omega), \qquad (4.148)$$

and

$$\bar{T}(\omega)\,\bar{R}(\omega)\,\underline{L}(\omega) = \bar{S}(\omega). \qquad (4.149)$$

With pulses of finite duration at the transmitter input with spectrum $\bar{S}_1(\omega)$, the corresponding transmitting filter characteristic is obtained from (4.76).

In Fig. 4.11, the optimum receiving-filter characteristic obtained from (4.147) is illustrated for the particular case $\kappa = 1$ and $\alpha = 8$.

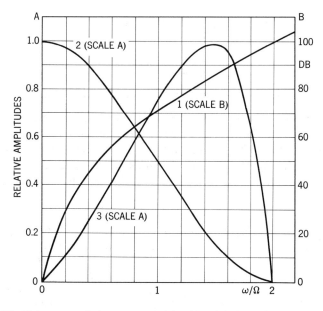

Figure 4.11 Pulse transmission over coaxial cable with flat noise power spectrum. 1. Assumed attenuation characteristic. 2. Desired raised cosine spectrum of received pulses. 3. Optimum receiving filter characteristic. Also transmitting filter characteristic with impulse transmission.

Binary pulse systems have an obvious advantage over multilevel pulse systems from the standpoint of simplified design of regenerative repeaters and increased tolerance to imperfections in the transmission medium. Nevertheless, it is of some interest to compare binary with multilevel pulse systems from the standpoint of repeater spacing, considering transmission impairments from random noise alone.

To this end it will be assumed that the signal is quantized into $L_r = 256$ levels. In accordance with (2.39) this is possible with certain combinations of r pulse amplitudes and m digits, as indicated in the following table.

r	=	2	4	16	256 amplitudes
m^*	=	8	4	2	1 digit
$L_r = r^m$	=	256	256	256	256 levels

Let N_0 be the average noise power in one baseband, in which case the noise power N_i in (4.144) is $N_i = mN_0$. Equation (4.144) can then be written

$$\frac{P_0}{N_0} = m\rho_r^0 e^{2\alpha_m} L^2(\alpha_m, \kappa), \qquad (4.150)$$

where

$\rho_r^0 = \rho^0$ for r pulse amplitudes, and

$\alpha_m = \alpha$ for bandwidth required with m digits.

Let $\alpha = \alpha_8 = 10$ nepers (87 dB) for $m = 8$ digits, in which case, for m digits $\alpha_m = 10(m/8)^{\frac{1}{2}}$ nepers. Let, further, the probability of error in a digit be $\epsilon = 10^{-6}$. The required ratios $\rho = \rho_r^0$ can then be determined from (4.118), using Fig. 3.2 for the error-function complement. For a raised cosine pulse spectrum, ($\kappa = 1$), the ratios P_0/N_0 are then as given in Table 2.

Table 2. *Relative Repeater Powers with Raised Cosine Pulse Spectrum*

r	=	2	4	16	256 amplitudes
m	=	8	4	2	1 digit
α_m	=	10	7.1	5	3.6 nepers
ρ_r^0	=	22	116	2000	128,000
$L^2(\alpha_m, 1)$	$\simeq$	1.02	0.43	0.33	0.33
P_0/N_0	$\simeq$	$8.5 \cdot 10^8$	$7.2 \cdot 10^7$	$1.4 \cdot 10^7$	$5.8 \cdot 10^7$
$(P_0/N_0)_{db}$	$\simeq$	109	79	72	78

The optimum number of pulse amplitudes in the foregoing case would thus be $r = 16$.

Let the average repeater output power be the same in all cases, and such that $P_0/N_0 = 109$ dB. For a binary system this corresponds to 87 dB transmission loss at the frequency Ω. The relative repeater spacings would then compare approximately as shown in Table 3.

Table 3. Repeater Spacings with $P_0/N_0 = 109\,dB$

No. of Pulse Amplitude:	2	4	16	256
Relative Spacings:	1	1.4	1.5	1.4

There is thus an insignificant repeater spacing advantage in $r = 16$ over $r = 4$ amplitude levels. Considerations of equalization and intersystem interference would favor $r = 4$ over a greater number of levels. Intersystem interference may even dictate the choice of $r = 2$ over a greater number, as demonstrated in Section 11.6 on the premise that the maximum capacity per pair is the basic criterion. A further consideration in favor of binary PCM is that it permits simple regenerative repeaters. With multilevel PCM it may be preferable to use a number of analog repeaters between regeneration points, because of the greater complexity of regenerative repeaters. This would entail more precise equalization and reduced tolerance to both additive random noise and intersystem interference.

In baseband systems, pulse distortion caused by low-frequency cutoff is a principal transmission limitation, as discussed in Sections 7.16 and 7.17. The pulse distortion caused by low-frequency cutoff can be largely avoided by "constrained bipolar" transmission, in exchange for a 3 dB increase in transmitter power, as discussed in Section 5.4. This method is presently used in some PCM systems for metallic pairs (Chapter V). With multilevel transmission, other methods of combating low-frequency cutoff would be required, such as double or vestigial-sideband amplitude modulation. These methods are more complicated and entail greater bandwidth, so that the transmission losses are increased. This may result in even shorter repeater spacings than with constrained bipolar baseband transmission.

In view of the foregoing considerations, the preferable method of digital transmission over metallic pairs would appear to be constrained bipolar transmission.

The transmission capacity of coaxial cable as limited solely by additive random noise is discussed in some detail elsewhere [13]. When the system is assumed to comprise both analog and regenerative repeaters in tandem, the formalization of the filter optimization problem is modified in some respects [14,15]. However, this does not have a significant bearing on the choice of binary versus multilevel transmission for metallic pairs.

4.14 Optimization against Impulse Noise

The various relations in the preceding sections apply for continual additive random noise, such as thermal noise. In some systems, appreciable additive impulse-type noise may be encountered during certain periods, owing to electromagnetic induction from external sources, such as atmospheric disturbances or transients in other systems. As will be shown presently, the optimum filter shapes are then the same as for continual random noise, that is, for analog systems where the basic criterion is minimum average interference power in the demodulated output. For digital systems, however, where error probability is to be minimized, the filter optimization problem is much more complex.

Let $V_i(t)$ represent a transient disturbance at the input to the receiving filter, with a spectrum $S_i(i\omega)$. The corresponding disturbance at the output of the receiving filter is then

$$V(t) = \frac{1}{2\pi} \int_{-\infty}^{\infty} S_i(i\omega) R(i\omega) e^{-i\omega t} d\omega. \qquad (4.151)$$

Let the duration of the disturbance be T. In view of the Fourier integral energy theorem, the average power over the interval T then becomes

$$\frac{1}{T} \int_0^T V^2(t) dt = \frac{1}{2\pi T} \int_{-\infty}^{\infty} |S_i(i\omega)|^2 |R(i\omega)|^2 d\omega$$

$$= \int_0^{\infty} n_i(\omega) r(\omega) d\omega, \qquad (4.152)$$

where $r(\omega) = |R(i\omega)|^2 = \bar{R}^2(\omega)$ and $n_i(\omega) = |S_i(i\omega)|^2/\pi T$.

Relation (4.152) applies for a single disturbance of any shape and conforms with (4.14) for the average noise power with continual random noise. When disturbances of various shapes are considered, $n_i(\omega)$ is the power spectrum of the ensemble of disturbances that can be encountered in a given situation. It follows that for any kind of disturbances the optimum shapes of transmitting and receiving filters are the same as for continual random noise, provided that the ratio of average interference power to average message power is to be minimized, as in analog systems.

In the case of digital systems, the error probability depends on the probability distribution of the amplitudes of the disturbances in the demodulator output. If the peak amplitudes coincided with the sampling instants, the probability distribution of the peaks would suffice for determination of error probability. Even so the optimization problem would be complicated, for the reason that the distribution of peak amplitudes can be modified by changes in the amplitude characteristic $\bar{R}(\omega)$ of the receiving filter, as well as by modification in the phase characteristic. A further complication is that the disturbances in general will not have the same wave shape and that their peaks will occur at random relative to the sampling instants.

Optimization against Impulse Noise

When the constraint (4.22) is imposed on the combination of transmitting and receiving filters, only a minor reduction in the peak amplitudes of the disturbances can be realized by departing from the optimum shapes determined on the premise of minimum average interference power, that is, the same shapes as for continual random noise. A more pronounced reduction in peak amplitudes can be realized by the introduction of phase distortion at the receiver, with complementary phase correction at the transmitter to avoid signal-pulse distortion. This will not affect the average interference power nor the average transmitter power. However, it will in general increase peak-transmitter power, owing to overlaps of transmitted pulses, so that possible limitation on peak-transmitter power is another consideration in this optimization problem. By way of illustration, the curves in Fig. 7.7 (Chapter 7) give an indication of the reduction in peak amplitudes afforded by introduction of linear delay distortion for disturbances with a raised cosine spectrum.

Such considerations apply as far as division of equalization between transmitting and receiving filters are concerned. Other measures may be required to reduce impulse noise, as discussed in Section 5.23 for digital systems.

5

Digital Transmission and Modulation Methods

5.0 General

A basic objective in digital systems design is to realize a high transmission or channel capacity in the most economical manner. The transmission capacity depends on the channel bandwidth as well as on the number of digital amplitudes that can be distinguished at the receiver with an acceptable specified error probability. The channel capacity, as limited solely by additive Gaussian noise, was determined in Chapter 4 for optimum baseband transmission and carrier-modulation methods, assuming ideal channels and detection methods. The transmission capacity obtained on these premises is dealt with extensively in the literature, but cannot be fully realized for several reasons to be outlined presently.

In actual systems, it is necessary to incorporate certain features to facilitate carrier-wave demodulation at the receiver, together with the extraction of a timing wave for sampling of the demodulated baseband-pulse train. This can be accomplished in various ways, all of which entail some reduction in channel capacity, partly because extra bandwidth or signal power may be required and partly because of certain unavoidable erratic deviations in the derived carrier and baseband timing waves.

Moreover, a low-frequency cutoff is encountered in most baseband facilities because of the need for transformers for increased transmission efficiency as well as for other reasons. To effectively reduce the adverse effects of a low-frequency cutoff it may be necessary to employ certain methods of baseband coding, pulse shaping, or subcarrier modulation, all of which entail some reduction in transmission capacity or an increase in signal power.

Finally, the transmission capacity will be limited by various sources of distortion beside additive random noise. Among them are unwanted amplitude and phase variations over the channel band, channel non-

General 173

linearity, random fluctuations in channel transmittance with time and intersystem interference, as dealt with separately in later chapters.

Among a multitude of digital systems possibilities, the preferable type depends, among other things, on certain inherent properties of the transmission medium under consideration, on the state of the device technology, and on compatibility with existing systems. Binary encoding enjoys a number of advantages in systems design and operation over multilevel transmission. For one thing, it simplifies the implementation of regenerative repeaters and various kinds of terminal equipment, such as carrier modulators and demodulators, and devices for timing-wave provision, coding and storing of messages, and automatic error checking and correction. For another, binary pulse transmission imposes less severe requirements on the transmission medium with respect to signal-to-noise ratio, amplitude and phase deviations over the channel band, low-frequency cutoff, nonlinearity, and transmission-level variations. For these reasons, binary rather than multilevel transmission is ordinarily the more practical and economical method for new facilities. This applies even for existing facilities designed primarily for voice or other analog transmission, where the signal-to-noise ratio is much higher than required for binary transmission, but where pronounced phase distortion may be encountered. With appropriate equalization, however, as in analog broadband channels intended for television or picture transmission, multilevel or multiphase transmission is ordinarily feasible and economical, and this also applies to properly equalized voice channels.

In some applications, it may be necessary or desirable to transmit pulses at a sub-binary rate in order to reduce error probability without an increase in signal power. This can be accomplished by various means considered herein, such as repeated binary transmission, pulse position modulation, orthogonal binary codes, error correcting codes, or by the use of codes that permit partial error detection in conjunction with partial retransmission. These various methods can be employed in the time or the frequency domain, or in combinations of both.

This chapter reviews the more important types of digital systems, together with their more significant characteristics and certain basic principles and problems in their implementation, such as the provision of a reference carrier and a timing wave at the receiver. Certain theoretical relations are presented on power spectra, timing wave deviations, error probabilities due to additive Gaussian noise, and on the transmission capacity as limited by Gaussian noise for linear time-invariant channels.

For further details in the derivations of most relations presented here and for other additional information, reference is made to the more directly applicable literature. At present there is a great volume of more or less pertinent papers in this field that can be found in periodicals on communication theory and technology. Much of the literature in recent years has been devoted to the unique problems encountered in data transmission over carrier-system channels designed for voice or other analog messages. This, however, is but one application of digital transmission and not a principal subject matter here.

5.1 Transmission and Modulation Methods

For the analysis of digital systems it is convenient to divide the time scale into a number of signaling intervals of duration T and to choose the midpoint of an assumed reference interval as the time origin. If t_0 is the time from the midpoint of any signaling interval, n, the time from origin is $t = t_0 + nT$. It is convenient to express this time as a number of signaling intervals by

$$x = \frac{t}{T} = \frac{t_0}{T} + n = x_0 + n. \tag{5.1}$$

The original baseband modulating wave is conveniently assumed to consist of a sequence of quantized rectangular pulses of duration $\delta \leq T$ and amplitude $a(n)$ in the interval n. Let the transmitting baseband filter have a pulse transmission characteristic $P_1(t)$. Within the reference interval $n = O$ the amplitude of the transmitted-pulse train is then given by the following summation, applying for $-\frac{1}{2} < x_0 < \frac{1}{2}$:

$$W_1(x_0) = \sum_{n=-\infty}^{\infty} a(n) \, P_1(x_0 - n). \tag{5.2}$$

In what follows, Σ will indicate a summation between $-\infty$ and ∞, unless otherwise indicated.

If the combined pulse-transmission characteristic of the channel is $P(t)$, then the received-pulse train is of the form.

$$W(x_0) = \sum a(n) \, P(x_0 - n). \tag{5.3}$$

The pulse train is sampled at $t_0 = O$, that is, $x_0 = O$. With ideal channels $P(-n) = O$, except for $n = O$, so that at the sampling instants,

$$W(O) = a(O), \tag{5.4}$$

where $a(O)$ can have a number of discrete values.

Direct baseband transmission, as just assumed, permits the maximum channel capacity per unit of bandwidth, but is impracticable for most transmission media. In a somewhat modified form designed to combat the effect of low-frequency cutoff, it has been implemented for new binary systems employing balanced pairs and coaxials. For nearly all existing analog transmission facilities, the only feasible method is to transmit digital sequences by modulation of a carrier in amplitude, phase, or frequency. In the following formulation it is assumed that sequences of carrier pulses modulated in the foregoing manner are applied to a bandpass channel. In the case of phase or frequency modulation, this is not the same modulation method as used in conventional analog systems, as discussed later.

At the modulator input, the baseband pulses are ideally of rectangular shape and duration $\delta \leq T$. The signal at the modulator output, or channel input, can then be assumed to be of the following form over a pulse interval n of duration T.

$$W_i(t) = a(n) \cos [\omega_0 t + u(n)t - \psi(n)]. \qquad (5.5)$$

For any value of n the amplitudes $a(n)$, the frequencies $u(n)$ and the phases $\psi(n)$ can have various discrete values. Each of these are ordinarily integral multiples of certain minimum amplitude, frequency, and phase differences. With amplitude modulation and digital phase modulation it is possible to transmit pulses over a channel band that need not exceed twice the baseband frequency. As shown in Chapter 1, this is also possible with binary FM. However, since FM is not as efficient as various other methods, most of the present discussion is confined to the case in which $u(n) = 0$, so that

$$W_i(t) = a(n) \cos [\omega_0 t - \psi(n)]. \qquad (5.6)$$

As before, let t_0 be the time from the midpoint of a selected pulse interval, $n = 0$, and let $x_0 = t_0/T$ be the time in pulse intervals. For $-\tfrac{1}{2} < x_0 < \tfrac{1}{2}$, the amplitude of the received wave can then be written in the form

$$W(x_0) = \sum a(n) \cos [\omega_0 t - \psi(n)] R(x_0 - n)$$
$$\quad - \sum a(n) \sin [\omega_0 t - \psi(n)] Q(x_0 - n) \qquad (5.7)$$
$$= \sum a(n) \cos [\omega_0 t + \varphi(x_0 - n) - \psi(n)] \bar{P}(x_0 - n). \qquad (5.8)$$

A corresponding expression applies for the wave at the output of the transmitting filter, with R_1, Q_1, and P_1 in place of R, Q, and P.

In (5.7) and (5.8), the received-pulse train is expressed in terms of the bandpass pulse transmission characterization of the channel, formulated in Section 1.3 and given by

$$P(x) = R(x) \cos \omega_0 t - Q(x) \sin \omega_0 t \qquad (5.9)$$
$$= \bar{P}(x) \cos [\omega_0 t + \varphi(x)], \qquad (5.10)$$

where $x = t/T$ is the time in pulse intervals.

The function $R(x)$ represents the inphase component of the carrier envelope, $Q(x)$ the quadrature component, and $\bar{P}(x)$ the resultant envelope, with the time x measured in sampling intervals T. The phase $\varphi(x)$ is given by

$$\varphi(x) = \tan^{-1} \frac{Q(x)}{R(x)}. \tag{5.11}$$

By appropriate channel shaping it is possible to avoid intersymbol interference at $x_0 = O$, so that all terms in the foregoing series vanish, except for $n = O$. The received wave is then

$$W(O) = a(O) \cos [\omega_0 t + \varphi(O) - \psi(O)] \bar{P}(O). \tag{5.12}$$

Special cases of (5.7) include symmetrical sideband-amplitude modulation, $\psi(n) = O$ and $Q = O$, vestigial sideband-amplitude modulation, $\psi(n) = O$, $Q \neq O$, and symmetrical sideband-phase modulation, $a(n) = $ constant, $Q = O$. Other possibilities are combinations of amplitude and phase modulation, in which for each $a(n)$ there may be different phases $\psi(n)$.

With the foregoing method of phase modulation, individual carrier pulses are transmitted, each having the same shape of the envelope as the baseband modulating pulses and a fixed peak amplitude. The phase of each carrier pulse will have one of several distinct values, depending on the peak amplitude of the baseband modulating pulses. The resultant carrier-pulse train will have a variable envelope and a continuously varying phase. However, at sampling instants the carrier-pulse train will have a fixed envelope and one of several distinct phases. With this method, it is possible to obtain carrier spectra of finite bandwidth, and to avoid intersymbol interference in transmission over bandlimited channels. This basic method can be implemented in various ways, some of which permit the same channel capacity as baseband transmission.

That method of phase modulation differs from the one used in conventional analog systems. With a modulating wave (5.2) the transmitted wave with conventional phase modulation is of the form

$$W_1(x_0) = \cos [\omega_0 t + c \sum a(n) P_1(x_0 - n)], \tag{5.13}$$

where c is a constant.

If the channel has adequate (in theory infinite) bandwidth the pulse train (5.2) can be recovered by frequency-discriminator detection and in turn the baseband pulse train (5.3) by post-detection, low-pass filtering.

It is also possible to employ two basically different methods of frequency modulation. One is to modulate the carrier frequency by the baseband modulating wave and to use frequency-discriminator detection as in conventional analog FM systems. The other is to transmit sequences of pulses at different frequencies, in a manner similar to that outlined earlier for phase modulated pulses. This method is ordinarily referred to as frequency-shift keying and can be used in conjunction with frequency-discriminator detection or multiple-filter detection, as discussed further in the next section. In much of the literature on error probability in digital carrier-modulation systems, the transmitted and received pulses are, for convenience, assumed to be rectangular in shape. In this extreme case, there is no distinction between the two basic methods just mentioned. However, for reasons discussed in Section 1.13, rectangular received pulses are only of minor technical interest and therefore not considered here.

In principle, the use of pure PM or FM as in analog systems entails carrier sideband spectra of infinite bandwidth, regardless of the shape of the baseband modulating pulses. Transmission distortion in the form of unwanted amplitude and phase modulation in the received wave will be encountered, owing to channel bandwidth limitation. Nevertheless, the circumstance that envelope fluctuations are avoided can be a definite advantage in applications to channels with such highly nonlinear repeaters as microwave transmitters, in which envelope fluctuations are converted into unwanted phase modulation in the transmitted wave. Another significant advantage, particularly in multilevel transmission, is that the problem of extraction of an accurate reference carrier is avoided with frequency-discriminator detection. In this respect, the latter method is similar to differential phase product demodulation, discussed further in the next section. These advantages are secured in exchange for a somewhat greater bandwidth than required with the methods to be considered herein.

5.2 Carrier Demodulation Methods

The optimum method of carrier demodulation is by synchronous or coherent detection. With this method the received wave would in general be applied to each of two product demodulators, as indicated in Fig. 5.1. In one of these the waves would be demodulated by a wave $\cos(\omega_0 t - \psi)$ and in the other by a wave $\sin(\omega_0 t - \psi)$. After elimination of high-frequency demodulation products by low-pass filtering, the two demodulator outputs become

$$U(O) = a(O)[\cos\theta \cos\psi(O) + \sin\theta \sin\psi(O)]\bar{P}(O), \quad (5.14)$$

$$V(O) = a(O)[\cos\theta \sin\psi(O) - \sin\theta \cos\psi(O)]\bar{P}(O), \quad (5.15)$$

where the amplitudes have, for convenience, been normalized and

$$\theta = \psi - \varphi(O). \quad (5.16)$$

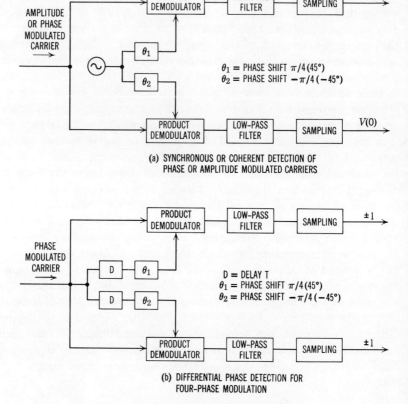

Figure 5.1 Carrier demodulation methods.

With proper choice of ψ such that $\theta = O$, the demodulator outputs become

$$U(O) = a(O) \cos \psi(O) \, \bar{P}(O), \qquad (5.17)$$

$$V(O) = a(O) \sin \psi(O) \, \bar{P}(O). \qquad (5.18)$$

From the two outputs it is now possible to determine both $a(O)$ and $\psi(O)$, that is, both the amplitude and the phase of the carrier at the sampling instants. In the special case of two-phase modulation, only a single demodulator would be required since $\psi(O) = O$ so that $V(O)$ vanishes.

With the above method of detection it is possible to transmit information over two carriers displaced 90° in phase, with two-phase modulation

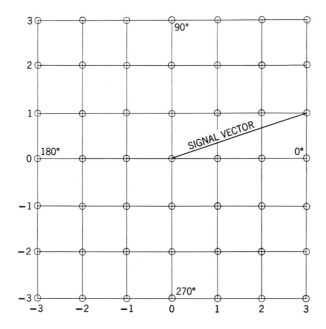

Figure 5.2 Quadrature carrier amplitude and phase modulation. With each carrier having 2 phases and 4 amplitudes as assumed above, the number of possible signals is $(2 \cdot 3 + 1)^2 = 49$ as indicated by the circles.

of each carrier. If the amplitude of one carrier is $x(O)$ and that of the other carrier is $y(O)$, the demodulator outputs are

$$U(O) = \pm x(O) \bar{P}(O), \tag{5.19}$$

$$V(O) = \pm y(O) \bar{P}(O). \tag{5.20}$$

With $\bar{P}(O)$ normalized to unity, the baseband outputs would be $U(O) = x(O) = O, \pm 1, \pm 2, \pm 3$ and $V(O) = y(O) = O, \pm 1, \pm 2$, etc. The amplitude of the modulated wave at sampling instants is in this case $a(O) = [x^2(O) + y^2(O)]^{1/2}$ and the phase is $\psi(O) = \tan^{-1} y(O)/x(O)$, as indicated in Fig. 5.2.

The foregoing method will be referred to here as orthogonal or quadrature carrier modulation, as contrasted with polar carrier modulation, indicated in Fig. 5.3. With the latter method, the amplitude, $a(n)$, and phase, $\psi(n)$, are modulated independently by the information. Two independent channels could thus be provided by separate amplitude and phase modulation.

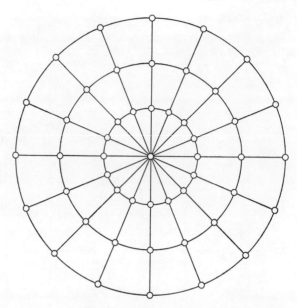

Figure 5.3 Combined amplitude and phase modulation with 4 amplitudes and 16 phases. The number of possible signals is 16 × 3 + 1 = 49 as indicated by circles.

The amplitude, $a(O)$, and the phase, $\psi(O)$, at sampling instants can in this case be derived from the two demodulator outputs (5.17) and (5.18). The amplitude is thus given by

$$a(O) = \frac{[U^2(O) + V^2(O)]^{1/2}}{\bar{P}(O)}, \qquad (5.21)$$

and the phase is obtained from

$$\psi(O) = \tan^{-1} \frac{V(O)}{U(O)}. \qquad (5.22)$$

Implementation to determine the amplitude and phase will be more complicated than that required with orthogonal carrier modulation. In the latter case, the two independent outputs are obtained directly from (5.19) and (5.20) as $x(O) = U(O)/\bar{P}(O)$ and $y(O) = V(O)/\bar{P}(O)$. In the particular case of four-phase modulation $x(O) = y(O)$ and the two methods coincide.

An alternate method of demodulation that has been employed in conjunction with phase modulation is differential phase detection [Fig. 5.1]. With this method the signal is applied to one pair of terminals of a product demodulator. The same pulse train delayed by one pulse interval T is applied to the other pair of terminals, with a suitable phase shift θ.

At sampling instants the normalized output of one demodulator after elimination of high-frequency demodulation products is, in place of (5.17), with $a(1) = a(O)$

$$U(O) = a^2(O) \cos [\psi(O) - \psi(1) - \theta] \bar{P}^2(O). \qquad (5.23)$$

The output of a second demodulator in which the phase shift in place of θ is $\theta + \pi/2$ becomes

$$V(O) = a^2(O) \sin [\psi(O) - \psi(1) - \theta] \bar{P}^2(O). \qquad (5.24)$$

The above two expressions apply in the absence of intersymbol interference and provided that the outputs are sampled at the midpoints of each pulse interval. It will be noted that with $\theta = O$ (5.23) and (5.24) conform with (5.17) and (5.18), except that the phase $\psi(O)$ is replaced by the phase difference $\psi(O) - \psi(1)$ between successive signal intervals.

5.3 Power Spectra of Pulse Trains

Relations in Section 5.1 for the transmitted and received pulse trains involve the amplitudes $a(n)$ of the rectangular input pulses in the various signal intervals together with the shapes $P_1(t)$ and $P(t)$ of the transmitted and received pulses. When these are known, the performance of the system in some important respects can be determined. To this end it is desirable to determine the power spectra of the transmitted pulse train, $W_1(t)$, and the received pulse train, $W(t)$. These power spectra are related to the probability distribution of the digit amplitudes $a(n)$ and the pulse shapes $P_1(t)$ and $P(t)$, or the corresponding pulse spectra $S_1(\omega)$ and $S(\omega)$. The conventional methods of determining power spectra are outlined in Appendix 2 and involve determination of the autocorrelation function, but more direct methods may be used in some simple cases for pulse trains.

The power spectra of transmitted pulse trains are of principal concern as regards possible intersystem interference, as considered further in Chapter 11. They are also of interest from the standpoint of deriving a reference carrier for synchronous demodulation. The power spectra of demodulated baseband pulse trains must be considered in connection with timing wave extraction for sampling and also as regards the effect of a low-frequency cutoff in baseband transmission facilities or terminations.

The power spectrum $W(\omega)$ in general comprises a component $W_c(\omega)$ that varies continuously with frequency and a discrete component $W_\delta(\omega)$ that consists of narrow spikes at certain frequencies. Thus

$$W(\omega) = W_c(\omega) + W_\delta(\omega). \qquad (5.25)$$

Let the rms deviation from the mean pulse peak amplitude be $\underline{A}$. From relations given in Section 4.7, it follows that the continuous one-sided power spectrum is given by

$$W_c(\omega) = \frac{\underline{A}^2}{\pi T} \bar{S}^2(\omega). \tag{5.26}$$

Here $\bar{S}(\omega) = |S(\omega)|$ is the amplitude of the spectrum of the pulse shape under consideration, that is, at the output of the transmitting or receiving filter, as the case may be.

Let A_1 be the mean pulse peak amplitude. The discrete components of the power spectrum is then given by the following series, which represents discrete components at those frequencies where $\bar{S}(n\Omega_r)$ does not vanish.

$$W_\delta(\omega) = \frac{A_1^2}{T^2} \sum_{n=0}^{\infty} \bar{S}^2(n\Omega_r)\delta(\omega - n\Omega_r), \tag{5.27}$$

where $\Omega_r = 2\pi/T$ is the radian repetition frequency. The various terms in this series represent the average power of each sine-wave component found by a Fourier series analysis of a periodic pulse train obtained with pulses $P(t)$ at uniform intervals T.

With r different amplitudes a_i of probabilities p_i the mean pulse peak amplitude is

$$A_1 = \sum_{i=1}^{r} a_i p_i, \tag{5.28}$$

and the variance from the mean is

$$\underline{A}^2 = \sum_{i=1}^{r} (A_1 - a_i)^2 p_i \tag{5.29}$$

In the particular case of binary transmission with amplitudes a_1 and a_2 with probabilities p_1 and $p_2 = 1 - p_1$

$$A_1 = a_1 p_1 + a_2(1 - p_1), \tag{5.30}$$

$$\underline{A}^2 = (a_1 - a_2)^2 p_1(1 - p_1). \tag{5.31}$$

With random bipolar pulses of equal amplitudes $a_2 = -a_1$ so that (5.26) and (5.27) become

$$W_c(\omega) = \frac{4a_1^2 p_1(1-p_1)}{\pi T} \bar{S}^2(\omega), \tag{5.32}$$

$$W_\delta(\omega) = \frac{a_1^2(1-2p_1)^2}{T^2} \sum_{n=0}^{\infty} \bar{S}^2(n\Omega_r)\delta(\omega - n\Omega_r) \tag{5.33}$$

Power Spectra of Pulse Trains

With unipolar transmission $a_2 = O$ and

$$W_c(\omega) = \frac{a_1^2 p_1(1 - p_1)}{\pi T} \bar{S}^2(\omega) \tag{5.34}$$

$$W_\delta(\omega) = \frac{a_1^2 p_1^2}{T^2} \sum_{n=0}^{\infty} \bar{S}^2(n\Omega_r)\delta(\omega - n\Omega_r). \tag{5.35}$$

Somewhat more general relations for binary systems are derived in reference [1].

As discussed in Chapter 7, in most transmission facilities a low-frequency cutoff is encountered between baseband inputs and outputs and may give rise to excessive pulse distortion. Transmission impairments on this account can be reduced by various means, among them the suppression of low-frequency components in the pulse trains. This can be accomplished by symmetrical sideband amplitude or phase modulation, as often used in data transmission over the voice channels of analog carrier systems, which have a cutoff near 200 c/s. An alternate and more efficient method in applications to new facilities designed specifically for binary transmission is "constrained bipolar" coding, as illustrated in Fig. 5.4. With this method the original unipolar binary pulse train is so converted that alternate pulses are reversed in polarity. The same effect is achieved by "dicode" transmission, in which the unipolar pulse train is delayed by one signal interval T and subtracted from the original unipolar pulse train.

With either method, the discrete component of the power spectrum vanishes. The continuous component with constrained bipolar transmission becomes [2]

$$W_c(\omega) = \frac{a_1^2 p_1(1 - p_1)}{\pi T} \frac{2(1 - \cos \omega T)\bar{S}^2(\omega)}{1 + 2(2p_1 - 1) \cos \omega T + (2p_1 - 1)^2}. \tag{5.36}$$

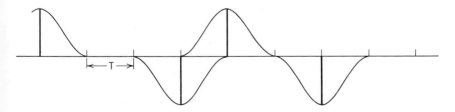

Figure 5.4 Constrained bipolar transmission for reduction of low-frequency components. Same as "on-off" transmission with reversal of polarity of successive pulses.

With appropriate changes in notation, the above relations can be translated to bandpass channels of adequately narrow bandwidth in relation to the midband frequency so that the approximations in Section 1.10 apply. Thus, with double-sideband amplitude modulation of a carrier $\cos \omega_0 t$, (5.26) and (5.27) apply with $u = \omega - \omega_0$ in place of ω, and the lower limit of the summation $-\infty$ in place of O. Accordingly

$$W_c(u) = \frac{A^2}{\pi T} \bar{S}^2(u), \tag{5.37}$$

$$W_\delta(u) = \frac{A_1^2}{T^2} \sum_{n=-\infty}^{\infty} \bar{S}^2(n\Omega_r) \delta(u - n\Omega_r). \tag{5.38}$$

At the carrier frequency $u = O$ and

$$W_\delta(O) = \frac{A_1^2}{T^2} \bar{S}^2(O). \tag{5.39}$$

Since there is complete correlation between spectral components at $-u$ and u the average power is twice the integral of the power spectra between $u = -\infty$ and ∞.

Relation (5.37) also applies to multiphase modulation of a carrier of amplitude $\underline{A}$. In this case the discrete component vanishes. To facilitate extraction of a carrier for demodulation, it may be desirable to inject a discrete component, for example, by recurrent transmission at uniform intervals of a carrier pulse of fixed phase, as discussed in Section 5.6.

The power spectra with frequency-shift keying will not be considered here since this method does not require carrier-wave extraction for demodulation. Comprehensive results have been published for binary [3] and multifrequency [4] keying of rectangular pulses.

5.4 Methods of Timing Wave Extraction

With any digital transmission method, the demodulated baseband pulse train must be sampled at intervals T. To this end, it is necessary to derive a periodic timing wave from the received pulse train. Such a timing wave can be extracted directly from the demodulated baseband wave with the aid of a tuned circuit and by other means, provided a discrete spectral component is present at a frequency k/T or $1/kT$, $k = 1, 2, 3, \ldots$. If such a component is not present, it may be provided without the need for extra bandwidth by appropriate processing of the demodulated baseband pulse train. This may consist of rectification or squaring, or by the triggering of impulses at zero crossings or at crossings with an appropriate slicing level.

Methods of Timing Wave Extraction

These methods can readily be implemented with binary and ternary coding, as discussed later.

After a timing wave of frequency $1/T$ or harmonics thereof is extracted, it is possible by appropriate clock circuits to divide the pulse train into digital sequences representing characters. To this end a special reference digit may be required to establish an appropriate origin for the digital sequences. In time-division systems, a framing digit is also required for synchronizations of the channels. The present discussion is concerned only with the provision of a basic timing wave of frequency $1/T$, corresponding to a radian frequency $2\pi/T$.

Let the demodulated baseband pulse train be applied to a bandpass extraction filter of narrow bandwidth Ω_x centered on the repetition frequency $\Omega_r = 2\pi/T$. The discrete spectral component will then give rise to a sine wave of amplitude proportional to $W_o(\omega)$ for $\omega = \Omega_r$. In addition a sine-wave component of variable amplitude and average power proportional to $\Omega_x W_c(\Omega_r)$. The ratio of the rms amplitude of the variable component to the mean amplitude of the derived timing wave becomes

$$\underline{\alpha}_x = \frac{\Omega_x W_c(\Omega_r)}{W_o(\Omega_r)} = \frac{\underline{A}}{A_1} \frac{\overline{S}(\Omega_r)}{\overline{S}(\Omega_r)} \left(\frac{T\Omega_x}{2\pi}\right)^{1/2}$$

$$= \frac{\underline{A}}{A_1} \left(\frac{\Omega_x}{\Omega_r}\right)^{1/2} \tag{5.40}$$

The last relation applies provided $S(\Omega_r)$ differs from zero, which is one necessary condition for extracting a timing wave. This will be the case when the pulse shape is such that a pulse train consisting of a long sequence of pulses of equal amplitude contains a ripple, as indicated in Fig. 5.5.

Another requirement for an acceptable derived timing wave is that $\underline{\alpha}_x \ll 1$, which imposes a certain condition on $\underline{A}/A_1$ for a given ratio Ω_x/Ω_r. When this condition is met it is possible to obtain a timing wave of

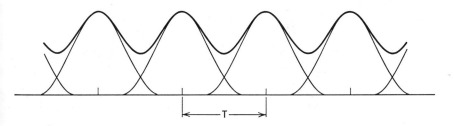

Figure 5.5 Pulse train ripple of fundamental frequency $1/T$ is required for direct timing wave extraction from received baseband wave.

constant amplitude with the aid of a limiter with a small probability that the timing wave will descend below the clipping level, owing to the effect of the variable component.

Direct extraction of a timing wave from the demodulated pulse train entails an increase in signal power and also in channel bandwidth beyond that required for transmission without intersymbol interference. This can be avoided by rectification or squaring of the demodulated pulse train, since this yields a power spectrum of increased bandwidth together with discrete spectral components at the repetition frequency or harmonics thereof. These methods have been used with binary systems [2]. An alternate procedure is to trigger impulses at zero crossings, as illustrated in Fig. 5.6 for a random bipolar train of pulses having a raised cosine spectrum. The same kind of impulse train can be triggered by random unipolar pulses with raised cosine spectrum at the intersection with a slicing level equal to half the peak pulse amplitude, as illustrated in Section 1.7.

Since impulses have a spectrum of great bandwidth compared to $\Omega_r = 2\pi/T$ the above unipolar pulse trains will have a discrete spectral component at $\omega = \Omega_r$ and at various harmonics. It is thus possible to extract a timing wave with a ratio $\underline{\alpha}_x$ as obtained from (5.40) with $(\underline{A}/A_1)^2 = (1 - p_1)/p_1$ in accordance with (5.30) and (5.31).

Relation (5.40) shows that in principle it is possible to reduce random amplitude fluctuation in the timing wave to any desired extent by reducing the bandwidth Ω_x of the extraction filter. This entails certain difficulties, owing to unavoidable variations in the midband frequency of the filter from the pulse-repetition frequency, or conversely. This will give rise to certain timing-wave deviations to be discussed in the next section for an extraction filter consisting of a simple resonator.

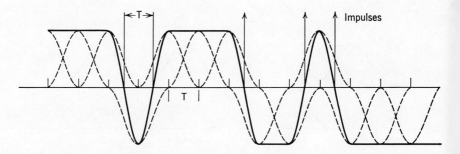

Figure 5.6 Generation of unipolar binary impulse train for timing wave extraction from random bipolar binary pulse train.

5.5 Timing Wave Deviations

Important considerations in the performance of digital systems are deviations in the timing wave and resultant deviations from the optimum sampling instants, usually called timing jitter. This is particularly important in systems employing a long chain of regenerative repeaters, because of the possibility of an excessive increase in the timing jitter along the repeater chain [5,6,7,8]. This can be avoided if adequate limitations are imposed on the allowable timing-wave deviations in a single repeater section. These deviations will be considered here on the assumption that a unipolar random impulse train is applied to a simple resonator of resonant frequency ω_0, which in general will differ from the impulse repetition frequency $\Omega_r = 2\pi/T$. As illustrated previously, such an impulse train can be triggered by the actual pulse train, and the relations given here then apply. When the other methods mentioned before are used, there will be additional timing deviations that are usually of lesser importance. These come about because the continuous part of the power spectrum will have a certain slope over the band of the resonator. This gives rise to additional phase modulation and resultant timing jitter.

Let the resonator or tuned circuit consist of a parallel combination of R, L and C. The resonant frequency is then

$$\omega_0 = (1/LC)^{1/2}, \tag{5.41}$$

and the loss constant is defined as

$$Q = \omega_0 RC. \tag{5.42}$$

The phase angle ψ of the resonator impedance at the frequency ω is given by

$$\tan \psi = Q\left(\frac{\omega}{\omega_0} - \frac{\omega_0}{\omega}\right). \tag{5.43}$$

With $\Delta\omega = \omega - \omega_0$ and $\Delta\omega/\omega_0 \ll 1$

$$\tan \psi \cong 2Q \frac{\Delta\omega}{\omega_0}. \tag{5.44}$$

The above and various relations that follow are derived in reference [5].

The timing wave derived from a random unipolar pulse train will contain two types of amplitude and phase deviations. The first type is a fixed amplitude reduction by a factor a_0 and a fixed phase displacement ψ, given by

$$a_0 = \cos \psi, \tag{5.45}$$

$$\psi = \tan^{-1}\left(\frac{2Q\Delta\omega}{\omega_0}\right). \tag{5.46}$$

This latter phase displacement gives rise to a fixed time displacement τ_0 in the sampling of pulse trains, given by

$$\tau_0 = \frac{T}{2\pi} \psi. \tag{5.47}$$

The second type consists of a random amplitude variation and a random phase variation with the rms values

$$\underline{\alpha}_x = \frac{A}{A_1} \left(\frac{\pi}{2Q} \frac{2-\psi^2}{2}\right)^{\frac{1}{2}} \frac{1}{\cos \psi}, \tag{5.48}$$

$$\bar{\varphi}_x = \frac{A}{A_1} \left(\frac{\pi}{4Q}\right)^{\frac{1}{2}} \frac{|\psi|}{\cos \psi}. \tag{5.49}$$

With a unipolar impulse train $\underline{A}/A_1 = [(1 - p_1)/p_1]^{\frac{1}{2}}$ in accordance with (5.30) and (5.31).

The above random phase deviations give rise to random timing-wave deviations, or timing jitter, having an rms value

$$\underline{\delta}_x = \frac{A}{A_1} \frac{T}{2\pi} \left(\frac{\pi}{4Q}\right)^{\frac{1}{2}} \frac{|\psi|}{\cos \psi}. \tag{5.50}$$

This is the rms deviation from an exact timing wave of period T.

If the applied impulse train has an initial small random timing jitter of rms value δ_0, then there will be an additional third random timing deviation in the derived timing wave of rms value

$$\delta'_x = \delta_0 \left(\frac{\pi}{Q}\right)^{\frac{1}{2}}. \tag{5.51}$$

The resultant rms timing jitter is obtained by a root-sum-square combination of $\underline{\delta}_x$ and δ'_x. There will also be a small additional amplitude variation which can be ignored. Relation (5.51) shows that if the received pulse train contains random timing jitter, owing to additive channel noise, the resultant jitter in the timing wave is smaller by the factor $(\pi/Q)^{\frac{1}{2}}$ and can be ignored when $Q \gg \pi$.

The fixed time displacement τ_0 will result in a reduced instantaneous amplitude of the pulse train sample, except for a pulse train consisting of a number of consequentive pulses of the same polarity so that the combined wave has a constant amplitude. For an isolated raised cosine pulse the amplitude at a sampling instant is reduced by a factor

$$\eta = \frac{1}{2}\left(1 + \cos \pi \frac{\tau_0}{T}\right)$$

$$= \frac{1}{2}(1 + \cos \psi/2). \tag{5.52}$$

By way of example, let $Q\Delta\omega/\omega_0 = 0.5$, corresponding to $\tan\psi = 1$ and $\psi = \pi/4$. In this case $\eta = 0.96$. With $Q = 100$, the corresponding required frequency precision is $\Delta\omega/\omega_0 = 0.005$. Hence the resonant frequency must be within 0.5% of the pulse repetition frequency. With the above values of Q, ψ, and $\Delta\omega/\omega_0$, the rms value of the random timing deviation obtained from (5.50) with $p_1 = 0.5$ is $\hat{\delta}_x = 0.015$. The random timing deviation will have a nearly Gaussian probability distribution in the range of interest. An instantaneous timing deviation four times the above rms value, or 0.06, would thus be rather improbable. When combined with the fixed timing displacement τ_0, the resultant instantaneous timing deviation will be about $0.5\tau_0$ or $1.5\tau_0$. With the latter value the amplitude of an isolated pulse at a sampling instant would be reduced by a factor of about 0.9.

In the above illustration the probability p_1 was taken as 0.5. In some applications p_1 may vary greatly, and may be much smaller than 0.5, as when a large number of channels are combined in time division and only a small fraction of the channels are active. The rms deviation $\hat{\delta}_x$ will then vary in proportion to $[(1 - p_1)/p_1]^{1/2}$. Constrained bipolar coding methods for reduction of low-frequency components have been devised, in which long sequences of zeros are avoided [9] so that amplitude fluctuations in the extracted timing wave and timing jitter are much reduced.

The above methods can be used with binary and ternary systems employing baseband transmission as well as amplitude or phase modulation. With four-phase modulation, the demodulated wave is split into two binary baseband waves, as indicated in Fig. 5.1, and a timing wave can be derived by the above methods for each binary train. Actually it may be more expedient to rectify or square the binary trains rather than to trigger impulses as assumed earlier.

With more complicated devices for timing-wave extraction, such as phase controlled oscillators, it is in effect possible to control and thus limit variations in the ratio $\Delta\omega/\omega_0$. This in essence permits the use of a larger Q so that both fixed and random timing deviations are reduced.

5.6 Derivation of Reference Carrier

With some methods of digital transmission by carrier modulation, a reference carrier for demodulation is not required. Thus envelope detection is feasible and practicable with amplitude modulation, and frequency-discriminator detection can be used with phase or frequency modulation. With phase modulation in conjunction with product demodulation, differential phase detection can be used, and is in some respects similar to frequency-discriminator detection. All of these methods entail some increase in signal-to-noise ratio as compared to ideal synchronous or coherent product demodulation with an exact carrier. However, with any practicable method of providing a carrier for product demodulation it is necessary to accept some increase in signal-to-noise ratio. This comes about because extra signal power may be required to facilitate carrier extraction, and partly because of unavoidable fixed and random deviations in the amplitude and phase of the derived carrier. These in turn give rise to amplitude and timing deviations in the demodulated baseband pulses. In addition it is necessary to consider deviations in the timing wave extracted from the baseband pulse train, as dealt with in the previous section.

A reference carrier can be derived from the received carrier wave with the aid of a resonator tuned to the carrier frequency or by a phase-controlled oscillator, provided that a discrete spectral component is present in the received wave at the carrier frequency or harmonics thereof. If it is not present, it may in certain cases be provided by squaring the received carrier wave, though this entails a fixed phase ambiguity that can be resolved by appropriate means. As in the previous section, it will be assumed that a simple resonator is used for carrier extraction in order to illustrate the amplitude and phase variation in the extracted carrier wave, as related to the loss constant Q of the resonator, the deviation $\Delta\omega$ in the resonant frequency from the carrier frequency ω_c and to the strength of the discrete spectral component used for carrier wave extraction.

When a strong enough discrete carrier-frequency component is present for extraction of a carrier wave, the latter will have a phase displacement ψ relative to the receiver carrier, as given by (5.44) or

$$\tan\psi \cong 2Q\left(\frac{\Delta\omega}{\omega_c}\right), \tag{5.53}$$

where $\Delta\omega = \omega_c - \omega_0$ is the difference between carrier and resonant frequency.

With this phase displacement, the demodulated baseband pulses will be reduced in amplitude by the factor $\cos\psi$, assuming double-sideband amplitude modulation with product demodulation. But there will be no reduction in the rms amplitude of additive random noise in the baseband output. Hence the average signal power must be increased by a factor $(\cos\psi)^{-2}$ for a given error probability.

In addition to the foregoing fixed phase displacement, the derived carrier will have random amplitude and phase variations. The ratio of the rms amplitude fluctuation to the mean amplitude of the derived carrier is given by (5.48) or

$$\underline{\alpha}_x = \frac{A}{A_1}\left(\frac{\pi}{2Q}\right)^{1/2}\frac{(1-\psi^2/2)^{1/2}}{\cos\psi}. \tag{5.54}$$

Derivation of Reference Carrier

The rms phase variation is in view of (5.49)

$$\bar{\varphi}_x = \frac{A}{A_1} \left(\frac{\pi}{4Q}\right)^{1/2} \frac{|\psi|}{\cos \psi}. \tag{5.55}$$

Let the transmitted signal have r equally probable amplitudes between $A_{\min}$ and $A_{\max}$. Relations (4.111), (4.112), and (4.113) then yield

$$\frac{A}{A_1} = \frac{A_{\max} - A_{\min}}{A_{\max} + A_{\min}} \left(\frac{r+1}{3(r-1)}\right)^{1/2}. \tag{5.56}$$

The rms amplitude and phase fluctuations (5.54) and (5.55) can thus be controlled by choice of the ratio $\eta = (A_{\max} - A_{\min})/(A_{\max} + A_{\min})$. For unipolar AM this ratio is 1 while for bipolar AM it becomes infinite if $A_{\min} = -A_{\max}$.

The amplitude fluctuations can be removed by a limiter at the resonator output, with insignificant probability that the amplitude will descend below the clipping level of the limiter, provided $\alpha_x < 0.1$. The performance is then determined by the fixed phase displacement ψ in conjunction with the random phase fluctuations of rms value (5.55).

By way of illustration, for unipolar AM, (5.55) becomes

$$\bar{\varphi}_x = \left(\frac{(r+1)\pi}{12(r-1)Q}\right)^{1/2} \frac{|\psi|}{\cos \psi}. \tag{5.57}$$

Let $Q = 100$, $r = 4$ and $\psi = \pi/6$, in which case $\Delta\omega/\omega_c = 0.0025$, $\alpha_x = 0.08$ and $\bar{\varphi}_x = 0.04$. The phase fluctuations will have a nearly Gaussian probability distribution in the range of interest. The probability is then about 10^{-4} that the phase error will be less than $\pi/6 - 4\bar{\varphi}_x = 0.36$ or greater than $\pi/6 + 4\bar{\varphi}_x = 0.68$ radian. The tolerable instantaneous-noise amplitude will be less than with ideal coherent detection by a factor between $\cos 0.68 = 0.78$ and $\cos 0.36 = 0.92$. The average value will be slightly less than $\cos \psi = 0.87$, which corresponds to an impairment in signal-to-noise ratio of about 1.3 dB compared to ideal coherent detection. This is comparable to the performance expected with the simpler method of envelope detection. Performance can be improved at increased cost by increasing Q and reducing $\Delta\omega/\omega_c$ so that ψ and $\bar{\varphi}_x$ are reduced. For example, with $Q = 200$ and $\Delta\omega/\omega_c = 0.0006$, $\psi = \pi/12$ and $\cos \psi = 0.97$.

The foregoing relations apply provided that the signal spectrum is symmetrical about the carrier frequency over the narrow band of the resonator. In vestigial sideband transmission the spectrum has a slope over this band, which gives rise to some additional random phase modulation. This can be offset by using a somewhat higher Q and greater frequency precision, and by such other measures as the suppression of spectral components in a narrow band at the carrier frequency with re-insertion at the receiver after carrier recovery [1].

In phase modulation systems, it is desirable to transmit carrier information in a manner that avoids amplitude modulation. With two-phase modulation, it is feasible to transmit a special carrier pulse in a fraction μ of the signal intervals, say after each group of digits representing a character. With equal probability of the two phases, it can be shown by using (5.30) and (5.31) that

$$\frac{\underline{A}^2}{A_1^2} = \frac{1-\mu^2}{\mu^2}. \qquad (5.58)$$

The resultant rms deviations in amplitude and phase can be obtained with (5.58) in (5.54) and (5.55). For reasonable values of μ, say $\mu = 0.1$ $\underline{A}/A_1$ becomes so large that a very high Q is required, together with high frequency precision, if excessive random fluctuations are to be avoided. Hence this method is unattractive.

A preferable method with two-phase modulation as well as bipolar AM is to rectify or square the received signal. The resultant second harmonic can be extracted with the aid of a resonator, the output of which is applied to a limiter and a frequency divider to obtain the desired carrier, except for a 180° phase ambiguity that can be resolved [1]. If ψ is the fixed phase error after frequency halving, then relation (5.53) is replaced by

$$\tan 2\psi = 2Q\left(\frac{\Delta\omega}{2\omega_c}\right) \qquad (5.59)$$

where $\Delta\omega = 2\omega_c - \omega_0$.

With two-phase modulation and carrier pulses of amplitude A, it turns out that for a raised cosine pulse shape the mean value of the squared signal is

$$A_1 = \frac{3A}{4}, \qquad (5.60)$$

and that the rms deviation from this mean is

$$\underline{A} = \frac{2^{1/2}A}{4}. \qquad (5.61)$$

Hence

$$\frac{\underline{A}}{A_1} = \frac{2^{1/2}}{3} = 0.47$$

The latter value compares with $\underline{A}/A_1 = 1$ obtained from (5.56) for unipolar AM, so that the random amplitude and phase fluctuations in the derived carrier are smaller.

In the preceding discussion, carrier extraction with the aid of a resonator was assumed. A preferable method when only minor frequency deviations can be tolerated is to use a phase-locked oscillator, particularly if carrier-frequency wander is encountered. A comprehensive analysis has been made of this method as applied to binary PSK with suppressed carrier and signal squaring [10]. In this reference, the error probability in the presence of Gaussian noise is determined for coherent detection of orthogonal codes by the method outlined in Section 5.19, on the premise of rectangular received pulses, so that $\underline{A}/A_1 = 0$. The results indicate that under the above conditions virtually the same performance as with an exact demodulating carrier can be realized, with reasonable parameters of the oscillator control loop.

5.7 Transmission Limitation by Gaussian Noise

The transmission capacity of idealized channels as limited solely by additive Gaussian noise inherent in the transmission medium is the one property of digital systems that can readily be determined and specified in terms of a single parameter; the signal-to-noise ratio at the receiver for a given error probability. For this reason, this channel capacity is dealt with extensively in the literature for various methods of digital coding, carrier modulation and detection. Although this channel capacity cannot be fully realized for a number of reasons mentioned before, it nevertheless serves a useful purpose as a reference for the performance of actual systems as determined by realistic theory or tests. Moreover, comparison of signal-to-noise ratios for a given transmission capacity and error probability also yields an approximate indication of relative signal powers required with various digital methods as actually implemented. Furthermore, such comparisons furnish an approximate criterion on relative performance with impulse noise rather than Gaussian noise.

In the sections that follow, the channel capacity, as limited by additive Gaussian noise alone, is considered for various conventional digital systems, assuming individual digit detection by instantaneous pulse-train sampling. This is the simplest and most readily implemented method and yields a performance that equals or surpasses that obtained by integration of the signal over an interval T, which is a feasible method for some highly idealized systems and is often assumed in mathematical treatises on communication theory. (See Sections 1.12 and 1.13.)

The probability of error can be determined with the aid of the probability densities or distributions of variations in the amplitude of a sine wave perturbed by Gaussian noise, as given in Chapter 3. The probability distributions of the above quantities are functions of the ratio of average sine-wave power to average noise power in an assumed noise band.

To obtain the probability of error in digital amplitude, phase, or frequency modulation systems, it is necessary as a first step to establish the maximum tolerable amplitude, phase, or frequency deviations before an error occurs. The error probability obtained after this first step is given in several publications, but applies only for certain extreme idealizations of channel transmittances of minor technical interest. One such extreme assumption discussed in Section 1.13 is rectangular received pulses, which entails infinite channel bandwidth and complicated detection procedures. Another is ideal flat channel filters with complete sharp cutoffs and a linear phase characteristic, which entails infinite transmission delay and other undesirable features, as discussed in Sections 1.5 and 1.6.

To determine the error probability with realistic idealizations of actual channels, it is also necessary to determine the average signal power at the receiver, which depends on the average transmitter power, the amplitude versus the frequency characteristic of the transmitting filter and that of the channel between transmitter and receiver. Moreover, the average noise power at the detector input must be determined, as related to the noise power spectrum at the input to the receiving filter and the amplitude versus frequency characteristic of this filter. Finally, after the above various quantities have been determined, it is possible to formulate the error probability as related to the various basic physical parameters of the digital system under consideration.

The above procedure is carried out in Chapter 4 for idealized baseband transmission and carrier modulation methods. As shown there, it is possible to determine optimum shapes of transmitting and receiving filters for a designated desired shape of received pulses, and thus to establish the maximum transmission rate for a given transmitter power and assigned error probability. The determination of optimum filter shapes is not considered in the sections that follow, which are concerned with transmission capacities and error probabilities as related to certain signal-to-noise ratios at the receiver.

5.8 Transmission Capacity of Baseband Systems

It will be assumed that the spectrum of received pulses at the output of the receiving filter has the properties indicated in Fig. 4.5, in which case pulses can be transmitted at intervals $T = 1/2B$ without intersymbol interference between pulse peaks. As shown in Section 4.10, the channel capacity is then

$$C = B \log_2 r^2, \tag{5.63}$$

where r is the number of pulse amplitudes.

This channel capacity is associated with a certain probability of an error in a digit, given by (4.118), or

$$\epsilon = \left(1 - \frac{1}{r}\right) \text{erfc} \left(\frac{3\rho/2}{r^2 - 1}\right)^{1/2}. \tag{5.64}$$

Here ρ is the ratio of the average instantaneous pulse-peak power at sampling instants to the average noise power, both at the output of the receiving filter.

The channel capacities obtained from (5.63) are given in Table 1, together with the ratios ρ obtained from (5.64) with $\epsilon = 10^{-4}$ and 10^{-6}

Table 1. *Baseband Channel Capacity versus Signal-to-Noise Ratio with r Digit Amplitudes and Instantaneous Decoding*

$r =$	2	4	8	16	32	64	
$C/B =$	2	4	6	8	10	12	
$\rho =$	13.7	92	400	1600	6400	25,600	($\epsilon = 10^{-4}$)
$\rho =$	22	116	500	2000	8000	32,000	($\epsilon = 10^{-6}$)

The ratios ρ apply for random bipolar transmission without intersymbol interference, with equal probabilities of positive and negative pulses of a given amplitude. The channel capacities are shown in Fig. 5.7 as a function of ρ for $\epsilon = 10^{-5}$.

In baseband systems, the attenuation in the transmission medium increases with frequency, and the problem is to determine the minimum required transmitter power for a specified ratio ρ and spectrum $\bar{S}(\omega)$ at the receiving filter output. This minimum transmitter power P_0 is obtained from (4.81) with $\rho = \rho^0$, which is the minimum required ratio of average signal power to average noise power at the output of the receiving filter. The corresponding optimum shapes of the receiving and transmitting filters are obtained from (4.72) and (4.74). The procedure is exemplified in Section 4.13 for transmission over metallic pairs.

Although baseband transmission media with constant attenuation $l(\omega) = l$ are not encountered, it may be noted that for constant attenuation and noise power density, the optimum ratio ρ for pulses with the type of spectrum indicated in Fig. 4.5 is given by the familiar relation

$$\rho^0 = \frac{P}{N} \tag{5.65}$$

where $P = P_0 l$ is the average signal power at the input to the receiving filter and N is the average noise power in a flat band B at the receiving filter input. The ratio ρ^0 in this particular case is the same as at the output of the receiving filter.

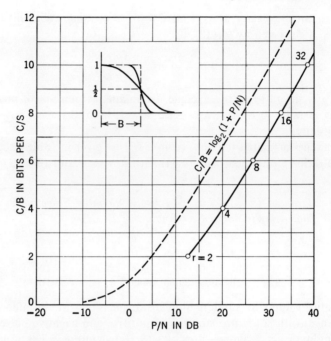

Figure 5.7 Channel capacities per unit of mean bandwidth B in presence of flat random noise of average power N in band B, for average received signal power P. Curves apply for random bipolar transmission with various numbers of amplitudes r and an error probability 10^{-5} per digit.

As discussed in Section 5.4, in binary systems it may be desirable to employ constrained bipolar transmission in order to reduce the adverse effects of a low-frequency cutoff. The average signal power is then half that with random bipolar transmission, since pulses are present on the average in half the signal intervals. On the other hand, the tolerable peak-noise amplitude before an error occurs is half that with random bipolar transmission. Hence the required signal-to-noise ratio is increased by a factor 2. The channel capacity with this method is

$$C = 2B, \tag{5.66}$$

and the associated probability of an error in a digit is

$$\epsilon = \mathrm{erfc}(\rho/4)^{1/2}. \tag{5.67}$$

A factor 2 as compared to (5.64) appears in (5.67) because an error in a digit falsifies the following transmission in a pulse train until a polarity reversal occurs. Each original digit error then gives rise to an average of two digit errors.

The above relations for error probability apply in applications to channels without a low-frequency cutoff. With a low-frequency cutoff, an increase in signal-to-noise ratio is required, as discussed in Sections 7.16 and 7.17.

5.9 Duobinary Baseband Transmission

In the previous section, pulse transmission without intersymbol interference at sampling instants was assumed. This is possible with an infinite variety of pulse spectra having the properties illustrated in Fig. 1.10 for low-pass and in Fig. 1.18 for bandpass channels. As indicated in Fig. 1.10, the mean bandwidth is $\Omega = 2\pi B$ and the maximum bandwidth is

$$B_{\max} = B\left(1 + \frac{B_x}{B}\right), \tag{5.68}$$

where $B_x = \omega_x/2\pi$.

Under the foregoing conditions, pulses can be transmitted without intersymbol interference at intervals $T = 1/2B$. When $B_x/B > 0$ it is possible to employ shorter intervals by accepting some intersymbol interference.

In the special case $B_x = B$, so that $B_{\max} = 2B$, satisfactory performance can be attained with appropriate decoding when pulses are transmitted at twice the binary rate, that is, at intervals $T' = T/2 = 1/4B$. In early transatlantic submarine telegraphy [1], this technique was known as "doubling the dotting speed" and in some recent applications to data transmission as "duobinary" transmission [11]. As will appear from the exposition that follows, this method embodies the second Nyquist criterion mentioned in Section 1.7 with a pulse spectrum bandwidth 2Ω.

For duobinary transmission the optimum pulse shape is as shown in Fig. 5.8 and of the form [1]

$$P(t) = \frac{2}{\pi} \frac{1}{1 - (4t/T)^2} \cos \frac{2\pi t}{T}, \tag{5.69}$$

where $T = 1/2B$.

When positive and negative pulses of the foregoing shape are transmitted at intervals $T/2$ to represent the digits 1 and -1, the possible amplitudes of the resultant pulse train at the sampling instant $t = T/4$ are as follows:

Previous Digit	Last Digit	Pulse Train Sample
-1	-1	-1
-1	$+1$	0
$+1$	-1	0
$+1$	$+1$	1

Samples $+1$ and -1 thus indicate the corresponding polarities of the last digit, while a zero indicates a polarity reversal in the last from the previous digit. With the aid of appropriate decoders, it is thus possible to recover the original binary sequence with digits at intervals $T/2$.

The spectrum of the above pulse (5.69) for $\omega < 2\pi/T$ is given by

$$\bar{S}(\omega) = \frac{T}{2} \cos \omega T/4. \tag{5.70}$$

For $\omega > 2\pi/T$, $\bar{S}(\omega) = 0$. The maximum bandwidth of this spectrum is $B_{\max} = 2B = 1/T$.

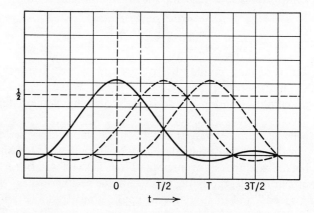

Figure 5.8 Optimum pulse shape for duobinary transmission with pulses at intervals $T/2$. Pulse train sampled at $t = T/4$ can have the amplitude $-1, 1,$ or $0,$ depending on polarities of adjacent pulses. For polarities shown the amplitude is 1.

As with the conventional methods, the optimum receiving filter has an amplitude characteristic proportional to $[\bar{S}(\omega)]^{1/2}$ for random noise with a flat power spectrum and a channel with assumed constant attenuation over the channel band. With this optimum receiving filter characteristic, the error probability becomes [12]

$$\epsilon = \tfrac{3}{4}\mathrm{erfc}\left[\frac{\pi}{4}\left(\frac{P}{4N}\right)^{1/2}\right] \tag{5.71}$$

where N and P are defined as in Section 5.8.

With this method, the original binary code is converted into a ternary code at the detector. It can be shown that for equal maximum bandwidths, equal-error probabilities and signal-to-noise ratios, it is possible to realize the same transmission capacity with conventional ternary transmission without intersymbol interference [12]. To this end the pulse spectra must be different, as indicated in Fig. 5.9.

The implementations and the performance of the duobinary technique as applied to baseband transmission and various methods of carrier modulation are discussed in further detail elsewhere [11]. In that article certain error correcting features are discussed and comparisons are made with conventional methods that employ a maximum bandwidth $B_{\max} = 2B$. While this bandwidth as well as a particular pulse shape is required for optimum performance with duobinary transmission, this is not essential with conventional methods in which intersymbol interference is absent at sampling instants. Hence the comparisons are not sufficiently exhaustive to be considered representative.

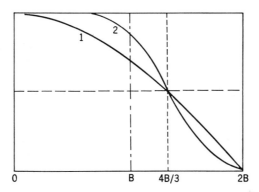

Figure 5.9 Pulse spectrum for duobinary coding, 1, and for conventional ternary coding, 2, for equal signal-to-noise ratios and about 5% greater transmission capacity with conventional ternary transmission.

5.10 Symmetrical Sideband Amplitude Modulation

In symmetrical sideband transmission, the carrier is at the midband frequency of a symmetrical channel of total mean bandwidth $B_i = 2B$ as indicated in Fig. 5.10.

With bipolar or two-phase AM, $\psi(n)$ in (5.6) is either O or π, so that (5.17) gives

$$U(O) = \pm a(O), \qquad (5.72)$$

while (5.18) yields $V(O) = O$.

Let q be the number of different amplitudes $a(O)$. With p phases the number of different signal states at a sampling instant is $r = pq$. Pulses can

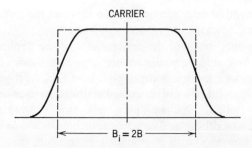

Figure 5.10 Sideband pulse spectrum in symmetrical sideband AM and PM.

in this case be transmitted at half the rate for baseband transmission, so that the channel capacity becomes

$$C = B_i \log_2 r = 2B \log_2 r. \tag{5.73}$$

With ideal synchronous detection and Gaussian channel noise, the amplitude-probability distribution of noise at the detector output is also Gaussian, as in baseband transmission, so that the error probability is the same as in baseband transmission with a signal-to-noise ratio ρ defined as in Section 5.8.

With $p = 2$ phases the error probability is the same as for bipolar baseband transmission as given by (5.64). In the particular case of $q = 1$, as in binary phase modulation, the channel capacity is

$$C = B_i = 2B, \tag{5.74}$$

and the error probability is

$$\epsilon = \tfrac{1}{2}\text{erfc}\,(\rho/2)^{1/2}. \tag{5.75}$$

With $p = 1$ and $q = 2$, as in unipolar binary AM, or on-off carrier modulation, the channel capacity is again given by (5.74), but the error probability is

$$\epsilon = \tfrac{1}{2}\text{erfc}\,(\rho/4)^{1/2}. \tag{5.76}$$

The optimum signal-to-noise ratio ρ^0 can be determined from (4.82) for a specified average transmitter power P_0, pulse spectrum $S(u)$, noise power spectrum $n_i(u)$ and transmission loss $l(u)$. Ordinarily ρ is specified, in which case the minimum transmitter power can be determined from (4.82) with $\rho = \rho^0$.

In the important case of a channel with flat loss $l(\omega) = l$ and constant noise power spectrum $n_i(\omega) = n_i$, the ratio ρ^0 is given by

$$\rho^0 = \frac{2P_i}{N_i} = 2\rho_i, \tag{5.77}$$

where $\rho_i = P_i/N_i$ and
  P_i = Average signal power at input to receiving filter;
  N_i = Average noise power in a flat band $B_i = 2B$ at input to receiving filter.

The channel capacity obtained from (5.73) is shown in Fig. 5.11 as a function of P_i.

With envelope detection there will be a certain optimum threshold level r_0 for the distinction between a mark and a space, which is greater than half the amplitude, a, of the carrier envelope during a mark. The error probability for various values of r_0/a can in this case be determined with the aid of (3.47).

The probability of an error in a space is obtained with $a = O$ and is

$$\epsilon_s = e^{-r_0^2/2\sigma_n^2}, \tag{5.78}$$

where $\sigma_n^2 = N_i$ = average noise power in the band B_i. The probability of an error in a mark is in a first approximation given by (3.49) and is

$$\epsilon_m = \tfrac{1}{2}\mathrm{erfc}\left(\frac{a - r_0}{2^{1/2}\sigma_n}\right). \tag{5.79}$$

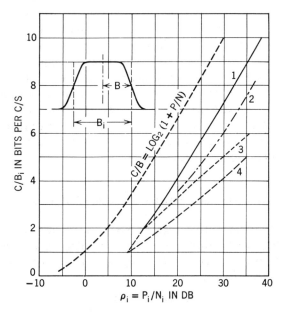

Figure 5.11 Channel capacities with various methods of carrier modulation with synchronous detection. 1. Two-phase vestigial sideband AM and quadrature two-phase double sideband AM. 2. Combined amplitude and multiphase modulation. 3. Multiphase double sideband modulation. 4. Two-phase double sideband AM.

With a probability $\frac{1}{2}$ of marks and spaces, the combined error probability is $\frac{1}{2}(\epsilon_s + \epsilon_m)$ or

$$\epsilon = \frac{1}{2}\left[e^{-r_0^2/2\sigma_n^2} + \frac{1}{2}\operatorname{erfc}\left(\frac{a - r_0}{2^{1/2}\sigma_n}\right)\right]. \tag{5.80}$$

Based on curves published by Bennett [13] the optimum ratio r_0/a is about 0.57 for $\epsilon = 10^{-2}$ and about 0.52 for $\epsilon = 10^{-7}$. For an error probability $\epsilon < 10^{-7}$, the required ratio ρ is greater than that obtained from (5.76) by about 0.5 dB.

In probabilistic literature on error probabilities with various methods of carrier modulation and detection, the transmitted and received pulses are ordinarily, for convenience, assumed to be rectangular in shape. This entails pulse spectra and channels of infinite bandwidth. Hence it is not appropriate to specify error probability in terms of a signal-to-noise ratio ρ_i defined as above. Instead, the error probability is in this case specified in terms of a ratio $\rho_i = E_i/n_i$ where E_i is the average energy per pulse and n_i the noise power density, both at the detector input. For rectangular pulses of duration T, $E_i = P_i T$ so that $\rho_i = P_i T/n_i = P_i/N_i$. The latter relation conforms with the present definition of ρ_i, since $T = 1/B_i$ for channels with the transmission-frequency characteristics assumed here. The above relations for error probability and those in later sections assume a flat noise-power spectrum, constant attenuation over the channel band and optimum division of channel shaping between transmission and receiving filters. On these premises the relations apply for an infinite variety of idealized channel transmittances and pulse shapes, including the extreme case of rectangular transmitted and received pulses and the other limiting case of a flat channel of bandwidth B_i. However, the solutions given in literature for these extreme cases are not directly applicable to communication channels without the supplemental derivation of optimum division of filter shaping indicated above.

5.11 Orthogonal Carrier and Vestigial Sideband Transmission

With bipolar or two-phase transmission on each of two carriers at quadrature with each other, over a channel of bandwidth $B_i = 2B$ the combined channel capacity is the sum of the two-channel capacities afforded by each carrier, or

$$C = 2B_i \log_2 r^2 = 4B \log_2 r^2. \tag{5.81}$$

The error probability for each channel is given by (5.75) and the combined error probability becomes

$$\epsilon = \operatorname{erfc}(\rho/2)^{1/2}. \tag{5.82}$$

PM with Synchronous Detection

The optimum ratio $\rho = \rho^0$ is obtained from (4.83). In the particular case of a channel with constant attenuation and noise power density, $\rho^0 = P_i/N_i = \rho_i$.

The channel capacity (5.81) is also afforded by two-phase vestigial sideband amplitude modulation with synchronous detection and is the same as for bipolar baseband transmission over a channel of bandwidth B_i. The channel capacity obtained from the above relations is shown in Fig. 5.11 as a function of ρ_i.

5.12 Phase Modulation with Synchronous Detection

For analog or digital transmission over channels with a rapid random variation in transmittance with time, it is desirable to employ methods that facilitate fast gain control. In this respect, phase and frequency modulation have a significant advantage over various methods of amplitude modulation, which is usually realized in exchange for some increase in signal-to-noise ratio, as considered here for time-invariant channels. (For further exposition see Section 10.2.)

With phase modulation $a(n)$ is constant in (5.6) and $\psi(n)$ takes on various discrete values. The channel bandwidth and optimum shaping is in this case the same as for two-phase amplitude modulation. With p phase positions the channel capacity per unit of bandwidth becomes

$$C = B_i \log_2 p. \tag{5.83}$$

The probability density $p(\theta)$ of phase deviations is given by (3.67) or

$$p(\theta) = \frac{1}{2\pi} e^{-\rho_i}(1 + \pi^{1/2}\alpha(1 + \text{erfc } \alpha)e^{\alpha^2}). \tag{5.84}$$

For the case of channels with constant attenuation and noise-power density, $\rho_i = P_i/N_i$ is the ratio of carrier power to noise power in a band $B_i = 2B$ and

$$\alpha = \rho_i^{1/2} \cos \theta. \tag{5.85}$$

With p phases, the probability of an error becomes

$$\epsilon_p = 1 - \int_{-\pi/p}^{\pi/p} p(\theta) \, d\theta \tag{5.86}$$

For $p = 2$, evaluation of (5.86) yields

$$\epsilon_2 = \tfrac{1}{2}\text{erfc } (\rho_i)^{1/2}, \tag{5.87}$$

and for $p = 4$

$$\epsilon_4 = \text{erfc}\left(\frac{\rho_i}{2}\right)^{1/2}. \tag{5.88}$$

Expressions (5.87) and (5.88) conform with (5.75) and (5.82), when applied to channels with constant attenuation and noise-power spectrum over the channel band as considered here.

For large values of ρ_i and small values of π/p, the following approximation applies (see Chapter 3).

$$\epsilon_p = \operatorname{erfc}\left(\rho_i^{\frac{1}{2}} \sin \frac{\pi}{p}\right) \tag{5.89}$$

This expression conforms with (5.88) for $p = 4$.

In Table 2 are given the ratio ρ_i obtained from the foregoing relation for various values of p, together with the corresponding channel capacity obtained from (5.83).

Table 2. *Channel Capacity with Phase Modulation and Ideal Synchronous Demodulation*

$p =$	2	4	8	16	32	64	
$C/B_i =$	1	2	3	4	5	6	
$\rho_i =$	8.8	17.6	49	200	800	3200	($\epsilon_p = 10^{-4}$)
$\rho_i =$	11	22	77	310	1240	5000	($\epsilon_p = 10^{-6}$)

The above channel capacities are shown in Fig. 5.11. For $p = 2$, the channel capacity conforms with those for bipolar AM with $q = 1$, and for $p = 4$ conform with that for quadrature two-phase AM with $q = 1$, or two-phase vestigial sideband transmission with $q = 1$.

With the above methods the optimum filter shapes are the same as illustrated in Fig. 4.7 for double-sideband systems.

The same channel capacity as with four-phase modulation can be obtained by duobinary coding and two-phase transmission. The error probability with the latter method is given by (5.71) with $\rho_i = P/2N$. For an error probability $\epsilon = 10^{-6}$, (5.71) yields $\rho_i = 38$ (15.8 dB) as compared with $\rho_i = 22$ (13.4 dB) for four-phase modulation. The latter method thus affords about a 2.4 dB advantage in signal-to-noise ratio.

5.13 Phase Modulation with Differential Phase Detection

With synchronous product demodulation as considered in the previous section it is necessary to extract a reference carrier at the receiver, as discussed in Section 5.6. This is avoided by the method of differential product demodulation mentioned in Section 5.2, which has a definite advantage in applications to rapidly time-varying channels. This advantage is realized in exchange for some increase in error probability for time-invariant channels as considered here, owing to the presence of noise in the demodulating wave.

With p phases, the maximum tolerable phase deviation $\hat{\theta}$ before an error occurs is $\hat{\theta} = \pi/p$. Let the probability of error with synchronous or coherent detection as a function of $\hat{\theta}$ be $P_e(\hat{\theta}) = P_e(\pi/p)$. If a phase deviation θ occurs in the demodulating pulse, then the tolerable phase deviation in the received pulse is $\pi/p - \theta$. The error probability for a particular θ is $P_e(\pi/p - \theta)$. The error probability, considering all possible values of θ, is

$$\epsilon_p = \int_{-\pi}^{\pi} p(\theta) P_e\left(\frac{\pi}{p} - \theta\right) d\theta, \tag{5.90}$$

where the probability density $p(\theta)$ is given by (5.84). In accordance with (5.86)

$$P_e\left(\frac{\pi}{p} - \theta_0\right) = 1 - 2 \int_0^{\pi/p - \theta_0} p(\theta) \, d\theta. \tag{5.91}$$

For $p = 2$, solution of (5.90) yields the following simple relation [14] for channels with constant attenuation and noise power density:

$$\epsilon_2 = \tfrac{1}{2} e^{-\rho_i}. \tag{5.92}$$

For small error probabilities, ρ_i is greater than with synchronous detection by about 1 dB.

For other values of p it is necessary to resort to approximations or to numerical integration of (5.90). For $p = 4$ the results of numerical integration are closely approximated by the following relation for $\epsilon_4 < 10^{-1}$

$$\epsilon_4 = \tfrac{1}{2} e^{-\rho_i/3}. \tag{5.93}$$

Thus for the same error probability, as with $p = 2$, a three-fold increase in signal-to-noise ratio is required.

For $p > 2$ and small error probabilities, the following approximate relations apply [15] for the factor by which the signal-to-noise ratio is greater than with synchronous detection

$$\lambda \cong \frac{\sin^2 \pi/p}{2 \sin^2 \pi/2p} \tag{5.94}$$

$$= 1.7 \ (2.3 \ \text{dB}) \quad \text{for} \quad p = 4$$

$$= 1.9 \ (2.8 \ \text{dB}) \quad \text{for} \quad p = 8$$

$$= 2 \ (3 \ \text{dB}) \quad \text{for} \quad p = \infty.$$

Thus relation (5.89) is replaced by the following approximation for $p > 2$:

$$\epsilon_p = \text{erfc}\left[\left(\frac{\rho_i}{\lambda}\right)^{1/2} \sin \pi/p\right]$$

$$= \text{erfc}\,[(2\rho_i)^{1/2} \sin \pi/2p]. \tag{5.95}$$

For $p = 4$, the latter relation conforms well with the results of numerical integration of (5.90) and the accuracy improves as p increases.

5.14 Binary FM with Dual Filter Detection

FM, like PM, facilitates rapid gain control, which is a distinct advantage in applications to rapidly time-varying channels, as discussed further in Section 10.2. Binary FM, or binary frequency-shift keying, as considered here, permits simpler detection methods than binary PM. However, some increase in bandwidth or signal-to-noise ratio, or both, is required for the same error probability as with PM.

In binary frequency-shift keying, the carrier frequency is changed from ω_1 to ω_2 over the duration T of a signal interval in order to indicate a change from space to mark. The signal can be detected in various ways; and the error probability as related to signal-to-noise ratio depends on the particular method and on the transmittances of the channels.

The simplest method from the standpoint of analysis is to assume two receiving filters centered on ω_1 and ω_2, as indicated in Fig. 5.12 with synchronous detection of the output of each filter. With this method, there would be two complementary channels in which marks and spaces are interchanged. The error probability as related to signal-to-noise ratio is the same as for one filter alone, on the premise that the frequency separation is sufficient to insure no correlation between noise in the two filters. The reason for this is that when the second filter is added, both the average signal power and the average noise power are doubled. The signal-to-noise ratio for a given error probability is thus the same as for on-off AM with synchronous detection and 3 dB greater than for two-phase transmission with synchronous detection. The error probability is thus given by (5.76). For channels with constant attenuation and noise power spectrum, (5.77) applies, and the following relation applies [16]:

$$\epsilon = \tfrac{1}{2}\text{erfc}\left(\frac{\rho_i}{2}\right)^{1/2}, \tag{5.96}$$

where ρ_i is the ratio of average signal power in one filter to the average noise power in one filter of bandwidth $2B$. This is the same as the ratio of

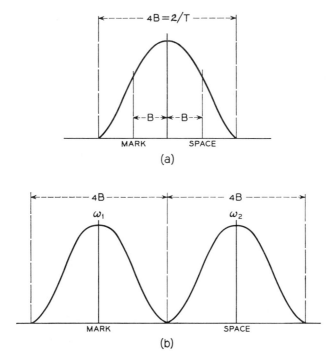

Figure 5.12 Channel bandwidth requirements in binary FM with frequency discriminator detection, a, and dual filter detection, b.

the combined power to combined noise power in the two filters. If synchronous demodulation were feasible, then the preferable method would of course be binary PM. Hence the foregoing method is of no practical significance.

An alternate method is to use envelope detection of each filter output and to combine the two detector outputs with reversal in polarity of one. The probability density of the resultant envelope is then in accordance with general relations given in Appendix 2.

$$p_{m-s}(r) = \int_0^\infty p_s(z - r) p_m(z) \, dz, \qquad (5.97)$$

where $p_m(z)$ is the probability density of the envelope z at the receiving filter output in the presence of a mark, and $p_s(z)$ the probability density in the presence of a space. The lower limit is zero, since negative values of z are not possible.

The probability that the combined envelope exceeds a specified amplitude r_1 is

$$P_{m-s}(r \geq r_1) = \int_{r_1}^{\infty} p_{m-s}(r) \, dr. \tag{5.98}$$

An error will occur when $r_1 \leq O$ such that the error probability is given by

$$\epsilon = \int_{-\infty}^{0} \int_{O}^{\infty} p_s(z - r) p_m(z) \, dz \, dr. \tag{5.99}$$

The probability density $p_m(z)$ is given by (3.44) and $p_s(z)$ is obtained from (3.44) with $a = O$. With these relations in (5.99) the following simple expression is obtained after evaluation of the integral [16]

$$\epsilon = \tfrac{1}{2} e^{-\rho_i/2}. \tag{5.100}$$

Comparison of (5.100) with (5.92) shows that this method entails a two-fold increase in signal-to-noise ratio and channel bandwidth, compared to two-phase modulation with differential phase detection.

With dual filter detection, the optimum filter characteristics are the same as discussed in Section 4.8 for double-sideband amplitude modulation.

5.15 Binary FM with Frequency Discriminator Detection

In binary FSK with dual filter detection and bandlimited pulses the envelope of the combined received carrier wave exhibits fluctuations except at sampling instants, in a similar manner as the received carrier with binary PSK. This can be a disadvantage in applications to channels with highly nonlinear repeaters, particularly when they exhibit pronounced AM-PM conversion (see Section 8.14). Such envelope fluctuations can be avoided by employing pure binary FM with frequency discriminator detection. It turns out that with optimum design the latter method is virtually equivalent as regards error probability to binary FSK with dual filter envelope detection. This applies for error probabilities in the range of practical interest, below 10^{-2} for equal carrier powers, channel bandwidths and pulse shapes.

By way of example, let the baseband pulses have a raised cosine spectrum of bandwidth $2B$ and let the semi-bandwidth of the bandpass channel be $4B$, as in Fig. 5.12 for dual filter detection. Optimum performance is in this case approached with cosine-shaped premodulation and post-detection low-pass filters, such that the carrier modulating pulses have a cosine spectrum. The maximum amplitude of a random train of modulating pulses is in this case about 1.6 times the amplitude at sampling instants during transmission of continuous sequences of marks or spaces.

For optimum performance it is necessary to use the maximum allowable frequency deviation with a channel bandwidth $8B$. This condition is approached with frequency deviations $\pm 1.25B$ for continuous marks and spaces, resulting in maximum deviation $\pm 2B$ for random pulse trains. With baseband pulses of bandwidth $2B$ the required semi-bandwidth of the bandpass channel is $4B$, in accordance with the conventional Carson rule.

The ratio of peak pulse power at sampling instant to average noise power at the output of the post-detection low-pass filter can be determined for frequency-discriminator detection, using relation (3.97) for frequency modulation caused by additive Gaussian noise. It turns out that for an error probability 10^{-10} the required carrier power would be 0.5 dB greater than with dual filter synchronous detection and for an error probability 10^{-2} would be about 1.5 dB greater. The required signal-to-noise ratio would be slightly less than for dual filter envelope detection and about 3 dB greater than for binary PSK with differential phase detection. Since the latter method also affords a two-fold reduction in channel bandwidth it would be preferable in most applications.

An alternate possibility with frequency-discriminator detection is to accept envelope fluctuations in exchange for a two-fold reduction in bandwidth compared to dual filter detection or pure binary FM. As discussed in Chapter 1, with frequency-discriminator detection the same bandwidth as for binary AM or PM suffices. For rectangular pulses of duration T at the channel input, and a raised cosine spectrum of the received wave at the detector input, the requisite amplitude characteristic of the channel is as shown in Fig. 1.25.

Determination of the optimum division of channel shaping and the corresponding minimum signal-to-noise ratio for a specified error probability is in this case rather difficult, particularly with a post-detection low-pass filter. Approximate evaluations indicate that about 4 dB greater signal-to-noise ratio would be required than for binary PM with synchronous detection, when no post-detection low-pass filter is used. (Chapter 1, reference 7.) Exact evaluations [17] indicate a 3 to 4 dB increase in the required signal-to-noise ratio over binary PM with synchronous detection for a variety of shapes of bandpass transmitting and receiving filters, giving the combined amplitude characteristic shown in Fig. 1.25.

In summary, the error probability for binary FM with frequency discriminator detection will be virtually the same as with dual filter envelope detection, such that the principal advantage of frequency discriminator detection is a two-fold reduction in channel bandwidth. Although binary FM entails a two-fold increase in signal-to-noise ratio over binary PM with differential phase detection, it is simpler to implement and may therefore be advantageous in applications to facilities with adequately high signal-to-noise ratio, such as channels designed primarily for voice or other analog transmission.

5.16 Sub-binary Transmission by Reduced Pulse Rate

In the previous analysis, transmission at the maximum rate for a given channel bandwidth was assumed, together with individual digit detection. On this premise, the minimum channel capacity, and the corresponding minimum signal-to-noise ratio is obtained with bipolar binary transmission. However, it is possible to realize a lower signal-to-noise ratio for a specified error probability, in exchange for a reduced channel capacity, in various ways.

The simplest method to this end is a k-fold reduction in the pulse transmission rate relative to the maximum permissible for the specified channel bandwidth. This results in a k-fold reduction in average signal power over the interval between pulses and in a corresponding reduction in channel capacity for a given error probability. With this method, the transmission arrangement remains unchanged except for the lower transmission rate. The reduction in the ratio of average signal power to average noise power is, however, somewhat illusory in that it is achieved by confining a given signal power to a single pulse interval.

An alternate arrangement that permits distribution of the signal power over a k-fold larger time interval is a k-fold reduction in noise power, with the aid of a filter at the detector input of bandwidth commensurate with the lower transmission rate. The signal power could then be reduced correspondingly for a given error probability, and the ratio of average signal power to average noise power in the channel band (not the filter band) would be reduced k times.

A third method would be to repeat each digit k times, with direct addition or integration of the received digits. The combined signal amplitude is increased by a factor k, and the combined rms amplitude of uncorrelated noise by a factor $k^{1/2}$. Hence the ratio of average signal power to average noise power is reduced by a factor k, for a given error probability. Instead of repeating each digit, simultaneous transmission could be made over k channels in frequency diversion multiplex.

With any one of the above three methods a k-fold reduction in signal-to-noise ratio is achieved in exchange for a k-fold reduction in channel capacity. For idealized channels and Gaussian noise the transmission capacity and error probability is in all cases the same as if a channel of bandwidth B/k were used. For $r = 2$, or bipolar binary baseband transmission, the channel capacity is, in view of (5.63),

$$C = \frac{2B}{k}, \tag{5.101}$$

and the error probability with baseband transmission becomes

$$\epsilon_k = \tfrac{1}{2}\mathrm{erfc}\left(\frac{k\rho}{2}\right)^{\!\!1/2}, \tag{5.102}$$

where ρ is defined as in Section 5.8.

It is important to note that the above equivalence of the three methods as regards error probability does not in general apply for channels with amplitude and phase distortion over the channel band. In this case the second method which employs a portion $1/k$ of the available band would ordinarily be preferable, because of the reduced distortion over this narrower band. Because of the resultant reduction in pulse distortion and intersymbol interference, the error probability with this method would ordinarily be less than with the other two. In the case of impulse noise, this method has the further advantage of a greater reduction in noise amplitudes with reduced bandwidth than for Gaussian noise as assumed in the foregoing comparison.

In the sections that follow, other methods that employ the full bandwidth are considered. Any advantage in signal-to-noise ratio they may have over the second method mentioned earlier in applications to ideal channels with Gaussian noise would not ordinarily apply for actual channels with attenuation and phase distortion, or with impulse noise.

5.17 Sub-binary Transmission by PPM

If a digit is represented by a pulse in one out of k possible positions in time, as in digital pulse position modulation, a k-fold increase in bandwidth is required as compared to k amplitudes in PAM. With baseband transmission and equal probabilities of all pulse positions, the channel capacity per unit of bandwidth B is then

$$C = \frac{2B}{k} \log_2 k. \tag{5.103}$$

The same relation applies if a pulse is transmitted over one out of k channels in frequency division multiplex. A special case of the latter method is binary FM with dual filter detection.

The above relation assumes pulses of one polarity in PPM. With both positive and negative polarities and k pulse positions, the channel bandwidth remains the same but the number of digital values that can be represented is $2k$. Hence with this more efficient method relation (5.103) is modified into

$$C = \frac{2B}{k} \log_2 2k. \tag{5.104}$$

With binary bipolar PPM the average signal power with a pulse in one out of k possible positions is smaller than for conventional bipolar transmission by a factor k. The tolerable peak-noise amplitude before an error occurs in a given pulse position is only half as great as with conventional bipolar transmission, since a distinction must be made between a negative pulse, no pulse, and a positive pulse. Hence for a given probability of error in a pulse, the required average signal-to-noise ratio ρ_k is related to the average signal-to-noise ratio ρ in conventional binary transmission by

$$\rho_k = \frac{4}{k} \rho. \tag{5.105}$$

The factor $1/k$ comes about because of the reduction in average signal power for the same channel band as in conventional binary transmission, and the factor 4 because of the two-fold reduction in tolerable peak noise amplitude.

In bipolar PPM k detectors would be required, one for each pulse position. It will be assumed that each detector has threshold levels of $-\frac{1}{2}$ and $\frac{1}{2}$. In $k - 1$ positions pulses would be absent, and the corresponding detector would give an erroneous indication if the peak-noise amplitude exceeded $\frac{1}{2}$, regardless of polarity. The one remaining detector would give an erroneous indication if the noise amplitude exceeded $\frac{1}{2}$ and had the appropriate polarity. In view of (5.105) and (5.102) the probability of false operation of one or more detectors would be

$$\epsilon_k = (k - 1 - \tfrac{1}{2}) \operatorname{erfc} \left(\frac{k\rho}{8}\right)^{1/2}. \tag{5.106}$$

If more than one detector indicated the presence of a pulse, the presence of an error would be known, but not the pulse position in error. Hence the error cannot be corrected, unless retransmission is used. Thus (5.106) gives the probability of error without retransmission. The proportion of erroneous indications involving more than one detector will be quite small, so that the reduction in error probability that could be realized by retransmission would be insignificant.

5.18 Sub-binary Transmission by Orthogonal Codes

In pulse-position modulation a single pulse is transmitted in one of k possible positions in the time or in the frequency domain. Instead of using a single pulse, k different combinations of pulses could be used, which would afford the same channel capacity. It is possible to select from the 2^k possible sequences of k digits, k sequences that have the orthogonal property indicated in Fig. 5.13. When the product of two such sequences is integrated over the duration of a sequence, the integral is zero, except when a sequence is multiplied by itself. The complement of an orthogonal code is obtained by reversing all polarities. An orthogonal code and its complement is referred to as a biorthogonal code, which corresponds to bipolar PPM. With biorthogonal codes the $2k$ possible combinations would be stored in k detectors. In the absence of noise a particular detector would then indicate 1 when its stored sequence is received, -1 for the reversed sequence of complement, and otherwise O. The channel capacities and ratios of average signal power to average noise power is the same as with bipolar PPM. The principal advantage over PPM is that for equal average signal power the peak signal power is smaller by a factor $(1/k)^{1/2}$.

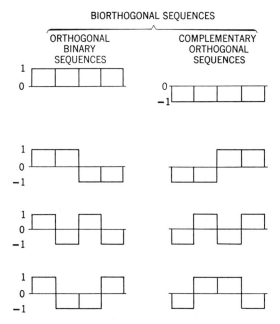

Figure 5.13 Orthogonal and biorthogonal binary sequences of 4 digits. Each sequence corresponds to a particular pulse position and polarity in PPM.

The signal-to-noise ratios and channel capacities obtained from the preceding relations (5.101), (5.63) and (5.106) are given in Table 3 for an error probability $\epsilon_k = 10^{-5}$.

Table 3 *Channel Capacities versus Signal-to-Noise Ratios with Subbinary Transmission Methods*

	Reduced Trans. Rate		Biorthogonal Codes and PPM	
k	ρ	C/B	ρ	C/B
1	18	2.0	72	2.0
2	9	1.0	40	2.0
4	4.5	0.5	22	1.5
8	2.2	0.25	11.6	1.0
16	1.1	0.125	6.3	0.62
32	0.55	0.062	3.2	0.375
64	0.275	0.031	1.7	0.22

Here ρ is the signal-to-noise ratio defined as in Section 5.8. The corresponding minimum transmitter power is determined in the same manner as outlined there.

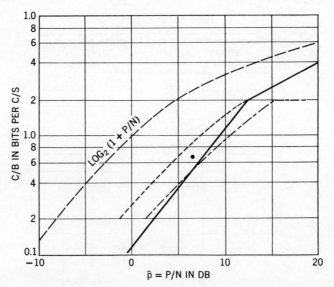

Figure 5.14 Channel capacities with various methods of sub-binary modulation and detection for error probability 10^{-5} in a decoded binary digit. ———— Transmission at reduced rate. – · – · – Biorthogonal codes and bipolar PPM with threshold detection. · · · · · Biorthogonal codes and bipolar PPM with correlation detection. • Bose-Chaudhuri (15,5) error correcting code.

Biorthogonal Codes—Correlation Detection

The foregoing channel capacities are compared in Fig. 5.14. It will be noted that for idealized channels and Gaussian noise as assumed here, biorthogonal codes and PPM have a slight advantage over transmission at reduced speed or by repetition for signal-to-noise ratios below about 10 dB.

However, as noted at the end of Section 5.16, in the case of actual channels with attenuation and phase distortion, the preferable method would ordinarily be transmission at a reduced rate with a k-fold reduction in channel bandwidth, or in the bandwidth of the receiving filter.

5.19 Biorthogonal Codes with Correlation Detection

In the previous section the error probability as given by (5.106) was based on threshold detection at each of the k detectors with biorthogonal codes or PPM. An alternate method is correlation detection, in which the above threshold levels of $\frac{1}{2}$ would be disregarded and instead the output of the k detectors compared. When all sequences are equally probable and have equal average power, the maximum detector output is obtained when the received sequence equals the stored sequence, with a probability of error owing to noise that is smaller than with individual detector threshold decision. The error probability is then reduced by selecting the detector with maximum output. This method is known as correlation or maximum likelihood detection, and is the optimum detection method in that it minimizes the probability of error.

With biorthogonal codes, only k correlation detectors are required for $2k$ code words, because the output is only reversed in polarity when the polarity of all the k digits are reversed. The probability that the correct detector gives the correct indication is given by [18]

$$1 - \epsilon_k = \int_0^\infty p(x)\, dx \prod_{i=1}^{k-1} \int_{-x}^{x} p(x_i)\, dx_i, \qquad (5.107)$$

where $p(x)$ is the probability density for the output of the correct detector, and $p(x_i)$ that of the other $k - 1$ detectors. With an output amplitude A in the absence of noise, and Gaussian noise of average power σ^2, relation (5.107) gives, for the probability of error,

$$\epsilon_k = 1 - \left(\frac{1}{2\pi\sigma^2}\right)^{1/2} \int_0^\infty e^{-(x-A)^2/2\sigma^2} \left[\mathrm{erf}\left(\frac{x}{2^{1/2}\sigma}\right)\right]^{k-1} dx. \qquad (5.108)$$

For $k = 1$ this integral yields relation (5.64) for $r = 2$.

The foregoing error probability has been determined by Viterbi [18] by computerized numerical integration. An approximate expression for the error probability with maximum likelihood detection has been derived by Gilbert [19], applying for a small error probability and various kinds of codes. As applied to biorthogonal codes, the approximate expression for $k \geq 2$ is

$$\epsilon_k \simeq k \operatorname{erfc} \left(\frac{k}{4} \rho\right)^{\frac{1}{2}}. \tag{5.109}$$

Comparison of (5.109) with (5.106) shows that in a first approximation, correlation detection gives a 3 dB advantage in signal-to-noise ratio over threshold detection.

The channel capacity as given by (5.104) is shown in Fig. 5.14 versus the signal-to-noise ratio ρ obtained from (5.109). For reasons noted at the ends of Sections 5.16 and 5.18, the foregoing advantage in signal-to-noise ratio over transmission at a reduced rate may not apply to actual channels with attenuation and phase distortion.

5.20 Sub-binary Transmission by Error-Correcting Codes

A reduction in signal-to-noise ratio in exchange for reduced channel capacity is possible by coding methods originated by Hamming [20] that facilitate detection and automatic correction of errors in one or more digits. These methods entail extra digits for error correction, so that the bandwidth required for a given information rate is increased, as are the noise amplitudes and the rate of errors prior to correction.

When a character represented by m binary digits is in error, the probabilities $p_0, p_1, p_2, \ldots$ that it contains random errors in 0, 1, 2, $\ldots$ digits, are given by the coefficients of the binominal expansion

$$(q + p)^m = p_0 + p_1 + p_2 + p_3 + \cdots, \tag{5.110}$$

where p is the probability of an error in a digit $q = 1 - p$ and

$$p_0 = q^m, \tag{5.111}$$

$$p_1 = mpq^{m-1}, \tag{5.112}$$

$$p_2 = \frac{m(m-1)}{2!} p^2 q^{m-2}, \tag{5.113}$$

$$p_3 = \frac{m(m-1)(m-2)}{3!} p^3 q^{m-3}. \tag{5.114}$$

Error-Correcting Codes

The probability that a character is correct, that is, that it contains an error in O digits is p_0, and the probability that it is in error is

$$p_e = 1 - p_0 = 1 - q^m = 1 - (1-p)^m \tag{5.115}$$

$$\simeq mp - \frac{m(m-1)}{2!}p^2 + \cdots, \quad \text{if } mp \ll 1. \tag{5.116}$$

Among various kinds of error-correcting codes, the more efficient appear to be Bose-Chaudhuri codes [21,22] that also have the property making them more readily implemented than most other types. With these codes the total number of digits m is related as follows to the maximum number of error-correction digits e and the number of digits d in which errors can be corrected:

$$m = 2^l - 1,$$

$$e \leq ld. \tag{5.117}$$

For $d = 1$ the foregoing relation gives various Hamming codes. By way of example, with $d = 1$ and $e = l = 3$, $m = 7$ and $n = m - e = 4$ information digits. With $d = 3$ and $l = 4$, $m = 15$ and $e \leq 12$. Actually only $e = 10$ error-correction digits are required in this particular case, which represents one of the more efficient codes.

The channel capacity with $n = m - e$ information digits becomes

$$\frac{C}{B} = \frac{2n}{m}\log_2 2 = \frac{2n}{m} = \frac{2}{k}, \tag{5.118}$$

where $k = m/n$ is the factor by which the bandwidth must be increased to maintain the same information rate as without coding.

When errors are corrected in d digits, the residual error probability in a character with n interaction digits is

$$\epsilon_k^{(n)} = p_{d+1} + p_{d+2} + \cdots$$

$$= \sum_{u=d+1}^{m} \binom{m}{u}p^u(1-p)^{m-u}$$

$$= E(m, d+1, p), \tag{5.119}$$

where $m = kn$ and E is the cumulative binomial probability distribution. (For tables of the latter see reference in Section 10.13.)

For $mp \ll 1$ the following approximation applies:

$$\epsilon_k^{(n)} \simeq \frac{m(m-1)\cdots(m-d)}{(d+1)!}p^{d+1}(1-p)^{m-d-1}. \tag{5.120}$$

The performance will be illustrated for a (15,5) code with $m = 15$ and $n = 5$. With this code $e = 10$, $k = 3$, and errors are corrected in $d = 3$ digits. In this case (5.120) gives

$$\epsilon_k^{(n)} = 1365 p^4 (1 - p)^{11}. \tag{5.121}$$

To conform with the comparison in Table 3 an error probability $\epsilon_k^{(5)} = 5 \cdot 10^{-5}$ will be assumed for a sequence of five information digits. In this case (5.121) yields $p = 1.7 \cdot 10^{-2}$. The corresponding signal-to-noise ratio is obtained from the relation $p = \frac{1}{2} \operatorname{erfc}(\rho/2)^{1/2}$ giving $\rho = 4.3$ or 6.3 dB, for $C/B = \frac{2}{3}$. The corresponding value of ρ for transmission by repetition or equivalent methods outlined before is about 1.5 dB greater, as indicated in Fig. 5.14.

A comprehensive comparison of some efficient error-correcting codes show improved performance with increasing length of the coded sequences [23,24]. Thus a particular (23,12) code with 12 information digits out of a total of 23, and $d = 3$, $k = 23/12$ affords about a 2.5 dB advantage for an error probability $1.2 \cdot 10^{-4}$ in 12 digits (10^{-5} per digit). A (73,45) code with $d = 4$ and $k = 73/45$ yields about a 3 dB advantage over repeated transmission, or no coding, for an error probability 10^{-5} per digit.

For reasons noted at the ends of Sections 5.16 and 5.18, the foregoing reductions in signal-to-noise ratio, as compared to transmission at a reduced rate, would not apply to actual channels with significant attenuation and phase distortion. Transmission at reduced rate with a reduction in the channel bandwidth or in the bandwidth of the receiving filter would also be the preferable method with impulse noise. It is also important to note that any of the foregoing methods that entail an increase in channel bandwidth for a reduction in signal-to-noise ratio would be at a disadvantage as regards the required transmitter power in applications to transmission media where the transmission loss increases with bandwidth, as in the case of baseband transmission over metallic circuits.

5.21 Binary Transmission with Feedback Error Correction

A more efficient means of error reduction, or of improving signal-to-noise ratio for a specified error rate, is to use error detection codes in conjunction with a feedback transmission path for partial error correction by retransmission. This method is feasible and has been used to obtain increased reliability in digital transmission where two-way channels are available, as in extensive communication networks. Since only a small increase in channel bandwidth is required for error detection, and re-

Feedback Error Correction

transmission, the problems associated with a large increase in bandwidth, as required with error-correcting codes, are largely avoided.

Comprehensive performance analyses of retransmission systems employing error detecting codes are given elsewhere [25,26]. The present exposition is confined to an approximate evaluation of the reduction in signal-to-noise ratio and in channel capacity when a single parity check digit [20] is used for detection of errors in a sequence of m digits. Each erroneous sequence of m digits would be retransmitted upon request, which may consist of a single or a few digits transmitted over the feedback channels. It is assumed here that comparatively few sequences of m digits would initially be in error, so that the time and the bandwidth required for request and retransmission of erroneous sequences can be disregarded.

By way of illustration of the above method, it will be assumed that a single parity check digit is inserted, which will facilitate detection of errors in an odd number of digits. With the approximation noted above, the channel capacity in this case is

$$\frac{C}{B} \simeq \frac{2(m-1)}{m} \log_2 2 = \frac{2(m-1)}{m}. \tag{5.122}$$

The error probability after the error correction is in a first approximation equal to the probability of two errors in m digits, or

$$\epsilon_m = p_2 = \frac{m(m-1)}{2!} p^2. \tag{5.123}$$

The permissible probability p of an error in a digit for a residual error probability ϵ_m in a sequence of m digits is thus

$$p = \left(\frac{2\epsilon_m}{m(m-1)}\right)^{1/2}. \tag{5.124}$$

Without error detection and correction the permissible probability p^0 of an error in a digit is

$$p^0 = \frac{\epsilon^0}{(m-1)}, \tag{5.125}$$

where ϵ^0 is the probability of error in a sequence of $m-1$ digits.

In Table 4 a comparison is made of the signal-to-noise ratio ρ required with error detection and feedback correction and the ratio ρ^0 with transmission without correction over a channel of bandwidth $B(m-1)/m$. It is assumed that after error correction the residual error probability e_m is the same as ϵ^0.

Table 4. *Signal-to-Noise Ratios ρ with Feedback Correction and ρ^0 without Correction*

m	6	12	24	48
C/B :	1.66	1.84	1.92	1.96
$\epsilon^0 = \epsilon_m$:	$5 \cdot 10^{-5}$	$11 \cdot 10^{-5}$	$23 \cdot 10^{-5}$	$47 \cdot 10^{-5}$
p :	$1.8 \cdot 10^{-3}$	$1.28 \cdot 10^{-3}$	$0.9 \cdot 10^{-3}$	$0.64 \cdot 10^{-3}$
ρ :	9.3	10	10.8	11.4
p^0 :	10^{-5}	10^{-5}	10^{-5}	10^{-5}
ρ^0 :	15	16.5	17.1	17.7
ρ^0/ρ :	1.62	1.65	1.58	1.54
	(2.1 dB)	(2.2 dB)	(2 dB)	(1.9 dB)

The ratio ρ is obtained from the relation $p = \frac{1}{2}\mathrm{erfc}\,(\rho/2)^{1/2}$ and the ratio ρ^0 from the relation $p^0 = \frac{1}{2}\mathrm{erfc}\,[m\rho^0/2(m-1)]^{1/2}$, which corresponds to (5.102) with $p^0 = \epsilon_k$ and $k = m/(m-1)$.

The foregoing comparisons show that a reduction in signal-to-noise ratio of about 2.2 dB is afforded without significant increase in channel bandwidth, about 8% for $m = 12$. With more than one parity check digit, the residual errors and the signal-to-noise ratio can be further reduced. To realize these improvements it is necessary to store the received information to facilitate correction of erroneous characters, which entails delays and complications in the implementation. Hence the above method precludes transmission of information in "real time."

5.22 Special Multiplexing Methods

In communication systems a number of independent signals are usually transmitted over a channel with common repeaters by frequency- or time-division multiplexing at the terminals. The conventional methods to this end are reviewed in Sections 2.13 and 2.14. With pulse transmission other methods are feasible, provided that certain constraints are placed on the pulse spectra and the transmission rates in the individual channels, as in the discussion that follows.

With the conventional method of frequency-division multiplexing, the signal spectra are confined to non-overlapping transmission bands, as indicated in Fig. 2.10, so as to avoid interchannel interference, except from amplifier nonlinearity. In pulse systems, it is possible to avoid both inter-symbol and interchannel interference with overlapping pulse spectra of finite bandwidth, provided that the various carrier sideband spectra and thus the transmission rates are the same. The transmission instants are synchronized in the various channels and the pulses are sampled at the midpoint of each signal interval [27]. With the integration methods dis-

Impulse Noise   **221**

cussed in Section 1.13, it is possible to avoid interchannel interference with identical overlapping sideband pulse spectra of infinite bandwidth, such as the familiar $\sin x/x$ spectrum [28]. However, some intersymbol interference is then encountered, owing to unavoidable elimination of some spectral components in transmission over bandlimited channels.

With the conventional method of time-division multiplexing, pulses of the same shape are used for all channels and they are sampled at the midpoint of each signal interval. If correlation integration is used, in accordance with (1.114), an alternate possibility is to use pulses of different shapes with orthogonal properties so as to avoid intersymbol interference. In the extreme case of pulses of infinite spectral bandwidth, the orthogonal pulses would consist of various combinations of rectangular pulses, as illustrated in Fig. 5.13 with a different objective in mind. It is possible to devise orthogonal pulses of finite bandwidth, as shown in references 10 and 11 of Chapter 1. However, these methods are chiefly of academic interest, since implementation is more complicated and performance less efficient than with the conventional methods of time-division multiplexing, for the same reasons as outlined in Section 1.13.

5.23 Impulse Noise

In some systems impulse or intermittent noise may be a more important source of transmission impairment than continual additive Gaussian noise, at least during certain periods, as in the case of occasional atmospheric interference (static) in certain wire and radio facilities. Impulse noise may also be encountered in the voice channels of some analog carrier systems, owing to interference from switching currents in local subscriber pairs in the same cable as the carrier system. This has been the source of an increased error rate in data transmission over such channels, particularly during busy periods with much switching activity. The increased error rate is also accompanied by other distributions of the intervals between errors than expected for additive Gaussian noise [29,30]. However, impulse noise is ordinarily of minor importance in broadband analog or digital systems employing such transmission media as coaxial pairs, shielded balanced pairs, wave guides or microwave radio paths.

In situations where impulse noise is important, it is dealt with in systems design as extraneous interference, rather than as noise inherent in the transmission medium, in a similar manner as outlined in Chapter 11 for intersystem interference. To enable the determination of transmission impairments and appropriate remedial measures, it is necessary to have adequate statistical information on the impulse noise under consideration.

This may include the power spectrum, the probability distribution of amplitudes and of intervals between impulses, together with data on the time of day or of the year during which impulse noise is most severe.

In multipair cable systems the conventional measures against impulse noise consist of adequate shielding of various kinds, the insertion of suppression filters at open-wire entrances, or a reduction in repeater spacings and thus increased signal power in sections where severe impulse noise is expected. By such measures it is possible to avoid excessive impulse noise, or to limit it to a tolerable small fraction of the total time.

In some situations it may be necessary to incorporate special remedial measures against impulse noise, as when channels originally designed for voice or other analog messages are used for digital transmission. An excessive error rate may then be encountered, even though the impulse noise does not cause excessive disturbances in voice transmission, where a basic design criterion is the energy of impulses that can be tolerated with a certain maximum frequency of occurrence, as established by subjective tests. Although this has some relation to error probability in digital systems, the error rate may be excessive, particularly if multilevel transmission is attempted over channels with appreciable phase distortion as usually encountered in voice channels [29]. In fact, the increased error rate may in large measure be attributable to this phase distortion and resultant intersymbol interference. Hence a partial remedial measure in such situations is adequate phase equalization.

Another remedial measure is the introduction of phase distortion at the receiver with complementary phase correction at the transmitter, as discussed in Section 4.14. This so called "smear-desmear" technique has its limitation, since the reduction of impulse-noise peaks is accompanied by increased peak signal power.

In binary systems, the error rate can be reduced with the aid of a wideband limiter ahead of the receiving filter, provided the impulses have a very short duration compared to the signal interval T. Under this condition, only impulses of amplitude much in excess of the signal pulses will give rise to errors, so that the foregoing kind of limiter may effect a significant reduction in error probability. However, when a number of voice channels are transmitted in frequency-division multiplex, as is usually the case in analog carrier systems, the combined signal at the receiver will be quite peaked, as discussed in Section 2.12. The limiter would in this case clip the combined signal and thereby cause excessive intermodulation distortion. This would preclude the use of a limiter or other nonlinear device ahead of the receiving filters.

For a given transmission capacity and signal power, there will be some difference in error probability among various digital modulation methods.

Impulse Noise

Ordinarily the amplitude probability distribution of impulse noise is such [30] that the spread in error rates among different modulation methods will be less than for additive Gaussian noise [1].

The principal problem with impulse noise is to determine its statistical properties, as noted earlier. When they are known, the error probabilities with various digital methods can readily be determined with engineering accuracy. A comprehensive mathematical formulation as well as some numerical results have been published for binary and quaternary PM with differential phase detection [31].

6

Attenuation and Phase Distortion in Analog Systems

6.0 General

In the analysis of transmission systems performance, it is convenient to first consider certain idealized channel transmittances, as in previous chapters, and to determine in turn the effect of certain departures from these idealizations. These departures are ordinarily referred to as channel distortion or characteristic distortion, and include two basic components: linear and nonlinear transmittance distortion. Linear distortion includes unwanted variations with frequency in the amplitude and phase characteristics of the channel that are independent of the amplitude of the transmitted wave. Nonlinear distortion relates to variations in amplitude and phase characteristics with the amplitude of the transmitted wave, and will be considered in a later chapter.

In dealing with transmission impairments from attenuation and phase distortion, it is convenient to consider at first the resultant modification in the received signals at the channel output, that is, the detector input. The resultant distortion in the demodulated baseband message wave depends on the modulation method, and is difficult to determine for nonlinear modulation methods such as FM or PM. With such nonlinear modulation methods, linear channel distortion will give rise to nonlinear distortion, or intermodulation distortion, in the demodulated baseband wave.

In the evaluation of transmission impairments from attenuation and phase deviations, it is necessary to consider the resultant distortion in the demodulated wave as well as the subjective effects of distortion, which depends both on the type of distortion and the nature of the message wave. For example, while the subjective effect of an echo is important in video reception, it can ordinarily be disregarded in audio reception, except for echoes of very long delays, as may be encountered in satellite systems. In video reception the subjective effect of an echo of given amplitude depends significantly on the echo delay. Erratic or random distortion will have a different subjective effect from an echo, for the same rms deviation in the received from the transmitted wave. These subjective effects can be established only by appropriate tests and are not considered here. A comprehensive bibliography on subjective effects in video and picture transmission is found in Reference 3 of Chapter 2.

In dealing with attenuation and phase deviations in a transmission medium, such as a coaxial line or a wave guide, it is customary and convenient to distinguish between coarse- and fine-structure deviations. The former can in general be compensated for by appropriate equalization. After such equalization, fine-structure deviations will remain, having an erratic variation with frequency over the channel band. Such fine-structure deviations are caused partly by a multiplicity of reflections or echoes in the transmission medium. However, even with a smooth medium, fine-structure deviations will remain when lumped networks are used for equalization.

Fine-structure irregularities in the transmission medium have two separate effects. One is a slight increase in attenuation, compared to that of a uniform medium, which varies uniformly with frequency and can be compensated for by appropriate equalization. The second and far more important effect as regards transmission impairments is a fine-structure deviation from the mean attenuation and phase characteristics, which varies in an erratic or virtually random manner with frequency. Both of these components depend on the attenuation and impedance characteristics of the medium as well as on the impedance irregularities. Because of the importance of the resultant attenuation and phase irregularities in broadband systems, they have been given much attention in the literature [1–5].

With the aid of measurements on a multitude of samples, statistics can be obtained on the impedance deviations, from which can be determined the rms values of the fine-structure deviations in the attenuation and phase characteristics, using relations in the foregoing references. Relations given here for various analog systems may in turn be employed to determine the resultant rms distortion in the demodulated baseband message wave.

6.1 Characterization of Transmittance Deviations

In the analysis of the effects of attenuation and phase distortion, it is convenient to consider the channel as composed of two transducers in tandem, as indicated in Fig. 6.1. The first of these would be an ideal

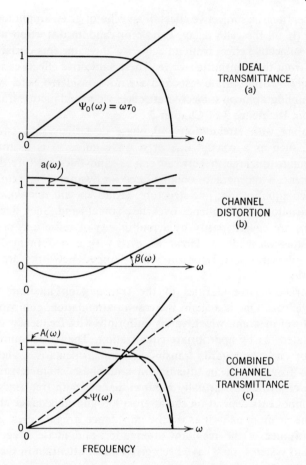

Figure 6.1 Combination of ideal transmittance, a, and channel distortion, b, into channel transmittance $c = a \cdot b$.

transducer with linear phase characteristic having the transmittance

$$T_0(i\omega) = A_0(i\omega)^{-i\omega\tau_0}. \tag{6.1}$$

The second transducer would represent the channel distortion and would have a transmittance as indicated in Fig. 6.1, given by

$$T_1(i\omega) = [1 + a(\omega)]e^{-i\beta(\omega)} \tag{6.2}$$

Here $a(\omega)$ represents the deviation from the mean amplitude characteristic 1 and $\beta(\omega)$ the deviation from the linear phase characteristic $\omega\tau_0$.

Characterization of Deviations

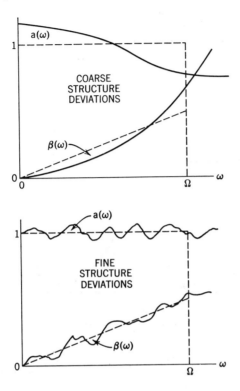

Figure 6.2 Coarse- and fine-structure deviations from constant amplitude characteristic and linear phase characteristic over signal band $\omega = 0$ to Ω.

In analog systems as considered here, limitations on tolerable transmission impairments impose the condition $a(\omega) \ll 1$, in which case the deviation in the attenuation characteristic becomes

$$\alpha(\omega) = \log_e [1 + a(\omega)] \simeq a(\omega). \tag{6.3}$$

Let a baseband channel have a bandwidth Ω, as indicated in Fig. 6.2. The deviation from the mean amplitude characteristic over the band Ω can then be represented by a cosine Fourier series as

$$a(\omega) = \sum_{n=1,2,3,\ldots} a_n \cos \frac{n\pi\omega}{\Omega}. \tag{6.4}$$

A corresponding relation applies for $\alpha(\omega)$ when $a(\omega) \ll 1$.

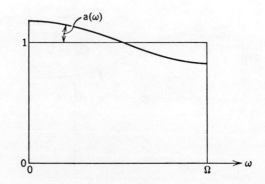

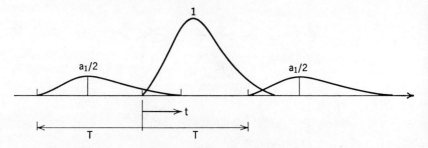

Figure 6.3 Signal echoes caused by cosine deviation $a(\omega) = a_1 \cos \omega T$ from constant amplitude characteristic. Signal spectrum is not shown but is confined to channel band 0 to Ω. $T = \pi/\Omega = 1/2B$.

The deviation from the linear phase characteristic can be represented by

$$\beta(\omega) = \sum_{n=1,2,3} \beta_n \sin \frac{n\pi\omega}{\Omega}. \tag{6.5}$$

In dealing with transmission impairments, from attenuation and phase distortion, it is convenient to distinguish between "coarse-structure" and "fine-structure deviations" as indicated in Fig. 6.2. If the term in (6.4) for $n = 1$ predominates, then there will be a coarse-structure deviation in the amplitude characteristic. This can be shown to give rise to two signal echoes of amplitudes $a_1/2$, as indicated in Fig. 6.3. If the term for $n = 1$ predominates in (6.5), two signal echoes of polarities $\mp \beta_1/2$ are produced, as indicated in Fig. 6.4. A combination of the above attenuation and phase distortion would in general result in two echoes of different amplitudes.

It is possible to use other than Fourier series representations of $\alpha(\omega)$ and $\beta(\omega)$, for example a series in ω^n or $(i\omega)^n$, $n = 1, 2, 3, \ldots$. However,

Signal Distortion from Deviations

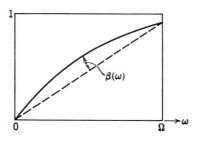

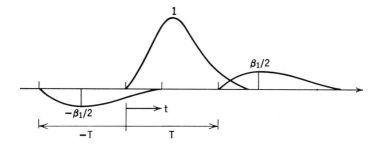

Figure 6.4 Signal echoes caused by sine deviation $\beta(\omega) = \beta_1 \sin \omega T$ from a linear phase characteristic. Signal spectrum is not shown but is confined within channel band 0 to Ω. $T = \pi/\Omega = 1/2B$.

Fourier series appear to have a distinct advantage from the standpoint of analysis and interpretation of analytical results.

6.2 Signal Distortion from Transmittance Deviations

The received wave in response to a transmitted wave $G_0(t)$ with spectrum $S_0(i\omega)$ is

$$G(t) = \frac{1}{2\pi} \int_{-\infty}^{\infty} S_0(i\omega) T(i\omega) e^{i\omega t} d\omega \tag{6.4}$$

$$= \frac{1}{2\pi} \int_{-\infty}^{\infty} S(i\omega)[1 + a(\omega)] e^{-i\beta(\omega)} e^{i\omega(t-\tau_0)} d\omega \tag{6.5}$$

$$= \frac{1}{2\pi} \int_{-\infty}^{\infty} S(i\omega) e^{\alpha(\omega) - i\beta(\omega)} e^{i\omega(t-\tau_0)} d\omega, \tag{6.6}$$

where $T(i\omega) = T_0(i\omega)T_1(i\omega)$ is the combined transmittance and $S(i\omega) = S_0(i\omega)T_0(i\omega)$ is the spectrum of a received wave $G_0(t)$ in the absence of channel distortion. Without loss of generality, it is permissible to take $\tau_0 = 0$, and this will be done in the expressions that follow.

The wave $G(t)$ at the channel output can be written

$$G(t) = G_0(t) + G_1(t), \tag{6.7}$$

in which $G_0(t)$ is the transmitted wave and

$$G_1(t) = \frac{1}{2\pi} \int_{-\infty}^{\infty} S(i\omega)[e^{\alpha(\omega) - i\beta(\omega)} - 1] e^{i\omega t}\, d\omega$$

$$= G_\alpha(t) + G_\beta(t) + G_{\alpha\beta}(t), \tag{6.8}$$

where

$$G_\alpha(t) = \frac{1}{2\pi} \int_{-\infty}^{\infty} S(i\omega)[e^{\alpha(\omega)} - 1] e^{i\omega t}\, d\omega, \tag{6.9}$$

$$G_\beta(t) = \frac{1}{2\pi} \int_{-\infty}^{\infty} S(i\omega)[e^{-i\beta(\omega)} - 1] e^{i\omega t}\, d\omega, \tag{6.10}$$

$$G_{\alpha\beta}(t) = \frac{1}{2\pi} \int_{-\infty}^{\infty} S(i\omega)[e^{\alpha(\omega)} - 1][e^{-i\beta(\omega)} - 1] e^{i\omega t}\, d\omega. \tag{6.11}$$

In the above relation

$$e^{\alpha(\omega)} - 1 = a(\omega). \tag{6.12}$$

It will be noted that in addition to the signal distortions $G_\alpha(t)$ and $G_\beta(t)$, owing to attenuation and phase deviations taken separately, a third component $G_{\alpha\beta}(t)$ is obtained, owing to interaction between attenuation and phase distortion. For values of α and β that would result in tolerable distortion G_α and G_β, the term $G_{\alpha\beta}$ can be disregarded, so that direct superposition is a legitimate approximation.

As discussed in Section 1.3, in dealing with bandpass channels it is ordinarily preferable to choose a reference frequency ω_0 within the passband, which usually would coincide with the carrier frequency. In this case the wave at the output of the bandpass channel, that is, the detector input, is of the general form

$$G(t) = X(t) \cos(\omega_0 t + \varphi_0) - Y(t) \sin(\omega_0 t + \varphi_0) \tag{6.13}$$

$$= \bar{Z}(t) \cos[\omega_0 t + \varphi_0 + \varphi(t)], \tag{6.14}$$

where $\bar{Z}(t)$ is the envelope, $\tan \varphi(t) = Y(t)/X(t)$ and

$$X(t) = \frac{1}{\pi} \int_{-\omega_0}^{\infty} S_0(u)\, A(u) \cos[ut - \psi_0(u) - \beta(u)]\, du \tag{6.15}$$

$$= \frac{1}{\pi} \int_{-\omega_0}^{\infty} S(u)[1 + a(u)] \cos[ut - \psi_0(u) - \beta(u)]\, du \tag{6.16}$$

$$Y(t) = \frac{1}{\pi} \int_{-\omega_0}^{\infty} S(u)[1 + a(u)] \sin[ut - \psi_0(u) - \beta(u)]\, du, \tag{6.17}$$

where $u = \omega - \omega_0$, $S_0(u)$ is the amplitude characteristic of the spectrum of the transmitted wave and $\psi_0(u)$ its phase. The amplitude of the spectrum in the absence of amplitude distortion is $S(u) = S_0(u)A_0(u)$.

It should be noted that in a baseband channel $A(\omega)$ must be an even function of ω and $\beta(\omega)$ an odd function. Corresponding restrictions do not apply to $A(u)$ and $\beta(u)$, each of which can have both an even and an odd component as functions of u, as indicated in Fig. 6.5.

With the aid of the preceding relations, it is possible to determine the effect of any transmittance deviations $\alpha(\omega)$ and $\beta(\omega)$, or $\alpha(u)$ and $\beta(u)$, when the shape of the transmitted wave or its spectrum is specified. Since these deviations can in general be represented by Fourier series, consideration will first be given to the effect of a single sinusoidal deviation in the amplitude and phase characteristics.

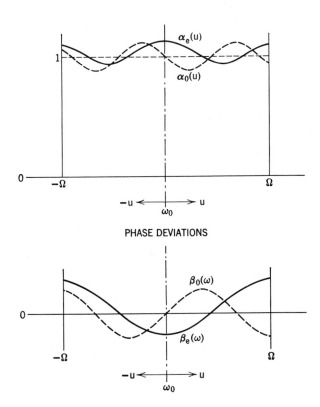

Figure 6.5 Even and odd amplitude and phase deviations in bandpass channels.

6.3 Sinusoidal Distortion and Echoes in Low-Pass Channels

Sinusoidal distortion is used here to designate a cosine variation in the amplitude versus the frequency characteristic, or a sine variation in the phase characteristic of a low-pass channel.

With a cosine variation in the amplitude characteristic $A(\omega)$, relation (6.9) becomes

$$e^{\alpha(\omega)} - 1 = a(\omega) = a \cos \omega T \qquad (6.18)$$

$$= \frac{a}{2}(e^{i\omega T} + e^{-i\omega T}).$$

The corresponding distortion obtained from (6.9) is

$$G_\alpha(t) = \frac{a}{2} G_0(t + T) + \frac{a}{2} G_0(t - T). \qquad (6.19)$$

As indicated in Fig. 6.3, the received wave thus contains two echoes of amplitudes $a/2$ relative to that of the principal wave $G_0(t)$ and displaced in time by $-T$ and T.

With cosine variation in the attenuation rather than the amplitude characteristic,

$$\alpha(\omega) = \alpha \cos \omega T.$$

The following expansions apply in terms of Bessel functions with imaginary arguments

$$e^{\alpha \cos \omega T} = I_0(\alpha) + 2I_1(\alpha) \cos \omega T + 2I_2(\alpha) \cos 2\omega T + \cdots, \qquad (6.21)$$

where

$$I_n(\alpha) = i^{-n} J_n(\alpha).$$

From (6.9), it follows with the aid of simple geometric relations that

$$G_\alpha(t) = [I_0(\alpha) - 1]G_0(t) + I_1(\alpha)[G_0(t + T) + G_0(t - T)]$$
$$+ I_2(\alpha)[G_0(t + 2T) + G_0(t - 2T)]$$
$$+ I_3(\alpha)[G_0(t + 3T) + G_0(t - 3T)]. \qquad (6.22)$$

For

$$\alpha \ll 1, \quad I_0(\alpha) \simeq 1 \quad \text{and} \quad I_1(\alpha) \simeq \frac{\alpha}{2},$$

so that

$$G_\alpha(t) \simeq \frac{\alpha}{2} G_0(t + T) + \frac{\alpha}{2} G_0(t - T). \qquad (6.23)$$

Let the modification in the phase characteristic be represented by

$$\beta(\omega) = \beta \sin \omega T. \tag{6.24}$$

The following Bessel function expansion then applies:

$$e^{-i\beta \sin \omega T} = J_0(\beta) - J_1(\beta)(e^{i\omega T} - e^{-i\omega T})$$
$$+ J_2(\beta)(e^{2i\omega T} + e^{-2i\omega T})$$
$$- J_3(\beta)(e^{3i\omega T} - e^{-3i\omega T}). \tag{6.25}$$

With (6.25) in (6.10) the resultant distortion becomes

$$G_\beta(t) = [J_0(\beta) - 1]G_0(t) - J_1(\beta)[G_0(t+T) - G_0(t-T)]$$
$$+ J_2(\beta)[G_0(t+2T) + G_0(t-2T)]$$
$$- J_3(\beta)[G_0(t+3T) - G_0(t-3T)]. \tag{6.26}$$

For $\beta \ll 1$

$$G_\beta(t) \simeq -\frac{\beta}{2} G_0(t+T) + \frac{\beta}{2} G_0(t-T). \tag{6.27}$$

With a modification of $\alpha \cos \omega T$ in the attenuation characteristic, the corresponding modification in the phase characteristic of a minimum phase shift network is in accordance with (1.33) $\alpha \sin \omega T$. The resultant transmittance thus becomes

$$T(i\omega) = T_0(i\omega) e^{\alpha(\cos \omega T - i \sin \omega T)}$$
$$= T_0(i\omega) \exp(\alpha e^{-i\omega T})$$
$$= T_0(i\omega)\left(1 + \alpha e^{-i\omega T} + \frac{\alpha^2}{2!} e^{-2i\omega T} + \frac{\alpha^3}{3!} e^{-3i\omega T} + \cdots\right). \tag{6.28}$$

From (6.28), it follows that in this case

$$G_1(t) = \alpha G_0(t-T) + \frac{\alpha^2}{2!} G_0(t-2T) + \frac{\alpha^3}{3!} G_0(t-3T) + \cdots. \tag{6.29}$$

For $\alpha \ll 1$, the echo becomes

$$G_1(t) \simeq \alpha G_0(t-T), \tag{6.30}$$

as indicated in Fig. 6.6. Relation (6.30) is also obtained by combining the echoes (6.19) and (6.27) with $\beta = \alpha$.

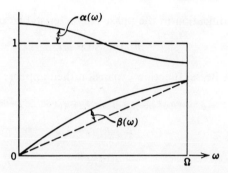

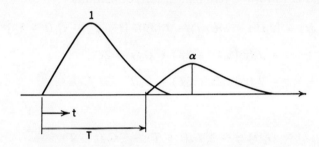

Figure 6.6 Signal echo caused by a combined small attenuation and phase deviation $\alpha(\omega) - i\beta(\omega) = \alpha(\cos \omega T - i \sin \omega T)$. Signal spectrum is not shown but is confined to channel band 0 to Ω.

6.4 Sinusoidal Distortion and Echoes in Bandpass Channels

In (6.16) and (6.17) let $\beta(\omega) = O$ and

$$a(u) = a_c \cos uT + a_s \sin uT. \tag{6.31}$$

For $X_1(t) = X(t) - X_0(t)$ and $Y_1(t) = Y(t) - Y_0(t)$, the following relations are obtained:

$$X_1(t) = \frac{a_c}{2}[X_0(t+T) + X_0(t-T)] + \frac{a_s}{2}[Y_0(t+T) - Y_0(t-T)], \tag{6.32}$$

$$Y_1(t) = \frac{a_c}{2}[Y_0(t+T) + Y_0(t-T)] - \frac{a_s}{2}[X_0(t+T) - X_0(t-T)]. \tag{6.33}$$

With $a(u) = O$ and

$$\beta(u) = \beta_c \cos uT + \beta_s \sin uT, \tag{6.34}$$

the first term in the expansion of the cosine and sine functions in (6.16) and (6.17) yield the following relations if β_c and $\beta_s \ll 1$

$$X_1(t) = \frac{\beta_c}{2}[Y_0(t+T) + Y_0(t-T)] - \frac{\beta_s}{2}[X_0(t+T) - X_0(t-T)], \tag{6.35}$$

$$Y_1(t) = -\frac{\beta_c}{2}[X_0(t+T) + X_0(t-T)] - \frac{\beta_s}{2}[Y_0(t+T) - Y_0(t-T)]. \tag{6.36}$$

A situation often encountered in radio channels is that of an echo of relative amplitude r transmitted over an alternate path with transmission delay T relative to the main path. In this case, the modified transmittance is

$$T(i\omega) = T_0(i\omega)[1 + re^{-i\omega T}]. \tag{6.37}$$

In this special case it also follows directly that the received wave (6.13) can be written

$$G(t) = G_0(t) + G_1(t), \tag{6.38}$$

where

$$\begin{aligned} G_1(t) &= rX(t-T)\cos[\omega_0(t-T) - \varphi_0] \\ &\quad - rY(t-T)\sin[\omega_0(t-T) - \varphi_0] \\ &= rX(t-T)\cos(\omega_0 t - \varphi_0)\cos\omega_0 T \\ &\quad + rX(t-T)\sin(\omega_0 t - \varphi_0)\sin\omega_0 T \\ &\quad - rY_0(t-T)\sin(\omega_0 t - \varphi_0)\cos\omega_0 T \\ &\quad + rY_0(t-T)\cos(\omega_0 t - \varphi_0)\sin\omega_0 T. \end{aligned} \tag{6.39}$$

Hence

$$X_1(t) = rX(t-T)\cos\omega_0 T + rY(t-T)\sin\omega_0 T \tag{6.40}$$
$$Y_1(t) = rY(t-T)\sin\omega_0 T + rX(t-T)\cos\omega_0 T. \tag{6.41}$$

The same result is obtained in a more cumbersome manner from the more general equations (6.31) through (6.36) by observing that the factor $1 + re^{-i\omega T}$ in (6.37) can be written

$$1 + re^{-i\omega T} = 1 + r\cos\omega T - ir\sin\omega T.$$

With $\omega = \omega_0 + u$, the following relations are obtained for the modifications in the amplitude and phase characteristics

$$\begin{aligned} a(u) &= r\cos(\omega_0 + u)T \\ &= r\cos\omega_0 T \cos uT - r\sin\omega_0 T \sin uT, \\ \beta(u) &= r\sin(\omega_0 + u)T \\ &= r\sin\omega_0 T \cos uT + r\cos\omega_0 T \sin uT. \end{aligned}$$

Thus in this particular case the factors in (6.31) and (6.34) are

$$a_c = r \cos \omega_0 T; \qquad a_s = -r \sin \omega_0 T,$$
$$\beta_s = r \sin \omega_0 T; \qquad \beta_s = r \cos \omega_0 T.$$
(6.42)

6.5 Signal Distortion with Fourier Series Transmittance Deviations

Let the amplitude or attenuation deviations be represented by a Fourier series in accordance with (6.4). From the results in the preceding sections it follows that the resultant distortion in the received wave is given by the echo series:

$$G_\alpha(t) = \sum_{n=1,2,3,\ldots} \frac{\alpha_n}{2} [G_0(t + nT) + G_0(t - nT)] \qquad (6.43)$$

With the deviations from a linear phase characteristic represented by (6.5), the resultant signal distortion becomes

$$G_\beta(t) = \sum_{n=1,2,3,\ldots} \frac{\beta_n}{2} [-G_0(t + nT) + G_0(t - nT)] \qquad (6.44)$$

In these relations

$$T = \frac{\pi}{\Omega} = \frac{1}{2B}, \qquad (6.45)$$

where $\Omega = 2\pi B$ is the bandwidth of the channel or the spectrum $S(i\omega)$ of the received wave in the absence of distortion.

With both amplitude and phase deviations, the combined signal distortion is, in a first approximation,

$$G_1(t) = G_\alpha(t) + G_\beta(t). \qquad (6.46)$$

In the case of bandpass channels $\alpha(u)$ and $\beta(u)$ would in accordance with (6.31) and (6.34) in general contain both cosine and sine terms. Expressions (6.4) and (6.5) would then be replaced by

$$\alpha(u) = \sum_{n=1,2,3,\ldots} \left(\alpha_{c,n} \cos \frac{n\pi\omega}{\Omega} + \alpha_{s,n} \sin \frac{n\pi\omega}{\Omega} \right), \qquad (6.47)$$

$$\beta(u) = \sum_{n=1,2,3,\ldots} \left(\beta_{c,n} \cos \frac{n\pi\omega}{\Omega} + \beta_{s,n} \sin \frac{n\pi\omega}{\Omega} \right). \qquad (6.48)$$

The resultant distortion can in this case be represented by sums of echoes of $X_0(t)$ and $Y_0(t)$, in accordance with rules that will follow from expressions given in Section 6.4.

6.6 Signal Distortion with Power Series Transmittance Deviations

Let the transmittance deviations be represented by

$$T(i\omega) = T_0(i\omega) \sum_{n=0,1,2,\ldots} c_k(i\omega)^k$$

$$= T_0(i\omega)[c_0 - c_2\omega^2 + c_2\omega^4 - \cdots] + i[c_1\omega - c_3\omega^3 + c_5\omega^5 - \cdots]. \quad (6.49)$$

The signal at the channel output is then

$$G(t) = \frac{1}{2\pi} \int_{-\infty}^{\infty} S(i\omega) \sum_{n=0,1,2,\ldots} c_k(i\omega)^k e^{i\omega t}\, d\omega$$

$$= \sum_{n=0,1,2,\ldots} c_k G_0^{(k)}(t)$$

$$= c_0 G_0(t) + c_1 G_0'(t) + c_2 G_0''(t) + \cdots. \quad (6.50)$$

The signal can thus be represented by the sum of an undistorted signal $c_0 G_0(t)$ and its various time derivatives $G_0^{(k)}(t)$.

Relation (6.49) represents a combined modification in the amplitude and phase characteristics of the channel. With $c_k = 0$ for $k = 1, 3, 5, \ldots$, only the amplitude characteristic is modified and in this case

$$G(t) = c_0 G_0(t) + c_2 G_0''(t) + c_4 G_0'''(t) - \cdots. \quad (6.51)$$

To conform with the result obtained by Fourier series representation (for $\beta_n = 0$), it is necessary that $c_0 = 1$. That is to say, c_0 must be the mean value of the amplitude characteristic over the band Ω of the signal spectrum. The remaining terms in (6.51) will then correspond to the series (6.43) for the signal distortion.

In the case of bandpass channels, the amplitude characteristic can in general be represented by

$$A(u) = A_0(u) \sum_{k=0,1,2,\ldots} c_k u^k \quad (6.52)$$

The solution for $X(t)$ and $Y(t)$ obtained from (6.16) and (6.17) can in this case be written in the form

$$X(t) = c_0 X_0(t) + c_1 Y_0'(t) - c_2 X_0''(t) - c_3 Y_0'''(t) + \cdots, \quad (6.53)$$

$$Y(t) = c_0 Y_0(t) - c_1 X_0'(t) + c_2 Y_0''(t) + c_3 X_0'''(t) - \cdots. \quad (6.54)$$

The departure from a linear phase characteristic can be represented by

$$\beta(u) = \sum_{k=1,2,\ldots} b_k u^k. \quad (6.55)$$

For adequately small values of b_k the first terms in the solution of the integrals (6.16) and (6.17) are

$$X(t) = X_0(t) - b_1 X_0'(t) - b_2 Y_0''(t) + b_3 X_0'''(t), \quad (6.56)$$

$$Y(t) = Y_0(t) - b_1 Y_0'(t) - b_2 X_0''(t) + b_3 Y_0'''(t). \quad (6.57)$$

It should be noted that for any value of b_1 the first two terms in an exact solution are replaced by $X_0(t - b_1)$ and $Y_0(t - b_1)$. For small values of b_1 the approximations represented by the first two terms apply.

With Fourier series representations, the coefficients α_n and β_n must be determined, and with power series representation the coefficients c_k and b_k. The preferable method will depend on the form of the deviations $\alpha(\omega)$ and $\beta(\omega)$ and also on the difficulties in evaluating the resultant signal distortion.

6.7 Message Distortion in Baseband Systems

In the previous sections, distortion at the receiver input was considered for transmitted waves $G_0(t)$ with specified amplitude spectra $S_0(i\omega)$. In analog systems where a variety of wave shapes are encountered it is customary to specify the power spectra of the message waves, or the related signals, as discussed in Chapter 2. When the power spectrum $W_0(\omega)$ of the transmitted signals is known, it is possible to determine the rms distortion in the received wave for any specified $\alpha(\omega)$ and $\beta(\omega)$.

For adequately small values of $\alpha(\omega)$ and $\beta(\omega)$, the following relations are obtained from (6.9) and (6.10) for the distortion at the input to the receiving filter:

$$G_\alpha(t) = \frac{1}{2\pi} \int_{-\infty}^{\infty} \alpha(\omega) S(i\omega) e^{i\omega t} \, d\omega, \quad (6.58)$$

$$G_\beta(t) = \frac{1}{2\pi} \int_{-\infty}^{\infty} -i\beta(\omega) S(i\omega) e^{i\omega t} \, d\omega. \quad (6.59)$$

By the Fourier integral energy relation theorem (see Appendix 1) it follows that

$$\int_{-\infty}^{\infty} G_\alpha^2(t) \, dt = \frac{1}{2\pi} \int_{-\infty}^{\infty} \alpha^2(\omega) \, d\omega, \quad (6.60)$$

with a corresponding relation for $G_\beta(t)$.

The foregoing relations apply for specific signals $G(t)$ with power spectrum $|S(i\omega)|^2$. The variance of the distortion over the duration T_1 of a particular signal $G_\alpha(t)$ is then

$$\frac{1}{T_1} \int_{-\infty}^{\infty} G_\alpha^2(t) \, dt = \frac{1}{2\pi T_1} \int_{-\infty}^{\infty} a^2(\omega) |S(i\omega)|^2 \, d\omega.$$

When all possible signals $G(t)$ are considered, the variances over an infinite time interval are given by the relations

$$\sigma_\alpha^2 = \int_0^\infty \alpha^2(\omega)\, W(\omega)\, d\omega, \tag{6.61}$$

$$\sigma_\beta^2 = \int_0^\infty \beta^2(\omega)\, W(\omega)\, d\omega, \tag{6.62}$$

where $W(\omega)$ is the one-sided power spectrum of the signals (see Appendix 2).

The ratio of the combined variance to the average power P at the input to the receiving filter is

$$\frac{\sigma^2}{P} = \frac{\int_0^\infty [\alpha^2(\omega) + \beta^2(\omega)] W(\omega)\, d\omega}{\int_0^\infty W(\omega)\, d\omega}. \tag{6.63}$$

In the latter relation $W(\omega) = W_0(\omega) A_0^2(\omega)$ is the power spectrum at the input to the receiving filter and $W_0(\omega)$ that of the transmitted signals. The ratio of distortion power D_m in the message wave at the output of the receiving filter to the average message power P_m is correspondingly

$$\frac{D_m}{P_m} = \frac{\int_0^\infty [\alpha^2(\omega) + \beta^2(\omega)] W_m(\omega)\, d\omega}{\int_0^\infty W_m(\omega)\, d\omega}, \tag{6.64}$$

where $W_m(\omega)$ is proportional to the power spectrum $s(\omega)$ of the message wave, so that $s(\omega)$ can be used in place of $W_m(\omega)$. The latter relation assumes that the transmitting and receiving filters are complementary, so that the combined transmittance is unity. Thus the distortion in the average wave is independent of the division of equalization between transmitter and receiver.

In the foregoing relations the upper integration limit ∞ can be replaced by the maximum frequency Ω_0 of the message spectrum. In the particular case of a flat message power spectrum $W_m(\omega) = P/\Omega_0$ and (6.65) yields

$$\frac{D_m}{P_m} = \frac{1}{\Omega_0} \int_0^{\Omega_0} [\alpha^2(\omega) + \beta^2(\omega)]\, d\omega$$

$$= \underline{\alpha}^2 + \underline{\beta}^2, \tag{6.65}$$

where $\underline{\alpha}$ and $\underline{\beta}$ are the rms deviations from the ideal attenuation and phase characteristics.

Relation (6.65) applies, regardless of the nature of $\alpha(\omega)$ and $\beta(\omega)$, provided that both are adequately small. In the case of a single coarse-structure cosine and sine variation $\alpha(\omega) = \alpha_1 \cos \omega \tau$ and $\beta(\omega) = \beta_1 \sin \omega \tau$, the echo preceding the main signal has the amplitude $\left(\dfrac{\alpha_1}{2} - \dfrac{\beta_1}{2}\right)$, and the echo following the mean signal has the amplitude $\left(\frac{\alpha_1}{2} + \frac{\beta_1}{2}\right)$. The combined echo power taken in relation to the original power is in this case

$$\left(\frac{\alpha_1}{2} - \frac{\beta_1}{2}\right)^2 + \left(\frac{\alpha_1}{2} + \frac{\beta_1}{2}\right)^2 = \frac{1}{2}\left(\alpha_1^2 + \beta_1^2\right) = \alpha_1^2 + \beta_1^2.$$

The same result is, in this case, obtained from (6.65). Hence the relation applies regardless of the relative polarities of the signal echoes.

In the case of fine-structure deviations, a large number of echoes will be encountered. The resultant distortion in the message will then have properties resembling those of random noise. This can be shown by assuming that the message is transmitted by sampling at intervals $T = 1/2B$. The instantaneous sample of the received message will then differ in amplitude from the amplitude in the absence of distortion by

$$\Delta(t) = \sum_{n=1}^{\infty} \left[A_{-n}(t)\left(\frac{\alpha_n}{2} - \frac{\beta_n}{2}\right) + A_n(t)\left(\frac{\alpha_{-n}}{2} + \frac{\beta_{-n}}{2}\right) \right]. \quad (6.66)$$

Here $A_{-n}(t)$ and $A_n(t)$ are instantaneous amplitudes of the signal samples at sampling instants $-nT$ and nT from the instant under consideration, $\alpha_n/2$, $\alpha_{-n}/2$, $\beta_n/2$, and $\beta_{-n}/2$ are the amplitudes of echoes at times $t = \pm nT$. With a sufficient number of terms, the resultant probability distribution of $\Delta(t)$ will approach that of random noise regardless of the probability distribution of $A_{-n}(t)$. Hence the message distortion resulting from fine structure deviations can be assumed to be similar to random noise.

6.8 Message Distortion in AM Carrier Systems

In the case of bandpass channels, the variances of the distortions in $X(t)$ and $Y(t)$ are given by

$$\sigma_X^2 = 2\int_{-\omega_0}^{\Omega_0} \{[\alpha_e^2(u) + \beta_o^2(u)]W_X(u) + [\alpha_o^2(u) + \beta_e^2(u)]W_Y(u)\}\, du, \quad (6.67)$$

$$\sigma_Y^2 = 2\int_{-\omega_0}^{\Omega_0} \{[\alpha_e^2(u) + \beta_o^2(u)]W_Y(u) + [\alpha_o^2(u) + \beta_e^2(u)]W_X(u)\}\, du, \quad (6.68)$$

where $W_X(u)$ and $W_Y(u)$ are the double-sided envelope power spectra of the random variables $X(t)$ and $Y(t)$. The subscripts e and o apply for even and odd symmetry components, as indicated in Fig. 6.5. The factor 2 comes about because of direct amplitude addition of the spectral components at $-u$ and u in the demodulator output, rather than power addition, as with two incoherent sideband power spectra.

With double-sideband transmission $W_Y = O$ and the lower limit can be replaced by $-\Omega_0$. Since, further, the integrands have even symmetry,

$$\sigma_X^2 = 4 \int_0^{\Omega_0} [\alpha_e^2(u) + \beta_0^2(u)] W_X(u) \, du, \tag{6.69}$$

$$\sigma_Y^2 = 4 \int_0^{\Omega_0} [\alpha_0^2(u) + \beta_e^2(u)] W_X(u) \, du. \tag{6.70}$$

With envelope detection, the carrier must be transmitted, and the effect of the component σ_Y^2 can then be neglected in comparison with the effect of variation in the in-phase component σ_X^2. Hence the ratio of message distortion power D_m to the message power becomes

$$\frac{D_m}{P_m} = \int_0^{\Omega_0} [\alpha_e^2(u) + \beta_0^2(u)] W_X(u) \, du \bigg/ \left(\int_0^{\Omega_0} W_X(u) \, du \right). \tag{6.71}$$

In these relations $W_X(u)$ is proportional to the message power spectrum $s(u)$, so that the latter can be used in place of $W_X(u)$.

With synchronous detection the quadrature component σ_Y^2 is eliminated in the demodulator output and (6.71) applies also in this case. It will be recognized that the above result is the same as for baseband transmission, since there $\alpha(\omega)$ is an even and $\beta(\omega)$ is an odd function of ω.

With quadrature double-sideband transmission and equal power in each of the two carriers at quadrature, $W_Y = W_X$. In this case (3.67) and (3.68) yield

$$\sigma_X^2 = \sigma_Y^2 = 2 \int_{-\Omega_0}^{\Omega_0} [\alpha_e^2(u) + \alpha_0^2(u) + \beta_e^2(u) + \beta_0^2(u)] W_X(u) \, du. \tag{6.72}$$

In this case the ratio of average message distortion to average message power is the same for each of the two quadrature channels. With synchronous detection this ratio is

$$\frac{D_m}{P_m} = \int_0^{\Omega_0} [\alpha_e^2(u) + \alpha_0^2(u) + \beta_e^2(u) + \beta_0^2(u)] W_X(u) \, du \bigg/ \left(\int_0^{\Omega_0} W_X(u) \, du \right), \tag{6.73}$$

where $W_X(u)$ can be replaced by $s(u)$. When the quantities $\alpha(u)$ and $\beta(u)$ are resolved into harmonic sinusoidal components, it follows that $\alpha_e^2(u) + \alpha_0^2(u) = \alpha^2(u)$ and $\beta_e^2(u) + \beta_0^2(u) = \beta^2(u)$. Hence

$$\frac{D_m}{P_m} = \int_0^{\Omega_0} [\alpha^2(u) + \beta^2(u)] W_X(u) \, du \bigg/ \left(\int_0^{\Omega_0} W_X(u) \, du \right), \tag{6.74}$$

where $W_X(u)$ can be replaced by $s(u)$.

In the case of single-sideband transmission, it follows from the form of (2.49) and (2.50) that with random modulation $W_Y(u) = W_X(u)$. (If the lower limits of the integrals had been $-\infty$, then $W_X = W$ and $W_Y = O$.) The lower limit in (6.67) and (6.68) is in this case O. With synchronous detection, relations (6.72) through (6.94) are thus obtained. It will be noted that (6.74) is the same as (6.64) for baseband transmission, except that ω is replaced by u.

Comparison of (6.74) and (6.71) shows that the ratios D_m/P_m are greater than with double-sideband transmission unless $\alpha_0(u)$ and $\beta_e(u) = O$. This comes about because the quadrature power spectrum in single-sideband transmission, and in quadrature double-sideband transmission, gives rise to a component of distortion σ_X^2 but does not contribute to the message power derived from $X(t)$.

6.9 Message Distortion in FM and PM Systems

With any nonlinear modulation method, such as FM, there is a complicated nonlinear relation between distortion of the signal wave at the detector input and the resultant message distortion in the demodulated wave. Because of this nonlinear relation, there is a component in the message power spectrum at any particular frequency which depends on the spectral components at other frequencies. This is referred to as intermodulation distortion, which will give rise to so-called interchannel crosstalk between voice channels in different frequency bands of the combined message wave. Other intermodulation effects discussed in Section 8.0 are also encountered in FM systems.

Exact formal solution of message distortion in FM systems is difficult except for adequately small values of $\alpha(u)$ and $\beta(u)$, so that the superposition theorem can be invoked with insignificant error. Even in this case solutions appear feasible only for modulation by a sine wave as considered in many publications, or for message waves having the properties of Gaussian noise, which is more representative of message waves in communication systems. For the latter case, a general formal solution has been derived by Rice for specified attenuation and phase deviations [6]. Specific relations together with numerical results have been published by Bennett, Curtis, and Rice for distortion owing to a small echo [7], and by Medhurst and Small for distortion caused by small sinusoidal transmittance deviations [8]. A somewhat different analytical procedure will be followed here in order to obtain specific relations for a number of important cases. A comprehensive analysis and bibliography for other cases than considered here has been published by Bedrosian and Rice [9].

Linear Amplitude Deviations in FM

To determine distortion at the output of the demodulator, it is necessary to return to (6.13). With a modification $\Delta_x(t)$ in $X(t)$ and $\Delta_y(t)$ in $Y(t)$ the resultant phase modulation is modified by $\epsilon(t)$ and the following relation applies:

$$\varphi(t) + \epsilon(t) = \tan^{-1} \frac{Y(t) + \Delta_y(t)}{X(t) + \Delta_x(t)}. \tag{6.75}$$

Without loss of generality it can for convenience be assumed that the amplitude of the carrier is $\bar{Z}(t) = E = 1$. In this case $X(t) = \cos \varphi(t)$ and $Y(t) = \sin \varphi(t)$, and solution of (6.75) yields the relation

$$\epsilon(t) = \tan^{-1} \frac{\Delta_y(t) \cos \varphi(t) - \Delta_x(t) \sin \varphi(t)}{1 + \Delta_x(t) \cos \varphi(t) + \Delta_y(t) \sin \varphi(t)}. \tag{6.76}$$

For adequately small values of Δ_x and Δ_y

$$\epsilon(t) \simeq \Delta_y(t) \cos \varphi(t) - \Delta_x(t) \sin \varphi(t). \tag{6.77}$$

The functions $\Delta_y(t)$ and $\Delta_x(t)$ will depend on the deviation in the amplitude and phase characteristics of the channel. A number of special cases of technical importance will now be considered.

6.10 Linear Amplitude Deviations in FM

A coarse-structure departure from a constant amplitude characteristic can often be represented by a small linear deviation. In this case (6.53) and (6.54) become, with $c_0 = 1$:

$$X(t) = X_0(t) + c_1 Y_0'(t), \tag{6.78}$$

$$Y(t) = Y_0(t) - c_1 X_0'(t). \tag{6.79}$$

Since $X_0(t) = \cos \varphi(t)$ and $Y_0(t) = \sin \varphi(t)$, the following relations apply:

$$\Delta_x(t) = c_1 Y_0'(t) = c_1 \varphi'(t) \cos \varphi(t),$$

$$\Delta_y(t) = -c_1 X_0'(t) = c_1 \varphi'(t) \sin \varphi(t).$$

With $\varphi = \varphi(t)$ and $\varphi' = d\varphi(t)/dt$, relation (6.76) yields

$$\epsilon(t) = \tan^{-1} c_1 \frac{\varphi' \sin \varphi \cos \varphi - \varphi' \sin \varphi \cos \varphi}{1 + c_1 \varphi'(\cos^2 \varphi + \sin^2 \varphi)} = 0 \tag{6.80}$$

The resultant amplitude modulation is $c_1 \varphi'(t)$, as represented by the second term in the denominator. Thus a linear deviation from a constant amplitude characteristic produces no distortion in the demodulated wave. In a first approximation, this also applies for a small linear deviation in the attenuation characteristic.

6.11 Quadratic Phase Deviations in FM

A linear phase deviation will give rise to a certain transmission delay but no distortion, and can thus be disregarded. A coarse-structure departure from a linear phase characteristic can often be approximated by a component of quadratic phase distortion, or linear delay distortion. (See Section 7.11.) For adequately small quadratic phase distortion (6.56) and (6.57) apply, so that

$$\Delta_x(t) = b_2 Y_0''(t) = -b_2[\varphi'' \cos \varphi - (\varphi')^2 \sin \varphi],$$

$$\Delta_y(t) = b_2 X_0''(t) = -b_2[\varphi'' \sin \varphi + (\varphi')^2 \cos \varphi].$$

Relation (6.76) in this case yields

$$\epsilon(t) = \tan^{-1} - \frac{b_2[\varphi'(t)]^2}{1 - b_2\varphi''(t)} \tag{6.81}$$

$$\simeq -b_2[\varphi'(t)]^2. \tag{6.82}$$

It is to be noted that these relations apply only for adequately small quadratic phase distortion, such that in (6.16) and (6.17) $\cos \beta(u) \simeq 1$, $\sin \beta(u) \simeq b_2 u^2$. The resultant amplitude modulation of the received carrier is $-b_2\varphi''(t)$.

The power spectrum of $\epsilon(t)$ will now be considered for the case in which $\varphi(t)$ and thus $\varphi'(t)$ have the properties of Gaussian random variables. In accordance with (6.82) the autocorrelation function of $\epsilon(t)$ is the same as for a square-law nonlinearity. From relations in Appendix 3, it follows that this autocorrelation function is

$$R_\epsilon(\tau) = b_2^2[R_{\varphi'}(O) + 2R_{\varphi'}(\tau)], \tag{6.83}$$

where $R_{\varphi'}(\tau)$ is the autocorrelation function of $\varphi'(\tau)$.

In accordance with relation (19) of Appendix 3, the one-sided power spectrum of the variation in $\epsilon(t)$ from its mean value is given by

$$\tilde{W}_\epsilon = b_2^2 \int_{-\omega_0}^{\infty} W_{\varphi'}(u) W_{\varphi'}(\omega - u) \, du. \tag{6.84}$$

The squared rms frequency deviation is in accordance with (4.51) of Chapter 4.

$$\underline{\Delta}^2 = \Delta_1^2 \int_0^{\Omega_0} s(\omega) t(\omega) \, d\omega. \tag{6.85}$$

Hence the power spectra of the frequency and the phase modulations are

$$W_{\varphi'}(\omega) = \Delta_1^2 s(\omega) t(\omega), \tag{6.86}$$

$$W_{\varphi}(\omega) = \Delta_1^2 s(\omega) t(\omega)/\omega^2. \tag{6.87}$$

From relations (6.84) through (6.87) it follows that the ratio of intermodulation distortion power to message power in a narrow band at ω is given by

$$\eta(\omega) = \frac{\tilde{W}_\epsilon(\omega)}{W_\varphi(\omega)}$$

$$= \beta_2^2 \underline{D}^2 a^2 \frac{\int_{-\omega_0}^{\infty} s(u)t(u)s(\omega - u)t(\omega - u) \, du}{s(\omega)t(\omega) \int_{0}^{\Omega_0} s(\omega)t(\omega) \, d\omega}, \quad (6.88)$$

where, as before, the lower limit $-\omega_0$ can, for practical purposes, be taken as $-\infty$ and

$$\underline{D} = \Delta/\Omega_0 = \text{rms deviation ratio},$$
$$a = \omega/\Omega_0, \quad \text{and} \quad \beta_2 = b_2 \underline{\Delta}^2.$$

In the particular case of a flat message power spectrum $s(\omega) = s$ of bandwidth Ω_0, the following relation is obtained from (6.88)

$$\eta(\omega) = \frac{\beta_2^2 \underline{D}^2 a^2}{t(a) \int_0^1 t(x) \, dx} \int_{a-1}^{1} t(x)t(a-x) \, dx. \quad (6.89)$$

When the power-transfer characteristic of the transmitting filter is of the form

$$t(x) = 1 + c\left(\frac{u}{\Omega_0}\right)^2, \quad (6.90)$$

the solution of (6.89) yields, after some manipulation, the relation [10]

$$\eta(\omega) = \beta_2^2 \underline{D}^2 G(c,a), \quad (6.91)$$

where

$$G(c,a) = \frac{3a^2}{(1 + ca^2)(3 + c)} F(c,a), \quad (6.92)$$

$$F(c,a) = 2 - a + \frac{2c + c^2 a^2}{3}[1 + (1-a)^3]$$

$$- \frac{c^2 a}{2}[1 - (1-a)^4] + \frac{c^2}{5}[1 + (1-a)^5]. \quad (6.93)$$

The function $G(c,a)$ is shown in Fig. 6.7 for pure FM and PM and for $c = 16$. The latter case approximates the type of frequency pre-emphasis ordinarily used in broadband systems. For pure FM, $c = 0$, the above relation conforms with one given by Rice [6].

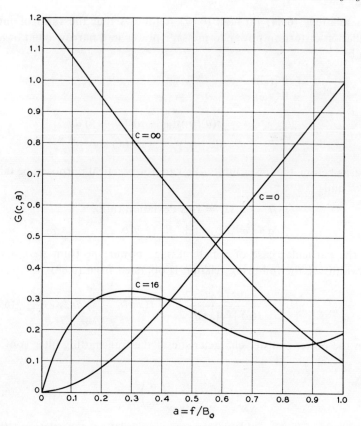

Figure 6.7 Function $G(c, a)$ for pure FM, $c = 0$, pure FM, $c = \infty$, and for $c = 16$.

6.12 Single-Echo Distortion in FM

Another form of distortion in FM and PM systems is that resulting from a single small echo, as caused by transmission over paths of different length owing to a reflection in the transmission medium. A solution for the resultant nonlinear distortion has been published by Bennett, Curtis, and Rice [7], for message waves having the properties of Gaussian random variables, such as thermal noise.

With a small echo of delay T, the phase error can be obtained from (6.77) with the aid of (6.40) and (6.41) by noting that $\Delta_x(t) = X_1(t)$ and $\Delta_y(t) = Y_1(t)$. Furthermore, with a phase modulation $\varphi(t)$ it follows that $X(t - T) = \cos \varphi(t - T)$ and $Y(t - T) = \sin \varphi(t - T)$. With these relations in (6.77) the latter yields

$$\epsilon(t) = r \cos [v(t) - \omega_0 T]$$
$$= r \cos \omega_0 T \cos v(t) + r \sin \omega_0 T \sin v(t), \qquad (6.94)$$

where

$$v(t) = \varphi(t - T) - \varphi(t). \qquad (6.95)$$

The autocorrelation function of $v(t)$ becomes

$$R_v(\tau) = \text{ave } [v(t)v(t + \tau)]$$
$$= \text{ave } [\varphi(t - T) - \varphi(t)][\varphi(t - T + \tau) - \varphi(t + \tau)]$$
$$= 2R_\varphi(\tau) - R_\varphi(\tau + T) - R_\varphi(\tau - T)]. \qquad (6.96)$$

Let $R_{\cos v}(\tau)$ and $R_{\sin v}(\tau)$ be the autocorrelation functions of $\cos v(t)$ and $\sin v(t)$. The autocorrelation function of $\epsilon(t)$ is then

$$R_\epsilon(\tau) = r^2[\cos^2 \omega_0 T R_{\cos v}(\tau) + \sin^2 \omega_0 T R_{\sin v}(\tau)]$$
$$= \frac{r^2}{2}[R_{\cos v}(\tau) + R_{\sin v}(\tau)] + \frac{r^2}{2} \cos 2\omega_0 T [R_{\cos v}(\tau) - R_{\sin v}(\tau)].$$
$$(6.97)$$

For a particular value of T, the autocorrelation function will thus depend on ω_0 and will vary rapidly with ω_0, since ordinarily $\omega_0 T \gg 2\pi$. The mean value of the last term in (6.97) is zero, so that the mean autocorrelation function is obtained by disregarding this term.

In Appendix 3 the following relations are derived:

$$R_{\cos v}(\tau) = e^{-R_v(0)} \cosh R_v(\tau), \qquad (6.98)$$

$$R_{\sin v}(\tau) = e^{-R_v(0)} \sinh R_v(\tau). \qquad (6.99)$$

Hence

$$R_\epsilon(\tau) = \frac{r^2}{2} e^{-R_v(0)} e^{R_v(\tau)}, \qquad (6.100)$$

where $R_v(\tau)$ is given by (6.96) in terms of $R_\varphi(\tau)$.

The corresponding one-sided power spectrum is

$$W_\epsilon(\omega) = \frac{2}{\pi} \int_0^\infty R_\epsilon(\tau) \cos \omega\tau \, d\tau \qquad (6.101)$$

$$= \frac{r^2}{\pi} e^{-R_v(0)} \int_0^\infty e^{R_v(\tau)} \cos \omega\tau \, d\tau. \qquad (6.102)$$

As discussed in Appendix 3 in dealing with nonlinear transformation, it is convenient to extract a discrete special component at $\omega = O$, which does not give rise to distortion. The remainder of the power spectrum is then

$$\tilde{W}_\epsilon(\omega) = \frac{r^2}{\pi} e^{-R_v(O)} \int_0^\infty (e^{R_v(\tau)} - 1) \cos \omega\tau \, d\tau. \qquad (6.103)$$

This power spectrum contains a component of the same shape as the power spectrum $W_v(\omega)$ of $v(t)$ defined by (6.95). When this component is brought out explicitly in (6.103), the latter relation can alternately be written

$$\tilde{W}_\epsilon(\omega) = W_\epsilon^{(1)}(\omega) + W_\epsilon^{(2)}(\omega), \qquad (6.104)$$

where the one-sided power spectra are given by

$$W_\epsilon^{(1)}(\omega) = \frac{r^2}{\pi} e^{-R_v(O)} \int_O^\infty R_v(\tau) \cos \omega\tau \, d\tau$$

$$= \frac{r^2}{2} e^{-R_v(O)} W_v(\omega), \qquad (6.105)$$

$$W_\epsilon^{(2)}(\omega) = r^2 F(\omega), \qquad (6.106)$$

$$F(\omega) = \frac{e^{-R_v(O)}}{\pi} \int_O^\infty [e^{R_v(\tau)} - 1 - R_v(\tau)] \cos \omega\tau \, d\tau. \qquad (6.107)$$

The ratio of intermodulation distortion power to average message power is a narrow band at ω and becomes

$$\eta(\omega) = r^2 \frac{F(\omega)}{W_\varphi(\omega)}. \qquad (6.108)$$

To determine this ratio it is necessary to first determine $R_\varphi(\tau)$ and in turn $R_v(\tau)$ from (6.96). The function $F(\omega)$ must next be determined from (6.107), which in general will entail numerical integration. A general integral for $R_\varphi(\tau)$ is given below, together with explicit expressions for pure FM and PM and a flat message power spectrum.

From (6.86) and (6.87) it follows that

$$W_\varphi(\omega) = \underline{\Delta}^2 \frac{s(\omega)t(\omega)/\omega^2}{\int_0^{\Omega_0} s(\omega)t(\omega)}. \qquad (6.109)$$

The corresponding autocorrelation function is

$$R_\varphi(\tau) = \underline{D}^2 \left[\int_0^{\Omega_0} \left(\frac{\Omega_0}{\omega}\right)^2 s(\omega)t(\omega) \cos \omega\tau \, d\omega \right] \Big/ \left[\int_0^{\Omega_0} s(\omega)t(\omega) \, d\omega \right], \qquad (6.110)$$

where

$$D = \Delta/\Omega_0 = \text{Rms deviation ratio.}$$

With pure FM, $t(\omega) = 1$ and (6.110) yields for a flat message power spectrum $s(\omega) = s$ of bandwidth Ω_0

$$R_\varphi(\tau) = D^2 \int_0^{\Omega_0} \frac{\Omega_0}{\omega^2} \cos \omega\tau \, d\omega \qquad (6.111)$$

Since $R_\varphi(O) = \infty$, it is necessary in evaluation of (6.107) to use the combination

$$R_\varphi(O) - R_\varphi(\tau) = D^2[xSi(x) + \cos x - 1], \qquad (6.112)$$

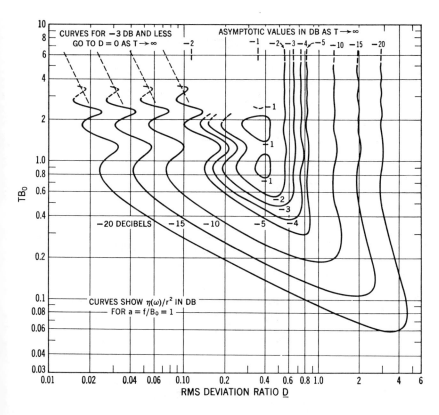

Figure 6.8 Intermodulation distortion at maximum baseband frequency in pure FM, $c = 0$. Curves show the ratio $\eta(\omega)/r^2$ at $\omega = 2\pi B_0 \cdot B_0 = $ maximum baseband frequency c/s, $T = $ echo delay in seconds. From reference 7, Bennett, Curtis, and Rice.

where $x = \Omega_0 \tau$ and Si is the sine integral function.

With pure PM, $t(\omega) = \omega^2$ and (6.110) gives

$$R_\varphi(\tau) = 3\underline{D}^2 \frac{\sin x}{x}. \tag{6.113}$$

Curves of the ratio $\eta(\omega)$ obtained from the equivalent of the foregoing expressions are given in reference [7] for pure FM, and the maximum baseband frequency $\omega = \Omega_0$. These are shown in Fig. 6.8. Similar curves have been published [11] for a frequency pre-emphasis $t(\omega)$ corresponding closely to (6.90) with $c = 16$. These curves are shown in Fig. 6.9.

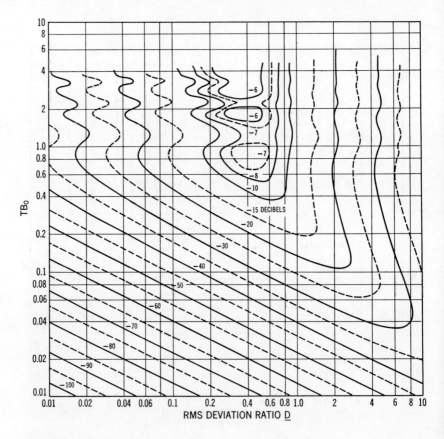

Figure 6.9 Intermodulation distortion at maximum baseband frequency in FM with pre-emphasis $t(f) = 1 + c(f/B_0)^2$, $c \cong 16$. Curves show the ratio $\eta(\omega)/r^2$ at $\omega = 2\pi B_0$. B_0 = maximum baseband frequency, $c/s.$, T = echo delay in seconds. From reference 11, Beach and Trecher.

6.13 Fine-Structure Transmittance Deviations in FM

Distortion, owing to fine-structure transmittance deviations, can be determined by an extension of the method used in Section 6.8 for AM systems.

In FM systems both $X(t)$ and $Y(t)$ are present in (6.13), such that they combine to a constant envelope $\bar{Z}(t)$. Hence $W_X(u)$ and $W_Y(u)$ are both present in (6.67) and (6.68), in which the limits are now $\pm \Omega$. The above relations show that in this case the in-phase and quadrature components of the distortion power spectrum are

$$N_X(u) = [\alpha_e^2(u) + \beta_0^2(u)]W_X(u) + [\alpha_0^2(u) + \beta_e^2(u)]W_Y(u), \quad (6.114)$$

$$N_Y(u) = [\alpha_e^2(u) + \beta_0^2(u)]W_Y(u) + [\alpha_0^2(u) + \beta_e^2(u)]W_X(u). \quad (6.115)$$

These are the double-sided envelope power spectra of distortion at the receiver input. The resultant error $\epsilon(t)$ in the demodulator output is obtained with the aid of (6.77). Let the four variables in the latter relation have zero mean value and let there be no correlation between the first and second products. This is a legitimate approximation for fine-structure deviations as assumed here. The autocorrelation function of $\epsilon(t)$ is then

$$R_\epsilon(\tau) = R_{\Delta_y}(\tau)\, R_X(\tau) + R_{\Delta_x}(\tau)\, R_Y(\tau), \quad (6.116)$$

and the corresponding one-sided baseband power spectrum at $u = \omega$ becomes

$$W_\epsilon(u) = \frac{1}{\pi}\int_{-\infty}^{\infty} R_\epsilon(\tau)\cos u\tau\, d\tau$$

$$= 2\int_{-\omega_0}^{\infty} [W_X(u - \lambda)\, N_Y(\lambda) + W_Y(u - \lambda)\, N_X(\lambda)]\, d\lambda, \quad (6.117)$$

where the last relation follows when $N_Y(u)$ and $N_X(u)$ are the power spectra corresponding to $R_{\Delta_x}(\tau)$ and $R_{\Delta_y}(\tau)$.

In view of (6.114) and (6.115), relation (6.117) can be written as

$$W_\epsilon(u) = 2\int_{-\omega_0}^{\infty} [\alpha_0^2(\lambda) + \beta_e^2(\lambda)][W_X(\lambda)\, W_X(u - \lambda) + W_Y(\lambda)\, W_Y(u - \lambda)]\, d\lambda$$

$$+ 2\int_{-\omega_0}^{\infty} [\alpha_e^2(\lambda) + \beta_0^2(\lambda)][W_Y(\lambda)\, W_X(u - \lambda) + W_X(\lambda)\, W_Y(u - \lambda)]\, d\lambda.$$

$$(6.118)$$

Here W_X and W_Y are the double-sided power spectra about the carrier frequency ω_0, and $W_\epsilon(u)$ is the corresponding one-sided baseband spectrum with $u = \omega$.

The ratio of distortion power at the frequency ω to message power $W_\varphi(\omega)$ becomes

$$\eta(\omega) = \frac{W_\epsilon(\omega)}{W_\varphi(\omega)} \tag{6.119}$$

The distortion power in the band Ω_0 in the demodulator output becomes

$$D_m = \int_0^{\Omega_0} \frac{\omega^2}{t(\omega)} W_\epsilon(\omega)\, d\omega. \tag{6.120}$$

The message power in the output is

$$P_m = \int_0^{\Omega_0} \frac{\omega^2}{t(\omega)} W_\varphi(\omega)\, d\omega. \tag{6.121}$$

The ratio of average distortion power to average message power in the band Ω_0 is then

$$\frac{D_m}{P_m} = \frac{1}{\Delta^2} \int_0^{\Omega_0} \frac{\omega^2}{t(\omega)} W_\epsilon(\omega)\, d\omega \frac{\int_0^{\Omega_0} s(\omega) t(\omega)\, d\omega}{\int_0^{\Omega_0} s(\omega)\, d\omega}. \tag{6.122}$$

A general solution for any variations $\alpha(u)$ and $\beta(u)$ becomes very complicated. Certain special cases will be considered next.

6.14 Distortion from Flat Fine-Structure Deviations in FM

Relation (6.118) applies for any particular channel when $\alpha_e(u)$, $\alpha_0(u)$, $\beta_e(u)$ and $\beta_0(u)$ are known as functions of the frequency $u = \omega - \omega_0$, provided that each has zero mean and varies rapidly with u, so that they can be regarded as fine-structure deviations. These quantities will vary among channels, and if they are averaged for a large number of channels there will be certain average values $\underline{\alpha}_e(u)$, $\underline{\alpha}_0(u)$, $\underline{\beta}_e(u)$ and $\underline{\beta}_0(u)$. The average distortion for a large number of channels can then be obtained by using these averages in (6.118). It will be assumed here that these averages do not vary with u and that they are related as follows:

$$\underline{\alpha}_0^2 + \underline{\beta}_e^2 = \underline{\alpha}_e^2 + \underline{\beta}_0^2$$

$$= \tfrac{1}{2}(\underline{\alpha}^2 + \underline{\beta}^2). \tag{6.123}$$

Flat Fine-Structure Deviations

In this case (6.116) simplifies to

$$W_\epsilon(u) = (\underline{\alpha}^2 + \underline{\beta}^2) \int_{-\omega_0}^{\infty} [W_X(\lambda) + W_Y(\lambda)][W_X(u-\lambda) + W_Y(u-\lambda)]\,d\lambda$$

$$= 4(\underline{\alpha}^2 + \underline{\beta}^2) \int_{-\omega_0}^{\infty} W(\lambda)\,W(u-\lambda)\,d\lambda \qquad (6.124)$$

$$= \frac{4(\underline{\alpha}^2 + \underline{\beta}^2)}{2\pi} \int_{-\infty}^{\infty} R_z^2(\tau) \cos u\tau\,d\tau \qquad (6.125)$$

$$= \frac{(\underline{\alpha}^2 + \underline{\beta}^2)}{2\pi} e^{-2R_\varphi(0)} \int_{-\infty}^{\infty} e^{2R_\varphi(\tau)} \cos u\tau\,d\tau. \qquad (6.126)$$

Relation (6.125) follows from (6.124) by virtue of transformation (18) in Appendix 3, when it is considered that $W_\epsilon(u)$ is a single-sided and $W(u)$ a double-sided power spectrum. Relation (6.126) is obtained by introducing eq. (32) of Appendix 3 for $R_z(\tau)$.

The power spectrum (6.126) conforms with that for a single small echo as given by (6.102) provided that

$$\underline{\alpha}^2 + \underline{\beta}^2 = r^2, \qquad (6.127)$$

and

$$2R_\varphi(\tau) = R_v(\tau). \qquad (6.128)$$

From (6.96) it follows that the latter relation can only be satisfied with $T \to \infty$, so that there is no correlation between the echo and the signal. This condition will be approximated for fine-structure deviations, since each sinusoidal component in the Fourier representation then corresponds to an echo of long delay.

For adequately small values of $R_\varphi(O)$, or of the rms deviation ratio $\underline{D}$, the phase modulation $\varphi(t)$ is related to the Y component of the envelope fluctuation by $\varphi(t) \simeq W_Y(u)$, while $W_X(u) \ll W_Y(u)$. Relations (6.67) and (6.68) for AM are hence modified into the following relations for PM:

$$\sigma_X^2 \simeq 2 \int_{-\omega_0}^{\Omega_0} [\alpha_0^2(u) + \beta_e^2(u)]\,W_Y(u)\,du, \qquad (6.129)$$

$$\sigma_Y^2 = 2 \int_{-\omega_0}^{\Omega_0} [\alpha_e^2(u) + \beta_0^2(u)]\,W_Y(u)\,du. \qquad (6.130)$$

With frequency-discriminator detection, distortion power in the demodulator output is virtually proportional to σ_Y^2. Hence relation (6.71) for AM is replaced by the following relation for adequately small values of the rms deviation ratio $\underline{D}$

$$\frac{D_m}{P_m} = \int_0^{\Omega_0} [\alpha_e^2(u) + \beta_0^2(u)]W_m(u)\,du \bigg/ \int_0^{\Omega_0} W_m(u)\,du, \qquad (6.131)$$

where the substitution $W_Y(u) = W_m(u)$ has been made. With this substitution (6.131) applies for any frequency pre-emphasis and conforms with relation (6.71) for double-sideband AM. In the particular case of flat fine-structure deviations, so that (6.123) applies

$$\frac{D_m}{P_m} = \tfrac{1}{2}(\underline{\alpha}^2 + \underline{\beta}^2) = \tfrac{1}{2}r^2. \tag{6.132}$$

For adequately large values of the rms deviation $\underline{D}$, the power spectrum can be approximated by relation (6.149) derived later. With this approximation (6.124) yields with $r^2 = \underline{\alpha}^2 + \underline{\beta}^2$

$$W_\epsilon(u) = \frac{r^2}{2\pi\underline{\Delta}^2} \int_{-\omega_0}^{\infty} e^{-u^2/2\underline{\Delta}^2} e^{-(u-\lambda)^2/2\underline{\Delta}^2} \, d\lambda$$

$$\simeq \frac{r^2}{2\pi^{1/2}\underline{\Delta}} e^{-u^2/4\underline{\Delta}^2}, \tag{6.133}$$

where the latter relation is obtained by change of variables to $z = \lambda - u/2$, when the lower limit is taken as $-\infty$.

When (6.133) and (6.109) are inserted in (6.119), the following relation is obtained with $a = u/\Omega_0$:

$$\eta(\omega) = \frac{r^2}{2\pi^{1/2}\underline{D}^3} a^2 e^{-a^2/4\underline{D}^2} \frac{1}{\Omega_0 s(\omega)t(\omega)} \int_0^{\Omega_0} s(\omega)t(\omega) \, d\omega. \tag{6.134}$$

With $s(\omega) = s$ and pure FM, and $t(\omega) = 1$, (6.134) yields

$$\eta(\omega) = \frac{r^2}{2\pi^{1/2}\underline{D}^3} a^2 e^{-a^2/4\underline{D}^2}. \tag{6.135}$$

With $s(\omega) = s$ and pure PM, $t(\omega) = \omega^2$ so that

$$\eta(\omega) = \frac{r^2}{6\pi^{1/2}\underline{D}^3} e^{-a^2/4\underline{D}^2}, \tag{6.136}$$

where r^2 is given by (6.127).

In Figs. 6.10 and 6.11 are shown curves of the foregoing ratios for large values of $\underline{D}$, together with curves obtained from relations and charts given in reference [7], for a single echo of amplitude r.

With (6.133) in (6.122) the following relation is obtained for the ratio of distortion to message power in a band Ω_0

$$\frac{D_m}{P_m} = \frac{r^2}{2\pi^{1/2}\underline{\Delta}^3} \frac{\int_0^{\Omega_0} s(\omega)t(\omega) \, d\omega}{\int_0^{\Omega_0} s(\omega) \, d\omega} \int_0^{\Omega_0} \frac{u^2}{t(u)} e^{-u^2/4\underline{\Delta}^2} \, du. \tag{6.137}$$

Flat Fine-Structure Deviations

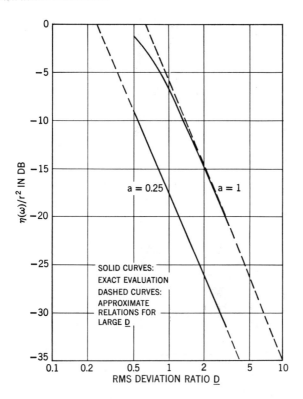

Figure 6.10 Intermodulation distortion in pure FM for an echo of relative amplitude r and long delay ($TB_0 \gg 1$). Curves also apply for flat fine-structure attenuation and phase deviations with $r^2 = \underline{\alpha}^2 + \underline{\beta}^2$.

With $s(\omega) = s$ and pure FM, $t(\omega) = 1$ and

$$\frac{D_m}{P_m} = \frac{r^2}{2\pi^{1/2}\underline{\Delta}^3} \int_0^{\Omega_0} u^2 e^{-u^2/4\underline{\Delta}^2} \, du$$

$$\simeq \left(\frac{1}{4\pi}\right)^{1/2} \frac{r^2}{3\underline{D}^3}. \qquad (6.138)$$

With $s(\omega) = s$ and pure PM, $t(\omega) = \omega^2$ and (6.137) yields

$$\frac{D_m}{P_m} = \frac{r^2}{6\underline{D}^2} \operatorname{erf}\left(\frac{1}{2\underline{D}}\right)$$

$$\simeq \frac{1}{4} \frac{r^2}{3\underline{D}^3}, \qquad (6.139)$$

where $r^2 = \underline{\alpha}^2 + \underline{\beta}^2$.

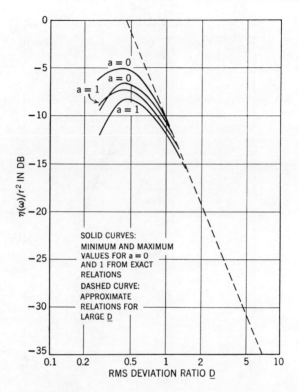

Figure 6.11 Intermodulation distortion in pure PM for an echo of relative amplitude r and long delay ($TB_0 \gg 1$). Curves also apply for flat fine-structure attenuation and phase deviations with $r^2 = \underline{\alpha}^2 + \underline{\beta}^2$.

In Fig. 6.12 are shown the ratios D_m/P_m obtained from (6.132) for $\underline{D} \ll 1$ together with the ratios obtained from (6.138) and (6.139) for large values of $\underline{D}$. The solid transition curve gives the approximate ratios for intermediate values of $\underline{D}$.

It should be noted that D_m includes both the linear and nonlinear distortion component, since both will produce distortion in broadband message transmission, although only the nonlinear component gives rise to interchannel interferences in frequency-division multiplex systems. Curve 1 also applies for double-sideband AM while the ratio for single-sideband transmission is O dB.

From Fig. 6.12 it can be concluded that FM and PM afford an advantage over double-sideband systems as regards message distortion from fine-structure attenuation and phase deviations. Moreover, FM and PM afford a greater reduction in such distortion than in distortion from ran-

Signal-Power Spectra in FM and PM

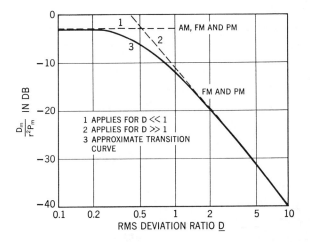

Figure 6.12 Distortion in broadband transmission by FM and PM caused by echo of relative amplitude r and long delay. Curves also apply for flat fine structure deviations with $r^2 = \underline{\alpha}^2 + \underline{\beta}^2$.

dom noise. By way of example let $\underline{D} = 0.6$, in which the required bandwidth would be greater than in double-sideband transmission by an approximate factor $1 + 4\underline{D} \simeq 3.4$. In this case $3\underline{D}^2 \simeq 1$ and there would be no advantage over double-sideband AM with suppressed carrier as regards random noise. However, in accordance with Fig. 6.12, FM and PM would afford about a 5 dB reduction in transmission distortion compared to double-sideband AM, in exchange for a 3.4-fold increase in bandwidth. The advantage over single-sideband transmission would be about 9 dB, with nearly a sevenfold increase in bandwidth.

6.15 Signal-Power Spectra in FM and PM

The signal power spectra in FM and PM are important in connection with distortion from attenuation and phase deviations as considered previously, and also as regards channel bandwidth requirements, to be discussed later. These power spectra are also important from the standpoint of intersystem interference, as dealt with in Chapter 11. Some basic integrals for the signal-power spectra are reviewed here, and curves are presented for the power spectrum with pure PM as obtained by numerical integration for a Gaussian modulating wave with a flat power spectrum. Power spectra with frequency modulation are discussed in further detail in references [12, 13, and 14], which also supply a comprehensive bibliography. With $X = \cos \varphi(t)$ and $Y = \sin \varphi(t)$, it follows from relations given in Appendix 3 that the corresponding autocorrelation functions are

$$R_X(\tau) = e^{-R_\varphi(O)} \cosh R_\varphi(\tau), \tag{6.140}$$

$$R_Y(\tau) = e^{-R_\varphi(O)} \sinh R_\varphi(\tau). \tag{6.141}$$

The autocorrelation function of the phase modulated carrier $Z(t) = \cos \varphi(t) \cos \omega_0 t - \sin \varphi(t) \sin \omega_0 t$ is

$$R_Z(\tau) = \tfrac{1}{2}[R_X(\tau) + R_Y(\tau)]$$

$$= \tfrac{1}{2} e^{-R_\varphi(O)} e^{R_\varphi(\tau)}. \tag{6.142}$$

The foregoing relations correspond to those given in Section 6.12 for a small echo, except that R_v is replaced by R_φ.

The corresponding double-sided power spectra are

$$W_X(u) = \frac{1}{\pi} \int_O^\infty R_X(\tau) \cos u\tau \, d\tau \tag{6.143}$$

$$W_Y(u) = \frac{1}{\pi} \int_O^\infty R_Y(\tau) \cos u\tau \, d\tau \tag{6.144}$$

$$W(u) = W_Z(u) = \frac{1}{\pi} \int_O^\infty R_Z(\tau) \cos u\tau \, d\tau$$

$$= \tfrac{1}{2}[W_X(u) + W_Y(u)]. \tag{6.145}$$

When the discrete spectral component at $u = O$ is placed outside the integrals, the latter take the forms

$$W_X(u) = \delta(u)e^{-R_\varphi(O)} + \frac{1}{\pi} e^{-R_\varphi(O)} \int_O^\infty [\cosh R_\varphi(\tau) - 1] \cos u\tau \, d\tau, \tag{6.146}$$

$$W_Y(u) = \frac{1}{\pi} e^{-R_\varphi(O)} \int_O^\infty \sinh R_\varphi(\tau) \cos u\tau \, d\tau, \tag{6.147}$$

$$W_Z(u) = \tfrac{1}{2}\delta(u)e^{-R_\varphi(O)} + \frac{e^{-R_\varphi(O)}}{2\pi} \int_O^\infty [\exp R_\varphi(\tau) - 1] \cos u\tau \, d\tau. \tag{6.148}$$

For very large values of $\underline{D}^2$, so that $R_\varphi(\tau) \to \infty$, it is permissible to use approximations of $R_\varphi(\tau)$ applying for small values of τ in (6.110). Thus, with $\cos \omega\tau - 1 \simeq -\omega^2\tau/2$, (6.110) gives

$$R_\varphi(\tau) - R_\varphi(O) \simeq \frac{\underline{\Delta}^2 \tau^2}{2}$$

Relations (6.146), (6.147) and (6.148) then yield the approximation

$$W_X(u) = W_Y(u) = W(u) \simeq \frac{1}{2\pi} \int_0^\infty e^{-\Delta^2\tau^2/2} \cos u\tau \, d\tau$$

$$= \frac{1}{(8\pi)^{1/2}\underline{\Delta}} e^{-u^2/2\underline{\Delta}^2}. \tag{6.149}$$

In most applications, the preferable method closely approximates pure PM, in which case (6.113) applies for a flat message power spectrum. For this case the sideband power spectrum represented by the integral term in (6.148) has been determined by numerical integration with the aid of a digital computer, for various ratios $a = u/\Omega_0$ and $\underline{D} = \underline{\Delta}/\Omega_0$. The resultant curves are shown in Fig. 6.13, and conform closely with approximation (6.149) for $\underline{D} \geq 2$. For $\underline{D} < 0.25$, the sideband power spectrum is closely approximated by $W_Y(u) = W_\varphi(u)$, as previously assumed arriving at relation (6.131).

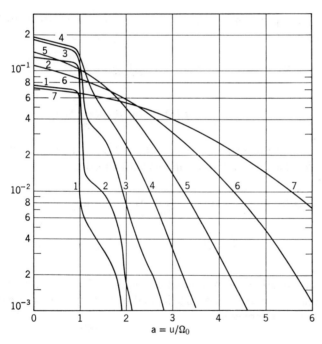

Figure 6.13 Sideband power spectrum in pure PM for Gaussian modulating wave with flat power spectrum of bandwidth Ω_0. Curves apply for carrier of unit amplitude.

Curve	1	2	3	4	5	6	7
$\underline{D}$	$2^{-3/2}$	2^{-1}	$2^{-1/2}$	1	$2^{1/2}$	2	$2^{3/2}$
Carrier power	.34	.24	.11	.025	.001	$3 \cdot 10^{-6}$	$2 \cdot 10^{-11}$
Sideband power	.16	.26	.39	.475	.499	$\sim .5$	$\sim .5$

6.16 Distortion from Channel Bandwidth Limitation in FM

Since the spectrum of an FM signal wave has infinite bandwidth, a restriction of the spectrum bandwidth will give rise to distortion. The resultant distortion will be considered here for an idealized channel with a constant amplitude characteristic and linear phase in the band $-\Omega < u < \Omega$, with complete cutoffs at $u = \pm\Omega$. It will further be assumed that distortion is adequately small so that the phase error $\epsilon(t)$ in the output can be determined from (6.77). The power spectrum of the phase error can then be obtained from (6.118) with the following relations for N_X and N_Y:

$$\frac{-\Omega < \lambda < \Omega}{N_X(\lambda) = O} \quad \frac{-\Omega > \lambda > \Omega}{N_X(\lambda) = W_X(\lambda)},$$

$$N_Y(\lambda) = O \quad N_Y(\lambda) = W_Y(\lambda).$$

The power spectrum of the phase error obtained from (6.118) is accordingly

$$W_\epsilon(u) = 2 \int_{-\omega_0}^{\infty} F(\lambda) \, W_Y(\lambda) \, W_X(u - \lambda) \, d\lambda$$

$$+ 2 \int_{-\omega_0}^{\infty} F(\lambda) \, W_X(\lambda) \, W_Y(u - \lambda) \, d\lambda, \quad (6.150)$$

where $F(\lambda) = O$ except for $|\lambda| > \Omega$ where $F(\lambda) = 1$. Taking the lower limit as $-\infty$, the above relation can be transformed into

$$W_\epsilon(u) = 2 \int_{\Omega}^{\infty} W_Y(\lambda)[W_X(u - \lambda) + W_X(u + \lambda)] \, d\lambda$$

$$+ 2 \int_{\Omega}^{\infty} W_X(\lambda)[W_Y(u - \lambda) + W_Y(u + \lambda)] \, d\lambda. \quad (6.151)$$

An explicit relation for $W_\epsilon(u)$ can be obtained for very large deviation ratios so that the spectra are approximated by (6.149). In this case the two integrals in (6.151) are equal and

$$W_\epsilon(u) = \frac{1}{\pi \Delta^2} \int_{\Omega}^{\infty} e^{-\lambda^2/2\Delta^2}(e^{(u-\lambda)^2/2\Delta^2} + e^{(u+\lambda)^2/2\Delta^2}) \, d\lambda$$

$$= \frac{1}{2\pi^{1/2}\Delta} e^{-u^2/4\Delta^2}\left(\operatorname{erfc}\left(\frac{\Omega}{2\Delta}\right) + \operatorname{erfc}\left(\frac{3\Omega}{2\Delta}\right)\right). \quad (6.152)$$

The second bracket term can be neglected in comparison with the first.

The ratio of average distortion power to average message power at the frequency ω is obtained from (6.119) and becomes

$$\eta(\omega) = \frac{a^2}{\pi^{1/2} \underline{D}^3} e^{-a^2/4\underline{D}^2} \operatorname{erfc}\left(\frac{\Omega/\Omega_0}{2\underline{D}}\right) \frac{\int_0^{\Omega_0} s(\omega) t(\omega) \, d\omega}{\Omega_0 t(\omega) s(\omega)}, \quad (6.153)$$

where $a = \omega/\Omega_0$ and $\underline{D} = \underline{\Delta}/\Omega_0$.

With pure FM, $t(\omega) = 1$. In this case (6.153) yields for a flat spectrum $s(\omega) = s$ of bandwidth Ω_0.

$$\eta(\omega) = \frac{a^2}{\pi^{1/2} \underline{D}^3} e^{-(a/2\underline{D})^2} \operatorname{erfc}\left(\frac{\Omega/\Omega_0}{2\underline{D}}\right). \quad (6.154)$$

With pure PM, $t(u) = \omega^2$, and the following relation is obtained for a flat power spectrum.

$$\eta(\omega) = \frac{1}{3\pi^{1/2} \underline{D}^3} e^{-a^2/4\underline{D}^2} \operatorname{erfc}\left(\frac{\Omega/\Omega_0}{2\underline{D}}\right). \quad (6.155)$$

The foregoing expressions are legitimate approximations for $\underline{D} \geq 2$. Since $a \leq 1$, the factor $\exp(-a^2/4\underline{D}^2)$ can then be replaced with 1. The average distortion ratio for the band Ω_0 is

$$\bar{\eta} = \frac{1}{\Omega_0} \int_0^{\Omega_0} \eta(\omega) \, d\omega$$

$$= \frac{1}{3\pi^{1/2} \underline{D}^3} \operatorname{erfc}\left(\frac{\Omega/\Omega_0}{2\underline{D}}\right), \quad (6.156)$$

which applies for both pure FM and PM.

With a power spectrum (6.133) the average power outside the channel band $-\Omega < u < \Omega$ is smaller than the total average signal power by the factor

$$\gamma = \operatorname{erfc}\left(\frac{\Omega/\Omega_0}{2^{1/2} \underline{D}}\right). \quad (6.157)$$

The above factors are given below as a function of Ω/Ω_0 for $\underline{D} = 2$

Factors $\bar{\eta}$ and γ for Pure FM and PM

$\Omega/\Omega_0 =$	2	4	6	8
$\Omega/\underline{D} =$	1	2	3	4
$10^4 \bar{\eta} =$	112	38	7	1
$10^4 \gamma =$	3300	450	25	0.35

The foregoing results show that when the message wave has the properties of Gaussian noise, distortion in wideband FM and PM is insignificant, provided that the semi-bandwidth Ω exceeds the rms frequency deviation by a factor 4. A greater bandwidth may be required in order to attain a nearly linear phase over the requisite band 2Ω, so as to avoid excessive distortion from phase deviations, particularly near the band edges.

For intermediate deviation ratios, evaluation of distortion with the aid of (6.151) would entail numerical integration and is not considered here. For deviation ratios $\underline{D} \leq 2^{-3/2} = 0.375$, curve 1 of Fig. 6.13 indicates that a channel semi-bandwidth $\Omega \simeq 2\Omega_0$ would suffice, provided that the phase characteristic is made virtually linear over the channel band 2Ω by appropriate phase equalization.

In accordance with the conventional Carson bandwidth rule, the required semi-bandwidth is $\Omega = \Omega_0(1 + \hat{D})$. The maximum deviation ratio in the foregoing case can be taken as $\hat{D} = 4\underline{D}$, for the reason that clipping of the modulating wave at 4 times the rms amplitude produces negligible distortion. On this premise, the Carson rule would yield $\Omega/\Omega_0 = 9$ for $\underline{D} = 2$ and $\Omega/\Omega_0 = 2.4$ for $\underline{D} = 0.375$. Thus the bandwidths obtained from the Carson rule are rather liberal. However, it must be realized that a principal problem with actual channels is the phase nonlinearity in the pass-band resulting from the cutoffs at the band edges. To avoid excessive intermodulation distortion, owing to phase nonlinearity, it is necessary to employ phase equalization, which is facilitated by using a liberal bandwidth as obtained from the Carson rule. Intermodulation distortion resulting from residual fine-structure attenuation and phase deviations can be evaluated from relations given in previous sections. Such distortion is ordinarily of greater importance than that caused by elimination of spectral components when the Carson rule bandwidth is used in systems design.

6.17 Distortion from AM-PM Conversion

As discussed further in Chapter 8, in microwave transmitters fluctuations in the envelope of a received wave are converted into unwanted phase modulation and resultant distortion in the transmitted wave. As shown in Chapter 8, this can be an important source of intermodulation distortion in multicarrier FM systems where pronounced envelope fluctuations are encountered. In single carrier FM systems, however, with representative microwave transmitters and appropriate equalization, distortion from AM-PM conversion can ordinarily be disregarded, as will be shown presently.

Distortion from AM-PM Conversion

Consider first a linear variation in the amplitude versus the frequency characteristic of the channel, as considered in Section 6.10. With a carrier of amplitude A the envelope of the received wave is then $A[1 + c_1\varphi'(t)]$. From relations given in Chapter 8 it follows that the resultant phase modulation, owing to AM-PM conversion, is

$$\psi_x(t) = k_0 A^2 [1 + c_1\varphi'(t)]^2$$
$$= 2k_0 S\{1 + 2c_1\varphi'(t) + c_1^2[\varphi'(t)]^2\}, \quad (6.158)$$

where $S = A^2/2$ is the average carrier power. The constant term can be disregarded, as well as the term in $[\varphi'(t)]^2$. The variable component of the unwanted phase modulation is thus

$$\tilde{\psi}_x(t) \simeq 4k_0 S c_1 \varphi'(t)$$
$$= 0.152\kappa c_1 \varphi'(t) \quad (6.159)$$

where κ is the AM-PM conversion factor in degrees/dB, which is shown in Fig. 8.10 for a particular traveling wave tube amplifier. With operation at saturation, $\kappa \simeq 3.5$ and $0.152\kappa \simeq 0.50$.

In accordance with (6.159) the demodulated message wave will contain an unwanted component proportional to its time derivative. This component must be considered in broadband transmission such as TV, but will not give rise to intermodulation distortion and resultant interchannel interference in transmission of channels in frequency-division multiplex. Hence in broadband transmission a limitation will be imposed on the tolerable value of c_1, although this is not the case in the absence of AM-PM conversion, since a linear amplitude versus frequency deviation does not then give rise to distortion (see Section 6.10).

Consider next a component of quadratic phase distortion, or linear delay distortion. The resultant amplitude modulation is then represented by the denominator of (6.81) and is $A[1 - b_2\varphi''(t)]$. The variable component of the unwanted phase modulation is in this case

$$\tilde{\psi}_x(t) = -0.152\kappa b_2 \varphi''(t). \quad (6.160)$$

The demodulated message wave in this case contains an unwanted component proportional to its second time-derivative. This component must be considered in broadband transmission, but will not give rise to interchannel interference in transmission of a number of channels in frequency division multiplex. As shown in Section 6.11, linear delay variation will cause intermodulation distortion, and this imposes a limitation on the maximum permissible value of b_2. With this limitation on b_2, unwanted phase modulation $\tilde{\psi}_x(t)$ given by (6.160) can ordinarily be disregarded in broadband transmission.

The above linear variations in a first approximation represent coarse-structure deviations in the amplitude and delay characteristic of the channel. Distortion from single echoes and fine-structure deviations will now be considered.

The phase error caused by a single small echo is given by (6.94). A similar derivation yields the following relation for the envelope of the received wave when the transmitted carrier has the amplitude A

$$\bar{z}(t) = A[1 + r \cos \omega_0 T \sin v(t) - r \sin \omega_0 T \cos v(t)]. \quad (6.161)$$

The variable component of the unwanted phase modulation in this case becomes

$$\tilde{\psi}_x(t) = 0.152 \kappa r [\cos \omega_0 T \sin v(t) - \sin \omega_0 T \cos v(t)]. \quad (6.162)$$

It follows by comparison with (6.94) that the average value of $\tilde{\psi}_x(t)$ is smaller than that of $\epsilon(t)$ by the factor 0.152κ. Hence distortion power, owing to AM-PM conversion, is smaller than that given in previous sections for a single echo and for fine-structure deviations by the factor $(0.152\kappa)^2$. The combined distortion is thus greater than in the absence of AM-PM conversion by the factor $1 + (0.152\kappa)^2$. For a traveling-wave tube amplifier with the characteristic shown in Fig. 8.10, this factor is at most 1.25, corresponding to 1 dB increase in distortion power.

In microwave systems, power transmitters such as traveling-wave tubes are separated by bandpass filters from the preceding modulators used for up conversion from intermediate frequencies. These filters introduce essentially quadratic delay distortion. In the preceding discussion it was assumed that phase equalization was used in conjunction with each filter at the input of the power transmitter, to the extent that only a linear component of delay distortion remained, together with small sinusoidal or fine-structure deviations. If phase equalization is not used with each filter, the quadratic delay distortion will give rise to amplitude modulation and resultant intermodulation distortion, owing to AM-PM conversion. This intermodulation distortion can be nearly eliminated by appropriate quadratic phase equalization at subsequent repeaters, provided they all have the same AM-PM conversion factors at the point of equalization. Otherwise a significant component of intermodulation distortion from AM-PM conversion may remain, in spite of apparently adequate equalization as determined by gain and phase deviation measurements that do not reflect the effect of AM-PM conversion as discussed in further detail in a paper by Cross [15]. The latter reference contains theoretical relations and numerical results on intermodulation distortion, including the effect of AM-PM conversion, for linear, quadratic, cubic, and quartic attenuation, and phase deviations.

6.18 Distortion in FM Systems with Power Series Transmittance Deviations

The objective in equalization of analog systems for broadband transmission is to remove coarse structure deviations in the transmittance, so that only small fine-structure deviations remain, as considered in Sections 6.13 and 6.14. This objective cannot be fully realized, so that in general small coarse-structure deviations will remain. In the preceding analysis, these residual coarse-structure deviations were assumed to consist of a single sinusoidal variation in the amplitude and phase characteristics, as considered in Sections 6.3 and 6.4, or alternately a linear deviation from a constant amplitude versus frequency characteristic and from a constant transmission delay, as dealt with in Sections 6.10 and 6.11. The latter can also be considered as a limiting case of a slow sinusoidal variation.

The foregoing are the simplest *a priori* assumptions that can be made regarding residual coarse-structure deviations in specifying equalization requirements for prospective systems. More precise information may be available for existing systems, as determined by measurements of gain and phase deviations. In the latter case, it is of course possible to determine intermodulation distortion by direct tests or, alternately, to undertake a more elaborate analysis based on a power series approximation of coarse-structure deviations. Relations have been published for power series deviations including four terms and a message wave with the properties of flat Gaussian noise [15,16]. The results are summarized in a tutorial exposition which also considers an AM-PM conversion factor that varies with frequency together with the effect of correlation between certain components of the combined distortion and presents illustrative numerical comparisons [17].

7

Attenuation and Phase Distortion in Digital Systems

7.0 General

Chapter 5 was concerned with the performance of various digital modulation methods in the presence of Gaussian noise, for channels with idealized transmission characteristics. An important consideration in many applications is the performance in the presence of phase distortion or equivalent envelope delay distortion, which may be appreciable in certain transmission facilities. An ideal amplitude spectrum of received pulses can be approached with the aid of appropriate terminal filters with gradual cutoffs, such that the associated phase characteristic is virtually linear. Nevertheless, pronounced phase distortion may be encountered in channels with sharp cutoffs outside the pulse spectrum band, as in frequency-division carrier-system channels designed primarily for voice transmission. Consideration will be given here to transmission impairments resulting from certain representative types of delay distortion in pulse transmission by various methods of carrier modulation and signal detection. In addition, an evaluation is made of transmission impairments, owing to low-frequency cutoff and fine-structure variations in the amplitude and phase characteristics of the channel. These transmission impairments are reflected in the need for increased signal-to-noise ratio at the detector input to compensate for the reduced noise margin.

The performance in pulse transmission by various carrier modulation and detection methods can be related to a basic function known as the carrier pulse transmission characteristic, introduced in Chapter 1. This basic function gives the shape of a single carrier pulse at the channel output, that is, the detector input, under ideal conditions or in the presence of the particular kind of transmission distortion under consideration. From this basic function can be determined the envelopes of carrier pulse trains at the detector input, together with the phase of the carrier within the envelope. The shape of demodulated pulse trains with various methods of carrier modulation and detection can, in turn, be determined for various combinations of transmitted pulses, together with the maximum transmission impairment from a specified type of channel imperfection.

Numerical values are given here for the carrier pulse transmission characteristics with linear and quadratic delay distortion, together with the maximum transmission impairments caused by these fairly representative forms of delay distortion with various methods of carrier modulation and signal detection. These include amplitude modulation with envelope and with synchronous detection, two-phase and four-phase modulation with synchronous detection and with differential phase detection and binary frequency modulation. In determining the effect of distortion, a raised cosine amplitude spectrum of the pulses at the detector input has been assumed in all cases, together with the minimum pulse interval permitted with this spectrum, ideal implementation of each modulation and detection method and optimum design from the standpoint of slicing levels and sampling instants at the detector output. These idealizations insure that only the effect of transmission distortion is evaluated and considered in comparing modulation methods, a condition that is difficult to realize with experimental rather than analytical comparisons.

As mentioned above, the present analysis involves a basic function common to all modulation methods, which in general would be determined with the aid of digital computers. This approach has certain advantages in comparison of modulation methods and from the standpoint of optimum system design, and has been used for the above modulation methods [1]. The procedure is fairly simple and yields direct relations for the maximum distortion at sampling instants in the case of baseband transmission and various optimum methods of amplitude and phase modulation with synchronous detection. However, with such methods as amplitude modulation with envelope detection, or two-phase and four-phase modulation with differential phase detection the procedure becomes laborious in that it is necessary to consider explicitly various digital sequences in order to determine maximum distortion. In these cases it may be expedient to use digital computer simulation of distortion with each modulation method rather than computer evaluation of the carrier pulse transmission characteristic. Computer simulation has been used for cosine and sine variations in transmission delay over the channel band, and combinations thereof, for binary AM with envelope detection [2] and four-phase modulation with differential phase detection [3].

Phase distortion in typical analog channels with sharp cutoffs precludes efficient digital transmission unless adequate phase equalization is provided, as for channels intended for television or picture transmission. A first step toward effective utilization of other channels for digital transmission, such as conventional voice channels, is the provision of adequate equalization. In the case of switched channels this would ordinarily entail some form of adaptive equalization, which is facilitated by tapped delay lines for reduction of intersymbol interference resulting from coarse-structure deviations [4].

7.1 Carrier Pulse Trains

In accordance with the discussion in Chapter 5, it will be assumed that a carrier pulse of rectangular or other suitable envelope is applied at the transmitting end of a bandpass channel. The received pulse with carrier frequency ω_0 can then be written in the general form

$$P(t) = \cos(\omega_0 t - \psi) R(t) + \sin(\omega_0 t - \psi) Q(t)$$
$$= \cos[\omega_0 t - \psi - \varphi(t)] \bar{P}(t). \tag{7.1}$$

In this relation R and Q are the envelopes of the in-phase and quadrature components of the received carrier pulse and $\bar{P}$ the resultant envelope. The time t is taken with respect to a conveniently chosen origin; as, for example, the midpoint of a pulse interval or the instant at which R or $\bar{P}$ reaches a maximum value.

Let carrier pulses be transmitted at intervals T, and let $t = t_0 + nT$ be the time from the midpoint of a selected interval. The following designation will be introduced for convenience

$$R(t_0 + nT) = R\left[T\left(\frac{t_0}{T} + n\right)\right] = R(x + n),$$
$$Q(t_0 + nT) = Q\left[T\left(\frac{t_0}{T} + n\right)\right] = Q(x + n), \tag{7.2}$$

where n is the time expressed in an integral number of pulse intervals of duration T and x the time in a fraction of a pulse interval.

Let $a(-n)$ and $\psi(-n)$ be the amplitude and phase of the carrier pulse transmitted in the nth interval prior to the interval O under consideration, and $a(n)$, $\psi(n)$ the corresponding quantities for the nth subsequent interval. The received pulse train in the interval $-T/2 < t < T/2$ is then

$$W_0(x) = \Sigma a(n) \cos[\omega_0 t - \psi(n)] R(x - n)$$
$$+ \Sigma a(n) \sin[\omega_0 t - \psi(n)] Q(x - n)$$
$$= \Sigma a(n) \cos[\omega_0 t - \psi(n) - \varphi(x - n)] \bar{P}(x - n), \tag{7.3}$$

where the summations are between $n = -\infty$ and $n = \infty$.

During the next interval, T to $2T$, the received wave is obtained by replacing $a(n)$ and $\psi(n)$ with $a(n + 1)$ and $\psi(n + 1)$ which is thus

$$W_1(x) = \Sigma a(n + 1) \cos [\omega_0 t - \psi(n + 1) - \varphi(x - n)] \bar{P}(x - n), \quad (7.4)$$

where t and x refer to midpoint of interval 1.

In pulse modulation systems, as considered herein, it is assumed that the modulating pulses are rectangular in shape and of duration equal to the pulse interval. For equal phases $\psi(n) = \psi$ of all the modulating pulses, (7.3) then becomes

$$W_0 = \cos(\omega_0 t - \psi)\Sigma a(n) R(x - n) + \sin(\omega_0 t - \psi)\Sigma a(n) Q(x - n). \quad (7.5)$$

When $a(n) = a$ is a constant the input is a continuous carrier, so that evaluation of (7.5) will give

$$W_0 = aA_0 \cos(\omega_0 t - \psi), \quad (7.6)$$

where A_0 is the amplitude of the transmission-frequency characteristic of the channel at $\omega = \omega_0$ and it is assumed in the determination of R and Q that the phase characteristic is zero at $\omega = \omega_0$, which is permissible without loss of generality.

When $R(t)$ and $Q(t)$ are determined from the channel transmission-frequency characteristic by the usual Fourier integral relations, the following relations apply for rectangular modulating pulses of duration T equal to the pulse interval:

$$A_0 = \sum_{n=-\infty}^{\infty} R(x - n), \quad (7.7)$$

$$0 = \sum_{n=-\infty}^{\infty} Q(x - n). \quad (7.8)$$

That is to say, for steady-state transmission of a carrier of unit amplitude the receiver carrier must have an amplitude A_0 and the same phase ψ as the transmitter carrier, since the phase characteristic of the channel has been taken as zero at $\omega = \omega_0$.

7.2 Slicing Levels and Noise Margins

The demodulated wave is related to the received carrier wave $W_0(x)$ in a manner that depends on the carrier modulation and detection method. In general, the demodulated wave at sampling instants may assume a number of different amplitudes. Let $U^{(s)}$ designate the demodulated wave for one particular amplitude or state a_s of the transmitted signal and $U^{(s+1)}$ the demodulated wave at a sampling instant for an adjacent amplitude or state a_{s+1} of the transmitted signal. There will then be a certain sequence of transmitted pulses for which a maximum value $U_{\max}^{(s)}$ is obtained, owing to intersymbol interference, and also a certain sequence resulting in a minimum value $U_{\min}^{(s+1)}$. If there is an equal probability of a_s and a_{s+1} and of positive and negative noise voltages, the optimum level for distinction between $U^{(s)}$ and $U^{(s+1)}$ is

$$L_0^{(s)} = \tfrac{1}{2}(U_{\min}^{(s+1)} + U_{\max}^{(s)}). \tag{7.9}$$

In the presence of $U^{(s)}$ the margin for distinction from $U^{(s+1)}$ is

$$M^{(s)} = L_0^{(s)} - U^{(s)}, \tag{7.10}$$

and in the presence of $U^{(s+1)}$ the margin for distinction from $U^{(s)}$ is

$$M^{(s+1)} = U^{(s+1)} - L_0^{(s)}. \tag{7.11}$$

The minimum margins are obtained with $U^{(s)} = U_{\max}^{(s)}$ and with $U^{(s+1)} = U_{\min}^{(s+1)}$ in (7.10) and (7.11). The minimum margins thus become

$$M_{\min}^{(s+1)} = M_{\min}^{(s)} = \tfrac{1}{2}(U_{\min}^{(s+1)} - U_{\max}^{(s)}). \tag{7.12}$$

For sequences of marks and spaces, or other signal patterns, such that the minimum margins for distinction between adjacent signal states are obtained, an error will occur if the noise voltage at the sampling instant exceeds $M_{\min}$ in amplitude and has the appropriate polarity. (Polarity is immaterial except for the two extreme signal states.) For other signal patterns, the tolerable amplitude of the noise voltage is greater. The value of $M_{\min}$ relative to the value in the absence of intersymbol interference thus gives the maximum transmission impairment. The average impairment obtained by considering various pulse-train patterns and the corresponding values of $M^{(s)}$ and $M^{(s+1)}$ will be less, as discussed in the next section.

7.3 Evaluation of Transmission Impairments

By way of illustration, it will be assumed that all values of M between $M_{\min}$ and a maximum value $M_{\max}$ are equally probable, and that the noise has a Gaussian amplitude distribution. With a given fixed value of M the probability of an error can be written as

$$p_e = \tfrac{1}{2} \operatorname{erfc}(aM), \tag{7.13}$$

where $\operatorname{erfc} = 1 - \operatorname{erf}$ is the error function complement and a is a factor that depends on the ratio of signal power to noise power.

Considering all noise margins between the limits just mentioned, the average error probability becomes

$$\bar{p}_e = \frac{1}{M_{max} - M_{min}} \cdot \frac{1}{2} \int_{M_{min}}^{M_{max}} \text{erfc}\,(aM)\,dM$$

$$= \frac{1}{2} \frac{1}{M_{max} - M_{min}}$$

$$\times \left(M_{max}\,\text{erfc}\,B - M_{min}\,\text{erfc}\,A + \frac{1}{a\sqrt{\pi}}(e^{-A^2} - e^{-B^2})\right), \quad (7.14)$$

where

$$A = aM_{min},$$
$$B = aM_{max}.$$

With

$$M_{max} = kM_{min},$$

(7.14) becomes

$$\bar{p}_e = \frac{1}{2}\frac{1}{k-1}\left(k\,\text{erfc}\,(kA) - \text{erfc}\,A + \frac{1}{A\sqrt{\pi}}(e^{-A^2} - e^{-k^2A^2})\right). \quad (7.15)$$

For $k = 1$, the latter expression conforms with (7.13).

The maximum error probability would be obtained by considering a fixed noise margin equal to M_{min} and would be

$$\hat{p}_e = \tfrac{1}{2}\,\text{erfc}\,A. \quad (7.16)$$

The error committed in assuming M_{min} can be determined by writing $\bar{p}_e$, as given by (7.15) in the form

$$\bar{p}_e = \tfrac{1}{2}\,\text{erfc}\,(cA), \quad (7.17)$$

where $c \geq 1$ is so chosen that (7.17) equals (7.15).

The average noise margin is then

$$\bar{M} = cM_{min}. \quad (7.18)$$

By way of numerical illustration, let A be so chosen that $\hat{p}_e$ as given by (7.16) in one case is 10^{-4} and in another case 10^{-5}. The results given in Table 1 are then obtained from (7.15) and (7.17).

As will be shown later, a factor $k = 3$ may correspond to a transmission impairment of about 6 dB based on the minimum noise margin, whereas the actual impairment would be 1.4 dB less for an error probability of 10^{-4}, and about 1 dB less for an error probability 10^{-5}. For an error probability of 10^{-6} or less, the error committed in evaluating transmission impairments on the basis of the minimum noise margin can be disregarded.

Table 1. Ratio $c = \bar{M}/M_{\min}$ for Equal Probability of All Noise Margins between $M_{\min}$ and $M_{\max} = kM_{\min}$ for Noise with a Gaussian Amplitude Distribution

$k =$	$A = 2.63; \hat{p}_e = 10^{-4}$			$A = 3.0; \hat{p}_e = 10^{-5}$		
	1	2	3	1	2	3
$\bar{p}_e$	10^{-4}	10^{-5}	5×10^{-6}	10^{-5}	1.5×10^{-6}	7.5×10^{-7}
c	1	1.15	1.17	1	1.1	1.12
c (in dB)	0	1.2	1.4	0	0.8	1.0

This also applies for greater error probabilities when the transmission impairment based on the minimum noise margin is small, in which case $k < 2$. Similar conclusions have been reached in an analysis based on certain other assumed probability distributions of intersymbol interference [5].

7.4 Synchronous AM and Two-Phase Modulation

Amplitude modulation can be used in conjunction with envelope detection and synchronous detection. The former method is simplest from the standpoint of implementation, but synchronous detection, also referred to as homodyne and coherent detection, affords an improvement in signal-to-noise ratio. Since synchronous detection is also the simplest method from the standpoint of analysis, it will be considered first.

Amplitude modulation, in general, implies several pulse amplitudes, and can be used with double-sideband and vestigial-sideband transmission. The particular case of bipolar binary AM with synchronous detection is equivalent to two-phase modulation.

With amplitude modulation and synchronous detection it is possible to transmit pulse trains on two carriers at quadrature with each other, and under certain idealized conditions to avoid mutual interference. The special case of bipolar binary AM on each of the two carriers is equivalent to four-phase modulation.

In AM systems $\psi(n) = \psi$ is constant and (7.3) becomes

$$W_0(x) = \cos(\omega_0 t - \psi)\Sigma a(n) R(x - n) + \sin(\omega_0 t - \psi)\Sigma a(n) Q(x - n).$$
(7.19)

As illustrated in Fig. 5.4 with synchronous detection, the received wave is applied to a product demodulator together with a demodulating wave $\cos(\omega_0 t - \psi)$. After the elimination of higher frequency demodulation products by low-pass filtering the demodulated baseband output becomes

$$U_0(x) = \Sigma a(n) R(x-n), \tag{7.20}$$

where a factor of one-half is omitted for convenience.

At a sampling instant $x = 0$, $U_0(0) = U$ becomes

$$U = a(0) R(0) + \sum_{n=1}^{\infty} [a(-n) R(n) + a(n) R(-n)]. \tag{7.21}$$

The following notation will be used:

$$r^+ = \sum_{n=1}^{\infty} [R^+(n) + R^+(-n)], \tag{7.22}$$

$$r^- = \sum_{n=1}^{\infty} [R^-(n) + R^-(-n)], \tag{7.23}$$

where R^+ designates positive values of R and R^- absolute values when R is negative.

Let there be l different amplitude levels, between a minimum amplitude $a_{\min}$ and a maximum amplitude $a_{\max}$. When a pulse of amplitude $a_s = a_s(0)$ is transmitted, the maximum value of (7.21) is

$$U_{\max}^{(s)} = a_s R(0) + a_{\max} r^+ - a_{\min} r^-. \tag{7.24}$$

For the next higher pulse amplitude $a_{s+1} = a_s + (a_{\max} - a_{\min})/(l-1)$, the minimum value of (7.21) is

$$U_{\min}^{(s+1)} = a_{s+1} R(0) + a_{\min} r^+ - a_{\max} r^-. \tag{7.25}$$

The minimum noise margin is, in accordance with (7.12),

$$M_{\min} = \frac{a_{\max} - a_{\min}}{2} \left(\frac{R(0)}{l-1} - r^+ - r^- \right). \tag{7.26}$$

In the absence of intersymbol interference, $r^+ = 0$, $r^- = 0$, and $R(0) = R^{(0)}(0)$, so that

$$M^0 = \frac{a_{\max} - a_{\min}}{2} \frac{R^0(0)}{l-1}. \tag{7.27}$$

The value of $M_{\min}$ as given by (7.26) is smaller than M^0 in the absence of intersymbol interference by the factor

$$\eta_{\min} = \frac{R(0)}{R^0(0)} \left[1 - (l-1) \frac{r^+ + r^-}{R(0)} \right]$$

$$= \frac{R(0)}{R^0(0)} \left[1 - (l-1) \frac{\bar{r}}{R(0)} \right], \tag{7.28}$$

where

$$\bar{r} = \sum_{n=1}^{\infty} [\bar{R}(-n) + \bar{R}(n)], \quad (7.29)$$

in which $\bar{R}$ designates the absolute value of R.

The factor $R(O)/R^o(O)$ represents the transmission impairment, owing to reduction in pulse amplitude at sampling instants. The summation term represents transmission impairments, owing to intersymbol interference.

Relation (7.28) applies regardless of the polarity of the transmitted pulses and for both symmetrical (double-sideband) and asymmetrical (vestigial-sideband) systems. The special case $l = 2$ and $a_{\min} = -a_{\max}$ represents binary bipolar AM, which can also be regarded as two-phase transmission.

7.5 Quadrature Carrier AM and Four-Phase Modulation

With synchronous detection, it is possible under certain ideal conditions to transmit signals on two carriers at quadrature without mutual interference. In general, however, the quadrature component in (7.19) will in this case give rise to interference and (7.21) is replaced by

$$U = a(O) R(O) + \sum_{n=1}^{\infty} [a(-n) R(n) + a(n) R(-n)]$$

$$+ b(O) Q(O) + \sum_{n=1}^{\infty} [b(-n) Q(n) + b(n) Q(-n)], \quad (7.30)$$

where $b(n)$ are the pulse amplitudes in the quadrature system.

For equal differences between maximum and minimum amplitudes in the two systems, that is, $a_{\max} - a_{\min} = b_{\max} - b_{\min}$, (7.28) is replaced by

$$\eta_{\min} = \frac{R(O)}{R^o(O)} \left\{ 1 - \frac{l-1}{R(O)} [Q(O) + \bar{r} + \bar{q}] \right\}, \quad (7.31)$$

where $\bar{r}$ is defined by (7.29), and similarly

$$\bar{q} = \sum_{n=1}^{\infty} [\bar{Q}(-n) + \bar{Q}(n)], \quad (7.32)$$

where $\bar{Q}$ designates the absolute values of Q.

In general the phase of the demodulating carrier can be so chosen that $Q(O) = O$, as is demonstrated later.

Expression (7.31) applies regardless of pulse polarities in the two quadrature systems. The special case of two binary bipolar AM systems, that is, $l = 2$ and $a_{min} = b_{min} = -a_{max} = -b_{max}$ can also be regarded as four-phase transmission.

When the spectrum of a received pulse at the detector input has even symmetry about the carrier frequency, and the phase characteristic has odd symmetry (even symmetry delay distortion), the quadrature components $Q(n)$ vanish. In this case (7.28) and (7.31) are identical, so that there is no mutual interference between pulse trains transmitted on two carriers at quadrature. In this special case, it is thus possible by quadrature carrier AM to realize a two-fold increase in pulse transmission rate, without increased intersymbol interference. An alternative means to the same end is to use vestigial sideband transmission, to be discussed later.

When the pulse spectrum at the detector input has even symmetry about the midband frequency and the phase characteristic has a component of even symmetry, that is, odd symmetry delay distortion, the in-phase and quadrature components both have even symmetry with respect to t. That is,

$$R(-t) = R(t); \quad Q(-t) = Q(t). \tag{7.33}$$

With synchronous detection, the phase of the demodulating carrier would preferably be so chosen that $Q(t)$ would vanish at $t = O$, since this would give the maximum amplitude of the demodulated pulse at a sampling constant, equal to $[R^2(O) + Q^2(O)]^{1/2}$. For purposes of analysis, it is therefore convenient to modify the phase angle so that the quadrature component vanishes at $t = O$. The modified quantities are related to R and Q by

$$R_0(t) = \frac{R(O)\,R(t) + Q(O)\,Q(t)}{[R^2(O) + Q^2(O)]^{1/2}}, \tag{7.34}$$

$$Q_0(t) = \frac{R(O)\,Q(t) - Q(O)\,R(t)}{[R^2(O) + Q^2(O)]^{1/2}}. \tag{7.35}$$

In the case of double-sideband transmission, (7.28) applies, with

$$\bar{r} = \bar{r}_0 = 2 \sum_{n=1}^{\infty} \bar{R}_0(n). \tag{7.36}$$

With quadrature, double-sideband transmission (7.31) applies, with $Q(O) = Q_0(O) = O$ and

$$\bar{q} = \bar{q}_0 = 2 \sum_{n=1}^{\infty} \bar{Q}_0(n), \tag{7.37}$$

where $\bar{R}_0$ and $\bar{Q}_0$ designate absolute values.

7.6 Vestigial Sideband AM with Synchronous Detection

In vestigial sideband AM, the carrier would be located as indicated in Fig. 7.1 at the frequency Ω from midband or at $-\Omega$. When $R(t)$ and $Q(t)$ are known for a carrier at midband, the corresponding quantities $R_c(t)$ and $Q_c(t)$ for a carrier at $\omega_c = \omega_0 + \Omega$ can be obtained by the method outlined in Section 1.3. Thus

$$R_c(t) = \cos \Omega t R(t) - \sin \Omega t Q(t), \quad (7.38)$$

$$Q_c(t) = \cos \Omega t Q(t) + \sin \Omega t R(t). \quad (7.39)$$

Let T be the pulse interval in double-sideband transmission, in which case the interval in vestigial sideband systems would be $T' = T/2$. When the carrier frequency is chosen, as above, $\Omega T' = \pi/2$. The following relations then apply at sampling instants mT', $m = 0, 1, 2, 3, \ldots$.

$$R_c(mT') = \cos\left(\frac{m\pi}{2}\right) R(mT') - \sin\left(\frac{m\pi}{2}\right) Q(mT'), \quad (7.40)$$

$$Q_c(mT') = \cos\left(\frac{m\pi}{2}\right) Q(mT') + \sin\left(\frac{m\pi}{2}\right) R(mT'). \quad (7.41)$$

When the pulse spectrum has even symmetry about the sideband frequency ω_0, and the phase characteristic has odd symmetry, $Q(t) = O$ and the following relations are obtained. That is, at even sampling points, $m = 0, 2, 4, 6, \ldots$,

$$\begin{aligned} R_c(mT') &= (-1)^{m/2} R(mT') = (-1)^{m/2} R(mT/2), \\ Q_c(mT') &= O \end{aligned} \quad (7.42)$$

At odd sampling points, that is, $m = 1, 3, 5, 7, \ldots$,

$$\begin{aligned} R_c(mT') &= O, \\ Q_c(mT') &= (-1)^{(m-1)/2} R(mT') = (-1)^{(m-1)/2} R(mT/2). \end{aligned} \quad (7.43)$$

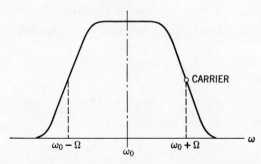

Figure 7.1 Carrier location in vestigial sideband transmission for pulse transmission at intervals $T' = T/2 = \pi/2\Omega = 1/4B$.

In accordance with the foregoing relations, at even sampling points only the in-phase components are present and are the same as in double-sideband AM. At odd sampling points the quadrature components are present, but are eliminated with synchronous detection and need not be considered.

In summary, when the amplitude spectrum at the detector input has even symmetry about the midband frequency, and the phase characteristic has odd symmetry, relation (7.28) applies for double-sideband AM, quadrature double-sideband AM, vestigial-sideband AM, as well as special cases thereof, such as two-phase and four-phase modulation.

When the pulse spectrum and phase distortion both have even symmetry about the midband frequency, relation (7.33) applies, or alternately (7.34) and (7.35). The in-phase component obtained from (7.40) is then given by the following relations in terms of $R_0(t)$ and $Q_0(t)$ as given by (7.34) and (7.35).

At even sampling points, $m = 0, 2, 4, \ldots,$

$$R_{c,0}(mT') = (-1)^m R_0(mT'). \tag{7.44}$$

At odd sampling points, $m = 1, 3, 5, \ldots,$

$$R_{c,0}(mT') = (-1)^{(m+1)/2} Q_0(mT'). \tag{7.45}$$

In this case (7.28) applies, with $R(O) = R_0(O)$ and

$$\bar{r} = 2 \sum_{m=2,4,6,\ldots} \bar{R}_0(mT') + 2 \sum_{m=1,3,5,\ldots} \bar{Q}_0(mT'). \tag{7.46}$$

7.7 Phase Modulation with Differential Phase Detection

In phase modulation with differential phase detection, the demodulator output would under ideal conditions depend on changes in carrier phase between two successive pulse intervals of duration T. In its simplest and ideal form, the signal with two-phase modulation would be applied to one pair of terminals of a product demodulator, while the signal delayed by a pulse interval T would be applied to the other pair. With four-phase modulation, two product demodulators are required, each with a delay network at one pair of terminals. In addition, a phase shift of 90° must be provided between all frequencies of the demodulating waves of the two demodulators, as indicated in Fig. 5.1. Such a phase shift over a frequency band can be realized in principle and closely approached with actual networks. The modulator outputs would be applied to low-pass filters of appropriate bandwidth for elimination of high-frequency demodulation products, and the output of these would be sampled at interval T.

With the foregoing method it is possible with ideal channel characteristics to avoid intersymbol interference at sampling instants, without the need of a wider channel band than required with synchronous detection. However, the two methods are not in all respects equivalent from the standpoint of bandwidth utilization. With synchronous detection the carrier frequency must be at least equal to the maximum baseband signal frequency, whereas with differential phase detection it must exceed twice the maximum baseband signal frequency. This requirement does not impose a limitation on bandwidth utilization with differential phase modulation provided that the midband frequency of the available channel is at least twice the lowest frequency, or that this condition is realized through frequency translation prior to demodulation.

Other implementations of differential phase detection than assumed herein have been used, but in principle these entail a wider channel band than with ideal synchronous detection. For example, the two demodulator inputs or outputs could be integrated over a pulse interval T with the aid of a narrow-band resonator tuned to the carrier frequency, and then be rapidly quenched before the next signal interval. When the channel bandwidth is limited, the phase of the demodulating carrier will then depend on the phases of the carrier during several pulse intervals. Thus some intersymbol interference from bandwidth limitation is encountered even in the absence of phase distortion, and the effect of phase distortion will be greater than that determined herein. However, excessive transmission impairments from bandwidth limitation and phase distortion can be avoided by appropriate techniques, as when a large number of narrow channels are provided within a common band of much greater bandwidth than that of the individual channels [6].

With differential phase detection, $W_0(x)$ as given by (7.3) is applied to one pair of terminals of a product demodulator, and $W_1(x)$ is given by (7.4) to the other pair with a suitable phase shift θ. The demodulator output is then, with $a(n)$ constant,

$$U(x) = \{\Sigma \cos [\omega_0 t - \psi(n) - \varphi(x - n) + \theta] \bar{P}(x - n)\}$$
$$\cdot \{\Sigma \cos [\omega_0 t - \psi(n + 1) - \varphi(x - n) \bar{P}(x - n)\}, \quad (7.47)$$

where, as before, the summations are between $n = -\infty$ and $n = \infty$.

After the elimination of high-frequency components present in (7.47) by low-pass filtering and omitting a factor of one-half, the resultant baseband output can be written

$$U(x) = \Sigma S_n(x) \bar{P}(x - n), \quad (7.48)$$

$$U(O) = \Sigma S_n(O) \bar{P}(-n), \quad (7.49)$$

PM with Differential Phase Detection

in which the summations are between $n = -\infty$ and ∞ and

$$S_n(x) = \sum_{m=-\infty}^{\infty} \bar{P}(x-m) \cos [\psi(-n) - \psi(-m+1)$$
$$+ \varphi(x-n) - \varphi(x-m) - \theta], \quad (7.50)$$

$$S_n(O) = \sum_{m=-\infty}^{\infty} \bar{P}(-m) \cos [\psi(-n) - (\psi-m+1)$$
$$+ \varphi(-n) - \varphi(-m) - \theta]. \quad (7.51)$$

The desired output is represented by the term in (7.49) for $n = O$ and is

$$U^o(O) = S_0(O)\bar{P}(O) \quad (7.52)$$

where, in accordance with (7.51)

$$S_0(O) = \sum_{m=-\infty}^{\infty} \bar{P}(-m) \cos [\psi(O) - \psi(-m+1) + \varphi(O) - \varphi(-m) - \theta]. \quad (7.53)$$

In the absence of intersymbol interference $\bar{P}(-m) = O$ for $m \neq O$ and

$$S_0^o(O) = \bar{P}(O) \cos [\psi(O) - \psi(1) - \theta] \quad (7.54)$$

so that (7.52) becomes

$$U^o(O) = \bar{P}^2(O) \cos [\psi(O) - \psi(1) - \theta]. \quad (7.55)$$

This expression is of the same form as for synchronous detection except that $\bar{P}(O)$ is replaced by $\bar{P}^2(O)$ and $\psi(O)$ by the phase change $\psi(O) - \psi(1)$ between two successive sampling instants.

It is convenient to write (7.51) in the form

$$S_n(O) = \sum_{m=-\infty}^{\infty} \bar{P}(-m) a_n(m), \quad (7.56)$$

where

$$a_n(m) = \cos (\psi_{n,m-1} + \varphi_{n,m} - \theta), \quad (7.57)$$

in which

$$\psi_{n,m-1} = \psi(-n) - \psi(-m+1), \quad (7.58)$$
$$\varphi_{n,m} = \varphi(-n) - \varphi(-m). \quad (7.59)$$

Here $\psi_{n,m-1}$ is the change in carrier phase between sampling intervals $-n$ and $-(m-1)$ while $\varphi_{n,m}$ is the phase difference resulting from distortion. The latter is given by

$$\varphi_{n,m} = \tan^{-1} \frac{Q(-n)}{R(-n)} - \tan^{-1} \frac{Q(-m)}{R(-m)}. \quad (7.60)$$

From (7.57) it will be noted that for $m = n + 1$ the value of $a_n(m)$ is always $a_n(n + 1) = \cos(\varphi_{n,n+1} - \theta)$. The corresponding component of $S_n(O)$ is $\cos(\varphi_{n,n+1} - \theta)\bar{P}(-n-1)$. The resultant component of $U(O)$ is a d–c or bias component that corresponds to the optimum slicing level as given by

$$L_0 = \sum_{n=-\infty}^{\infty} \cos(\varphi_{n,n+1} - \theta)\,\bar{P}(-n-1)\,\bar{P}(n). \qquad (7.61)$$

The minimum noise margin in the presence of a mark is

$$\eta_{\min}^{(1)} = U_{\min}^{(1)} - L_0, \qquad (7.62)$$

and the maximum noise margin is

$$\eta_{\max}^{(1)} = U_{\max}^{(1)} - L_0. \qquad (7.63)$$

Here $U_{\min}^{(1)}$ and $U_{\max}^{(1)}$ are the minimum and maximum values of $U(O)$ obtained from (7.49) with $a_0(O) = 1$, corresponding to a pulse of unit amplitude.

With two-phase modulation $\psi_{n,m-1}$ can have the values O or $\pm \pi$. With $\theta = O$ the values of $a_n(m)$ are in this case

$$a_n(m) = \cos \psi_{n,m-1} \cos \varphi_{n,m} + \sin \psi_{n,m-1} \sin \varphi_{n,m}$$

$$= \pm \cos \varphi_{n,m}. \qquad (7.64)$$

Thus $a_0(O) = \pm 1$ and a mark corresponds to $a_0(O) = 1$.

With four-phase modulation, the carrier phase change $\psi_{n,m-1}$ can have the values $\psi_{n,m-1} = O, \pm \pi/2, \pm \pi,$ and $\pm 3\pi/2$. With $\theta = \pi/4$ $a_n(m)$ can be as follows after multiplication by a normalizing factor $\sqrt{2}$

$$a_n(m) = \cos \psi_{n,m-1}(\cos \varphi_{n,m} + \sin \varphi_{n,m})$$

$$+ \sin \psi_{n,m-1}(\sin \varphi_{n,m} - \cos \varphi_{n,m}) \qquad (7.65)$$

Hence

$$a_n(m) = \pm(\cos \varphi_{n,m} + \sin \varphi_{n,m}),$$

or $\qquad (7.66)$

$$a_n(m) = \pm(\sin \varphi_{n,m} - \cos \varphi_{n,m}).$$

Thus $a_0(O) = \pm 1$ and $a_0(O) = 1$ for a mark.

To determine the noise margins from (7.62) and (7.63), it is necessary to evaluate $U_{\min}^{(1)}$ and $U_{\max}^{(1)}$. To this end various digital sequences are assumed, all with $a_0(O) = 1$ but with the other values of $a_n(m)$, depending on the sequence under consideration. For each sequence a value of $S_n(O)$ is

In-Phase and Quadrature Envelopes

determined from (7.56) and in turn the corresponding value of $U^{(1)}$ from (7.49). From these calculations the minimum and maximum values of $U^{(1)}$ can be determined, and in turn $\eta^{(1)}_{\min}$ and $\eta^{(1)}_{\max}$ from (7.62) and (7.63). This laborious procedure will not be discussed further here, and is preferably carried out with the aid of a digital computer. (This would be one type of computer simulation of distortion but not that used in references [2] and [3].)

7.8 Basic Relations for In-Phase and Quadrature Envelopes

The functions $R(t)$ and $Q(t)$ in (7.1) can be determined from relations (6.15) and (6.17) or

$$R(t) = \frac{1}{\pi} \int_{-\omega_0}^{\infty} S(u) A_1(u) \cos[ut - \psi_0(u) - \beta(u)] \, du \qquad (7.67)$$

$$Q(t) = -\frac{1}{\pi} \int_{-\omega_0}^{\infty} S(u) A_1(u) \sin[ut - \psi_0(u) - \beta(u)] \, du \qquad (7.68)$$

where $S(u) = S_0(u) A_0(u)$ is the desired amplitude spectrum of the pulses at the channel output, or detector input, and $\psi_0(u)$ is the phase function of the spectrum. (The opposite sign of the phase $\varphi(t)$ has been chosen here, hence the negative sign for $Q(t)$.) The amplitude distortion introduced by the channel is $A_1(u)$ and the phase distortion $\beta(u)$. In general the desired amplitude spectrum will have even symmetry about the channel midband frequency, in which case $\psi_0(u)$ varies linearly with u and can be disregarded or included by appropriate choice of the time origin at the receiver.

A raised cosine pulse spectrum will be assumed here, in which case

$$S(-u) = S(u) = \frac{T}{2} \cos^2 \frac{\pi u}{4\Omega}, \qquad (7.69)$$

where $\Omega = \pi/T$ is the mean bandwidth to each side of the midband frequency.

The corresponding time function with $\beta(u) = O$ is

$$R(t) = \frac{2}{\pi} \int_{O}^{\infty} S(u) \cos ut \, du$$

$$= \frac{\sin 2\Omega t}{2\Omega t [1 - (2\Omega t/\pi)^2]}. \qquad (7.70)$$

7.9 Linear Amplitude versus Frequency Deviation

As discussed in Chapter 6, it is convenient to distinguish between coarse- and fine-structure deviations in the transmission-frequency characteristic of a channel. As a first approximation, a coarse-structure deviation in the amplitude characteristic can be represented by a linear deviation, in which case

$$A_1(u) = 1 + cu. \tag{7.71}$$

In this case (7.67) and (7.68) yield for spectra $S(u)$ with even symmetry

$$R(t) = \frac{2}{\pi} \int_0^\infty S(u) \cos ut\, du \tag{7.72}$$

$$Q(t) = -\frac{2c}{\pi} \int_0^\infty S(u) u \sin ut\, du$$

$$= c\, dR(t)/dt. \tag{7.73}$$

For a raised cosine spectrum, (7.72) gives (7.70) and (7.73) yields the relation

$$\frac{Q(t)}{R(O)} = 2c\Omega \frac{\cos 2\Omega t}{2\Omega t[1 - (2\Omega t/\pi)^2]}$$

$$- 2c\Omega \frac{\sin 2\Omega t}{(2\Omega t)^2[1 - (2\Omega t/\pi)^2]^2} \tag{7.74}$$

$$= O \quad \text{for} \quad t = O$$

$$= \pm \frac{c\Omega}{3\pi} \quad \text{for} \quad t = \pm T$$

$$= \pm \frac{c\Omega}{30\pi} \quad \text{for} \quad t = \pm 2T.$$

With $c\Omega = 0.25$, the maximum value of $A_1(u)$ is 1.5 at $u = 2\Omega$ and the minimum value is 0.5 at $u = -2\Omega$. In this case distortion at the two first sampling points is about $\pm 2.5\%$ and $\pm 0.25\%$ of the peak pulse amplitude. The coarse-structure deviations from a constant amplitude characteristic will ordinarily be much smaller than just assumed. Hence when this distortion can be approximated by a linear deviation, the resultant intersymbol interference at sampling instants can for practical purposes be disregarded. As shown in Chapter 6 (Section 6.10), a linear deviation from a constant amplitude characteristic does not give rise to distortion in FM or PM analog systems.

7.10 Quadratic Delay Distortion

Quadratic delay distortion is in theory approached near midband of a flat bandpass channel with sharp cutoffs, such as a carrier system voice channel, and approximates the type of delay distortion often encountered in channels without phase equalization.

The phase distortion is in this case given by

$$\beta(u) = cu^3, \qquad (7.75)$$

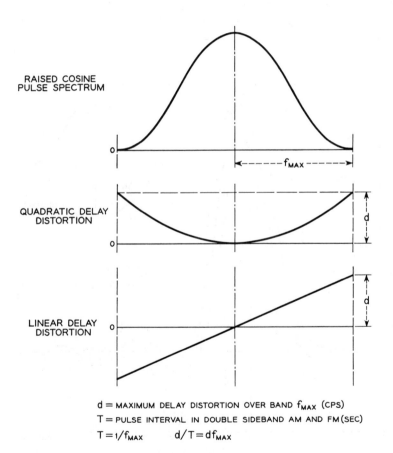

Figure 7.2 Parameter designations for quadratic and linear delay distortion with raised cosine pulse spectrum at detector input. Maximum frequency $f_{\max} = 2B = 1/T$.

and the corresponding delay distortion is

$$\beta'(u) = 3cu^2. \tag{7.76}$$

In this case $Q(t) = O$ for a spectrum with even symmetry. For a raised cosine spectrum (7.69) relation (7.67) can be brought into the form

$$R(t) = \frac{4}{\pi} \int_0^{\pi/2} \cos^2 x \cos(ax - bx^3)\, dx, \tag{7.77}$$

$$a = 4\frac{t}{T}, \qquad b = \frac{16}{3\pi^2}\frac{d}{T}.$$

The ratio t/T is the time measured in pulse intervals and the ratio of d/T the maximum delay distortion measured in pulse intervals, with d defined as in Fig. 7.2.

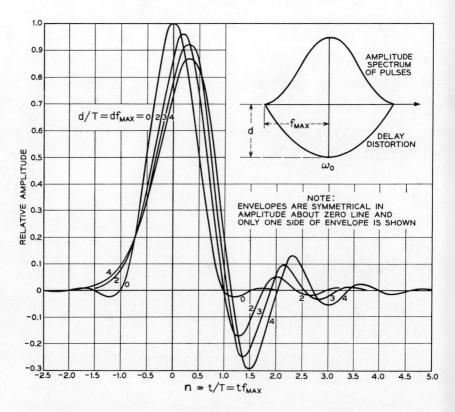

Figure 7.3 Function $R(n)$ for raised cosine pulse spectrum and quadratic delay distortion. Maximum frequency $f_{\max} = 2B = 1/T$.

Table 2. Function $R(n)$ for Raised Cosine Spectrum and Quadratic Delay Distortion

$n = t/T$	d/T				
	0	1	2	3	4
−3	0	−0.0006	0.0025	0	0
−2	0	−0.0013	0.0011	0.0017	0.0028
−1	0	0.0467	0.0756	0.0891	0.0986
0	1	0.9633	0.8795	0.7956	0.7336
1	0	−0.0341	−0.0098	0.0827	0.2045
2	0	0.0196	0.0543	0.0655	0.0142
3	0	0.0044	0.0020	−0.0231	−0.0584
4	0	0.0014	−0.0014	−0.0087	−0.0037
5	0	0.0006	−0.0008	−0.0022	0.0040
$\sum_{-3}^{5} R(n)$	1.0	1.0000	1.0004	1.0006	0.9957

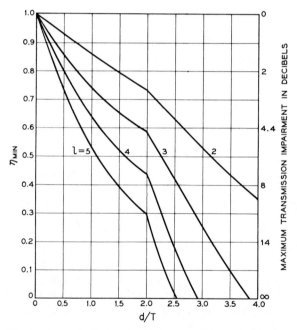

Figure 7.4 Maximum transmission impairments with raised cosine spectrum and quadratic delay distortion. Curves apply for: (a) double sideband AM systems employing synchronous or envelope detection, and baseband systems; (b) vestigial sideband AM systems and quadrature double sideband systems with synchronous detection; (c) for $l = 2$, two-phase and four-phase systems with synchronous detection.

Numerical evaluation of (7.76) has yielded the values given in Table 2 for integral values of n. (A more comprehensive table is given in reference [1] for other than integral values of n.)

With exact evaluation of $R(n)$ the summation $\Sigma R(n)$ between $n = -\infty$ and $n = \infty$ should equal 1.

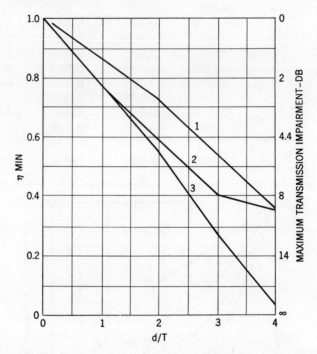

Figure 7.5 Maximum transmission impairments with raised cosine pulse spectrum and quadratic delay distortion for: 1. Two-phase and four-phase systems with synchronous detection. 2. Two-phase systems with differential phase detection. 3. Four-phase systems with differential phase detection.

The corresponding carrier pulse transmission characteristic is shown in Fig. 7.3.

In Fig. 7.4 are shown the factors $\eta_{\min}$ obtained from relations in Sections 7.4 through 7.6 for $l = 2, 3, 4,$ and 5 amplitude levels. These curves also apply to baseband transmission, when the baseband spectrum and delay distortion is as shown for $u > O$.

In Fig. 7.5 are shown the factors $\eta_{\max}$ for two-phase and four-phase modulation with differential phase detection, obtained by the procedure outlined in Section 7.7. In the same figure is shown, for comparison, the noise margin with synchronous detection.

A comparison is made in Fig. 7.6 of the various binary methods considered earlier, together with binary AM with envelope detection and binary FM with frequency-discriminator detection. The curves for the two latter methods were obtained by a procedure carried out in further detail in reference [1].

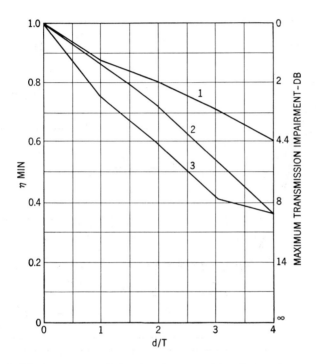

Figure 7.6 Maximum transmission impairments with raised cosine spectrum and quadratic delay distortion for: 1. Binary FM with frequency discriminator detection. 2. Binary PM with synchronous detection and binary AM with synchronous or envelope detection. 3. Binary PM with differential phase detection.

7.11 Linear Delay Distortion

As mentioned previously, quadratic delay distortion is approached near midband of a flat bandpass channel with sharp cutoffs. After phase equalization of such channels, a component of linear delay distortion may remain, owing to inexact equalization. Linear delay distortion is also approximated when a narrow bandpass channel is established to one side of midband of a flat band-pass channel with sharp cutoffs. Delay distortion encountered in troposcatter channels as a result of frequency-selective fading can also in a first approximation be represented by linear delay distortion, as shown in Chapter 9.

Phase distortion is in this case given by

$$\beta(u) = cu^2, \tag{7.78}$$

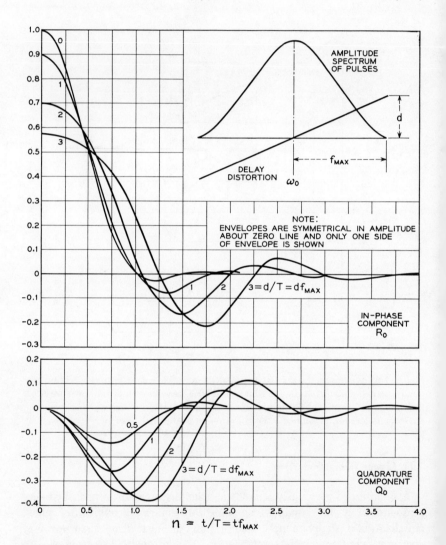

Figure 7.7 Functions $R_0(n) = R_0(-n)$ and $Q_0(n) = Q_0(-n)$ for raised cosine pulse spectrum at detector input and linear delay distortion. Maximum frequency $f_{\max} = 2B = 1/T$.

Linear Delay Distortion

and the corresponding delay distortion is

$$\beta'(u) = 2cu. \tag{7.79}$$

For a raised cosine pulse spectrum at the detector input, relations (7.67) and (7.68) can in this case be transformed into

$$R(-t) = R(t) = \frac{4}{\pi} \int_0^{\pi/2} \cos^2 x \cos ax \cos bx^2 \, dx, \tag{7.80}$$

$$Q(-t) = Q(t) = \frac{4}{\pi} \int_0^{\pi/2} \cos^2 x \cos ax \sin bx^2 \, dx, \tag{7.81}$$

where

$$a = 4\frac{t}{T}, \quad b = \frac{4}{\pi}\frac{d}{T},$$

in which the delay d is defined as in Fig. 7.2.

The corresponding values of $R_0(t)$ and $Q_0(t)$ obtained from (7.34) and (7.35) are given in Table 3 and are shown in Fig. 7.7.

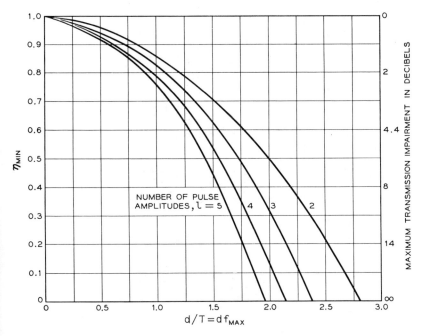

Figure 7.8 Maximum transmission impairments with raised cosine spectrum and linear delay distortion in double sideband AM systems with synchronous detection. Curve for $l = 2$ also applies for two-phase modulation with synchronous detection.

Table 3. Functions $R_0(t/T)$ and $Q_0(t/T)$ for Raised Cosine Pulse Spectrum and Linear Delay Distortion

t/T	$d/T = 0$		$d/T = 0.5$		$d/T = 1$		$d/T = 2$		$d/T = 3$	
	R_0	Q_0	R_0	Q_0	R_0	Q_0	R_0	Q_0	R_0	Q_0
0	1.00	0	0.9712	0	0.8943	0	0.6975	0	0.5712	0
±0.25	0.8488	0	0.8324	−0.0395	0.7879	−0.0667	0.6646	−0.0641	0.5614	−0.0344
±0.50	0.5000	0	0.5056	−0.1138	0.5204	−0.1967	0.5484	−0.2120	0.5188	−0.1317
±0.75	0.1698	0	0.1817	−0.1421	0.2169	−0.2577	0.3356	−0.3385	0.4131	−0.2628
±1.00	0	0	−0.0015	−0.0973	0.0012	−0.1933	0.0792	−0.3473	0.2293	−0.3629
±1.25	−0.0243	0	−0.0392	−0.0263	−0.0754	−0.0687	−0.1118	−0.2247	0.0044	−0.3588
±1.50	0	0	−0.0132	0.0130	−0.0511	0.0171	−0.1604	−0.0529	−0.1714	−0.2288
±1.75	0.0081	0	0.0063	0.0111	−0.0042	0.0293	−0.0910	0.0595	−0.2153	−0.0392
±2.00	0	0	0.0044	−0.0023	−0.0144	0.0058	0.0005	0.0693	−0.1284	0.0978
±2.25	−0.0037	0	−0.0017	−0.0047	0.0060	−0.0079	0.0391	0.0195	−0.0010	0.1154
±2.50	0	0	−0.0017	0.0007	−0.0039	−0.0036	0.0209	−0.0171	0.0669	0.0454
±2.75	0.0020	0	0.0007	−0.0023	−0.0032	0.0026	0.0056	−0.0160	0.0498	−0.0243
±3.00	0	0	0.0008	−0.0002	0.0014	0.0020	−0.0107	0.0009	−0.0000	−0.0377
±3.25	−0.0012	0							−0.0234	−0.0100
±3.50	0								−0.0115	0.0127
±3.75	0.0007								−0.0040	0.0177

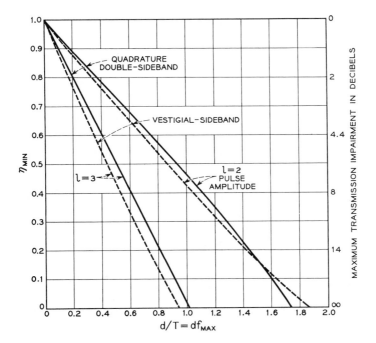

Figure 7.9 Maximum transmission impairments with raised cosine spectrum and linear delay distortion in vestigial sideband systems and quadrature double sideband systems with synchronous detection.

In Figs. 7.8 and 7.9 are shown the factors $\eta_{\min}$ obtained from relations in Sections 7.4 through 7.6 and carried out in further detail in reference [1].

For two-phase and four-phase modulation with differential phase detection, the factors $\eta_{\min}$ determined by the procedure outlined in Section 7.7 are as shown in Fig. 7.10. In the same figure are also shown the factors for binary AM with envelope detection and binary FM with frequency discriminator detection obtained from reference [1].

7.12 Fine-Structure Deviations

Coarse-structure deviations in the amplitude and phase characteristics, such as those considered previously, can in general be removed by appropriate equalization. After such equalization, fine-structure deviations will remain, as indicated in Fig. 6.2 for a baseband channel.

In accordance with the discussion in Section 6.7, the power spectrum of distortion at the input of the receiving filter would be $[\alpha^2(\omega) + \beta^2(\omega)]W(\omega)$,

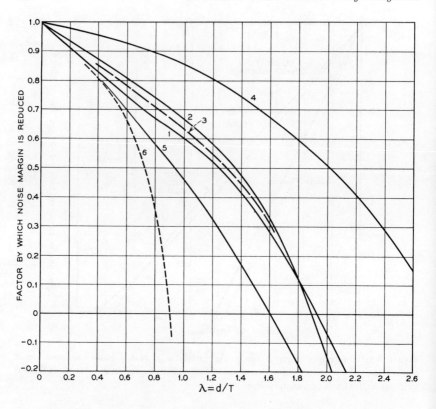

Figure 7.10 Maximum transmission impairments with raised cosine spectrum and linear delay distortion for: 1. Binary AM with envelope detection. 2. Binary FM with frequency discriminator detection. 3. Binary PM with differential phase detection. 4. Binary PM with synchronous detection. 5. Four-phase modulation with synchronous detection. 6. Four-phase modulation with differential phase detection.

where in the present case $W(\omega)$ is the power spectrum of the pulse train at the input to the receiving filter. In accordance with the notation in Chapter 4, the latter power spectrum is $s(\omega)l(\omega)t(\omega)$. With random noise of power spectrum $n_i(\omega)$ at the input to the receiving filter, the combined equivalent noise power spectrum becomes

$$n_e(\omega) = n_i(\omega) + [\alpha^2(\omega) + \beta^2(\omega)]s(\omega)l(\omega)t(\omega). \qquad (7.82)$$

With a receiving filter having a power transmittance $r(\omega) = 1/l(\omega)t(\omega)$, the average equivalent noise power at the output of the receiving filter becomes

$$N_e = N + D. \qquad (7.83)$$

The ratio of average equivalent noise power to average pulse-train power at the receiving filter output is

$$\frac{N_e}{P} = \frac{N}{P} + \frac{D}{P}, \qquad (7.84)$$

where

$$P = \int_0^{\Omega_0} s(\omega)\, d\omega, \qquad (7.85)$$

$$N = \int_0^{\Omega_0} \frac{n_i(\omega)\, d\omega}{l(\omega) t(\omega)}, \qquad (7.86)$$

$$D = \int_0^{\Omega_0} [\alpha^2(\omega) + \beta^2(\omega)]\, s(\omega)\, d\omega. \qquad (7.87)$$

For a specified pulse amplitude spectrum and corresponding power spectrum $s(\omega)$, and a given average transmitted power, the ratio N/P can be minimized by the procedure given in Chapter 4 for analog and digital pulse systems. However, the ratio D/P will remain fixed.

For reasons discussed in Section 6.7, transmission distortion owing to fine-structure deviations will have an amplitude probability distribution approaching that of Gaussian noise. Hence the maximum error probability is obtained by using an equivalent signal-to-noise $\rho_e = P/N_e$ in place of the ratio ρ in the relations for error probability given in Section 5.8. By way of illustration, let the fine-structure deviations for a large ensemble of channels have a flat power spectrum over the band Ω_0, in which case

$$D/P = \underline{\alpha}^2 + \underline{\beta}^2, \qquad (7.88)$$

where $\underline{\alpha}$ and $\underline{\beta}$ are the rms deviations from the ideal amplitude and phase characteristic. It will further be assumed that $N/P = D/P$, in which case the ratio ρ_e must have the values shown for ρ in Table 1 of Chapter 5. The corresponding values for $2\rho_e = (\underline{\alpha}^2 + \underline{\beta}^2)^{-1}$ are now given in Table 4 for an error probability 10^{-4}, together with $\underline{\alpha}$ and $\underline{\beta}$ on the premise that $\underline{\alpha} = \underline{\beta}$.

It is evident that with more than 8 pulse amplitudes the requirements imposed on tolerable fine-structure deviations become quite severe. In most broadband systems these fine-structure deviations place a limitation on transmission quality in analog systems and on the channel capacity in digital systems, that may be more important than additive random noise.

In the case of double-sideband systems, relation (7.82) is replaced by

$$n_e(u) = n_i(u) + 2[\alpha_e^2(u) + \alpha_0^2(u)]\, s(u) l(u) t(u)$$

$$\simeq n_i(u) + [\alpha^2(u) + \beta^2(u)]\, s(u) l(u) t(u), \qquad (7.89)$$

Table 4. *Maximum Permissible rms Fine-Structure Deviations for Various Numbers of Pulse Amplitudes r, for a Digit Error Probability* 10^{-4}

r	ρ_e	$\overline{\underline{\alpha}^2 + \underline{\beta}^2}$	$\underline{\alpha}$ and $\underline{\beta}$	$\underline{\alpha}$ dB	$\underline{\beta}$ deg.
2	13.7	27.4	0.135	1.15	7.7
4	92	184	0.053	0.46	3
8	400	800	0.025	0.21	1.4
16	1600	3200	0.0125	0.10	0.7
32	6400	12800	0.006	0.05	0.33
64	25600	51200	0.003	0.025	0.17

where the factor 2 comes about for reasons mentioned in Section 6.8 and the second approximate relation would ordinarily apply. For fine-structure deviations with a flat power spectrum, (7.88) is in accordance with (6.71) replaced by $D/P = \frac{1}{2}(\underline{\alpha}^2 + \underline{\beta}^2)$. For a given error probability, the tolerable rms deviations $\underline{\alpha}$ and $\underline{\beta}$ over the band $2\Omega_0$ would be greater than those given in Table 4 by a factor $2^{1/2}$.

In quadrature double-sideband systems, (6.72) applies. With $\alpha_e^2(u) + \alpha_0^2(u) = \alpha^2(u)$ and $\beta_e^2(u) + \beta_0^2(u) = \beta^2(u)$ the following relation applies:

$$n_e(u) = n_i(u) + 2[\alpha^2(u) + \beta^2(u)] s(u)l(u)t(u). \qquad (7.90)$$

In this case the tolerable variances $\underline{\alpha}^2$ and $\underline{\beta}^2$ are half as great as for double sideband transmission.

In the particular case of binary transmission on each of two carriers at quadrature, or four-phase transmission, a two-fold increase in channel capacity as limited by random noise can be realized in exchange for a two-fold increase in average signal power. The above relations show that in order to realize this two-fold increase in channel capacity in the presence of fine-structure deviations, it is also necessary to reduce $\underline{\alpha}^2$ and $\underline{\beta}^2$ to half the values permissible with double-sideband transmission.

7.13 Digital Transmission by Pure PM

As mentioned in Section 5.1, there are two basic methods of digital phase or frequency modulation. The method considered in Chapter 5 and in previous sections of this chapter assumed transmission of sequences of carrier pulses of different phases, in conjunction with synchronous or differential phase detection. With this method the phase modulation is accompanied by envelope fluctuations, that is, amplitude modulation, but

Digital Transmission by Pure PM

the required bandwidth is no greater than with double sideband AM. The other basic method is to use pure PM in conjunction with frequency discriminator detection, as in conventional analog PM systems. This method in principle entails carrier sideband spectra of infinite bandwidth, and some unwanted amplitude and phase modulation will arise from channel bandwidth limitation. However, the circumstance that amplitude modulation is largely avoided with this method may be a significant advantage in applications to microwave channels with highly nonlinear repeaters, in which amplitude modulation is converted into unwanted phase modulation in the received wave. The preferable method will depend on repeater nonlinearity and on the channel bandwidth required to avoid excessive unwanted amplitude and phase modulations.

To determine channel bandwidth requirement it is convenient to first consider the sideband spectrum with pure PM. With this method (7.3) for a carrier pulse train is replaced by

$$W_0(t) = \cos\left(\omega_0 t + \sum_{n=-\infty}^{\infty} a(n)\psi_0(t-T)\right), \quad (7.91)$$

where $\psi_0(t) = \theta_1 P_0(t)$ is a unit phase modulation function.

$P_0(t)$ represents the shape of a baseband modulating pulse of peak amplitude $P_0(O) = 1$ and $a(n)$ is the amplitude of a pulse during the nth pulse intervals. In a symmetrical system with r levels, $a(n) = \pm 1, \pm 3, \ldots, \pm(r-1)$. With r levels the maximum phase excursion is $\pm 2\hat{\theta}$, from $-\hat{\theta} = -(r-1)\theta_1$ to $\hat{\theta} = (r-1)\theta_1$, or conversely. Hence the maximum phase excursions are

$$\pm 2\hat{\theta} = 2(r-1)\theta_1. \quad (7.92)$$

The phase excursion θ_1 per unit amplitude cannot exceed π/r, that is,

$$\theta_1 \leq \frac{\pi}{r}. \quad (7.93)$$

Thus

$$\pm 2\hat{\theta} < 2\pi \frac{r-1}{r}. \quad (7.94)$$

The optimum choice of θ_1 and thus $\hat{\theta}$ depends on the sideband spectra and distortion arising from the elimination of sideband components in actual channels.

Owing to the nonlinear operation involved in (7.91), the transmitted wave $W_0(t)$ will have a spectrum of infinite bandwidth, for any kind of baseband modulating wave. It may be possible to determine the power spectrum of the waves $W_0(t)$ by considering all possible pulse sequences. With appropriate kinds of power spectra it would in principle be possible to determine average distortion in demodulated pulse trains owing to channel bandwidth restriction, by the same procedure as outlined in Chapter 6 for analog PM systems. However, in efficient digital systems pronounced average distortion is always encountered, but only distortion at sampling instants is of any consequence. By appropriate control of pulse distortion it is possible to avoid intersymbol interference at sampling instants in linear systems.

In FM and PM the bandwidth of the sideband spectrum increases with increasing phase or frequency deviation. Hence an appropriate criterion for channel bandwidth requirement would appear to be distortion at sampling instants for the maximum phase or frequency excursion during a signal interval of duration T. A convenient approximation that facilitates numerical evaluation is to assume a long sequence of alternately positive and negative modulating pulses, as indicated in Fig. 7.11. In this case (7.91) can be written

$$W_0(t) = \cos\left(\omega_0 t + \varphi + (r-1)\sum_{n=-\infty}^{\infty}(-1)^n \psi_0(t - nT)\right). \quad (7.95)$$

With ideal modulating pulses $P_0(t)$ and $\psi_0(t)$ are even functions of t, and (7.95) can then be resolved into a series of the following form:

$$W_0(t) = \cos(\omega_0 t + \varphi)\sum_{k=0,2,4,\ldots}^{\infty} A(k)\cos\frac{k\pi t}{T}$$

$$- \sin(\omega_0 t + \varphi)\sum_{k=1,3,5,\ldots}^{\infty} B(k)\cos\frac{k\pi t}{T} \quad (7.96)$$

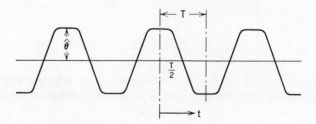

Figure 7.11 Phase variations with periodic pulse train. Maximum phase excursions from 0 with r pulse amplitudes are $\pm \hat{\theta} = \pm(r-1)\theta_1$.

Bandwidth Limitation with Pure PM

The coefficients $A(k)$ and $B(k)$ depend on the maximum phase excursion and on the shape of $P_0(t)$, as will be demonstrated next for raised cosine modulating pulses.

With raised cosine modulating pulses

$$P_0(t) = \cos^2 \frac{\pi t}{2T} \qquad |t| < T$$

$$= 0 \qquad |t| > T, \qquad (7.97)$$

and (7.95) takes the form

$$W_0(t) = \cos\left(\omega_0 t + \varphi + \hat{\theta} \cos \frac{\pi t}{T}\right), \qquad (7.98)$$

where $\hat{\theta}$ is given by (7.92).

The conventional expansion into sideband components yields the following series, in which φ is for convenience omitted.

$$W_0(t) = J_0(\hat{\theta}) \cos \omega_0 t$$

$$- J_1(\hat{\theta})\left[\sin\left(\omega_0 + \frac{\pi}{T}\right)t + \sin\left(\omega_0 - \frac{\pi}{T}\right)t\right]$$

$$- J_2(\hat{\theta})\left[\cos\left(\omega_0 + \frac{2\pi}{T}\right)t + \cos\left(\omega_0 - \frac{2\pi}{T}\right)t\right]$$

$$+ J_3(\hat{\theta})\left[\sin\left(\omega_0 + \frac{3\pi}{T}\right)t + \sin\left(\omega_0 - \frac{3\pi}{T}\right)t\right]$$

$$- \text{etc.} \qquad (7.99)$$

or

$$W_0(t) = \left[J_0(\hat{\theta}) - 2J_2(\hat{\theta}) \cos \frac{2\pi t}{T} + \cdots\right] \cos \omega_0 t$$

$$- 2\left[J_1(\hat{\theta}) \cos \frac{\pi t}{T} - J_3(\hat{\theta}) \cos \frac{3\pi t}{T} + \cdots\right] \sin \omega_0 t. \qquad (7.100)$$

In this case relation (7.96) applies with

$$A(0) = J_0(\hat{\theta}); \qquad A(2) = -2J_2(\hat{\theta}); \qquad A(4) = 2J_4(\hat{\theta}), \ldots,$$
$$B(1) = 2J_1(\hat{\theta}); \qquad B(3) = -2J_3(\hat{\theta}); \qquad B(5) = 2J_5(\hat{\theta}), \ldots.$$

7.14 Distortion from Bandwith Limitation with Pure PM

With a periodic modulating wave as assumed previously, the envelope and the phase of the received wave are given by the following relations when spectral components for $k > m$ are eliminated:

$$1 + \alpha(t) = \left\{ \left[\sum_{k=0,2,4,\ldots}^{m-1} A(k) \cos \frac{\pi k t}{T} \right]^2 + \left[\sum_{k=1,3,5,\ldots}^{m} B(k) \cos \frac{\pi k t}{T} \right]^2 \right\}^{1/2}, \tag{7.101}$$

$$\hat{\theta} \cos \frac{\pi t}{T} + \beta(t) = \tan^{-1} \frac{\sum_{k=1}^{m} B(k) \cos \frac{\pi k t}{T}}{\sum_{k=0}^{m-1} A(k) \cos \frac{\pi k t}{T}}, \tag{7.102}$$

where $\alpha(t)$ is the envelope fluctuation and $\beta(t)$ the phase error, and m is assumed odd. For an even m the upper limits are interchanged.

At a sampling instant $t = O$,

$$\alpha(O) = \left\{ \left[\sum A(k) \right]^2 + \left[\sum B(k) \right]^2 \right\}^{1/2} - 1, \tag{7.103}$$

$$\beta(O) = \frac{\tan^{-1} \sum B(k)}{\sum A(k) - \hat{\theta}} \tag{7.104}$$

where the summations are as before.

In the particular case of raised cosine modulating pulses, (7.100) applies. The relations to be given next are then obtained for various semi-bandwidths B_c of the channel, by observing that the first harmonic sideband component of amplitude $J_1(\hat{\theta})$ has a frequency $1/2T$, the second harmonic component a frequency $1/T$, etc.

For $1/2T < B_c < 1/T$:

$$\alpha(O) = (J_0^2(\hat{\theta}) + 4J_1^2(\hat{\theta}))^{1/2} - 1, \tag{7.105}$$

$$\beta(O) = \tan^{-1} \frac{2J_1(\hat{\theta})}{J_0(\hat{\theta})} - \hat{\theta}. \tag{7.106}$$

For $1/T < B_c < 3/2T$:

$$\alpha(O) = \{[J_0(\hat{\theta}) - 2J_2(\hat{\theta})]^2 + 4J_1^2(\hat{\theta})\}^{1/2} - 1, \tag{7.107}$$

$$\beta(O) = \tan^{-1} \frac{2J_1(\hat{\theta})}{J_0(\hat{\theta}) - 2J_2(\hat{\theta})} - \hat{\theta}. \tag{7.108}$$

For $3/2T < B_c < 2/T$:

$$\alpha(O) = \{[J_0(\hat{\theta}) - 2J_0(\hat{\theta})]^2 + 4[J_1(\hat{\theta}) - J_3(\hat{\theta})]^2\}^{1/2} - 1, \tag{7.109}$$

$$\beta(O) = \tan^{-1} \frac{2[J_1(\hat{\theta}) - J_3(\hat{\theta})]}{J_0(\hat{\theta}) - 2J_2(\hat{\theta})} - \hat{\theta}. \tag{7.110}$$

Linear Delay Distortion with Pure PM

From the foregoing relations, the numerical results now given in Table 5 are obtained.

Table 5. *Distortion with Periodic Raised Cosine Modulating Wave*

	$\hat{\theta} = \pi/4$		$\hat{\theta} = \pi/2$		$\hat{\theta} = 3\pi/4$	
	$\alpha(O)$	$\beta(O)$	$\alpha(O)$	$\beta(O)$	$\alpha(O)$	$\beta(O)$
$\frac{1}{2T} < B_c < \frac{1}{T}$	0.119	-0.079	0.228	-0.394	0.059	-0.809
$\frac{1}{T} < B_c < \frac{3}{2T}$	0.012	0.015	0.134	0.023	0.34	-0.125
$\frac{3}{2T} < B_c < \frac{2}{T}$	-0.002	0.0016	-0.004	0.027	0.066	0.096

The semi-bandwidth B_c required to avoid excessive transmission impairment depends on the number of pulse amplitudes r. In a binary system $\theta_1 \leq \pi/2$ in accordance with (7.93) and $\hat{\theta} \leq \pi/2$. The phase error $\beta(O)$ must be significantly smaller than θ_1. With B_c slightly in excess of $1/T$ the phase error is $\beta(O) = 0.023 \simeq 0.015\,\theta_1$. This corresponds to the same bandwidth as required with double-sideband AM and a raised cosine pulse spectrum.

With $r = 4$, $\theta_1 \leq \pi/4$, and $\hat{\theta} \leq 3\pi/4$. In this case $\beta(O) = -0.125 \simeq -0.166\,\theta_1$ with $B_c = 1/T$ and $\beta(O) = 0.096 \simeq 0.12\,\theta_1$ with $B_c = 3/2T$. Thus about a 1.5-fold increase in channel bandwidth over double-sideband AM would be required to avoid excessive distortion.

The above channel bandwidths are based on the maximum values of θ_1 and $\hat{\theta}$. With $r = 4$ it would be possible to use $\hat{\theta} = \pi/2$, in which case $\theta_1 = \pi/6$ and $\beta(O) = 0.023 \simeq 0.04\,\theta_1$. Distortion from channel bandwidth limitation would then be reduced, but the maximum tolerable phase error from all sources would be reduced from $\pi/4$ to $\pi/6$, that is, by a factor $\frac{2}{3}$.

The optimum choice of θ_1 also depends on transmission impairments owing to phase distortion, which will diminish as θ_1 and thus $\hat{\theta}$ is reduced. The important case of quadratic phase distortion is considered in the next section.

7.15 Effect of Linear Delay Distortion With Pure PM

The maximum effect of phase distortion, like channel bandwidth limitation, is encountered with the maximum phase excursion, and will be considered here for quadratic phase distortion as represented by (7.78). Let $\Omega = \pi/T$ and let the phase distortion at the frequency Ω be γ, in which case $c\Omega^2 = \gamma$. In conformity with the notation used previously, let delay distortion at the frequency 2Ω be d. The following relation then applies:

$$\gamma = c\Omega^2 = \frac{d}{T}\frac{\pi}{4}. \qquad (7.111)$$

Relation (7.100) for a periodic raised cosine pulse train is then modified into

$$W_0(t) = [J_0(\hat{\theta}) - 2J_2(\hat{\theta})\cos(2\Omega t + 4\gamma)$$
$$+ 2J_4(\hat{\theta})\cos(4\Omega t + 16\gamma) - \cdots]\cos\omega_0 t$$
$$- 2[J_1(\hat{\theta})\cos(\Omega t + \gamma) - J_3(\hat{\theta})\cos(3\Omega t + 9\gamma) + \cdots]\sin\omega_0 t$$

$$(7.112)$$

The phase error $\beta(O)$ at sampling instant is obtained from the relation

$$\hat{\theta} + \beta(O) = \tan^{-1}\frac{2[J_1(\hat{\theta})\cos\gamma - J_3(\hat{\theta})\cos 9\gamma + \cdots]}{J_0(\hat{\theta}) - 2J_2(\hat{\theta})\cos 4\gamma + 2J_4(\hat{\theta})\cos 16\gamma + \cdots}. \qquad (7.113)$$

The solutions of the latter equation for certain special cases are:
For $\gamma = O$, or $d/T = O$, $\beta(O) = O$
For $\gamma = \pi/4$ or $d/T = 1$:

$$\beta(O) = [\tan^{-1}(2^{-\frac{1}{2}}\sin\hat{\theta}) - \hat{\theta}] \qquad (7.114)$$

For $\gamma = \pi/2$ or $d/T = 2$:

$$\beta(O) = -\hat{\theta} \qquad (7.115)$$

In the absence of distortion the difference between adjacent phases is θ_1 and as the presence of distortion is $\theta_1 + \beta(O)$. The following ratio thus gives an approximate indication of the minimum noise margin

$$\eta_{\min} = \frac{\theta_1 + \beta(O)}{\theta_1}. \qquad (7.116)$$

For $r = 2$, $\hat{\theta} = \theta_1 = \pi/2$, so that in accordance with (7.115) and (7.116), $\eta_{\min} = O$ for $d/T = 2$. For $r = 4$ and $\hat{\theta} = 3\pi/4$, $\theta_1 = \pi/4$. In this case (7.116) yields for $d/T = 2$, $\eta_{\min} = -2$. This indicates that $\eta_{\min} = O$ is obtained for d/T substantially less than 2.

The ratio $\eta_{\min}$ obtained from (7.113) and (7.116) is shown in Fig. 7.12 as a function of d/T for $\hat{\theta} = \pi/2$ and $3\pi/4$. These curves apply for a periodic pulse train, and it is possible that a greater distortion may be

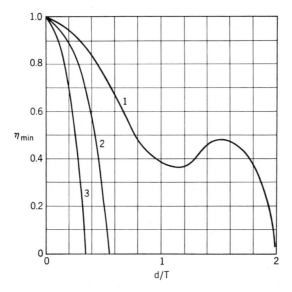

Figure 7.12 Maximum transmission impairments from linear delay distortion with pure PM and raised cosine pulses. 1. Two-phase modulation with $\hat{\theta} = \pi/2$. 2. Four-phase modulation with $\hat{\theta} = \pi/2$. 3. Four-phase modulation with $\hat{\theta} = 3\pi/4$.

obtained under other conditions for certain values of d/T. It may be noted that for pure binary PM $\eta_{\min} = O$ for $d/T = 2$. This conforms closely with curve 3 in Fig. 7.10 for two-phase modulation with differential phase detection. However, for pure quaternary PM the values of $\eta_{\min}$ are significantly smaller than given by curve 6 in Fig. 7.10 for four-phase modulation with differential phase detection, even with $\hat{\theta} = \pi/2$. This is to be expected because of the greater bandwidth of the spectrum with pure PM and the smaller value of θ_1 in (7.116).

7.16 Pulse Distortion from Low-Frequency Cutoff

Most communication channels involve a number of modulation steps and have a low-frequency cutoff between baseband inputs and outputs. This may be caused by transformers or by RC coupled amplifiers in the baseband sections, or it may result from elimination of the carrier and one sideband as required for maximum efficiency in transmission over bandpass channels. Such low-frequency cutoffs do not give rise to transmission impairments in analog transmission of waves in which low-frequency components are absent or unimportant, and may have a significant advantage from the standpoint of elimination of low-frequency interference.

It is possible to determine rms distortion in the received pulse train caused by removal of low-frequency components by the same method as used in Section 7.12 for fine structure deviations. (Reference 6 of Chapter 1.) With an adequately high ratio of upper to lower channel cutoff frequencies, this rms distortion can be kept small enough for random pulse trains to be transmitted without excessive error probability. A comprehensive analysis has been made of error probabilities with random pulse trains in the presence of Gaussian noise and low-frequency removal, for conventional binary and multilevel systems together with duobinary systems and multilevel extensions thereof. [7]

Although random pulse trains may be transmitted with an acceptable average error rate under the conditions just mentioned, an additional requirement for most pulse systems is that it be possible to transmit any conceivable digital sequence without excessive error probability. This requirement cannot be met with a low-frequency cutoff in the channels, for the reason that a great many long pulse sequences will contain pronounced low-frequency components that will be suppressed in transmission. This causes pronounced distortion and an excessive error probability for these particular sequences, even though the error probability averaged over all possible pulse patterns in a random pulse train may be adequately low.

It is possible to avoid an excessive error probability for all digital sequences by certain methods to be discussed in Section 7.17, in exchange for a moderate increase in the required signal-to-noise ratio. To illustrate the effect of low-frequency cutoff and facilitate determination of the effectiveness of remedial measures, it is convenient to first consider the resultant distortion for a single transmitted pulse, in a similar manner as for other sources of distortion dealt with previously.

In baseband pulse transmission a low-frequency cutoff gives rise to pulse distortion and resultant intersymbol interference, owing partly to elimination of low-frequency components and partly to the phase distortion over the transmission band that accompanies a low-frequency cutoff. The latter is ordinarily by far the more important source of pulse distortion and will be considered here.

Let the amplitude characteristic of the channel be given by

$$A(\omega) = A_0(\omega) \quad \text{for} \quad 0 < \omega < \omega_0 \quad (7.117)$$

$$= A_1 \quad \text{for} \quad \omega > \omega_0. \quad (7.118)$$

Pulse Distortion from Low-Frequency Cutoff

The corresponding phase characteristic of a minimum phase shift transducer is then in accordance with (1.52)

$$\psi^0(\omega) = \frac{1}{\pi} \int_{-\infty}^{\infty} \frac{\ln A(u)}{\omega - u} du$$

$$= \frac{1}{\pi} \int_{-\infty}^{\infty} \frac{\ln A_1}{\omega - u} du + \frac{1}{\pi} \int_{-\omega_0}^{\omega_0} \frac{\ln (A_0(\omega)/A_1)}{\omega - u} du. \quad (7.119)$$

The first term yields a linear component $\omega\tau$ and the second term a phase distortion component

$$\psi_0(\omega) = -\frac{1}{\pi} \int_{-\omega_0}^{\omega_0} \frac{\alpha_0(u)}{\omega - u} du, \quad (7.120)$$

where $\alpha_0(u)$ is the attenuation in nepers in the low-frequency cutoff range, as given by

$$\alpha_0(u) = -\ln \frac{A_0(\omega)}{A_1}. \quad (7.121)$$

For $\omega \gg \omega_0$ it is permissible to use the approximation $\omega - u = \omega$ in (7.120), in which case

$$\psi_0(\omega) \simeq -\frac{2}{\pi} \frac{\omega_0}{\omega} \underline{\alpha_0}, \quad (7.122)$$

where

$$\underline{\alpha_0} = \frac{1}{\omega_0} \int_0^{\omega_0} \alpha_0(u) \, du \quad (7.123)$$

is the average attenuation in the low-frequency cutoff range.

By way of illustration, let

$$A_0(\omega) = e^{-\beta(|\omega_0/\omega|^{1/2} - 1)}, \quad (7.124)$$

and let $A(\omega) = 1$ for $|\omega| > \omega_0$. In this case

$$\alpha_0(\omega) = -\beta\left(1 - \left|\frac{\omega_0}{\omega}\right|^{1/2}\right), \quad (7.125)$$

and (7.123) yields $\underline{\alpha_0} = \beta$. Thus (7.122) becomes

$$\psi_0(\omega) \simeq -\frac{2\beta}{\pi} \frac{\omega_0}{\omega}. \quad (7.126)$$

The exact phase distortion obtained with (7.124) in (7.120) is

$$\psi_0(\omega) = \frac{\beta}{\pi}\left(\ln\left|\frac{1+x}{1-x}\right| - \left(\frac{1}{x}\right)^{1/2} \ln\left|\frac{1+x^{1/2}}{1-x^{1/2}}\right| - 2\left(\frac{1}{x}\right)^{1/2} \tan^{-1}\left(\frac{1}{x}\right)^{1/2}\right), \quad (7.127)$$

where $x = \omega/\omega_0$.

For $x = 1$, the phase distortion obtained from (7.126) is in this case in error by a factor 0.70, but for $x \geq 2$ the error is insignificant.

Let the pulses at the channel input have a spectrum $S(\omega)$. The shape of a received pulse for a channel with a phase characteristic $\psi(\omega)$ is then

$$P(t) = \frac{1}{\pi} \int_0^\infty A(\omega) \, S(\omega) \cos\left[\omega t - \psi(\omega)\right] d\omega. \tag{7.128}$$

In the particular case of a raised cosine pulse spectrum of mean bandwidth Ω and maximum bandwidth 2Ω, (7.128) becomes

$$P(t) = \frac{1}{\Omega} \int_0^{2\Omega} A(\omega) \cos^2 \frac{\pi \omega}{4\Omega} \cos\left[\omega t - \psi(\omega)\right] d\omega. \tag{7.129}$$

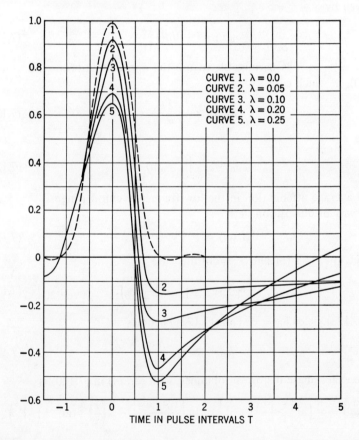

Figure 7.13 Received pulses with low-frequency cutoff for transmitted pulses with raised cosine spectrum. The parameter λ is given by (7.131).

When ω_0/Ω is adequately small, say less than 0.05, the contribution to the integral between $\omega = O$ and ω_0 can, for practical purposes, be disregarded. That is to say, elimination of the spectrum between O and ω_0 is of secondary importance compared to the effect of the resultant phase distortion in the band above ω_0. With this approximation, the following relation applies for the shape of a received pulse after transmission over a channel with phase distortion $\psi_0(\omega)$:

$$P(t) = \frac{1}{\Omega} \int_{\omega_0}^{2\Omega} \cos^2 \frac{\pi\omega}{4\Omega} \cos\left[\omega t - \psi_0(\omega)\right] d\omega. \tag{7.130}$$

In Fig. 7.13 are shown the pulse shapes obtained by numerical integration when $\psi_0(\omega)$ is given by (7.127), for various values of the parameter

$$\lambda = \frac{2}{\pi} \frac{\omega_0}{\Omega} \alpha_0 = \frac{2}{\pi} \frac{\omega_0}{\Omega} \beta. \tag{7.131}$$

The curves in Fig. 7.13 apply without excessive error for any kind of low-frequency cutoff characteristic when the average attenuation below ω_0 is α_0 as obtained from (7.123). More precisely, they apply for the specific cutoff characteristic (7.124) in terms of the attenuation $\beta = \alpha_0$.

7.17 Low-Frequency Cutoff with Dicode Transmission

There are various methods of compensating in part for the pulse distortion caused by low-frequency cutoff and resultant zero wander. One method employs "clamping" circuits activated by pulses of known amplitude transmitted at appropriate fixed intervals. These control pulses, and in turn the zero line will be distorted by channel noise, resulting in about 3 dB impairment in signal-to-noise ratio. In addition, there may be a significant cumulative effect, owing to the pulse distortion in the interval between control pulses. Another method is to use "quantized negative feedback," in which the long negative pulse tail resulting from low-frequency cutoff is compensated for by transmission of a pulse of the same shape as this tail but of opposite polarity. This entails appropriate design of the feedback arrangement for each application, depending on the shape of the low-frequency cutoff characteristic.

A preferable method for most applications is to use constrained bipolar binary transmission or dicode transmission, as discussed in Chapter 5. The maximum residual effect of low-frequency cutoff with this method will be determined here on the basis of the pulse shapes shown in Fig. 7.13.

When a mark is transmitted, and the pulses have a negative tail as shown in Fig. 7.13, the minimum amplitude of a received pulse train at a sampling instant is the same as the peak amplitude $P(O)$ in Fig. 7.13. Thus

$$U^{(1)}_{\min} = P(O). \qquad (7.132)$$

When a space is transmitted, the maximum amplitude of a pulse train is obtained with a negative pulse in the preceding pulse position and is

$$U^{(0)}_{\max} = -P(T) = \bar{P}(T), \qquad (7.133)$$

where $\bar{P}(T)$ is the absolute instantaneous amplitude of the pulse at $t = T$.

The optimum slicing level is $L = \frac{1}{2}[U^{(1)}_{\min} + U^{(0)}_{\max}]$ and the minimum noise margin is

$$M_{\min} = \frac{1}{2}[U^{(1)}_{\min} - U^{(0)}_{\max}]$$
$$= \frac{1}{2}[P(O) - \bar{P}(T)]. \qquad (7.134)$$

The noise margin in the absence of a low-frequency cutoff is $M^{(O)} = \frac{1}{2}P^O(O)$. The tolerance to noise is thus modified by the factor

$$\eta = \frac{M_{\min}}{M^{(O)}} = \frac{[P(O) - \bar{P}(T)]}{P^O(O)}. \qquad (7.135)$$

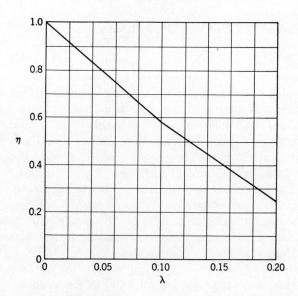

Figure 7.14 Maximum transmission impairments from low-frequency cutoff for pulses with a raised cosine spectrum, as a function of λ as given by (7.131).

In Fig. 7.14 the factor η is shown as a function of the parameter λ given by (7.131). In accordance with the discussion in Section 7.3, this factor applies with greater accuracy the smaller the error probability.

The curve in Fig. 7.14 applies provided no phase equalization is used to compensate for the phase distortion in the transmission band caused by the low-frequency cutoff. When $\eta \ll 1$ it is necessary to employ adequate phase equalizations, to avoid excessive transmission impairment.

8

Intermodulation Distortion in Nonlinear Channels

8.0 General

The steady-state transfer characteristic or transmittance of a channel is a complex quantity $A(\omega)e^{-i\psi(\omega)}$, where the amplitude characteristic $A(\omega)$ and the phase characteristic $\psi(\omega)$ in general exhibit some unwanted variations with frequency ω over the channel band. Distortion, owing to such variations, was dealt with in Chapter 6. In addition, these characteristics will ordinarily vary to some extent with the amplitude of the input signal. In conventional circuit theory, the latter type of departure is referred to as nonlinearity. Unlike variations in the attenuation and phase characteristics with frequency, nonlinearity will give rise to harmonic distortion and intermodulation distortion when the envelope of the signal varies, as in baseband and carrier amplitude modulation systems. However, with nonlinear modulation methods, like FM and PM intermodulation distortion is encountered even with a linear channel characteristic, owing to unwanted variations in the attenuation and phase characteristics with frequency as shown in Chapter 6.

In the transmission of broadband message waves, intermodulation distortion gives rise to the same kind of noiselike distortion as considered earlier for linear modulation methods. When the signal wave contains a number of message channels in frequency-division multiplex, intermodulation distortion causes interchannel interference not encountered with the previous linear modulation methods. In the particular case of voice channels this effect is known as interchannel crosstalk, which is usually noiselike but may contain an intelligible component. As another example, in color video transmission two channels are involved, one carrying the broadband signal representing the brightness or luminance, the other carrying a narrow-band signal containing the color or chrominance information. Interference can then be encountered in the color channel,

General 309

owing particularly to intermodulation between the color carrier and the low-frequency components of the luminance signal. This interference causes so-called differential amplitude and phase distortion of the color carrier, which may result in excessive fluctuations in the hue and intensity of the colors unless appropriate measures are taken. These may consist of pre-emphasis, in which the low-frequency components are reduced in the transmitted luminance signal and restored at the receiver. Such transmission impairments are also encountered when intermodulation distortion is caused by attenuation and phase deviations in FM systems, as dealt with in Chapter 6.

This chapter is concerned with conventional methods of determining distortion resulting from a nonlinear transmittance in the presence of amplitude variations in the signal, as encountered in baseband and carrier amplitude modulation systems and in multicarrier frequency or phase modulation systems. Most of the exposition applies to analog systems, where a high signal-to-noise ratio is required so that only minor distortion can be tolerated. However, nonlinear distortion may also need consideration in certain digital systems, as discussed at the end of this chapter.

In early investigations of intermodulation distortion in frequency-division carrier systems [1–2], the signal in each voice channel was represented by a sine wave of equivalent average power. Various modulation products were determined, as resulting from second and third harmonic distortion and from intermodulation of two or more sine waves. This is a rather cumbersome method, particularly as applied to frequency-division multiplex systems with a large number of voice channels. In the latter case the combined signals that are averaged over sufficiently long intervals will have statistical properties similar to those of random noise. This facilitates a simpler determination of average intermodulation distortion with the aid of methods and techniques based on an equivalent uniform signal power spectrum, or so-called equivalent "noise loading" that do not specifically consider various intermodulation products.

Transmission distortion, owing to a nonlinear characteristic, will not be entirely nonlinear, but will contain a component of the same shape as the signal that does not cause interference, and must therefore be excluded. The linear and nonlinear components of distortion power and of distortion spectra can be determined from the autocorrelation function. To determine the autocorrelation function of the received signal it is necessary to evaluate a double integral involving the channel transfer characteristic and the joint probability density function for two signal amplitudes occurring at different times, as shown in Appendix 2. For signals having a Gaussian probability distribution, as is the case for random noise, the autocorrelation function has been determined by evaluation of the integrals for a nonlinear characteristic that is represented by power series [3]. For Gaussian signals, evaluation of the integrals for power series and other representations of the nonlinearity is ordinarily facilitated by the transform method introduced by Bennett and Rice [4], and outlined in Appendix 3. The general form of solution thus obtained for the autocorrelation function has been given by Rice [5] and is developed in further detail in more recent tutorial publications [6–7]. The solution is in the form of an infinite power series of the correlation function of the input signal, with coefficients that depend on the Fourier transform of the transfer characteristic. For signals with the properties of random noise, solutions obtained by this method have been published for the output autocorrelation function and power spectrum for certain transfer characteristics, such as a power law characteristic [6–7] and an ideal linear limiter characteristic [7–8].

In the analysis of transmission-systems performance, it is convenient to distinguish between wide-band channels or signal spectra that comprise a large number of octaves, and narrow-band channels or spectra of small bandwidth in relation to the midband frequency. A distinct advantage of the transform method is that the solution for wide-band spectra with a Gaussian probability distribution is conveniently extended to narrow-band spectra with a Rayleigh probability distribution of the envelope.

In addition to distortion from nonlinearity in the amplitude characteristic $A(\omega)$ it may be necessary to consider distortion due to variations in the phase characteristic $\psi(\omega)$ with the amplitude of the signal. These variations give rise to unwanted phase modulation in the transmitted wave by AM-PM conversion and to resultant intermodulation distortion. Such AM-PM conversion is of particular importance in microwave transmitters, such as klystrons and traveling wave tubes. The resultant intermodulation distortion may need consideration even in single-carrier microwave FM systems, owing to small unavoidable envelope fluctuations, as discussed in Section 6.17, and may be of first importance in single-sideband microwave transmission [9] and in multicarrier FM systems [10] where pronounced envelope fluctuations are encountered.

In band limited nonlinear channels the received wave will differ from the transmitted wave partly because of variations with frequency in the amplitude and phase characteristics and partly because of the nonlinearity. In most analog transmission systems only minor transmission distortion can be tolerated. It is then permissible to determine transmission impairments from attenuation and phase distortion on the premise of a linear transducer, as in Chapter 6, and to evaluate separately transmission impairments owing to nonlinearity, assuming a nonlinear transducer of unlimited bandwidth (zero memory) in which the transmittance does not vary with frequency. This is a basic simplifying premise underlying conventional methods of determining intermodulation distortion that will also be used here.

8.1 Analytical Representation of Nonlinear Transmittance

Let the input to the transducer be $z(t)$, which may designate a sine wave, a combination of sine waves, or a random variable characterized by certain statistical parameters. As mentioned previously, for purposes of determining intermodulation distortion owing to nonlinearity, the transducer will be assumed to have unlimited bandwidth and a transmittance that does not vary with frequency. The signal $z(t)$ will have a restricted bandwidth and frequency components of distortion outside the transmission band will be eliminated by filters at the receiver. For this reason the assumption of unlimited bandwidth of the nonlinear element in the channel is a legitimate approximation in application to bandlimited channels.

On the foregoing premise, the output wave $v(t)$ will be a function $g[z(t)] = g(z)$ of the amplitude of the input wave only, but not of its spectrum, and can be represented as

$$v(t) = g[z(t)] \tag{8.1}$$

To facilitate analysis, it is in general necessary to approximate $g(z)$ by an explicit expression. To this end $g(z)$ is often represented by a power series as

$$g(z) = g_e(z) + g_o(z) \tag{8.2}$$

where the even and odd functions are

$$g_e(z) = a_0 + a_2 z^2 + a_4 z^4 + \cdots, \tag{8.3}$$

$$g_o(z) = a_1 z + a_3 z^3 + a_5 z^5 + \cdots. \tag{8.4}$$

In accordance with McLaurin's theorem, the coefficients a_k are related to $g(z)$ by

$$a_k = \frac{1}{k!} \left[\frac{d^k g(z)}{dz^k} \right]_{z=0} \tag{8.5}$$

When $g(z)$ is given by an explicit function, the latter relation facilitates solution by the above power series expansion. However, if $g(z)$ is in the form of a curve derived from measurements, it would hardly be feasible to determine a_k from the kth derivative of $g(z)$ at $z = 0$, since this would entail unusual experimental accuracy. Direct determination of the instantaneous transfer characteristic $g(z)$ would entail measurements with an input consisting of impulses or unit step functions of various amplitudes. This would in general be unfeasible because distortion caused by bandwidth limitation in actual transducers may exceed that resulting from nonlinearity. For this reason, the instantaneous transfer characteristic is ordinarily determined from measured harmonic or intermodulation distortion with an input of two or more sine waves, as discussed later.

In most analog systems only a minor nonlinearity can be tolerated, such as represented by the quadratic term in (8.3) or the cubic term in (8.4). Determination of intermodulation distortion is then fairly simple. With an actual power series representation, determination of distortion is rendered complicated for the reason that in addition to distortion, owing to each term $a_k z^k$, it is necessary to consider distortion from cross-products of the various terms. Moreover, at high input power the distortion becomes extremely sensitive to the amplitudes of the higher coefficients a_k in the expansion, which must be determined with great accuracy. For this reason the foregoing expansion is in effect applicable only when $g(z)$ is an explicit function. In the latter case, the transform method mentioned in the introduction is ordinarily more convenient.

Some transducers are almost linear below a certain maximum input amplitude c, but have nearly constant output for higher amplitudes. A convenient idealization in this case for a transmittance with even symmetry is a linear rectifier with the characteristic

$$\begin{aligned} g_e(z) &= O & \text{for } z < O \\ &= \alpha z & \text{for } O < z < c \\ &= \alpha c & \text{for } z > c \end{aligned} \quad (8.6)$$

For a transmittance with odd symmetry a convenient idealization is a linear limiter as represented by

$$\begin{aligned} g_0(z) &= \alpha z & -c < z < c \\ &= -\alpha c & z < -c \\ &= \alpha c & z > c \end{aligned} \quad (8.7)$$

8.2 Steady-State Sine-Wave Transmittance

Let the input be a sine wave

$$z(t) = A \cos \omega_0 t. \quad (8.8)$$

The output is then of the general form

$$v(t) = g[A \cos \omega_0 t] \quad (8.9)$$

$$= \sum_{m=0,1,2,\ldots}^{\infty} A_m \cos m\omega_0 t. \quad (8.10)$$

Steady-State Sine-Wave Transmittance

Of principal interest in applications to bandpass channels is the amplitude A_1 of the fundamental component, since the d–c component and the various harmonics would be eliminated by filters. This component can be determined by Fourier series analysis of (8.9) with respect to ω_0.

With $\omega_0 t = x$ the following relation is thus obtained for a transmittance with odd symmetry:

$$g_1(A) = \frac{1}{\pi}\int_{-\pi}^{\pi} g_0(A\cos x)\cos x\, dx$$

$$= \frac{4}{\pi}\int_{0}^{\pi/2} g_0(A\cos x)\cos x\, dx. \tag{8.11}$$

For a transmittance with even symmetry only a d–c component and even harmonics will be present.

By way of example, let $g_0(z) = a_k z^k = a_k A^k (\cos x)^k$, $k = 1, 3, 5, \ldots$. In this case (8.11) yields

$$g_1(A) = a_k A^k \frac{4}{\pi}\int_0^{\pi/2}(\cos x)^{k+1}\, dx$$

$$= a_k C(k) A^k, \tag{8.12}$$

where

$$C(k) = \frac{k!}{\left(\dfrac{k-1}{2}\right)!\left(\dfrac{k+1}{2}\right)!\,2^{k-1}}. \tag{8.13}$$

When (8.4) applies, the amplitude of the fundamental thus becomes

$$g_0(A) = A_1 = a_1 C(1)A + a_3 C(3) A^3 + a_5 C(5) A^5 + \cdots, \tag{8.14}$$

where $C(1) = 1$, $C(3) = 3/4$, $C(5) = 5/8$, $C(7) = 35/64$, etc. It will be noted that the sine-wave transmittance differs from the instantaneous transmittance by the factor $C(k)$ in the term $a_k A^k$. The above factors $C(k)$ given by (8.13) appear in later relations on intermodulation distortion in narrow bandpass channels.

For a transducer with the limiter characteristic (8.7) the transmittance $g_1(A)$ is for $|A| < c$:

$$g_1(A) = A_0 = \alpha A, \tag{8.15}$$

while for $|A| > c$ (8.11) yields

$$g_1(A) = A_0(A) = \frac{4\alpha}{\pi}\left(c\int_0^{x_1}\cos x\, dx + A\int_{x_1}^{\pi/2}\cos^2 x\, dx\right)$$

$$= \alpha c \frac{2}{\pi}\left(\left(1 - \frac{c^2}{A^2}\right)^{1/2} + \frac{A}{c}\sin^{-1}\frac{c}{A}\right) \tag{8.16}$$

$$= \alpha c \frac{4}{\pi} \quad \text{for } A/c = \infty.$$

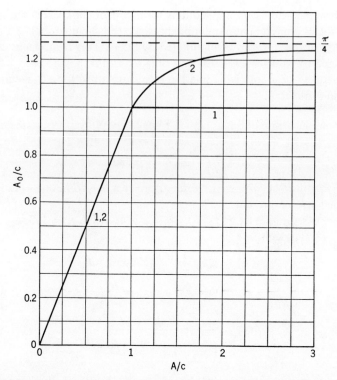

Figure 8.1 Instantaneous transmittance, 1, and sine-wave transmittance, 2, of a linear limiter.

The limit x_1 is determined from the condition $A \cos x_1 = c$ or $x_1 = \cos^{-1} c/A$. In the interval $x = O$ to $x = x_1$ the amplitude $A \cos x$ exceeds c while in the interval x_1 to $\pi/2$ the amplitude is below c.

The above sine-wave transmittance $g_1(A)$ of a linear limiter is shown in Fig. 8.1 as a function of A/c.

8.3 Output Signal and Distortion Powers

To determine the effect of intermodulation distortion in communication systems, it is in general necessary to obtain the power spectrum of the distortion. This may entail rather elaborate and difficult analysis, even for the simplest case of Gaussian signals, to be discussed later. In many cases the total average distortion power affords a satisfactory approximate criterion, and can be obtained by simpler analytical procedures, as outlined here.

Output Signal and Distortion Powers

Let the input be a random variable with a probability $p(z)$ that z is between z and $z + dz$. The average output power is then

$$P_0 = S_0 + D_0 = \int_{-\infty}^{\infty} g^2(z) p(z) \, dz, \quad (8.17)$$

where S_0 is the average undistorted signal power and D_0 the average distortion power.

The function $g(z)$ will contain a component $g_l(z)$ linearly related to z, that is, $g_l(z) = cz$. The coefficient c will depend on the characteristic $g(z)$ and on the shape of the input function z. When a large number of random signals with probability density $p(z)$ are considered, the variance of the output power with respect to an average linear component $\underline{c}z$ is given by

$$P_0 - \underline{c}^2 z^2 = \int_{-\infty}^{\infty} [g(z) - \underline{c}z]^2 p(z) \, dz$$

$$= \int_{-\infty}^{\infty} [g^2(z) - 2\underline{c}zg(z) + \underline{c}^2 z^2] p(z) \, dz. \quad (8.18)$$

The minimum value of the integral is obtained with $\underline{c}$ so chosen that the derivative with respect to $\underline{c}$ vanishes. This gives

$$\underline{c} \int_{-\infty}^{\infty} z^2 p(z) \, dz = \int_{-\infty}^{\infty} zg(z) p(z) \, dz, \quad (8.19)$$

or

$$\underline{c} = \frac{1}{S} \int_{-\infty}^{\infty} zg(z) p(z) \, dz, \quad (8.20)$$

where S is the average input signal power given by

$$S = \int_{-\infty}^{\infty} z^2 p(z) \, dz. \quad (8.21)$$

The linear component of the average output power is

$$S_0 = \underline{c}^2 S = \frac{1}{S} \left(\int_{-\infty}^{\infty} zg(z) p(z) \, dz \right)^2. \quad (8.22)$$

The nonlinear part of the output, or the nonlinear distortion power, becomes

$$D_0 = \int_{-\infty}^{\infty} g^2(z) p(z) \, dz - S_0. \quad (8.23)$$

Let $g(z)$ be resolved into an even and an odd function in accordance with (8.2) and let probability density be an even function, that is

$$p(-z) = p(z).$$

Relations (8.22) and (8.23) can then be written

$$S_0 = \frac{2}{S} \int_0^\infty z g_0(z) \, p(z) \, dz \qquad (8.24)$$

$$D_0 = 2 \int_0^\infty g_0^2(z) \, p(z) \, dz + 2 \int_0^\infty g_e^2(z) \, p(z) \, dz \qquad (8.25)$$

The first relation shows that for signals with the foregoing even probability density, the output power depends only on the odd component of the transmittance. The second relation shows that the distortion power resulting from the two components can be determined separately and combined by direct addition.

In the case of narrow bandpass channels, let $p(\bar{z})$ be the probability density of the envelope fluctuations and $g_1(\bar{z})$ the sine-wave transmittance as obtained from (8.11) with $A = \bar{z}$. Relations (8.22) and (8.23) are then replaced by

$$\bar{S}_0 = \frac{1}{\bar{S}} \left(\int_0^\infty \bar{z} g_1(\bar{z}) \, p(\bar{z}) \, d\bar{z} \right)^2, \qquad (8.26)$$

$$\bar{D}_0 = \int_0^\infty g_1^2(\bar{z}) \, p(\bar{z}) \, d\bar{z} - \bar{S}_0, \qquad (8.27)$$

where $\bar{S} = 2S$ is the average envelope power of the input signal, with corresponding definitions $\bar{S}_0 = 2S_0$ and $\bar{D}_0 = 2D_0$.

The average input and output powers can alternately be expressed as follows:

$$S = \lim_{T \to \infty} \frac{1}{T} \int_0^T z^2(t) \, dt$$

$$= \int_0^\infty W(\omega) \, d\omega \qquad (8.28)$$

$$= \int_{-\infty}^\infty z^2 p(z) \, dz;$$

$$S_0 + D_0 = \lim_{T \to \infty} \int_0^T g^2(t) \, dt$$

$$= \int_0^\infty W_0(\omega) \, d\omega \qquad (8.29)$$

$$= \int_{-\infty}^\infty g^2(z) \, p(z) \, dz,$$

where

$W(\omega)$ = power spectrum of input waves,
$W_0(\omega)$ = power spectrum of output waves
$= \underline{W}_0(\omega) + \underline{\underline{W}}_0(\omega)$,
$\underline{W}_0(\omega) = W(\omega)S_0/S$ = linear component,
$\underline{\underline{W}}_0(\omega)$ = nonlinear component of output power spectrum.

8.4 Autocorrelation Functions and Power Spectra

In most problems involving a nonlinear channel it is desirable to determine not only the total output signal and distortion powers just shown, but also the power spectrum of the output. To this end it is necessary to introduce more elaborate statistical methods as discussed in Appendix 2. These in effect introduce one or more statistical parameters that specify the expected time variation in the input waves. Equations (8.28) and (8.29) are replaced by the following relations

$$R(\tau) = \lim_{T \to \infty} \frac{1}{T} \int_0^T z(t)\, z(t + \tau)\, dt$$

$$= \int_0^\infty W(\omega) \cos \omega\tau\, d\omega + C \tag{8.30}$$

$$= \int_{-\infty}^\infty \int_{-\infty}^\infty uv\, p(u,v)\, du\, dv,$$

$$R_0(\tau) = \lim_{T \to \infty} \frac{1}{T} \int_0^T g(t)\, g(t + \tau)\, dt$$

$$= \int_0^\infty W_0(\omega) \cos \omega\tau\, d\omega + C_0 \tag{8.31}$$

$$= \int_{-\infty}^\infty \int_{-\infty}^\infty g(v)\, g(u)\, p(u,v)\, du\, dv.$$

It will be noted that $R(O) = S$ and $R_0(O) = S_0 + D_0$.

The constants C and C_0 may be required to avoid infinite values of $R(\tau)$ or $R_0(\tau)$ for certain forms of spectra for which the integrals do not converge. For example, let the power spectrum vary as $W(\omega) = 1/\omega^2$ so that $R(\tau) = \infty + C$. Since only the variation in $R(\tau)$ with τ is important, the constant C is, in the above case, chosen as $C = -R(O)$. The factor $\cos \omega\tau$ is then replaced by $(\cos \omega\tau - 1)$ and $R(\tau)$ is finite.

The linear component of $R_0(\tau)$ can ordinarily be determined more readily by the single integral for S_0 in Section 8.3 and is

$$\underline{R}_0(\tau) = R(\tau)S_0/S. \qquad (8.32)$$

When this component is subtracted, the resultant nonlinear component of the autocorrelation function becomes

$$\underline{\underline{R}}_0(\tau) = R_0(\tau) - \underline{R}_0(\tau). \qquad (8.33)$$

When $R(\tau)$ and $R_0(\tau)$ are known, the power spectra are obtained from the relations

$$W(\omega) = \frac{2}{\pi}\int_0^\infty R(\tau)\cos\omega\tau\, d\tau, \qquad (8.34)$$

$$W_0(\omega) = \frac{2}{\pi}\int_0^\infty R_0(\tau)\cos\omega\tau\, d\tau, \qquad (8.35)$$

$$\underline{\underline{W}}_0(\omega) = \frac{2}{\pi}\int_0^\infty \underline{\underline{R}}_0(\tau)\cos\omega\tau\, d\tau, \qquad (8.36)$$

$$\underline{W}_0(\omega) = \frac{S_0}{S} W(\omega). \qquad (8.37)$$

Evaluation of the double integrals in (8.30) or (8.31) is in general quite difficult, even for Gaussian probability functions $p(u,v)$, except for some simple forms of nonlinearity.

8.5 Solution by Transform Method

For signals with the statistical properties of Gaussian noise, evaluation of the double integral (8.31) is ordinarily facilitated by the transform method referred to before. With this method the transfer characteristic $g(z)$ is represented by an appropriate contour integral as

$$g(z) = \frac{1}{2\pi}\int_C f(iw)\, e^{iwz}\, dw, \qquad (8.38)$$

where the contour C may be between $-\infty$ and ∞ as for a Fourier integral, or may follow some other path depending on $f(iw)$. The latter function is the transform of $g(z)$.

The joint probability density $p(u,v)$ is also represented by its Fourier transform, referred to as its joint characteristic function. As outlined further in Appendix 3, the general solution for $R_0(\tau)$ is of the following form for an even junction $g_e(z)$:

$$R_0(\tau) = \sum_{k=0,2,4} \frac{h_{0k}^2 R^k(\tau)}{k!}, \qquad (8.39)$$

Solution by Transform Method

while for odd functions $g_0(z)$

$$R_0(\tau) = \sum_{k=1,3,5} \frac{h_{0k}^2 R^k(\tau)}{k!}, \tag{8.40}$$

where

$$h_{0k} = \frac{i^k}{2\pi} \int_C f(iw) w^k e^{-\sigma^2 w^2/2} \, dw. \tag{8.41}$$

The foregoing equations give the total autocorrelation function, and apply for both wide-band and narrow-band signals. In the particular case of a narrow bandpass spectrum, relation (3.21) applies for the input correlation function, so that in the above relations

$$R(\tau) = \bar{R}(\tau) \cos \omega_0 \tau \tag{8.42}$$

$$R^k(\tau) = \bar{R}^k(\tau) \cos^k \omega_0 \tau. \tag{8.43}$$

Only the component of (8.40) in $\cos \omega_0 \tau$ is ordinarily of interest, since the spectra about higher harmonics are eliminated by relatively narrow bandpass filters in the output. For each value of k in (8.40) this fundamental or pass-band component is obtained by replacing $R^k(\tau)$ by

$$R_1^k(\tau) = \bar{R}^k(\tau) \, C(k) \cos \omega_0 \tau, \tag{8.44}$$

where $C(k)$ is given by (8.13) and $\bar{R}(\tau)$ is the envelope of the autocorrelation function of the bandpass power spectrum. This autocorrelation function is obtained from relations (3.22) through (3.25) with $W_n(u) = W(u)$, the bandpass power spectrum of the input signal. In the important case of a power spectrum with even symmetry about the midband frequency

$$\bar{R}(\tau) = \int_{-\infty}^{\infty} W(u) \cos u\tau \, du. \tag{8.45}$$

With (8.44) and (8.13) in (8.40), the output correlation function with respect to the pass-band becomes

$$R_0(\tau) = \cos \omega_0 \tau \sum_{k=1,3,5}^{\infty} \frac{h_{0k}^2 \bar{R}^k(\tau)}{2^{k-1} \left(\frac{k-1}{2}\right)! \left(\frac{k+1}{2}\right)!}. \tag{8.46}$$

An advantage of the transform method is that the solution for low-pass channels is readily extended to narrow bandpass channels by insertion of the factor $C(k)$ in the summation, as above. An alternate and more cumbersome procedure would be to determine the envelope transfer characteristic from (8.11) and the corresponding function $f(iw)$ together with the characteristic function of the probability density of the envelope.

It is to be noted that the foregoing relations for bandpass channels apply when there is no correlation between the spectral components, as with Gaussian bandpass spectra. In the case of double-sideband AM systems, there is complete correlation between the two sideband spectra, and in this case the same relations as for wideband transmission apply, except for a translation in frequency.

In the following sections various important special cases will be considered.

8.6 Square-Law Nonlinearity

Let
$$g(z) = a_0 + a_1 z + a_2 z^2, \tag{8.47}$$

and
$$z(t) = A_0 + A_1 \cos x_1 + A_2 \cos x_2 + A_3 \cos x_3, \tag{8.48}$$

where $x_j = \omega_j t$.

With (8.48) in (8.47) the output wave becomes
$$v(t) = v_0(t) + v_1(t) + v_2(t), \tag{8.49}$$

where
$$v_0 = a_0 A_0 + \frac{a_2}{2}(A_1^2 + A_2^2 + A_3^2),$$

$$v_1 = a_1(A_1 \cos x_1 + A_2 \cos x_2 + A_3 \cos x_3),$$

$$v_2 = \frac{a_2}{2}(A_1^2 \cos 2x_1 + A_2^2 \cos 2x_2 + A_3^2 \cos 2x_3)$$
$$+ a_2 A_1 A_2 [\cos(x_1 + x_2) + \cos(x_1 - x_2)]$$
$$+ a_2 A_2 A_3 [\cos(x_2 + x_3) + \cos(x_2 - x_3)]$$
$$+ a_2 A_1 A_3 [\cos(x_1 + x_3) + \cos(x_1 - x_3)].$$

Here v_0 is a d–c component and v_1 a linear signal component while v_2 represents square law distortion, usually referred to as second-order distortion. In baseband systems square-law distortion will ordinarily be greatest at the lower frequencies. With bandpass transmission all distortion components fall outside the passband, as indicated in Fig. 8.2, provided that the highest sine-wave frequency is less than twice the lowest frequency. The adverse effects of square-law distortion can thus be avoided by appropriate choice of transmission band.

In the foregoing case of three sine waves, the frequencies and amplitudes of the second-order intermodulation products are obtained by

Square-Law Nonlinearity

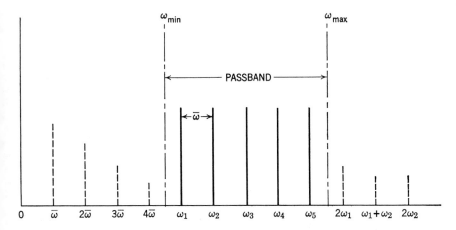

Figure 8.2 Second-order intermodulation products with 5 equally spaced sine waves. All products are outside pass-band if $\omega_{max} < 2\omega_{min}$.

taking the various possible combinations of two sine waves. This applies with any number of sine waves. Of particular interest in connection with frequency-division multiplex systems is a number of equally spaced sine waves, as indicated in Fig. 8.3, of amplitudes $A(js)$ at the frequency js. The amplitude of the second-order intermodulation product at the frequency is then becomes

$$v_2(is) = a_2 \sum_{j=0}^{n} A(js) A(is - js) \qquad (8.50)$$

At the frequency is the average distortion power is $\frac{1}{2}v_2^2(is) = W_2(is)$. With $W(is) = \frac{1}{2}A^2(js)$ and $W(is - js) = \frac{1}{2}A^2(is - js)$ the following relation is

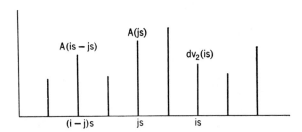

Figure 8.3 Second-order intermodulation products with sine waves at uniform frequency separation s. Voltage component at is owing to intermodulation of sine waves at frequencies js and $is - js$ is $dv_2(is) = A(js)A(is - js)$.

obtained for $W_2(is) = \tfrac{1}{2}v_2^2(is)$:

$$W_2(is) = \tfrac{1}{2}a_2^2 \sum_{j=0}^{n} A^2(js)\, A^2\,(is - js) \qquad (8.51)$$

$$= a_2^2 2 \sum_{j=0}^{n} W(js)\cdot W(is - js).$$

In the limit $s \to O$, the following continuous power spectrum is obtained with $is = \omega$ and $js = \omega_1$

$$W_2(\omega) = a_2^2 2 \int_O^\infty W(\omega)\, W(\omega - \omega_1)\, d\omega. \qquad (8.52)$$

This conforms with relation (19) on Appendix 3, obtained by the transform method outlined in Section 8.5.

8.7 Cube-Law Nonlinearity—Multiple Sine-Wave Transmission

Let

$$g(z) = a_0 + a_1 z + a_2 z^2 + a_3 z^3. \qquad (8.53)$$

With an input wave (8.48), the output resulting from the first three terms in (8.53) is then given by (8.49). The cubic term $a_3 z^3$ gives rise to the following additional linear component:

$$v_3 = \tfrac{3}{4} a_3 (A_1 (A_1^2 + 2A_2^2 + 2A_3^2) \cos x_1 + A_2 (A_2^2 + 2A_1^2 + 2A_3^2) \cos x_2$$
$$+ A_3 (A_3^2 + 2A_1^2 + 2A_2^2) \cos x_3), \qquad (8.54)$$

and also generates the following cube-law distortion component

$$y_3 = \tfrac{1}{4} a_3 (A_1^3 \cos 3x_1 + A_2^3 \cos 3x_2 + A_3^3 \cos 3x_3)$$

$$+ \tfrac{3}{4} a_3 \{ A_1^2 A_2 [\cos (2x_1 + x_2) + \cos (2x_1 - x_2)]$$

$$+ A_1^2 A_3 [\cos (2x_1 + x_3) + \cos (2x_1 - x_3)]$$

$$+ A_2^2 A_1 [\cos (2x_2 + x_1) + \cos (2x_2 - x_1)]$$

$$+ A_2^2 A_3 [\cos (2x_2 + x_3) + \cos (2x_2 - x_3)]$$

$$+ A_3^2 A_1 [\cos (2x_3 + x_1) + \cos (2x_3 - x_1)]$$

$$+ A_3^2 A_2 [\cos (2x_3 + x_2) + \cos (2x_3 - x_2)] \}$$

$$+ \tfrac{3}{2} a_3 A_1 A_2 A_3 [\cos (x_1 + x_2 + x_3) + \cos (x_1 + x_2 - x_3)$$

$$+ \cos (x_1 - x_2 + x_3) + \cos (x_1 - x_2 - x_3)]. \qquad (8.55)$$

Cube-Law Nonlinearity—Multiple Sines

The cube-law distortion $y_3(t)$ is ordinarily referred to as third-order distortion. It should be noted that "second," "third," and "higher" order distortion as applied to intermodulation products are conventional engineering definitions. In contrast to the usual mathematical definition of order, these designations do not imply that the products are less important the higher the order.

With any number of sine waves the amplitudes and frequencies of the third-order distortion components are obtained by taking the various combinations of three sine waves in accordance with rules that follow from (8.55).

An important case from the standpoint of frequency division multiplex systems with uniformly spaced channels is that of a number of equally spaced sine waves. With unmodulated sine waves the various modulation products will have frequencies that coincide with those of the sine waves, in addition to the third harmonics that can be eliminated by filters. With four equally spaced sine waves of equal amplitudes the third order modulation products will be as indicated in Fig. 8.4.

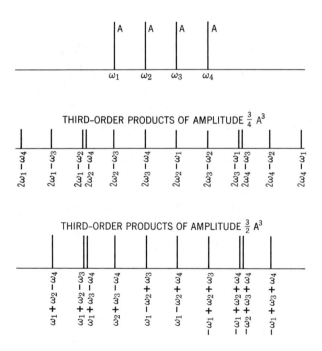

Figure 8.4 Third-order intermodulation products with input of four equally spaced sine waves of equal amplitudes.

For any number of equally spaced sine waves the number of third-order intermodulation products can be determined from relations given by Bennett [1]. With n sine waves, let $Q_n(m)$ designate the number of products of the general type $2x_1 - x_2$, $2x_2 - x_1$, etc., that coincides with the sine wave in position m in relation to the innermost sine wave. Thus, in Fig. 8.4, $m = 0$ for the second sine wave, $m = -1$ for the first, and $m = 1$ for the third. Correspondingly, let $P_n(m)$ designate the number of products of the type $x_1 \pm x_2 \pm x_3$. The number of products at these three inner sine-wave frequencies are then as given in Table 1.

Table 1. *Number of Third-Order Products*

n	$Q_n(-1)$	$Q_n(0)$	$Q_n(1)$	$P_n(-1)$	$P_n(0)$	$P_n(1)$
2	1	0	1	0	0	0
4	1	1	1	1	2	2
6	2	2	2	6	7	7
8	3	3	3	14	15	15
12	5	5	5	40	40	40
∞	$\frac{n}{2} - 1$	$\frac{n}{2} - 1$	$\frac{n}{2} - 1$	$\frac{3}{8}n^2$	$\frac{3}{8}n^2$	$\frac{3}{8}n^2$

Let all sine waves have equal amplitudes A. In accordance with (8.55) the amplitude of each third-order product of frequency $2x_1 - x_2$, $2x_2 - x_1$, etc., is then $3a_3A^3/4$ while that of each product of frequency $x_1 \pm x_2 \pm x_3$ is $3a_3A^3/2$. When all possible phases of the sine waves are considered, the average powers of the intermodulation products at a particular frequency is proportional to the numbers $Q_n(m)$ and $P_n(m)$. The combined distortion power at the sine-wave position m is thus

$$d_0 = a_3^2 \left(\left(\frac{3}{4}\right)^2 Q_n(m) + \left(\frac{3}{2}\right)^2 P_n(m) \right) A^6. \tag{8.56}$$

At any frequency, the linear component (8.54) becomes $\frac{3}{4}a_3(2n - 1)A^3$, so that the average undistorted output power at any frequency is

$$s_0 = \left(a_1 + a_3 \frac{3}{4}(2n - 1)A^2 \right)^2 A^2. \tag{8.57}$$

The ratio of average distortion power d_0 to average output power s_0 in one sine wave is thus

$$\frac{d_0}{s_0} = a_3^2 \frac{\left(\frac{3}{4}\right)^2 Q_n(m) + \left(\frac{3}{2}\right)^2 P_n(m)}{\left(a_1 + 3a_3 \frac{2n-1}{2n} S \right)^2} \cdot \left(\frac{2S}{n}\right), \tag{8.58}$$

where the substitution $S = nA^2/2$ has been made. For $n \to \infty$, $Q_n(m) \to \frac{n}{2} - 1$ and $P_n(m) \to 3n^2/8$ so that

$$\frac{d_0}{S_0} = a_3^2 \frac{27}{8} \frac{S^2}{(a_1 + 3a_3 S)^2}. \tag{8.59}$$

In the foregoing relations sine waves of equal amplitudes were assumed. Relations (8.56) through (8.59) also apply when each sine wave varies in amplitude, provided the average power is the same for all sine waves. The ratio (8.59) is obtained by a different derivation in the Section 8.9 for a bandpass channel and a Gaussian signal with a flat-power spectrum.

8.8 Cube-Law Nonlinearity—Gaussian Wideband Signals

With any probability density $p(z)$ of the input wave, the undistorted output power S_0, together with the total average distortion power D_0 can be obtained from relations given in Section 8.3. In the particular case of Gaussian wideband signals, the spectrum comprises a large number of octaves and the probability density of the instantaneous amplitude is then

$$p(z) = \left(\frac{1}{2\pi}\right)^{1/2} \frac{1}{\sigma} e^{-z^2/2\sigma^2}, \tag{8.60}$$

where $\sigma^2 = S =$ average signal power.

With

$$g(z) = g_0(z) = a_1 z + a_3 z^3, \tag{8.61}$$

the following basic function is involved in solutions of (8.22) and (8.23)

$$G(x) = \int_{-\infty}^{\infty} z^x p(z)\, dz$$

$$= S^{x/2} \left(\frac{2^x}{\pi}\right)^{1/2} \Gamma\left(\frac{x+1}{2}\right). \tag{8.62}$$

The latter relation yields for various values of x:

$$x = 2 \quad 4 \quad 6 \quad 8$$
$$G(x) = S \quad 3S^2 \quad 3 \cdot 5 S^3 \quad 3 \cdot 5 \cdot 7 S^4$$

Relations (8.22) and (8.23) become

$$S_0 = \frac{1}{S}[a_1 G(2) + a_3 G(4)]^2$$
$$= S[a_1 + 3a_3 S]^2, \tag{8.63}$$

$$D_0 = a_3^2 \left(G(6) - \frac{1}{S} G^2(4)\right)$$
$$= a_3^2[3 \cdot 5 - 3^2]S^3 = 6a_3^2 S^3. \tag{8.64}$$

Thus
$$\frac{D_0}{S_0} = a_3^2 \frac{6S^2}{(a_1 + 3a_3 S)^2}. \tag{8.65}$$

The distortion-power spectrum can be obtained with the aid of the autocorrelation function (8.40). For a cube-law nonlinearity, this function is determined as shown in Appendix 3 and becomes

$$R_3(\tau) = \underline{R}_3(\tau) + \underset{\sim}{R}_3(\tau), \tag{8.66}$$

$$\underline{R}_3(\tau) = 9a_3^2 R(\tau) R^2(O) \tag{8.67}$$

$$\underset{\sim}{R}_3(\tau) = 6a_3^2 R^3(\tau). \tag{8.68}$$

The autocorrelation function for the undistorted output power becomes

$$\underline{R}_0(\tau) = [a_1 + 3a_3 R(O)]^2 R(\tau) \tag{8.69}$$

while the autocorrelation function for the distortion is

$$\underset{\sim}{R}_0(\tau) = \underset{\sim}{R}_3(\tau) = 6a_3^2 R^3(\tau). \tag{8.70}$$

With $R(O) = S$, $\underline{R}_0 = S_0$ and $\underset{\sim}{R}_0(O) = D_0$, relations (8.69) and (8.70) conform with (8.63) and (8.64).

In the particular case of a flat signal-power spectrum of bandwidth Ω, $W(\omega) = W = S/\Omega$ and (8.30) yields

$$R(\tau) = S \frac{\sin \Omega \tau}{\Omega \tau}. \tag{8.71}$$

The power spectrum of the nonlinear distortion is, in accordance with (8.36) and (8.70),

$$\underset{\sim}{W}_0(\omega) = 6a_3^2 S^3 \frac{2}{\pi} \int_O^\infty \left(\frac{\sin \Omega \tau}{\Omega \tau}\right)^3 \cos \omega \tau \, d\tau$$

$$= \frac{6a_3^2 S^3}{\Omega} \frac{2}{\pi} \int_O^\infty \left(\frac{\sin x}{x}\right)^3 \cos ax \, dx$$

$$= 6a_3^2 S^2 W M_3(a), \tag{8.72}$$

where $a = \omega/\Omega$ and

$$M_m(a) = \frac{2}{\pi} \int_O^\infty \left(\frac{\sin x}{x}\right)^m \cos ax \, dx$$

$$= \frac{m}{2^{m-1}} \sum_{i=0,1,2,\ldots} \frac{(-1)^i (m - 2i + a)^{m-1}}{i!(m-i)!} \tag{8.73}$$

The final term in the series is the last term for which $m - 2i + a > O$ and for which $i < m$.

For $0 < a < 1$:
$$M_3(a) = \frac{3}{2^2}\left(\frac{1}{3!}(3+a)^2 + \frac{1}{2!}(1+a)^2\right)$$
$$= \tfrac{1}{4}(3-a)^2. \tag{8.74}$$

For $1 < a < 3$:
$$M_3(a) = \frac{1}{4}\left(3 - a^2 + \frac{3}{2}(1-a)^2\right), \tag{8.75}$$

and for $a > 3$, $M_3(a) = 0$. The ratio of distortion powers to output-signal power in a narrow band at ω becomes

$$\alpha(\omega) = \frac{3a_3^2 S^2(3-a^2)}{2(a_1 + 3a_3 S)^2}; \quad \text{for} \quad 0 < a < 1 \tag{8.76}$$

$$= \frac{3a_3^2 S^2\left(3 - a^2 + \frac{3}{2}(1-a)^2\right)}{2(a_1 + 3a_3 S)^2}; \quad \text{for} \quad 1 < a < 3 \tag{8.77}$$

$$= 0 \quad \text{for} \quad a > 3.$$

For $\omega = 0$ (8.76) gives

$$\alpha(0) = \frac{9}{2}\frac{a_3^2 S^2}{(a_1 + 3a_3 S)^2}. \tag{8.78}$$

This ratio is smaller than D_0/S_0 as given by (8.65) by a factor 3/4. Hence the ratio D_0/S_0 affords a safe engineering approximation.

The ratio $\alpha(\omega)/\alpha(0)$ is shown in Fig. 8.5 as a function of $a = \omega/\Omega$.

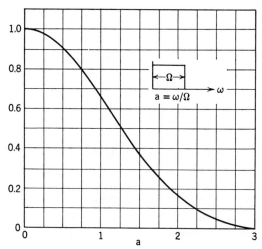

Figure 8.5 Normalized power spectrum of third-order intermodulation distortion for Gaussian low-pass signals with flat power spectrum.

8.9 Cube-Law Nonlinearity—Gaussian Narrowband Signals

The preceding relations apply for baseband channels and double-sideband bandpass channels, in which there is a complete correlation between the two sideband spectra of the signals. When there is no correlation between spectral components of the narrow-band signal spectrum, as in the case of Gaussian narrow-band noise, the undistorted output power and total average distortion power is obtained from (8.26) and (8.27) with a Rayleigh probability density of the envelope, as given by

$$p(\bar{z}) = \frac{\bar{z}}{\sigma} e^{-\bar{z}^2/2\sigma^2}, \tag{8.79}$$

where $\sigma^2 = S = \bar{S}/2$.

The function (8.62) is now replaced by

$$\bar{G}(x) = \frac{1}{\sigma^2} \int_0^\infty \bar{z}^{(x+1)} e^{-\bar{z}^2/2\sigma^2} \, d\bar{z}$$

$$= \bar{S}^{x/2} \left(\frac{x}{2}\right)! \tag{8.80}$$

From (8.26)

$$\bar{S}_0 = \frac{1}{\bar{S}} \left(\bar{a}_1 \bar{G}(2) + \frac{1}{\bar{S}} \bar{a}_3 \bar{G}(4)\right)^2$$

$$= \bar{S}[a_1 C(1) - 2! a_3 C(3) \bar{S}]^2$$

$$= \bar{S}(a_1 + 3a_3 S)^2 \tag{8.81}$$

From (8.27)

$$\bar{D}_0 = \bar{a}_3^2 \left[\bar{G}(6) - \frac{1}{\bar{S}} \bar{G}^2(4)\right]$$

$$= a_3^2 C^2(3) [\bar{S}^3 3! - \bar{S}^3 (2!)^2]$$

$$= a_3^2 \frac{9}{2} \bar{S} S^2. \tag{8.82}$$

The undistorted output power thus conforms with (8.63) for a low-pass channel while the total average distortion power (8.82) in the pass-band is less than for a low-pass channel as given by (8.64).

In accordance with (8.44) and (8.46) the autocorrelation function for the output signal is obtained by multiplying the first term in (8.66) by $C(1) = 1$ and the second term by $C(3) = 3/4$. Thus

$$R_1(\tau) = a_3^2 3[C(1) 3 \bar{R}(\tau) \bar{R}^2(0) + C(3) 2 \bar{R}^3(\tau)] \cos \omega_0 \tau. \tag{8.83}$$

Cube-Law—Gaussian Narrowband Signals

When the power spectrum of the signal has even symmetry about ω_0, $\bar{R}(\tau)$ is obtained from (8.45). For a flat signal-power spectrum $W(u) = W$ of semi-bandwidth Ω, $W = S/2\Omega$ and

$$\bar{R}(\tau) = \frac{S}{2\Omega} \int_{-\Omega}^{\Omega} \cos ut \, dt$$

$$= S \frac{\sin \Omega \tau}{\Omega \tau}. \tag{8.84}$$

This is the same relation (8.71) as for a baseband signal of average power S and bandwidth Ω.

The total distortion power is obtained from the second term in (8.83) with $\tau = 0$. With $\bar{R}(0) = S$ and $\cos \omega_0 \tau = 1$, the total distortion power becomes $D_0 = a_3^2 6 \cdot (3/4) S^3$ in conformance with (8.82). The distortion-power spectrum varies with $u = \omega - \omega_0$ in the same manner as the spectrum for a low-pass channel varies with ω, but is smaller by a factor $3/4$. The ratio $\alpha(u)$ is thus smaller than $\alpha(\omega)$ as given by (8.76) and (8.77) by the factor $3/4$ and becomes with $a = |u/\Omega|$ and for $0 < a < 1$:

$$\alpha(u) = \frac{9}{8} \frac{a_3^2 S^2 (3 - a^2)}{(a_1 + 3a_3 S)^2}, \tag{8.85}$$

$$\alpha(O) = a_3^2 \frac{27}{8} \frac{S^2}{(a_1 + 3a_3 S)^2}. \tag{8.86}$$

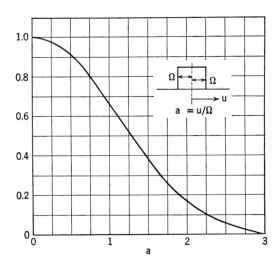

Figure 8.6 Normalized power spectrum of third-order intermodulation distortion for narrowband Gaussian signals with flat power spectrum.

The last relation conforms with (8.59), obtained in a different manner. The ratio (8.86) is smaller than the ratio D_0/S_0 obtained from (8.81) and (8.82) by the same factor 3/4 as in the case of a baseband channel.

The ratio $\alpha(u)/\alpha(O)$ is shown in Fig. 8.6 as a function of $|u/\Omega|$.

8.10 Power Series Nonlinearity

The preceding analytical procedure for a cubic nonlinearity can be extended to a power series transmittance. To determine the combined distortion it then becomes necessary to consider the cross-products of the various power series terms. This results in analytical complications, as shown next for a transmittance of the form (8.4) with a nonlinearity

$$h(z) = \sum_{m=3,5,7}^{n} a_m z^m. \tag{8.87}$$

Relations (8.22) and (8.23) can then be written

$$S_0 = S\left(1 + \frac{1}{S}\int_{-\infty}^{\infty} zh(z)\,p(z)\,dz\right)^2 \tag{8.88}$$

$$D_0 = \int_{-\infty}^{\infty} [z + h(z)]^2 p(z)\,dz - S_0$$

$$= \int_{-\infty}^{\infty} h^2(z)\,p(z)\,dz - \frac{1}{S}\left(\int_{-\infty}^{\infty} zh(z)\,dz\right)^2. \tag{8.89}$$

When $h(z)$ is of the form (8.87), the following relations apply

$$S_0 = S\left(1 + \frac{1}{S}\sum_{i=3,5,7,\ldots}^{n} a_i G(i+1)\right)^2 \tag{8.90}$$

$$D_0 = \sum_{i=3,5,7,\ldots}^{n}\sum_{j=3,5,7,\ldots}^{n} a_i a_j G(i+j) - \frac{1}{S}\left(\sum_{i=3,5,7,\ldots}^{n} G(i+1)\right) \tag{8.91}$$

where

$$G(x) = \int_{-\infty}^{\infty} z^x p(x)\,dx.$$

The linear component of the autocorrelation function becomes

$$\underline{R}_0(\tau) = \frac{S_0}{S} R(\tau), \tag{8.92}$$

and the nonlinear component is given by

$$\mathcal{R}_0(\tau) = \sum_{i=3,5,7,\ldots}^{n}\sum_{j=3,5,7,\ldots}^{n} H(i,j), \tag{8.93}$$

where

$$H(i,j) = \int_{-\infty}^{\infty} \int_{-\infty}^{\infty} u^i v^j p(u,v) \, du \, dv \qquad (8.94)$$

In the particular case of Gaussian signals $G(x)$ is given by (8.62). A solution for this case can be obtained [3] by direct evaluation of (8.94). Solution of (8.94) by the transform method results in the following relation

$$H(i,j) = H(j,i) = i!j! \sum_{k=1,3,5}^{\infty} \frac{R(O)^{r-k} R^k(\tau)}{2^{r-k} k! \left(\frac{i-k}{2}\right)! \left(\frac{j-k}{2}\right)!}, \qquad (8.95)$$

where $r = (i + j)/2$. The last term in the series is that for $i - k$ or $j - k = 0$.

By way of illustration

$$H(1,1) = R(\tau); \quad H(3,3) = 3[3R(\tau) R^2(O) + 2R^3(\tau)]$$

$$H(5,5) = 15[15R(\tau) R^4(O) + 40R^3(\tau) R^2(O) + 8R^5(\tau)]$$

$$H(3,5) = 15[3R(\tau) R^3(O) + 4R^3(\tau) R(O)]$$

For $n = 7$, the above formulations (8.90) and (8.91) in conjunction with (8.62) yield the relations

$$S_0 = S(1 + 3a_3 S + 15a_5 S^2 + 105a_7 S^3)$$

$$D_0 = 6a_3^2 S^3 + 720a_5^2 S^5 + 124110a_7^2 S^7 + 120a_3 a_5 S^4$$

$$\qquad + 1260 a_3 a_7 S^5 + 837900 a_5 a_7 S^7. \qquad (8.97)$$

For $n = 5$, (8.93) in conjunction with (8.95) gives

$$R_0(\tau) = a_3^2 6R^3(\tau) + a_5^2[600R^3(\tau)R^2(O) + 120R^5(\tau)] + 120 a_3 a_5 R^3(\tau) R(O)$$

$$\quad = [6a_3^2 + 120 a_3 a_5 R(O) + 600 a_5^2 R^2(O)]R^3(\tau) + 120 a_5^2 R^5(\tau). \qquad (8.98)$$

For $\tau = O$, $R_0(O) = D_0$ in conformance with (8.97) with $a_7 = O$.

The power spectrum of the nonlinear distortion is obtained from (8.36). In the particular case of a flat-power spectrum $W = S/\Omega$ the same procedure as in Section 8.9 results in the relation

$$W_0(\omega) = WR^2(O)[6a_3^2 + 120 a_3 a_5 R(O) + 600 a_5^2 R^2(O)]M_3(a)$$

$$\qquad + 120 W a_5^2 R^4(O) M_5(a), \qquad (8.99)$$

where $R(O) = S$ and $M_m(a)$ is given by (8.73).

The foregoing relations apply for wideband signal spectra and also for symmetrical narrow-band spectra in which there is complete correlation between the two sideband spectra, as in double-sideband transmission with a Gaussian baseband modulating wave. When there is no correlation between the spectral components of a narrow-band modulating wave, as with a narrow-band Gaussian signal, the relations for the autocorrelation function can be obtained by the procedure outlined previously. The following relations are thus obtained for $n = 5$.

$$S_0 = S(1 + a_3 S + 15 a_5 S^2)^2, \tag{8.100}$$

$$D_0 = S\left(\frac{9}{2} a_3^2 S^2 + 90 a_3 a_5 S^3 + 525 a_5 S^4\right), \tag{8.101}$$

$$R_0(\tau) = \left(\frac{9}{2} a_3^2 + 90 a_3 a_5 S + 450 a_5^2 S^2\right) \bar{R}^3(\tau) + 75 a_5^2 \bar{R}^5(\tau). \tag{8.102}$$

For $\tau = 0$, $\bar{R}(O) = S$ and $R_0(O) = D_0$ in conformance with (8.101).

For a flat narrow-band signal with power spectrum W of semi-bandwidth Ω the power spectrum of the nonlinear distortion becomes, in place of (8.99) for a low-pass channel

$$W_0(u) = W\bar{R}^2(O)\left(\frac{9}{2} a_3^2 + 90 a_3 a_5 S + 450 a_5^2 S^2\right) M_3(a) + 75 W S^4 M_5(a) \tag{8.103}$$

where $a = |u/\Omega|$ and $M_m(a)$ is defined by (8.73).

8.11 Linear Limiter Distortion—Sine-Wave Transmission

The preceding power series representations are convenient for adequately small nonlinearities within the range of the maximum signal amplitudes of significant probability. For pronounced nonlinearities such that saturation is approached, it is necessary to use an infinite power series with appropriate convergence at saturation. In this case a linear limiter simulation of the transfer characteristic is much more convenient, and ordinarily affords the requisite engineering accuracy.

Let the transmittance be given by (8.7) and for convenience let $\alpha = 1$. Relations (8.22) and (8.23) then become

$$S_0 = \frac{1}{S}\left(2 \int_0^c z^2 p(z)\, dz + 2c \int_c^\infty z p(z)\, dz\right)$$

$$= \frac{1}{S}[S - G_2(c) + c G_1(c)]^2, \tag{8.104}$$

$$D_0 = 2 \int_0^c z^2 p(z)\, dz + 2c^2 \int_c^\infty p(z)\, dz - S_0$$

$$= S - G_2(c) + c^2 G_0(c) - S_0, \tag{8.105}$$

where S is the average signal power and

$$G_0(c) = 2\int_c^\infty p(z)\,dz, \tag{8.106}$$

$$G_1(c) = 2\int_c^\infty zp(z)\,dz \tag{8.107}$$

$$G_2(c) = 2\int_c^\infty z^2 p(z)\,dz. \tag{8.108}$$

With a single sine-wave input (8.8), the probability density is

$$p(z) = 0 \qquad |z| > A$$

$$= \frac{1}{\pi}(A^2 - z^2)^{1/2} \qquad |z| < A \tag{8.109}$$

and the above functions become $G_0 = 1 - \frac{2}{\pi}\sin^{-1} y_1$, $G_1 = \frac{2A}{\pi}(1-y_1^2)^{1/2}$ and $G_2 = \frac{Ac}{\pi}(1-y_1^2)^{1/2} + \frac{A^2}{2}\sin^{-1} y_1$, where $y_1 = c/A$. In this case $S_0 = S$ and $D_0 = 0$ for $A < c$, while for $A \geq c$ the following relations apply

$$S_0 = \frac{2c^2}{\pi^2}\left((1-y_1^2)^{1/2} + \frac{1}{y_1}\sin^{-1} y_1\right)^2 \tag{8.110}$$

$$D_0 = c^2\left\{\frac{y_1^2}{\pi}\sin^{-1} y_1 - \frac{y_1}{\pi}(1-y_1^2)^{1/2}\right.$$

$$\left. + 1 - \frac{2}{\pi}\sin^{-1} y_1 - \frac{2}{\pi^2}\left[(1-y_1^2)^{1/2} + \frac{1}{y_1}\sin^{-1} y_1\right]^2\right\} \tag{8.111}$$

where

$$y_1 = c/A = (S_c/2S)^{1/2} \tag{8.112}$$

$S_c = c^2$ is the clipping power and $S = A^2/2$ the average sine-wave power.

The ratio S_0/S_c is shown in Fig. 8.7 as a function of S/S_c. In the same figure is shown the same ratio S_0/S_c for two sine waves of equal amplitudes and combined average power $S = A^2$ and for an infinite number of sine waves of combined input power S. The latter two curves are obtained from relations that follow.

With n sine waves of equal amplitudes A, the probability density is given by the relation [11]

$$p(z) = \frac{2}{\pi}\int_0^\infty J_0^n(Au)\cos uz\,du. \tag{8.113}$$

For $n > 1$ $p(z)$ can be expressed in terms of elliptic integrals, and solution of (8.104) and (8.105) is no longer possible in terms of elementary functions, except in the limiting case of $n = \infty$ when $p(z)$ is the Gaussian

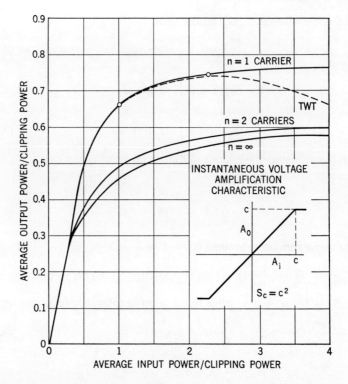

Figure 8.7 Power transmission characteristic of linear limiter with input of $n = 1, 2$ and ∞ number of sine waves of equal amplitudes. Dashed curve represents traveling wave tube characteristic shown in Fig. 8.10 with $S_c = -4$ dBm.

probability density (8.60). For the case $n = 2$ the average output power and the third-order intermodulation products are of particular interest since measurements of the latter facilitate determination of the parameter c or the clipping power S_c that is involved in simulations of actual repeaters by the foregoing limiter representation. The average output power and third-order intermodulation products can be obtained from a general formulation given by Bennett [12] for an asymmetrical limiter, or biased rectifier. From this general solution the relations that follow are obtained for a limiter as considered here.

For a linear rectifier as represented by (8.6), and equal amplitudes A of the two input sine waves, the amplitude of each fundamental in the output is

$$A_{O1} = \alpha A \left\{ 1 + \frac{2}{\pi^2} \int_\theta^\pi [(1 - u^2)^{1/2} - u \cos^{-1} u] \cos y \, dy \right\}, \quad (8.114)$$

where $u = y_1 + \cos y$ and $\theta = \cos^{-1}(1 - y_1)$, and $y_1 = c/A$.

The amplitude of the third-order products of frequencies $2\omega_1 - \omega_2$ and $2\omega_2 - \omega_1$ are

$$A_{21} = \frac{2\alpha A}{3\pi^2} \int_\theta^\pi (1 - u^2)^{3/2} \cos y \, dy. \tag{8.115}$$

For a linear limiter as represented by (8.7), the amplitude of each fundamental in the output is

$$A_0 = 2(A_{01} - \alpha A) \tag{8.116}$$

$$= \frac{8}{\pi^2} \alpha c \quad \text{for} \quad y_1 = c/A = 0,$$

and the amplitude of each third-order intermodulation product is

$$A_3 = 2A_{21}. \tag{8.117}$$

In Fig. 8.7 is shown the ratio $A_0^2/c^2 = S_0/S_c$ versus $A^2/c^2 = S/S_c$.
In Fig. 8.8 is shown the distortion ratio $\alpha_2^2 = (A_3/A_0)^2$ as a function of $y_1^2 = c^2/A^2 = S_c/S$. The curves in these figures are obtained from curves of A_{01} and A_{21} given in Bennett's paper [12].

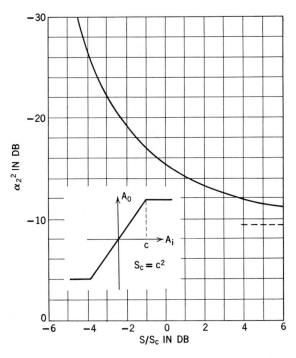

Figure 8.8 Third-order intermodulation ratio α_2^2 for linear limiter with input of two sine waves of equal amplitudes.

In the limiting case $n = \infty$, $p(z)$ is given by (8.60) and relations (8.106), (8.107) and (8.108) become

$$G_0(z) = \operatorname{erfc}(y),$$

$$G_1(z) = \left(\frac{2}{\pi}\right)^{1/2} S^{1/2} e^{-y^2},$$

$$G_2(z) = S \operatorname{erf}(y) + (SS_c)^{1/2} \left(\frac{2}{\pi}\right)^{1/2} e^{-y^2},$$

where

$$y = (c^2/2\sigma^2)^{1/2} = (S_c/2S)^{1/2}.$$

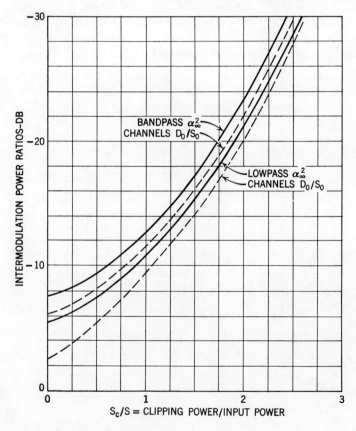

Figure 8.9 Intermodulation ratios D_0/S_0 and α_∞^2 for linear limiter with input of Gaussian noise with flat power spectrum.

From (8.107) and (8.108)

$$S_0 = S[\text{erfc}(y)]^2 \qquad (8.118)$$

$$= \frac{2}{\pi} S_c \quad \text{for} \quad y = 0,$$

$$D_0 = S\left\{\text{erfc}(y)[\text{erf}(y) + 2y^2] - \frac{2y}{\pi^{1/2}} e^{-y^2}\right\}. \qquad (8.119)$$

For $y = 0$, or $S/S_c = \infty$,

$$D_0/S_0 = \frac{\pi}{2} - 1 \qquad (8.120)$$

For $y \gg 1$, or $S/S_c \ll 1$

$$D_0/S_0 \simeq \frac{1}{\pi^{1/2} y^3} e^{-y^2}. \qquad (8.121)$$

The ratio S_0/S_c obtained from (8.118) is shown in Fig. 8.7.

In Fig. 8.9 is shown the ratio D_0/S_0 obtained from (8.118) and (8.119), together with some related ratios derived in the sections that follow.

8.12 Linear Limiter Distortion—Gaussian Wideband Signals

With a limiter transmittance (8.7) the transform method yields the following relation for the autocorrelation function [7] by steps outlined in Appendix 3:

$$R_0(\tau) = \underline{R_0(\tau)} + \underline{\underline{R}}(\tau), \qquad (8.122)$$

$$\frac{\underline{R_0(\tau)}}{R(\tau)} = \rho(\tau)[\text{erf}(y)]^2 \qquad (8.123)$$

$$\frac{\underline{\underline{R_0(\tau)}}}{R(\tau)} = \frac{4}{\pi} e^{-2y^2} \sum_{j=1,2,3,\ldots}^{\infty} \frac{[\rho(\tau)]^{2j+1} H_{2j-1}^2(y)}{2^{2j}(2j+1)!}, \qquad (8.124)$$

where as before $y = (S_c/2S)^{1/2}$ and

$$\rho(\tau) = \frac{R(\tau)}{R(0)} \qquad (8.125)$$

The Hermite polynomial in (8.124) is defined as

$$H_m(y) = (-1)^m e^{y^2} \frac{d^m}{dx^m} e^{-y^2}$$

$$= 2^m y^m - 2^{m-1} \frac{m(m-1)}{2} y^{m-2}$$

$$+ 2^{m-2} \frac{m(m-1)(m-2)(m-3)}{2 \cdot 4} y^{m-4} -, \ldots \qquad (8.126)$$

In an alternate definition [13] that is often used y^2 in (8.126) replaced by $y^2/2$.

The function $e^{-y^2/2}H_m(y)$ has been tabulated [14] for $m = 0$ to 11 for $y = 0$ to 8.

For $\tau = 0$, $\rho(\tau) = 1$ and the total output as obtained from (8.122) is

$$\frac{R_0(O)}{R(O)} = [\text{erf}(y)]^2 + \frac{4}{\pi} e^{-2y^2} \sum_{j=1}^{\infty} \frac{H^2_{2j-1}(y)}{2^{2j}(2j+1)!}. \tag{8.127}$$

The first term represents the linear component of the output power and conforms with (8.128). The remainder represents the nonlinear component of distortion, and should conform with (8.119) as obtained with the simpler direct method.

For the limiting case of $y = 0$, the following relation is obtained

$$\frac{R_0(\tau)}{R(\tau)} = \frac{4}{\pi} y^2 \left(\rho(\tau) + \frac{1}{3!} \rho^3(\tau) + \frac{3^2}{5!} \rho^5(\tau) + \frac{(3 \cdot 5)^2}{7!} \rho^7(\tau) + \cdots \right)$$

$$= \frac{4}{\pi} y^2 \sin^{-1} \rho(\tau). \tag{8.128}$$

For $\tau = 0$, $\rho(\tau) = 1$ and $\sin^{-1} \rho(\tau) = \pi/2$. In this case (8.128) gives

$$\frac{R_0(O)}{R(O)} = \frac{(S_0 + D_0)}{S} = 2y^2. \tag{8.129}$$

The linear component of the output power is given by the first term in (8.128) and is $S_0/R(O) = y^2 4/\pi$. The remaining components represent nonlinear distortion, which is given by $D_0/S = 2y^2 - (4/\pi)y^2$. The ratio $D_0/S_0 = \pi/2 - 1$ conforms with (8.120).

The distortion power spectra obtained from (8.36) and (8.37) are

$$W_0(\omega) = W(\omega)[\text{erf}(y)]^2, \tag{8.130}$$

$$W_0(\omega) = \frac{4R(O)}{\pi} e^{-2y^2} \sum_{j=1}^{\infty} \frac{L_{2j+1} H^2_{2j-1}(y)}{2^{2j}(2j+1)!}, \tag{8.131}$$

where $W(\omega)$ is the power spectrum of the input signal, $R(O) = S$ and

$$L_m = L_m(\omega) = \frac{2}{\pi} \int_0^\infty [\rho(\tau)]^m \cos \omega\tau \, d\tau. \tag{8.132}$$

In the particular case of a signal of average power S and a flat power spectrum of bandwidth Ω $R(\tau)$ is given by (8.71). In this case the following relation is obtained

$$\frac{W_0(\omega)}{\underset{\sim}{W}_0(\omega)} = \left(\frac{4}{\pi} e^{-2y^2} \sum_{j=1}^{\infty} \frac{M_{2j+1}(a)}{2^{2j}(2j+1)!} H_{2j-1}^2(y)\right) \cdot \left(\text{erf}(y)\right)^{-2} \quad (8.133)$$

where W and M_m are defined as in Section 8.8.

The maximum distortion is obtained for $\omega = 0$ corresponding to $a = 0$. In this case the following values of M_m apply:

m:	1	2	3	5	7	>7
M_m:	1	1	$\frac{3}{4}$	$\frac{115}{192}$	$\frac{5887}{11520}$	$\left(\frac{6}{\pi m}\right)^{1/2}$

The ratio $\alpha_\infty^2 = \underset{\sim}{W}(0)/W(0)$ is shown in Fig. 8.9 as a function $c^2/\sigma^2 = S_c/S$. The notation α_∞^2 is used to indicate intermodulation distortion, owing to amplification nonlinearity with an infinite number of sine waves.

8.13 Linear Limiter Distortion—Gaussian Narrowband Signals

In the case of narrow bandpass channels expression (8.124) is modified by the introduction of the factor $C(k)$ defined by (8.13) and becomes

$$\frac{R_0(\tau)}{R(\tau)} = \bar{\rho}(\tau)[\text{erf}(y)]^2 + \frac{4}{\pi} e^{-2y^2} \sum_{j=1,2,\ldots}^{\infty} \frac{C(j)[\bar{\rho}(\tau)]^{2j+1}}{2^{2j}(2j+1)!} H_{2j-1}^2(y) \quad (8.134)$$

where $\bar{\rho}(\tau) = \bar{R}(\tau)/R(0)$ and

$$C(j) = \frac{(2j+1)!}{2^{2j}(j+1)!j!}. \quad (8.135)$$

The latter expression for $C(j)$ is obtained by introducing $k = 2j + 1$ in (8.13) so that $k = 3, 5, \ldots$, for $j = 1, 2, 3, \ldots$, respectively.

In view of (8.135)

$$\frac{R_0(\tau)}{R(\tau)} = \bar{\rho}(\tau)[\text{erf}(y)]^2 + \frac{4}{\pi} e^{-2y^2} \sum_{j=1,2,\ldots}^{\infty} \frac{[\bar{\rho}(\tau)]^{2j+1} H_{2j-1}^2(y)}{4^{2j}(j+1)!j!}. \quad (8.136)$$

For $\tau = 0$ the latter ratio becomes $(S_0 - D_0)/S$ so that

$$D_0/S_0 = \left(\frac{4}{\pi} e^{-2y^2} \sum_{j=1,2}^{\infty} \frac{H_{2j-1}^2(y)}{4^{2j}j!(j+1)!}\right) \left(\text{erf}(y)\right)^{-2} \quad (8.137)$$

which is shown in Fig. 8.9 as a function of S/S_c.

In place of (8.133), the ratio of distortion to signal power in a narrow band at u becomes

$$\frac{W_0(u)}{\underline{W}_0(u)} = \left(\frac{4}{\pi}e^{-2y^2}\sum_{j=1}^{\infty}\frac{M_{2j+1}(a)H_{2n-1}^2(y)}{4^{2j}j!(j+1)!}\right)\left(\text{erf}(y)\right)^{-2} \quad (8.138)$$

where $M_m(a)$ is defined as in Section 8.8 with $a = |u/\Omega|$.

The ratio $a_\infty^2 = W_0(O)/\underline{W}_0(O)$ is shown in Fig. 8.9.

The curves in Fig. 8.9 show that for a low-pass channel the ratio of distortion power to signal power is somewhat greater than for a bandpass channel. Furthermore, that the ratio D_0/S_0 is somewhat greater than $W_0(O)/\underline{W}(O)$, so that it is a safe approximation to use the simpler relation D_0/S_0, as was the case for power series representation of a nonlinear transmittance.

In the particular case of "hard limiting," $y = (S_c/2S)^{1/2} = O$. For this case, numerical results on distortion with 3 and more unmodulated carriers of generally different amplitudes has been given in a paper by Shaft [15] which also contains references to prior literature on the effect of hard limiting.

8.14 AM-PM Conversion Factors

In addition to distortion from nonlinearity in the amplitude characteristic $A(\omega)$ it may also be necessary to consider distortion due to variations in the phase characteristic $\psi(\omega)$ with the amplitude of the signal. Fluctuations in the phase characteristic with the amplitude of the received wave will result in unwanted phase modulation in the transmitted wave and resultant intermodulation distortion. This effect may be important in microwave transmitters, such as traveling wave tubes and klystrons. This is illustrated in Fig. 8.10, which shows the power transfer characteristic of a traveling wave tube transmitter as measured with a single sine-wave input, together with the AM-PM conversion factor measured with a small sinusoidal amplitude modulation. For adequately small input powers the conversion factor is nearly proportional to the input power or to the squared envelopes, and this appears to be a general rule if not a law. For input powers near saturation, the conversion factor in Fig. 8.10 approaches a limiting value, but other types of behavior are also encountered [16,17] in which the conversion factor reaches a maximum value for input powers near saturation and then diminishes.

The sine-wave power amplification characteristic shown in Fig. 8.10 can be closely approximated by a linear limiter with $S_c = -4$ dBm. With

AM-PM Conversion Factors

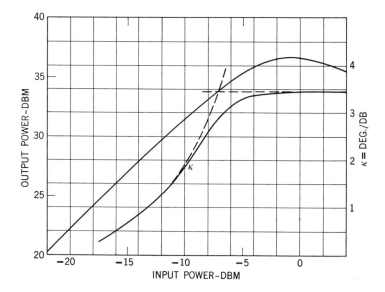

Figure 8.10 Single sine wave power transfer characteristic and AM-PM conversion factor κ of traveling wave tube transmitter.

input powers $S > S_c$, intermodulation distortion is caused principally by amplification nonlinearity rather than AM-PM conversion, while for $S \ll S_c$ amplification nonlinearity can be disregarded as a minor contributor in comparison with AM-PM conversion. On this basis it is a legitimate approximation to consider the effect of AM-PM conversion only at input powers $S \ll S_c$, in the range where the AM-PM conversion factor is essentially proportional to input power or to the squared envelope. This results in certain simple relations as shown next.

The AM-PM conversion factor shown in Fig. 8.10 was determined with an input

$$z(t) = A(1 + m \cos \omega_x t) \cos \omega_0 t, \tag{8.139}$$

with $m \ll 1$. The corresponding one-ohm normalized input power is

$$S = \frac{A^2}{2}\left(1 + \frac{m^2}{2}\right) \simeq \frac{A^2}{2}, \tag{8.140}$$

and the phase modulation becomes

$$\varphi_x(t) = k_0 \bar{z}^2(t) \simeq k_0 A^2 (1 + 2m \cos \omega_x t)$$
$$= 2k_0 S (1 + 2m \cos \omega_x t). \tag{8.141}$$

The peak deviation of the variable component of $\varphi_x(t)$ from the mean value is

$$\hat{\varphi}_x = 4k_0 Sm. \qquad (8.142)$$

The conversion factor expressed in degrees/dB, as in Fig. 8.10 is

$$\kappa = \frac{\hat{\varphi}_x \times 180/\pi}{20 \log_{10}(1+m)} = \frac{4k_0 Sm}{8.69m} \frac{180}{\pi}$$

$$= 26.6 k_0 S$$

or

$$k_0 = 0.038\kappa/S \qquad (8.143)$$

This relation applies in the range where the conversion factor $\kappa = \kappa(S)$ increases in proportion to S.

8.15 Intermodulation Distortion by AM-PM Conversion

The measured AM-PM conversion factors are virtually independent of the frequency ω_x of the amplitude modulation in the pass-band of the amplifier. Hence it is legitimate to assume that the following relation applies for the phase modulation $\psi_x(t)$ resulting from any envelope fluctuations $\bar{z}(t)$ that are caused by signal frequency components.

$$\psi_x(t) = k_0 \bar{z}^2(t). \qquad (8.144)$$

This assumes the square law relation mentioned above which is closely approximated at adequately low input powers.

Let the repeater input consist of n sine waves of amplitudes A_i as represented by

$$z(t) = \sum_{i=1}^{n} A_i \cos[\omega_0 t + u_i t + \varphi_i(t)], \qquad (8.145)$$

where ω_0 is conveniently taken as the midband frequency, $u_i = \omega_i - \omega_0$ is the frequency from midband of carrier i and $\varphi_i(t)$ its phase modulation. The above relation can alternately be written

$$z(t) = \bar{z}(t) \cos[\omega_0 t + \psi_0(t)], \qquad (8.146)$$

where the envelope $\bar{z}(t)$ is given by

$$\bar{z}^2(t) = \left(\sum_{i=1}^{n} A_i \cos(u_i t + \varphi_i)\right)^2 + \left(\sum_{i=1}^{n} A_i \sin(u_i t + \varphi_i)\right)^2, \qquad (8.147)$$

Intermodulation by AM-PM Conversion

and the phase function by

$$\tan \psi_0(t) = -\frac{\sum_{i=1}^{n} A_i \sin(u_i t + \varphi_i)}{\sum_{i=1}^{n} A_i \cos(u_i t + \varphi_i)}, \tag{8.148}$$

where $\varphi_i = \varphi_i(t)$.

Let the envelope fluctuation give rise to a phase modulation $\psi_x(t)$. The output wave is then in place of (8.146)

$$z_0(t) = \bar{z}_0(t) \cos[\omega_0 t + \psi_0(t) + \psi_x(t)]$$

$$= \sum_{i=1}^{n} A_i \cos[\omega_0 t + u_i t + \varphi_i(t) + \psi_x(t)]. \tag{8.149}$$

The latter relation shows that the phase modulation $\psi_x(t)$ appears on each of the transmitted carriers.

The transmitted sine wave of frequency ω_i becomes

$$z_i(t) = A_i \cos[\omega_i t + \varphi_i(t) + \psi_x(t)]$$
$$= A_i \cos[\omega_i t + \varphi_i(t)] \cos \psi_x(t) - A_i \sin[\omega_i t + \varphi_i(t)] \sin \psi_x(t). \tag{8.150}$$

For small values of $\psi_x(t)$ the following approximation applies

$$z_i(t) = A_i \cos[\omega_i t + \varphi_i(t)] - A_i \sin[\omega_i t + \varphi_i(t)] \psi_x(t), \tag{8.151}$$

The first component represents the signal wave and the second component is an interference wave

$$x_i(t) = A_i \sin[\omega_i t + \varphi_i(t)] \psi_x(t), \tag{8.152}$$

In accordance with (8.144), $\psi_x(t)$ is proportional to $\bar{z}^2(t)$. In determining $\bar{z}^2(t)$ it suffices to consider a signal consisting of three sine waves, as was done in Section 8.6 for square law nonlinearity. The contribution from the first squared bracket term in (8.147) is then of the same form as $v_2(t)$ in (8.49). The contribution from the second squared bracket term is of the same form as $v_2(t)$ except for a reversal of sign and cancellation of all terms except those in $x_1 - x_2$, $x_2 - x_3$, and $x_1 - x_3$. The amplitudes of the latter terms are thus doubled, so that for three sine waves (8.144) becomes with $2k_0$ in place of a_2

$$\psi_x(t) = 2k_0[A_1 A_2 \cos(x_1 - x_2) + A_2 A_3 \cos(x_2 - x_3) + A_1 A_3 \cos(x_1 - x_3)]. \tag{8.153}$$

With the latter relation in (8.152) the resultant interference becomes

$$x_1(t) = k_0 A_i A_1 A_2 [\sin(x_i + x_1 - x_2) + \sin(x_i - x_1 + x_2)]$$
$$+ k_0 A_i A_2 A_3 [\sin(x_i + x_2 - x_3) + \sin(x_i - x_2 + x_3)]$$
$$+ k_0 A_i A_1 A_3 [\sin(x_i + x_1 - x_3) + \sin(x_i - x_1 + x_3)]. \tag{8.154}$$

Since three sine waves are assumed, it is necessary that $A_i = A_3$ and $x_i = x_3$ in the first term, $A_i = A_1$ and $x_i = x_1$ in the second term, and $A_i = A_2$ and $x_i = x_2$ in the third term. With these substitutions the combined interference for three sine waves becomes

$$u_3(t) = x_1(t) + x_2(t) + x_3(t)$$
$$= \underline{\mu}_3 + \underset{\sim}{\mu}_3, \qquad (8.155)$$

where the linear component is

$$\underline{u}_3(t) = k_0[A_1(A_2^2 + A_3^2) \sin x_1 + A_2(A_1^2 + A_3^2) \sin x_2$$
$$+ A_3(A_1^2 + A_2^2) \sin x_3], \qquad (8.156)$$

and the nonlinear component is

$$\underset{\sim}{u}_3(t) = k_0[A_1^2 A_2 \sin(2x_1 - x_2) + A_2^2 A_1 \sin(2x_2 - x_1)$$
$$+ A_2^2 A_3 \sin(2x_2 - x_3) + A_3^2 A_1 \sin(2x_3 - x_2)$$
$$+ A_1^2 A_3 \sin(2x_1 - x_3) + A_3^2 A_1 \sin(2x_3 - x_1)]$$
$$+ 2k_0 A_1 A_2 A_3 [\sin(x_1 + x_2 - x_3) + \sin(x_1 - x_2 + x_3)$$
$$+ \sin(x_2 - x_1 + x_3)]. \qquad (8.157)$$

Comparison with (8.55) shows that the intermodulation products in the passband are the same as for a cubic amplitude nonlinearity, but at quadrature with the latter. Thus the relations for third-order intermodulation due to a cubic amplitude nonlinearity applies, with the substitution

$$k_0 = \tfrac{3}{4} a_3. \qquad (8.158)$$

In the simplest case of an input consisting of two sine waves of equal amplitudes, the following relation applies for the amplitudes of the third-order intermodulation products of frequency $2\omega_1 - \omega_2$ or $2\omega_2 - \omega_1$ taken in relation to the amplitude $A_1 = A_2 = A$ of each sine wave

$$\beta_2 = k_0 S \qquad (8.159)$$

where $S = A^2$ is the average input power. The latter relation is obtained from (8.157) or from (8.55) with the substitution (8.158). In view of (8.143), the following relation applies

$$\beta_2 = 0.038\kappa \qquad (8.160)$$

where $\kappa = \kappa(S)$ is the conversion factor in degrees/dB as shown in Fig. 8.10 and it is assumed that the input power is confined to the range in which the square law relation (8.144) is approximated.

Intermodulation by AM-PM Conversion

When the signal consists of a large number of sine waves of equal amplitudes, or alternately random noise with a flat power spectrum, relation (8.86) applies for the ratio of average distortion power density to signal power density at midband. With the substitution (8.158) the following relation is obtained

$$\beta_\infty^2 = 6(k_0 S)^2$$
$$= 8.8 \cdot 10^{-3} \kappa^2 \qquad (8.161)$$

where (8.143) has been introduced in the last relation, which applies in the range where $\kappa = \kappa(S)$ follows the square law relation.

In Fig. 8.11 are shown the ratios β_2^2 and β_∞^2 obtained from the above equations based on the square law relation. In the same figure are shown the corresponding ratios obtained on the approximate basis that (8.160) and (8.161) also apply outside the range of the square-law variation,

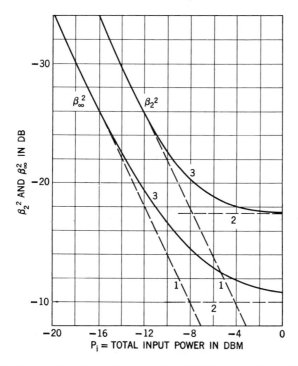

Figure 8.11 Intermodulation ratios owing to AM-PM conversion. 1. Curves based on square law AM-PM conversion. 2. Upper bound on distortion for constant AM-PM conversion factor $\kappa = 3.4$ degrees/dB. 3. Approximate ratios for AM-PM conversion factor shown in Fig. 8.10.

provided κ is taken as the measured conversion factor shown in Fig. 8.10. This should present an upper bound on the interference from AM-PM conversion.

8.16 Combined Intermodulation with AM-PM Conversion

The combined intermodulation distortion resulting from amplification nonlinearity in conjunction with AM-PM conversion will be considered here for the same two limiting cases as earlier of two sine waves and an infinite number of sine waves, or random noise with flat power spectrum.

Since intermodulation products resulting from AM-PM conversion are at quadrature with those caused by amplification nonlinearity, the combined ratio of intermodulation product power to the average power of each sine wave becomes

$$\gamma_2^2 = \alpha_2^2 + \beta_2^2. \tag{8.162}$$

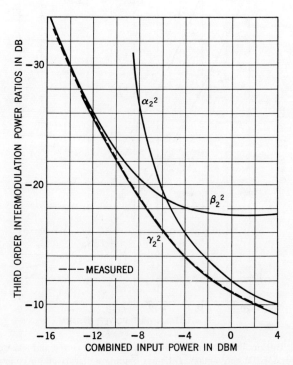

Figure 8.12 Combined intermodulation from amplification nonlinearity and AM-PM conversion for two sine waves of equal amplitudes. $S_c = -4$ dBm.

In Fig. 8.12 is shown the ratio α_2^2 obtained from the curves in Fig. 8.8 with $S_c = -4$ dBm, together with the ratio β_2^2 shown in Fig. 8.11 and the combined ratio γ_2^2. The latter ratio conforms closely with the results of measurements of third-order intermodulation products made on the particular traveling wave tube amplifier with the characteristics shown in Fig. 8.10. The curves in Fig. 8.12 show that for low-input powers the third-order intermodulation is caused principally by AM-PM conversion, whereas for high-input powers amplification nonlinearity is a principal contributor.

When the signal has the properties of Gaussian noise with a flat power spectrum, the combined intermodulation ratio becomes

$$\gamma_\infty^2 = \alpha_\infty^2 + \beta_\infty^2. \tag{8.163}$$

In Fig. 8.13 are shown the α_∞^2, β_∞^2, and γ_∞^2, when the clipping power is

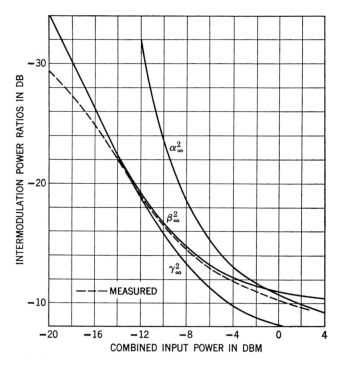

Figure 8.13 Combined intermodulation from amplification nonlinearity and AM-PM conversion. The greater measured intermodulation for small inputs arise from intermodulation extraneous to the traveling wave tube. The smaller measured intermodulation for high inputs may be due to a sharp reduction in the AM-PM conversion factor at inputs above 3 dBm.

taken as -4 dBm in adjusting the repeater amplification characteristic in Fig. 8.10 to that in Fig. 8.7 for a limiter. The curves of Fig. 8.13 also apply for single-sideband modulation of one or more carriers by random noise. The theoretical curves do not conform as well with measurements as for an input of two carriers. An examination of the test arrangement indicates that the greater measured distortion for low inputs can be attributed to extraneous intermodulation products contributed by a pre-amplifier. A smaller measured distortion would be expected for high inputs, for the reason that the AM-PM conversion factor will not approach a constant value but will diminish rapidly and even become negative [16,17] for higher inputs than shown in Fig. 8.10. Moreover, intermodulation products of higher order than the third will be encountered, with relatively lower spectral density near midband. The actual ratio γ_∞^2 would thus conform with the theoretical curve for inputs below -12 dBm and with the measured curve for higher inputs. This modified ratio is shown in Fig. 8.16 for $n = \infty$.

8.17 Intermodulation Distortion versus Number of Phase Modulated Carriers

For third-order intermodulation, the ratio γ_n^2 for n sine waves can be obtained by considering the number of third-order intermodulation products as previously discussed in Section 8.7. From relations given there it follows that for unmodulated sine waves

$$\lambda(n) = \frac{\gamma_n^2}{\gamma_\infty^2} = \frac{\left(\frac{3}{4}\right)^2 Q_n(O) + \left(\frac{3}{2}\right)^2 P_n(O)}{\left(\frac{3}{2}\right)^2 \frac{3}{8} n^2}$$

or

$$\lambda(n) = \frac{8}{3n^2}\left(P_n(O) + \frac{1}{4} Q_n(O)\right) \tag{8.164}$$

This factor is shown in Fig. 8.14.

The above factor applies for equally spaced unmodulated sine waves or carriers of equal average power. Intermodulation products resulting from any form of nonlinearity will then coincide with the various frequencies. Because of unavoidable departures from the nominal carrier frequencies, particularly with orbiting satellite repeaters as a result of Doppler shift, intermodulation products will fall at some frequency from each carrier. If this frequency is within the band of the baseband modulating waves, intermodulation noise will be present in the demodulated waves. When the baseband wave consists of a number of voice channels in frequency-

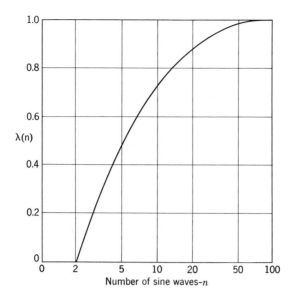

Figure 8.14 Factor $\lambda(n)$ for third-order intermodulation distortion with n equally spaced sine waves of equal amplitudes.

division multiplex, the minimum baseband frequency with conventional multiplexing techniques is about 12 kc. Hence intermodulation distortion in the limiting case of unmodulated carriers would be of no concern if the carrier frequency departures from the nominal values were well below 12 kc.

In the other extreme, consider a number of equally spaced carriers, each fully modulated by a Gaussian signal with a flat-power spectrum. The power spectrum of intermodulation distortion will then be a maximum at midband of the central carrier groups, and will be nearly flat over the transmission band of each group. With single- or double-sideband amplitude modulation, the amplitude probability distribution of intermodulation distortion will be Gaussian regardless of the number of carrier groups and the same is for an infinite number of sine waves, that is, $\gamma_n = \gamma_\infty$.

With phase modulation, the probability distribution of the envelope is the same as for unmodulated carriers, as is the total intermodulation distortion power. However, the power spectrum of distortion depends on the degree of modulation. In the important case of a very large frequency deviation ratio, the signal-power spectrum of each modulated carrier will approach a Gaussian shape, as shown in Chapter 6 and illustrated in Fig. 8.15. The combined spectrum of third-order intermodulation will then be nearly flat as indicated in the figure. It is thus a legitimate approximation with four or more carrier groups to determine third-order intermodulation by assuming an equivalent flat signal power spectrum as indicated in Fig. 8.15. With this approximation the factor $\lambda(n)$ given by (8.164) applies for any degree of phase or frequency modulation.

When the factor $\lambda(n)$ is applied to the curve in Fig. 8.13 for γ_∞^2, the approximate curves in Fig. 8.16 are obtained for $n = 4, 6, 8,$ and 12 phase or frequency modulated carriers. At low inputs, the factor $\lambda(n)$ applies, while at saturation the factor must approach unity. The curves in Fig. 8.16 are obtained by drawing approximate transition curves.

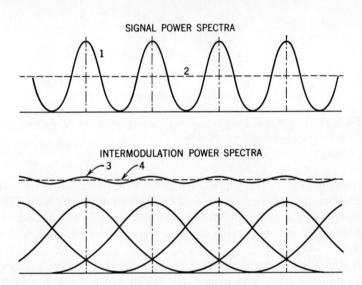

Figure 8.15 Approximate evaluation of intermodulation distortion in multicarrier FM with full modulation. 1. Signal power spectrum at repeater input. 2. Equivalent flat power spectrum of equal power. 3. Third-order intermodulation distortion at repeater output. 4. Power spectrum of intermodulation distortion based on equivalent flat power spectrum 2.

With two carrier groups all intermodulation products will fall outside the transmission band of each group when the carriers are unmodulated, regardless of the carrier separation. With modulated carriers, intermodulation distortion depends on the carrier separation in relation to the signal power spectrum, and can in principle be avoided by adequate carrier separation. The carrier separation required to this end for a given signal power spectrum depends on the kind of intermodulation products that need to be considered. Third-order intermodulation products from one carrier group will not fall into the band of the other carrier group provided that the carrier separation exceeds twice the bandwidth of the signal power spectrum of each group. Under this condition higher order intermodulation products are also avoided.

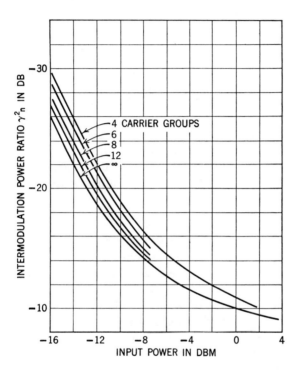

Figure 8.16 Approximate intermodulation ratio γ_n^2 for amplifier with single sine wave power transfer characteristic and AM-PM conversion factor shown in Fig. 8.10.

Approximate evaluation of intermodulation distortion by the foregoing procedure entails measurements of the single sine-wave amplification characteristic, of intermodulation products with a twin carrier input and of the AM-PM conversion factor. To secure more reliable data for engineering purposes it is necessary to perform tests on the type of amplifier in question, which should preferably include measurements with various numbers of carriers under consideration, as exemplified by the published data for a somewhat different traveling-wave tube transmitter [18].

8.18 Combination of Intermodulation Distortion and Random Noise

Let μ^2 be the ratio of average noise power in band B of each carrier group to the average carrier power, taken at the receiver input. If γ_n^2 is the ratio of average intermodulation power in band B at the transmitter output the same ratio applies at the receiver input. The ratio of combined average interference power in band B to the average power of each carrier at the receiver input is then

$$\eta^2(n) = \mu^2 + \gamma_n^2 \tag{8.165}$$

With linear modulation methods, such as double or single sideband AM, the resultant combined average interference power in the demodulated baseband output is independent of the probability distribution of the intermodulation distortion and can be derived from the same relations as those for random noise at the receiver input. In the particular case of full amplitude modulation of each carrier by Gaussian noise the combined interference in band B at the receiver input will have the same probability distribution as Gaussian noise, regardless of the number of carrier groups. In this case $\gamma_n^2 = \gamma_\infty^2$, and the following relation applies for any number of carriers

$$\eta^2(n) = \mu^2 + \gamma_\infty^2 \tag{8.166}$$

With analog or digital FM or PM it is necessary in determining distortion or errors in the demodulated output to consider the probability distribution of the combined interference in band B at the receiver input. The combined wave at the transmitter input, consisting of n phase modulated carriers, will then have a probability distribution that depends on n and will be less peaked than for Gaussian noise, that is, the probability of substantially exceeding the rms value is smaller. This also applies for intermodulation distortion in band B and for the combined interference at the receiver input. On this account the relations between input and output signal-to-noise ratios for analog FM and PM will be modified in the nonlinear region below the threshold in such a manner that output distortion is less than for Gaussian input noise. However, for operation above the threshold the relation between input and output signal-to-noise ratios is essentially linear and the probability distribution of the input wave is of only minor importance.

For these reasons the relations given in Chapter 5 for FM and PM apply to operation above the threshold with $\rho_i = 1/\eta^2$ where $\eta^2 = \eta^2(n)$ is given by (8.165), and depends on the number of carrier groups. Below the threshold the relations discussed in Section 2.12 for Gaussian noise give an upper bound on output distortion, the difference being insignificant for more than 10 carriers.

For binary or multilevel digital FM or PM an upper bound on error probability is obtained from relations given in Chapter 5 for Gaussian input noise, with $\rho_i = 1/\eta^2$. With more than 10 carriers a fairly close approximation is obtained except for very small error probabilities. Exact relations for any number of carriers have not been derived because of the difficulties in accurate determination of the average value of intermodulation distortion, not to mention its probability distribution and that of the combined interference in band B at the receiver.

In the above relations μ and γ_n are functions of the repeater input power S or corresponding output power. An increase in S is accompanied by a reduction in μ^2 and an increase in γ_n^2. Hence there is a certain minimum value of η^2 and corresponding optimum repeater input or output power.

8.19 Output Distortion in Multicarrier FM Systems

Multicarrier FM systems using a common amplifier for several independently modulated carriers are of particular interest for simultaneous satellite communication between several pairs of ground stations. With pronounced satellite transmitter nonlinearity, such as that exemplified in Fig. 8.10, intermodulation distortion at the transmitter output, or receiver input, would preclude efficient single- or double-sideband operation near saturation. With FM, intermodulation distortion in the demodulator output, like random noise, can be held within tolerable limits in exchange for increased bandwidth by using an adequately large frequency deviation ratio.

For ratios $\eta^2 < 10^{-1}$ conventional FM theory for high signal-to-noise ratios applies as discussed in Section 4.6. For a flat noise and distortion-power spectrum, the following relations are obtained for the ratio $\eta_0^2(f)$ of average combined fluctuation and distortion power $n_0(f)$ in a narrow band at the frequency f to the average baseband signal power $s_0(f)$ in the same narrow band at the detector output.

For pure FM:

$$\eta_0^2(f) = \frac{n_0(f)}{s_0(f)} = \frac{\eta^2}{2\underline{D}^3}\left(\frac{f}{B_0}\right)^2 = \frac{\eta_1^2}{\underline{D}^2}\left(\frac{f}{B_0}\right)^2. \qquad (8.167)$$

For pure PM:

$$\eta_0^2(f) = \frac{n_0(f)}{s_0(f)} = \frac{\eta^2}{6\underline{D}^3} = \frac{\eta_1^2}{3\underline{D}^2}, \qquad (8.168)$$

where η is given by (8.165) and

$$\eta_1^2 = \eta^2 \cdot B_0/B, \qquad (8.169)$$

and

B_0 = Bandwidth of baseband modulating wave and post detection low-pass filters,
D = rms deviation ratio = Δ/B_0,
Δ = rms frequency deviation.

Thus η_1^2 is the ratio of combined average noise and distortion power in the band B_0 at the detector input to average carrier power per group.

With the aid of negative feedback in the FM demodulator it is possible to realize a reduction in the threshold carrier-to-noise ratio, that is, to increase η^2 for a given minimum allowable output message-to-noise ratio. As mentioned in Section 2.12, the effectiveness of such feedback diminishes with increasing bandwidth of the message wave and is insignificant at a certain maximum bandwidth because of the adverse effect of intermodulation distortion in the feedback circuit. For this reason negative feedback will be somewhat more effective the greater the number of carriers. This will tend to compensate for the greater intermodulation distortion as the number of carriers is increased in multicarrier FM systems if FM demodulators with negative feedback are used.

8.20 Intelligible Interference in Multicarrier FM Systems

Intermodulation as considered previously gives rise to noiselike or unintelligible interference in the various carrier groups of a multicarrier FM system. In addition, it is possible to have coherent or intelligible interference between message waves in different carrier groups. To this end it is necessary to have a linear deviation in the amplitude versus frequency characteristic of the channel preceding the common amplifier with AM-PM conversion. As shown in connection with (6.80), when the phase modulation of a carrier is $\varphi(t)$, a linear amplitude versus frequency deviation modifies the amplitude of the carrier by the factor $1 + c_1\varphi'(t)$. The resultant intelligible interference will be considered here for two carrier groups.

With two carrier groups, the squared envelope (8.147) is modified into

$$\bar{z}^2(t) = A_1^2[1 + c_1\varphi_1'(t))^2 + A_2^2(1 + c_1\varphi_2'(t)]^2$$
$$+ 2A_1A_2[1 + c_1\varphi_1'(t))(1 + c_1\varphi_2'(t)]$$
$$\cdot \cos[(\omega_1 - \omega_2)t + \varphi_1(t) - \varphi_2(t)]. \quad (8.170)$$

The resultant phase modulation is in accordance with (8.144) $\Psi_x(t)^2 = k_0\bar{z}^2(t)$. This phase modulation will contain a principal term representing intermodulation distortion, obtained with $c_1 = 0$, and an incremental term $\Delta\Psi_x(t)$ which for $c_1 \ll 1$ becomes

$$\Delta\Psi_x(t) \simeq 2c_1k_0[A_1^2\varphi_1'(t) + A_2^2\varphi_2'(t)]$$
$$+ 2c_1k_0A_1A_2\varphi[\varphi_1'(t) + \varphi_2'(t)]$$
$$\cdot \cos[(\omega_1 - \omega_2)t + \varphi_1(t) - \varphi_2(t)]. \quad (8.171)$$

This incremental phase modulation will appear on both carriers and give rise to corresponding terms in the demodulator output. Thus, let the modulating wave of each carrier group consist of a sum of sine waves of various frequencies s_it. For this case

$$\varphi_1(t) = m_1 \Sigma a_i \sin s_it, \quad (8.172)$$

$$\varphi_2(t) = m_2 \Sigma b_i \sin s_it, \quad (8.173)$$

where m_1 and m_2 represent modulation indices. The first two terms in (8.171) then give rise to the following incremental baseband output in carrier groups 1 and 2:

$$\Delta V_1 = 2c_1k_0A_2^2 \frac{m_2}{m_1} \Sigma b_is_i \cos s_it, \quad (8.174)$$

$$\Delta V_2 = 2c_1k_0A_1^2 \frac{m_1}{m_2} \Sigma a_is_i \cos s_it. \quad (8.175)$$

The corresponding interference ratios for a message wave in a narrow band at the frequency s_i is thus

$$X_1 = 2c_1k_0A_2^2 \frac{m_2}{m_1} \frac{b_i}{a_i} s_i, \quad (8.176)$$

$$X_2 = 2c_1k_0A_1^2 \frac{m_1}{m_2} \frac{a_i}{b_i} s_i. \quad (8.177)$$

The foregoing intelligible interference arises from the first two terms in (8.171). The remainder of this expression represents phase modulation of each of the two carriers by a sine wave of frequency $(\omega_1 - \omega_2)t$, which has both phase modulation $\varphi_1(t) - \varphi_2(t)$ and amplitude modulation $\varphi_1'(t) + \varphi_2'(t)$. The resultant frequency components in the baseband output fall outside the band of the post-detection low-pass filter, and hence need not be considered.

In accordance with (8.176) and (8.177), the amplitude of intelligible interference increases in proportion to the frequency position s_i of the narrow bands (usually voice channels) in the baseband output of each carrier group. With different modulation indexes m_1 and m_2, interference depends on the ratio m_1/m_2 and m_2/m_1 but for equal indexes is independent of the modulation index, that is, the frequency deviation ratio.

With two carrier groups, as assumed above, intelligible crosstalk is a maximum, while intermodulation distortion is quite small. As the number of carrier groups is increased, intelligible interference diminishes, for the reason that the maximum frequency s_i per group is reduced, as well as the average signal power $A_1^2/2$ per group. By contrast, intermodulation distortion increases and, unlike intelligible interference, cannot be reduced by reducing the slope c_1 of the amplitude versus frequency distortion. Hence intermodulation distortion is usually a primary consideration with more than two carrier groups.

A more detailed analysis of intelligible crosstalk with two or more carrier groups is given elsewhere [19], together with experimental data for a traveling-wave tube with the AM-PM conversion factor shown in Fig. 8.10.

8.21 Nonlinear Distortion in Digital Systems

One of the advantages often cited for digital systems is that channel nonlinearities will not give rise to transmission impairments. This is true only of certain idealized binary systems often assumed in simplified theoretical expositions and not actually used. For example, in a binary on-off or bipolar system with rectangular pulses, a nonlinear transmitter will not deform the pulse shape. However, this pulse shape entails infinite channel bandwidth if intersymbol interference is to be avoided. With any practicable pulse shape that permits finite bandwidth the amplitude of the transmitted pulse train will vary continually. However, in the absence of intersymbol inteference it will have a fixed amplitude at each sampling instant. When such a binary pulse train is applied to a nonlinear amplifier the transmitted pulse train will have infinite bandwidth. The received pulse train will in this case have a fixed amplitude at each sampling instant only if the channel has infinite bandwidth. Otherwise some intersymbol interference will be encountered, owing to the combination of channel nonlinearity and bandwidth limitation. However, this effect is ordinarily so small that it does not preclude efficient binary transmission by time-division multiplex methods, even with pronounced channel nonlinearity.

The situation is different with binary transmission over each of a number of channels in frequency division multiplex with a common nonlinear amplifier. In this case, the combined wave will have pronounced amplitude fluctuations and intermodulation distortion may be an important consideration when the transmitters have pronounced nonlinearity. For example, with multicarrier transmission by binary PM, and a transmitter with the power transfer characteristic illustrated in Fig. 8.10, the ratio γ_n^2 would be about as shown in Fig. 8.16. The approximate ratio η^2 can in this case be obtained from relation (8.165). The corresponding approximate error probability with synchronous carrier demodulation can be determined from (5.87) with $\rho_i = 1/\eta^2$ and from (5.92) with differential phase detection. From the curves in Fig. 8.16 it is evident that even in the absence of additive random noise, so that $\rho_i = 1/\gamma_n^2$, it would be necessary to operate below saturation to avoid an excessive error probability.

With multilevel transmission by amplitude or phase modulation, the tolerable distortion may be so small that the permissible number of levels may be limited by distortion from nonlinearity rather than by additive random noise. This may be true even in the case of multilevel transmission by pure FM or PM so that envelope fluctuations are avoided. To this end it is necessary to have transmitters and repeaters with infinite bandwidth. Because of repeater bandwidth limitation, envelope fluctuations will be encountered in the received wave, together with resultant distortion owing to amplification nonlinearity and AM-PM conversion. A case in point would be multilevel transmission over a microwave radio relay designed for analog PM with a small deviation ratio and hence a high signal-to-noise ratio at repeater inputs. With repeaters having the amplification nonlinearity and AM-PM conversion factor illustrated in Fig. 8.10, and operated close to saturation, distortion would be an important consideration in determining the maximum number of levels with acceptable error probabilities.

9
Random Multipath and Troposcatter Transmittances

9.0 General

In preceding chapters, time-invariant linear and nonlinear channels were assumed, a condition hardly encountered in nature. When the channel parameters vary significantly and in an unpredictable manner with time, the conventional methods of determining the response to signals must be modified in various respects. In exact formulations this results in appreciable complications in analytical procedures and also in certain conceptual difficulties [1,2].

However, in efficient communication systems the fluctuations in the transmittances must of necessity be slow in relation to those in the signals. It is then a legitimate approximation to assume that the transmittances are essentially time-invariant over an appreciable number of signal intervals T. This corresponds to a form of "quasi-stationary" analysis, in which distortion is determined by conventional analytical methods for the various shapes of time-invariant transmittances that can be encountered.

This form of analysis is, of course, tacitly implied in dealing with very slow and small variations that do not pose analytical problems, such as those resulting from temperature fluctuations in metallic circuits. It is also a legitimate analytical approximation for radio transmission facilities that exhibit very pronounced and rapid random fluctuations in transmission properties. Among all present broadband radio facilities, troposcatter channels exhibit the more pronounced fluctuations and will be discussed further herein.

The transmittance variations in troposcatter facilities resemble those encountered in idealized random multipath channels with an infinity of transmission paths. In such channels each path has a different but fixed length and transmission delay. However, the amplitude of the signal component received over each path fluctuates at random, since it depends on reflections in the transmission medium that vary with time, owing to structural changes caused by atmospheric turbulence. When a steady-state sine wave is transmitted, the received wave will consequently exhibit variations in its envelope and phase, known as fading.

The transmission-frequency characteristic of a multipath channel can be determined by simultaneous transmission of a multiplicity of sine waves at uniform frequency spacing. The envelope and phase of each received wave will vary with frequency, as illustrated in Fig. 9.1. At a particular instant the transmission-frequency characteristic may be as indicated in Fig. 9.1a and at a later instant as shown in Fig. 9.1b. Such variations, known as frequency-selective fading, give rise to distortion in the spectrum of the received signals with resultant transmission impairments that depend on the modulation method.

Certain pertinent statistical properties of random multipath channels can be obtained from relations given in Chapter 3 for narrowband Gaussian variables. These relations involve the autocorrelation function of the time variations in the two quadrature components of the random variable. In application to random multipath channels the formal solution in terms of the autocorrelation function is the same as given in Chapter 3 for random noise. However, the autocorrelation function now relates to changes with time or with frequency in the channel transmittances. In dealing with time variations in the channel transmittance at a particular frequency, the time-autocorrelation function is related to the rapidity of changes in the heterogeneities in the transmission medium that are responsible for multipath transmission. In considering variations with

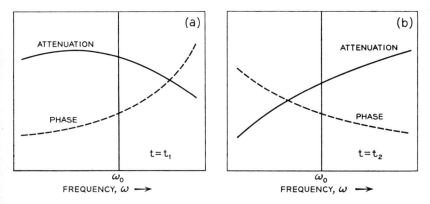

Figure 9.1 Illustrative variations in attenuation and phase characteristics at two instants t_1 and t_2.

frequency observed simultaneously over brief intervals, as in sweep-frequency measurements, the frequency-autocorrelation function is related to the distribution in space of power over the various transmission paths, or to the corresponding distribution in terms of the transmission delays encountered in the various paths. These distributions, and thus the autocorrelation function, depend on the size and shape of the common antenna volume, which is related to the antenna beam pattern. In a first approximation a Gaussian autocorrelation function is obtained for the variation with frequency [3].

When the foregoing two autocorrelation functions are known, it is possible to determine the probability distribution of the first, second, and higher derivatives of the envelope and phase variations with respect to time and frequency, as well as other statistics that are pertinent to the behavior of transmission systems, such as the distribution of the duration of fades of various depths [4].

The variations in transmittance will give rise to distortion of received signals, with resultant transmission impairments of various kinds, depending on the modulation method. At present, frequency modulation is used for transmission of voice channels in frequency-division multiplex over troposcatter links. With this method, the transmittance variations with frequency give rise to intermodulation distortion. Approximate relations are given here for such intermodulation distortion as related to various basic systems parameters, and a comparison is made with the results of comprehensive measurements.

The liability to deep signal fades and resultant excessive transmission impairments can be significantly reduced by various methods of diversity operation over several independently fading channels. The performance of various methods depends on the resultant probability distributions of the combined signal-to-noise ratios. These are given here for various methods of diversity transmission over independently fading channels with a Rayleigh probability distribution of received signal power.

In digital systems the transmittance variations give rise to pulse distortion and resultant intersymbol interference so that the error probability is increased. The various basic statistical relations presented here are used in Chapter 10 to derive specific equations and curves for the error probability with binary AM, FM and PM, as related to certain basic system parameters.

9.1 Transmittance Function

Let a sine wave of frequency ω be transmitted, and let $\omega = \omega_0 + u$, as indicated in Fig. 9.2, where ω_0 is a conveniently chosen reference fre-

Transmittance Function

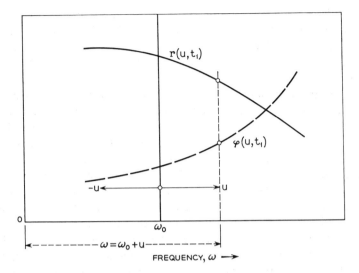

Figure 9.2 Illustrative dependence of amplitude and phase of transmittance with frequency u from a reference frequency ω_0 at a specified time t_1.

quency. In complex notation the received wave is then of the general form

$$e(u,t) = r(u,t) \exp[-i\varphi(u,t)] \exp(i\omega t). \qquad (9.1)$$

The channel transmittance is then

$$T(u,t) = r(u,t) \exp[-i\varphi(u,t)]. \qquad (9.2)$$

The following general relations apply

$$r(u,t) = [U^2(u,t) + V^2(u,t)]^{1/2} \qquad (9.3)$$

$$\varphi(u,t) = \tan^{-1}\frac{V(u,t)}{U(u,t)}. \qquad (9.4)$$

With random multipath transmission, $U(u,t)$ and $V(u,t)$ contain a very large number of components received over the multiplicity of paths. At a given frequency $u = u_1$ each component can be considered a Gaussian random variable of t, owing to the great number of components and the variation in each component with t. At a given time $t = t_1$, the components U and V can be considered Gaussian random variables of u. The contribution from each path will vary periodically with frequency. With a large number of different paths the combined periodic variation with frequency becomes rather complicated. As the number of paths approaches infinity, so will the frequency interval between repetition of the periodic pattern. Hence the variation over a finite frequency interval can be considered random. This will also be the case with a finite number of time-varying paths, provided that the average is taken for a large number of instants $t = t_1, t_2, t_3, \ldots$, etc., with adequate intervals between these instants.

On the above premises $U(u,t)$ and $V(u,t)$ will be Gaussian random variables of u and t and in addition both will have the same statistical parameters, so that no distinction need be made between U and V. Under this condition the envelope $r(u,t)$ and the phase function $\varphi(u,t)$, will have the statistical properties of narrowband Gaussian variables discussed in Chapter 3, as regards variations with t for a fixed u and conversely. This applies as regards relatively rapid variations from the mean or the median values. However, it should be noted that in actual multipath channels there will be a fluctuation in the mean or the median values taken over a convenient and meaningful interval, such as an hour or a day. These fluctuations may contain a systematic diurnal and a seasonal component as well as an unpredictable component, and can only be determined experimentally, as discussed later for troposcatter channels. By contrast, the statistical properties of the rapid fluctuations considered here can be derived on the sole premise of time-variant Gaussian random multipath transmission.

9.2 Envelope and Phase Distribution

On the preceding premises, all phases $\varphi(u,t)$ are equally probable as functions of u or of t and have a probability density $1/2\pi$. The probability density of the envelope follows the Rayleigh law as functions of u or of t. As shown in Chapter 3, the probability that the envelope exceeds a specified value r_1 is

$$P(r \geq r_1) = \exp\left(\frac{-r_1^2}{\bar{r}^2}\right), \tag{9.5}$$

where $\bar{r}$ is the rms amplitude of the envelope or the transmittance taken over an appropriate time interval.

The average received envelope power is in this case $\bar{r}^2 = \bar{S} = 2S$, where S is the average carrier power, that is, the average power within the envelope. The probability that the received envelope power at any instant exceeds a specified value $\bar{S}_1 = 2S_1$ is

$$P(S > S_1) = \exp\left(\frac{-\bar{S}_1}{\bar{S}}\right) = \exp\left(\frac{-S_1}{S}\right). \tag{9.6}$$

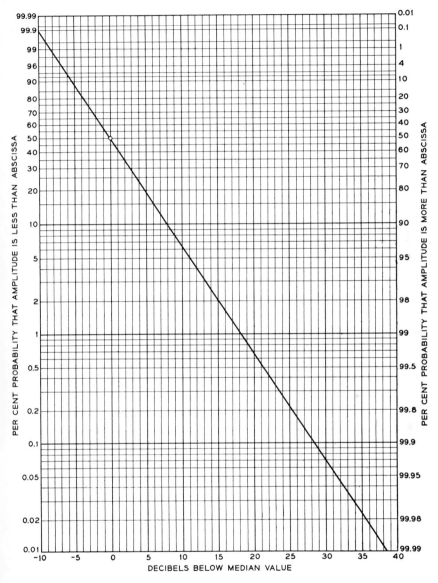

Figure 9.3 Rayleigh probability distribution of rapid fluctuations in envelope of a received carrier owing to multipath propagation.

The median value S_m of S is obtained from $P(S \geq S_m) = \frac{1}{2}$, which gives $S_m = S \ln 2$. Hence, in terms of the median value

$$P(S \geq S_1) = \exp\left(-\frac{S_1}{S_m} \ln 2\right). \tag{9.7}$$

The distribution represented by (9.6) or (9.7) is shown in Fig. 9.3.

The foregoing distribution of rapid fades is to be distinguished from the distribution of slow variations in the envelope. In the case of troposcatter channels the latter have a log-normal distribution, as discussed in Section 9.12.

In order to determine various other statistical properties of transmittance, the autocorrelation functions of $U(u,t)$ or $V(u,t)$, with respect to u and t are required.

9.3 Time Autocorrelation Functions

The time autocorrelation functions are related to the rapidity of changes in the heterogeneities in the transmission medium that are responsible for multipath transmission. The mean rate at which these changes occur may vary from hour to hour or from day to day, and there will be a certain probability distribution of the mean rates, as discussed in Section 9.13 for troposcatter links. With the aid of the autocorrelation function determined experimentally for appropriate periods, such as an hour, it is possible to determine various related statistical properties of rapid fluctuations relative to the mean or the median for the period under consideration.

Let $R(\tau)$ be the autocorrelation function of variations in $U(u,t)$ or $V(u,t)$ with t. The corresponding double-sided power spectrum is then

$$W(\gamma) = \frac{1}{\pi}\int_0^\infty R(\tau) \cos \gamma\tau \, d\tau, \tag{9.8}$$

where γ is used to designate the radian frequency of spectral components to avoid confusion with the frequency ω of the transmitted wave.

The autocorrelation function $R(\tau)$ or the corresponding power spectrum $W(\gamma)$ of the components U and V cannot be determined as readily by measurements as the autocorrelation function $R_r(\tau)$ of the envelope. The latter is related to $R(\tau) = \bar{R}(\tau)$ by (3.29) or

$$R_r(\tau) = R(O)\{2E[\kappa(\tau)] - [1 - \kappa^2(\tau)]\,K[\kappa(\tau)]\}, \tag{9.9}$$

where

$$\kappa(\tau) = \frac{R(\tau)}{R(O)}, \tag{9.10}$$

E = complete elliptic integral of the second kind,

K = complete elliptic integral of the first kind.

Frequency Autocorrelation Function

For $\tau = 0$, $R_r(O) = 2R(O)$. Hence the autocorrelation coefficient of the envelope can be written

$$\kappa_r(\tau) = E[\kappa(\tau)] - \tfrac{1}{2}[1 - \kappa^2(\tau)] K[\kappa(\tau)]. \tag{9.11}$$

With the aid of (9.11), the autocorrelation coefficient $\kappa(\tau)$ of each quadrature component can be determined from measurements of $\kappa_r(\tau)$.

In the particular case of a Gaussian autocorrelation function

$$R(\tau) = R(O)e^{-\gamma_0^2 \tau^2/2}, \tag{9.12}$$

the corresponding symmetrical power spectrum obtained from (9.8) is

$$W(\gamma) = R(O)\left(\frac{1}{2\pi\gamma_0^2}\right)^{1/2} e^{-\gamma^2/2\gamma_0^2}, \tag{9.13}$$

where $R(O) = \sigma^2$ is the average envelope power of each component U and V, so that the rms amplitude of the envelope is $\sigma_r = 2^{1/2}\sigma$.

The equivalent semi-bandwidth $\bar{\gamma}$ of a flat symmetrical power spectrum of spectral density $W(O) = \left(\dfrac{1}{2\pi\gamma_0^2}\right)^{1/2} R(O)$ is obtained from the relation

$$W(O)\bar{\gamma} = W(O)\int_0^\infty e^{-\gamma^2/2\gamma_0^2}\, d\gamma,$$

which yields

$$\bar{\gamma} = \left(\frac{\pi}{2}\right)^{1/2}\gamma_0 \simeq 1.25\,\gamma_0. \tag{9.14}$$

9.4 Frequency Autocorrelation Function

The frequency autocorrelation function depends on the distribution of received signal power among the various transmission paths of different length or transmission delay. Let $\nu = \omega_2 - \omega_1 = u_2 - u_1$ designate the frequency difference and $R(\nu)$ the autocorrelation function with respect to u of the components $U(u,t)$ and $V(u,t)$. The corresponding symmetrical power spectrum is

$$W(\delta) = \frac{1}{\pi}\int_0^\infty R(\nu)\cos\nu\delta\, d\delta. \tag{9.15}$$

Here $W(\delta)$ is the power spectral density for paths with transmission delay δ from the mean delay.

The corresponding autocorrelation coefficient of the envelope is in place of (9.11)

$$\kappa_r(\nu) = R(O)\{E[\kappa(\nu)] - \tfrac{1}{2}[1 - \kappa^2(\nu)] K[\kappa(\nu)]\}, \tag{9.16}$$

where E and K are elliptic integrals as defined previously and

$$\kappa(\nu) = R(\nu)/R(O). \tag{9.17}$$

For a Gaussian frequency autocorrelation function

$$R(\nu) = R(O)e^{-\delta_0^2 \nu^2/2}, \tag{9.18}$$

the corresponding symmetrical power spectrum is

$$W(\delta) = R(O)\left(\frac{1}{2\pi\delta_0^2}\right)^{1/2} e^{-\delta^2/2\delta_0^2}. \tag{9.19}$$

If the signal power is distributed equally among paths with delay differences δ varying between $-\Delta$ and Δ, the spectral density is

$$\begin{aligned} W(\delta) &= \frac{R(O)}{2\Delta}; \quad \text{for} \quad -\Delta < \delta < \Delta \\ &= O \quad \text{for} \quad \delta > |\Delta|. \end{aligned} \tag{9.20}$$

The corresponding frequency autocorrelation function is

$$\begin{aligned} R(\nu) &= \frac{R(O)}{2\Delta}\int_{-\Delta}^{\Delta} \cos \nu\delta \, d\delta \\ &= R(O)\frac{\sin \nu\Delta}{\nu\Delta} \end{aligned} \tag{9.21}$$

The autocorrelation coefficient $\kappa_r(\nu)$ obtained with $\kappa(\nu) = \sin \nu\Delta/\nu\Delta$ in (9.16) is shown in Fig. 9.4.

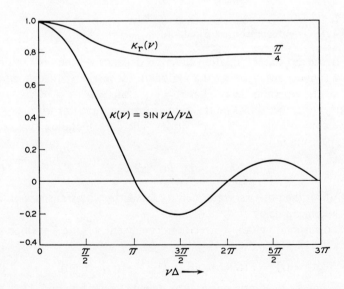

Figure 9.4 Frequency autocorrelation coefficient $\kappa_r(\nu)$ of envelope for autocorrelation coefficient $\kappa(\nu)$ of components U and V.

9.5 Distribution of Envelope Time Derivative $r'(t)$

The time derivative $r'(t)$ of the envelope of a received sine wave, or the time derivative of the transmittance, is of interest in determination of errors that may occur in narrow-band digital transmission by amplitude modulation, owing to fading variations over a pulse interval. When the time autocorrelation function $R(\tau)$ is known, or the corresponding power spectrum $W(\gamma)$, it is possible to determine the probability distribution of $r'(t)$. As shown in Chapter 3, the probability distribution of r' follows the normal law

$$P(|r'| > k\sigma') = \text{erfc}(k/2^{1/2}), \qquad (9.22)$$

or

$$P(|r'| > k\sigma'_r) = \text{erfc}(k), \qquad (9.23)$$

where

$$\sigma' = 2^{-1/2}\sigma'_r = \left(\int_{-\infty}^{\infty} W(\gamma)\gamma^2\, d\gamma\right)^{1/2}. \qquad (9.24)$$

Here $W(\gamma)$ is the power spectrum of each envelope component $X(t)$ or $Y(t)$ or of the sine wave within the envelope $r(t)$.

For a Gaussian power spectrum (9.13), relation (9.24) yields

$$\sigma' = \gamma_0 R^{1/2}(O) = \gamma_0 \sigma$$
$$\simeq 0.8\,\bar{\gamma}\sigma \qquad (9.25)$$

where $\bar{\gamma}$ is the equivalent semi-bandwidth (9.14).

For a flat-power spectrum of semi-bandwidth $\hat{\gamma}$ the spectral density is $R(O)/2\hat{\gamma}$ and (9.24) gives

$$\sigma' = \hat{\gamma}[R(O)/3]^{1/2} = \hat{\gamma}\sigma/3^{1/2}$$
$$\simeq 0.58\,\hat{\gamma}\sigma. \qquad (9.26)$$

With $\bar{\gamma} = \hat{\gamma}$ the average power $R(O) = \sigma^2$ is the same with both power spectra. Relations (9.25) and (9.26) show that under this condition σ' is somewhat greater for a Gaussian than for a flat power spectrum.

9.6 Distribution of Envelope Frequency Derivative $\dot{r}(u)$

The derivative $\dot{r}(u) = dr(u)/du$ represents a component of linear slope in the amplitude versus frequency characteristic of a channel, as indicated in Fig. 9.5. This linear component affords a first approximation to coarse

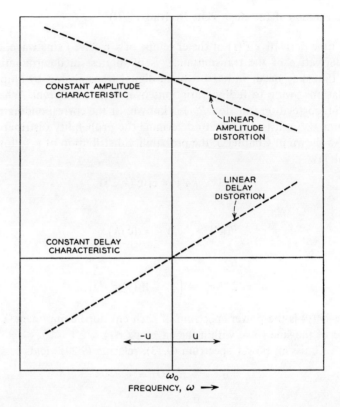

Figure 9.5 First approximations to random departures from constant amplitude and delay characteristics are represented by linear variations with frequency.

structure amplitude versus frequency deviations, which may be important from the standpoint of performance in broadband transmission.

When the frequency autocorrelation function $R(\nu)$ is known, or the corresponding power spectrum $W(\delta)$, the probability distribution of $\dot{r}$ can be determined, in the same manner as that for $r'(t)$ in the preceding section. Thus

$$P(|\dot{r}| \geq k\dot{\sigma}) = \mathrm{erfc}\,(k/2^{1/2}), \tag{9.27}$$

or

$$P(|\dot{r}| \geq k\dot{\sigma}_r) = \mathrm{erfc}\,(k), \tag{9.28}$$

where

$$\dot{\sigma} = 2^{-1/2}\dot{\sigma}_r = \left(\int_{-\infty}^{\infty} W(\delta)\delta^2\,d\delta\right)^{1/2} \tag{9.29}$$

Distribution of Envelope Delay

For a Gaussian power spectrum (9.19), (9.25) is replaced by

$$\dot{\sigma} = \delta_0 R^{1/2}(O) = \delta_0 \sigma. \tag{9.30}$$

For a flat power spectrum (9.20), (9.26) is replaced by

$$\dot{\sigma} = \Delta\sigma/3^{1/2} \simeq 0.58\,\Delta\sigma. \tag{9.31}$$

9.7 Distribution of Instantaneous Frequency $\varphi'(t)$

The probability distribution of the time derivative $\varphi'(t)$ of the phase function is of interest in connection with errors that may occur in narrow-band digital transmission by phase modulation, owing to phase variations over a pulse interval. When $R(\tau)$ or the corresponding $W(\gamma)$ is known, the probability distribution of $\varphi'(t)$ can be determined.

In accordance with (3.87) the mean frequency deviation is

$$\varphi'_m = \frac{\sigma'}{\sigma}, \tag{9.32}$$

where $\sigma = R(O)$ and σ' is given by (9.24).

In accordance with (3.85) the probability that $|\varphi'|$ exceeds φ'_1 is

$$P(|\varphi'| > \varphi'_1) = 1 - \frac{k}{(1+k^2)^{1/2}}, \tag{9.33}$$

where $k = z_1 = \varphi'_1/\varphi'_m$. In view of (9.32)

$$P(|\varphi'| > k\varphi'_m) = 1 - \frac{k}{(1+k^2)^{1/2}}. \tag{9.34}$$

For $k \gg 1$

$$P(|\varphi'| > k\varphi'_m) \simeq \frac{1}{2k^2}. \tag{9.35}$$

For a Gaussian power spectrum (9.13)

$$\varphi'_m = \gamma_0 \simeq 0.8\,\bar{\gamma}. \tag{9.36}$$

For a flat-power spectrum of semi-bandwidth $\hat{\gamma}$

$$\varphi'_m = \hat{\gamma}/3^{1/2} = 0.58\,\hat{\gamma}. \tag{9.37}$$

9.8 Distribution of Envelope Delay $\dot{\varphi}(u)$

The frequency derivative of the phase function represents the envelope delay. This delay will vary with time, so that in digital transmission at uniform intervals T the intervals between received pulses will vary. The probability distribution of $\dot{\varphi}(u)$ is obtained in a similar way as shown for $\varphi'(t)$ in the preceding section.

Thus

$$P(|\dot{\varphi}| \geq k\dot{\varphi}_m) = 1 - \frac{k}{(1 + k^2)^{1/2}} \quad (9.38)$$

$$\simeq \frac{1}{2k^2} \quad \text{for} \quad k \gg 1, \quad (9.39)$$

where

$$\dot{\varphi}_m = \frac{\dot{\sigma}}{\sigma}, \quad (9.40)$$

in which $\sigma^2 = R(O)$ and $\dot{\sigma}$ is given by (9.29).

For a Gaussian power spectrum (9.19)

$$\dot{\varphi}_m = \delta_0, \quad (9.41)$$

while for a flat-power spectrum (9.20)

$$\dot{\varphi}_m = \Delta/3^{1/2} \simeq 0.58\,\Delta \quad (9.42)$$

9.9 Distribution of Time Derivative $\varphi''(t)$

The second time derivative of the phase variations corresponds to the time derivative of the instantaneous frequency. The probability distribution of $\varphi''(t)$ is of interest in connection with errors that may occur in narrowband digital transmission by frequency modulation, as a result of frequency changes over a pulse interval. The probability distribution of $\varphi''(t)$ is obtained from relation (3.112), or the approximate relation (3.115).

Thus

$$P\left[|\varphi''| \geq k\left(\frac{\sigma'}{\sigma}\right)^2\right] = H(k), \quad (9.43)$$

where $H(k)$ is the function shown in Fig. 3.10 for a flat power spectrum and in Fig. 9.6 for a Gaussian and for a flat-power spectrum.

For a Gaussian power spectrum (9.13)

$$\left(\frac{\sigma'}{\sigma}\right)^2 = \gamma_0^2. \quad (9.44)$$

For a flat-power spectrum of semi-bandwidth $\hat{\gamma}$,

$$\left(\frac{\sigma'}{\sigma}\right)^2 = \frac{\hat{\gamma}^3}{3} \quad (9.45)$$

Distribution of $\ddot{\varphi}(u)$

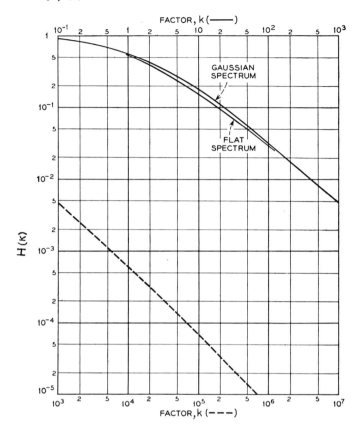

Figure 9.6 Function $H(k)$ relating to probability distributions of $\varphi''(t)$ and $\ddot{\varphi}(u)$.

For $k \geq 100$, the following approximation applies

$$P\left[|\varphi''| > k\left(\frac{\sigma'}{\sigma}\right)^2\right] \simeq \frac{2}{\pi k}\left[1 + \ln\left(\frac{k}{2} + 1\right)\right]. \tag{9.46}$$

9.10 Distribution of Frequency Derivative $\ddot{\varphi}(u)$

The frequency derivative $\ddot{\varphi}(u)$ represents a component of linear delay distortion as indicated in Fig. 9.5, which gives rise to intermodulation distortion in broadband transmission by FM and causes pulse distortion in broadband digital transmission with any modulation method. The probability distribution of $\ddot{\varphi}$ is obtained in the same manner as that for φ'' in the preceding section.

Thus

$$P\left[|\ddot{\varphi}| > k\left(\frac{\dot{\sigma}}{\sigma}\right)^2\right] = H(k), \qquad (9.47)$$

where $H(k)$ is the function shown in Fig. 9.6. For $k \geq 100$

$$H(k) \simeq \frac{2}{\pi k}\left[1 + \ln\left(\frac{k}{2} + 1\right)\right]. \qquad (9.48)$$

For a Gaussian power spectrum (9.19)

$$\left(\frac{\dot{\sigma}}{\sigma}\right)^2 = \delta_0^2. \qquad (9.49)$$

For a flat-power spectrum (9.20)

$$\left(\frac{\dot{\sigma}}{\sigma}\right)^2 = \frac{\Delta^2}{3}. \qquad (9.50)$$

9.11 Troposcatter Channels

In tropospheric scatter transmission beyond the horizon, narrow-beam transmitting and receiving antennas are used in a frequency range from about 400 to 10,000 megacycles. Various publications differ in their views regarding the exact mechanism of the reflections in troposcatter links, but they appear to agree that they occur as a result of heterogeneities within the common antenna volume indicated in Fig. 9.7.

Various properties of troposcatter channels have been dealt with in several publications [3–8]. These properties include the expected average

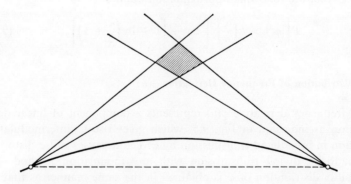

Figure 9.7 Illustrative antenna beams and common antenna volume.

Observed Transmission Loss Variation 373

path loss and systematic seasonal variations from the average, together with the probability distributions of slow and rapid fluctuations from the mean loss. Other important properties from the standpoint of systems design and performance are the distribution of duration of fades and the fading rapidity or rate. The above properties relate to transmittance variations with time at a particular frequency. In addition observations have been made of variations with frequency by sweep-frequency measurements and by measurements of the frequency autocorrelation function of the envelope.

In the following sections the results of observations are reviewed briefly.

9.12 Observed Transmission Loss Variation of Troposcatter Channels

On troposcatter links there is a certain average transmission loss over a year, which depends on the length of the link, on the properties of the terrain, and on climatic conditions. Experimental data indicate that there will be systematic monthly and seasonal departures from this yearly average, owing principally to slow temperature changes. The average loss during a winter month may thus be up to 20 dB greater than the average during a summer month. That is, the departure in transmission loss from the yearly mean may be ± 10 dB.

During each month there will be a more or less random fluctuation in the hourly average loss from the mean of the month. This fluctuation has been found to be almost independent of frequency and seems to be associated with the variations in average refraction of the atmosphere and resultant variation in the bending of beams. This fluctuation in the hourly average loss relative to the monthly average has been found to follow closely the log-normal law. That is to say, let the monthly median loss be

$$\alpha_m = -\ln \bar{r}_m^2, \tag{9.51}$$

and the hourly average loss be

$$\alpha = -\ln \bar{r}^2 \tag{9.52}$$

where $\ln = \log_e$, $\bar{r}$ is the rms amplitude of the envelope over an hour, and $\bar{r}_m$ is the monthly median of all the hourly rms amplitudes.

The probability that the average hourly loss exceeds a specified value $\alpha_1 = -\ln \bar{r}_1^2$ is then given by

$$P(\alpha \geq \alpha_1) = \frac{1}{2}\left(1 - \operatorname{erf} \frac{\alpha_1 - \alpha_m}{\sqrt{2}\sigma_\alpha}\right), \tag{9.53}$$

where erf is the error function and σ_α the standard deviation in transmission loss expressed in nepers, when α and α_m are expressed in nepers as earlier. For links 100 to 200 miles in length, a representative value of σ_α appears to be about 0.9 neper (8 dB).

In addition to the above slow variations in the average hourly loss, there will be more rapid fluctuations in the envelope $r(u,t)$, owing to changes in the multipath transmission structure caused principally by winds. This type of fluctuation follows the Rayleigh distribution law, discussed in Section 9.2.

9.13 Observed Time Autocorrelation of Troposcatter Channels [6]

Observations of the correlation function $R(\tau)$ of rapid fluctuations in transmission loss indicate a nearly Gaussian shape (9.12). The correlation function will change from hour to hour and day to day, depending on the velocity of winds. There will thus be a certain median autocorrelation function and corresponding median power spectrum. Measurements indicate that these median values depend on the antenna beamwidths and that the fading rate is not quite proportional to the carrier frequency. Furthermore, there can be appreciable departures from the median values. From measurements of the median number of fades per minute, the median value of γ_0 can be determined. The observations indicate that for a particular antenna arrangement, $\gamma_0 \simeq 0.1$ c/s at 460 mc and about 1.3 c/s at 4110 mc. The corresponding equivalent semi-bandwidth $\bar{\gamma}$ of a flat-power spectrum is thus $\bar{\gamma} \approx 0.125$ c/s, or 0.8 radian/sec at 460 mc and $\bar{\gamma} \approx 1.6$ c/s or 10 radians/sec at 4110 mc. The measurements further indicate that there is a probability of about 0.01 that the fading rate exceeds the median by a factor of about 7 at 460 mc and a factor of about 3.5 at 4110 mc.

9.14 Differential Delay Δ of Troposcatter Channels

Exact determination of the equivalent maximum departure from the mean transmission delay requires consideration of the beam patterns as affected by scattering. On the approximate basis of equivalent beam angles α, the following relation applies, with notation as indicated in Fig. 9.8

$$\Delta \approx \frac{L}{v}\frac{\alpha+\beta}{2}\left(\theta + \frac{\alpha+\beta}{2}\right), \tag{9.54}$$

Differential Delay

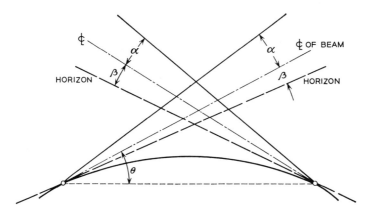

Figure 9.8 Definition of antenna beam angles α, take-off angle β and chord angle θ to midbeam. With different angles at the two ends, the mean angles are used in expressions for Δ. In applications to actual beams, α would be the angle to the 3-dB loss point.

where $\beta \leq \alpha$, v is the velocity of propagation in free space, L is the length of the link, and

$$\theta = \frac{L}{2R} = \frac{L}{2R_0 K}, \qquad (9.55)$$

where R_0 is the radius of the earth and the factor K is ordinarily taken as 4/3.

The equivalent beam angle α from midbeam to the 3-dB loss point depends on the free-space antenna beam angle α_0 and on the effect of scatter, which is related in a complex manner to α_0 and the length L, or alternately θ. Narrow-beam antennas as now used in actual systems are loosely defined by $\alpha_0 \leq 2\theta/3$. For these $\alpha \approx \alpha_0$ on shorter links, while on longer links $\alpha > \alpha_0$ owing to beam-broadening by scatter. Analytical determination of α for longer links appears difficult, and only limited experimental data are available at present.

By way of numerical example, let $L = 170$ miles and $K = 4/3$, in which case $\theta = 0.016$ radian. Since $\alpha_0 = 0.004$ radian $\ll 2\theta/3$, it is permissible to take $\alpha = \alpha_0$. With $\beta \approx \alpha_0$, (9.54) gives $\Delta = 0.08 \times 10^{-6}$ second.

For small wide-beam antennas with $\alpha_0 > \theta/3$, the effective beam angle depends principally on the scatter process, even for links of relatively short lengths, say 100 miles. In this case the distribution of power within the beam may be such that the error involved in the use of an equivalent beam angle α in accordance with (9.54) is greater than for narrow-beam antennas.

9.15 Observed Frequency Autocorrelation of Troposcatter Channels

In Fig. 9.9 is indicated the shapes of the envelope versus frequency variations that can be obtained from (9.3) when the components U and V are random variables. These fluctuations will vary with time but will have the characteristic shapes indicated in the figure, which resemble shapes obtained in sweep-frequency measurements on a link of the length for which the above value of Δ applies.

A better indication of the adequacy of the present idealized troposcatter model is obtained by comparing the autocorrelation coefficient of the envelope as given by (9.16) with the correlation coefficient derived from observations. In Fig. 9.10 is shown the theoretical coefficient for $\Delta = 0.08 \times 10^{-6}$ second together with coefficients obtained from four experimental runs. However, a much greater number of experimental runs under various atmospheric conditions would be required to establish a representative average autocorrelation coefficient.

The foregoing results apply for narrow-beam antennas. With wide-beam antennas, the beam power pattern depends on the scatter process rather than on the antenna beam pattern. Measurements have been made of the correlation function $\kappa(\nu)$ of each component U and V, on a 100-mile link employing wide-beam antennas, at about 5000 mc. [8]. These indicate median values of about 0.36, 0.08, 0.05 and 0.05 for frequency separations of 1.5, 3, 4.5, and 6 mc. During 1 percent of the time the correlation coefficients exceeded 0.66, 0.35, 0.29, and 0.27 at the above frequency separations.

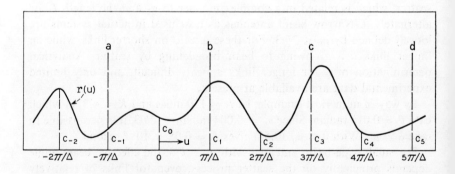

Figure 9.9 Illustrative envelope vs frequency characteristic in sweep-frequency measurements with a radian frequency sweep from $-\pi/\Delta$ to π/Δ from the carrier frequency. The envelope variations during a sweep might be like that in any of the intervals a-b, b-c, c-d, etc.

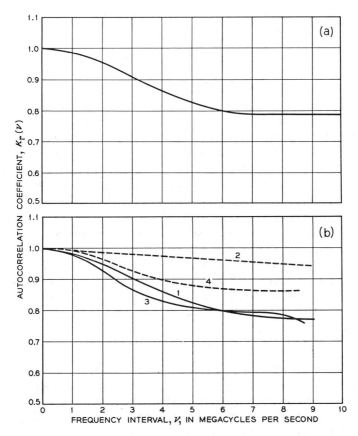

Figure 9.10 Theoretical versus observed frequency correlation functions of envelope. a. Autocorrelation function for random multipath transmission with $\Delta = .08 \cdot 10^{-6}$ second. b. Autocorrelation functions given in Fig. 7 of reference (6) and derived from measurements of envelope variation with narrow-beam antennas with the above Δ on four different days over a period of about 8 weeks. Curve 2 considered very unusual.

9.16 Measured versus Predicted Intermodulation in FM Systems

As shown in Chapter 6, attenuation and phase deviations over the transmission band will give rise to intermodulation distortion in FM systems. Comprehensive measurements have been made of such distortion in four troposcatter links ranging in length from 185 to 440 miles [9,10]. A comparison has been made elsewhere [11] of measured median intermodulation distortion with a first-order prediction of median intermodulation distortion based on random multipath transmission. The results are reviewed briefly here.

The amplitude and phase characteristics as a function of u at any time t_0 can in general be represented by a power series as

$$A(u,t_0) = a_0 + a_1 u + a_2 u^2 + a_3 u^3 + \cdots, \tag{9.56}$$

$$\varphi(u,t_0) = b_0 + b_1 u + b_2 u^2 + b_3 u^3 + \cdots. \tag{9.57}$$

In accordance with the results in Section 6.10, a linear deviation $a_1 u$ from a constant amplitude characteristic will not give rise to intermodulation distortion. It can also be shown that the term in $a_2 u^2$ will cause distortion proportional to the second time derivative of the modulating wave, which does not contribute new frequency components and hence does not constitute intermodulation distortion. Accordingly, the principal contributor to intermodulation distortion at adequately low frequency u will be the term in $b_2 u^2$. Intermodulation distortion, owing to this term, is given by (6.91) for adequately small values of the rms frequency deviation $b_2 \underline{\Delta}^2$ so that the assumptions underlying (6.91) apply.

In the intermodulation measurements mentioned above a pre-emphasis was used that corresponded closely with that for $c = 16$ in Fig. 6.7. Furthermore, the measurements were made at the top frequency $\omega = \Omega_0$ so that $a = 1$. Hence in this case $G(c,a) = G(16,1) \simeq 0.19$ and (6.91) becomes

$$\eta(\Omega_0) = 0.19 b_2^2 \underline{\Delta}^2 \Omega_0^2 \tag{9.58}$$

In accordance with (9.57) as $u \to O$, $\ddot{\varphi}(u) = 2b_2$. The probability distribution of $\ddot{\varphi}(u)$ and thus of b_2 is obtained from Fig. 9.6, which gives the probability that $\ddot{\varphi}(O)$ or $2b_2$ exceeds $\Delta^2/3$ by a factor k. The median value corresponding to $H(k) = 0.5$ is obtained with $k \simeq 1.2$. The median value $(2b_2)_{\frac{1}{2}}$ is accordingly $1.2\ \Delta^2/3$ so that the median b_2 is

$$(b_2)_{\frac{1}{2}} \simeq \frac{1.2}{6} \Delta^2 \simeq 0.2\ \Delta^2. \tag{9.59}$$

With (6.59) in (6.58) the median value of η becomes

$$\eta_{\frac{1}{2}}(\Omega_0) = 7.6 \times 10^{-3} \Delta^4 \underline{\Delta}^2 \Omega_0^2. \tag{9.60}$$

With $\underline{\Delta} = \underline{D}\Omega_0$ and $\Omega_0 = 2\pi B_0$

$$\eta_{\frac{1}{2}}(B_0) \simeq 11.4\ \Delta^4 \underline{D}^2 B_0^4, \tag{9.61}$$

where Δ is the equivalent maximum departure from the mean transmission delay, in seconds, $\underline{D}$ is the rms frequency deviation ratio and B_0 is the maximum baseband frequency in c/s.

The basic parameters of the systems on which the measurements were made are given in reference [10] and are summarized in Table 1. In this table α_0 is the free-space antenna beam angle from midbeam to the 3-dB loss point.

Table 1. Basic Parameters of Troposcatter Test Systems in Caribbean (A) and in Arctic (B,C,D)

System	A	B	C	D
Length, miles	185	194	340	440
Radio frequency, mc	725	900	900	800
Antenna/diameter, ft	60, 60	30, 60	120, 120	120, 120
α_0 (radian)	0.0115	0.017	0.0058	0.0058
θ (radian)	0.015	0.016	0.031	0.034
Δ_0 (microsecond)	0.12	0.21	0.185	0.255

The value Δ_0 of Δ given in the table was calculated with $\alpha = \alpha_0$, rather than the actual beam angle with scatter. Systems A, B, C, and D correspond to paths 1, 2, 4, and 3 in reference [10].

In Fig. 9.11 are shown the measured ratios $\eta_{1/2}(B_0)$ in dB as a function of the rms frequency deviation DB_0 for different bandwidths B_0 of the baseband signals.

In the same figure are shown median values of intermodulation noise obtained from (9.61) for each case, based on values Δ_m of Δ that afford the best average approximation to the measurements. The latter values are somewhat greater than Δ_0, as indicated in Table 2.

A ratio Δ_m/Δ_0 or $\alpha_m/\alpha_0 > 1$ is to be expected owing to beam-broadening by scatter, and the above ratios appear reasonable in the light of present knowledge. Thus, if the actual angles α were known so that Δ could be determined, it appears plausible that satisfactory conformance with observed intermodulation noise would be obtained.

Table 2. Ratio Δ_m/Δ_0

System	A	B	C	D
Length, miles	185	194	340	440
Δ_0, microsecond	0.12	0.21	0.185	0.255
Δ_m, microsecond	0.12	0.25	0.25	0.55
Δ_m/Δ_0	1.0	1.2	1.35	2.15
α_m/α_0	1.0	1.1	1.35	2.15

It should be noted that the simple relation (9.61) applies only for small values of the rms frequency deviation and of the quadratic component in (9.57). In a more exact analysis it is necessary to include additional terms in (9.57) and to use relations for intermodulation distortion that are valid also for large rms frequency deviations. As shown elsewhere [11], inter-

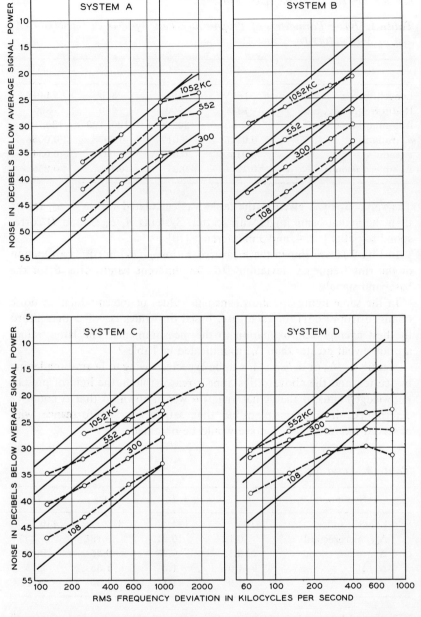

Figure 9.11 Dashed curves show measured median intermodulation distortion at top frequency with modulation by flat random noise of the indicated bandwidth. Solid curves are theoretical values based on Δ_m as given in Table 2.

modulation distortion resulting from the quadratic term in (9.57) will increase less rapidly with the frequency deviation than indicated by (9.61) and will tend to approach the observed values for large frequency deviations. The inclusion of additional terms in (9.57) will result in a further reduction in intermodulation distortion for large rms frequency deviations, for the reason that these terms will limit phase distortion [11].

By the same procedure as used earlier to derive the median values of $\eta_{1/2}$, it is possible to derive values of η that are exceeded with any designated probability. The probability distribution of intermodulation distortion can thus be determined, as shown in further detail elsewhere [11]. Comparison with observations indicate that the probability of exceeding the median value by a specified factor is less than predicted by the foregoing first-order theory, for the reasons mentioned earlier that it does not apply for large phase distortion encountered with a small probability.

Exact theoretical evaluation of intermodulation distortion for an autocorrelation function (9.21) of frequency-selective fading is unfeasible, but would no doubt yield reasonable conformance with observations, also for large frequency deviations.

9.17 Diversity Transmission Objectives and Methods

The objective in simultaneous or selective diversity transmission over several time-varying channels is to reduce the liability to excessively deep fades. This can be accomplished in part by increasing the combined transmitter power over that employed with a single channel. But the more important advantage is derived from the circumstance that deep fades occur but rarely during the same intervals on two or more paths, provided the variations in transmission loss are substantially uncorrelated.

Let the average transmitter power for each of m diversity channels be the same as for a single channel, so that there is an m-fold increase in total transmitter power. With one receiver per channel and synchronous demodulation of each signal, it is well known that with time invariant channels the ratio of combined average message power to combined average noise power is

$$\rho_m = m\rho_1, \tag{9.62}$$

where ρ_1 is the ratio for one channel and it is assumed that all channels have the same transmission loss and the same average noise power at the receiver inputs.

When the transmission losses vary simultaneously and in the same manner in all channels, relation (9.62) is modified into

$$\bar\rho_m = m\bar\rho_1 \qquad (9.63)$$

where $\bar\rho_m$ and $\bar\rho_1$ are now the mean ratios over very long intervals.

If the transmission losses in the m channels have the same mean values and the same probability distributions but vary independently, the ratio $\bar\rho_m$ depends on the method used in controlling the gain of the m receivers and combining their outputs. With the optimum method to be discussed in the next section, maximal ratio combining, relation (9.63) again applies.

It should be noted that (9.62) and (9.63) are the same relations as obtained with an m-fold increase of transmitter power in a single channel. Thus, at best, there is no advantage in diversity transmission as regards the average signal-to-noise $\bar\rho_m$ and with some methods there is some disadvantage. The inherent advantage resides in the reduced probability of deep fades when the fluctuations in transmission loss in the m channels are substantially uncorrelated. With independent variations and Rayleigh probability distributions, the probability of deep fades is substantially reduced, even without an m-fold increase in transmitter power, owing to the small probability of simultaneous fading in several channels, as shown later.

In the literature on diversity transmission it has been customary to assume that each of several channels has the same transmitter power as for single-channel transmission. Thus the effect of increased total transmitter power is included in relations for diversity transmission probability distributions and average signal-to-noise ratio, though often this is not explicitly stated. This may possibly be attributed to the circumstance that the simplest method is to use a single transmitter with several receivers, in which case the combined received signal power very nearly increases in proportion to the number of receivers without increase in transmitter power. Moreover, if several transmitters operating at different frequencies are provided, there may be no significant cost penalty in using the same power for each as with single-channel transmission, and this could also apply for several antennas operating at the same frequency. Finally, the formulas for the probability distributions of the signal-to-noise ratios are simplified by the above assumption, as are the relations for error probabilities in digital transmission given in Chapter 10. Hence it will be assumed that with m-diversity transmission a single receiver is used for each channel or link and that there is an m-fold increase in total signal power. With r receivers per channel the combined signal-to-noise ratios and their probability distribution is obtained by replacing m with mr in the relations that follow.

In diversity transmission, either space, frequency, or time diversity can be used. With space diversity it is also possible to use both horizontal and vertical polarization to double the number of diversity paths at a given frequency. In some situations combinations of the above various methods may be used for increased security and protection against jamming.

Optimum performance, from the standpoint of the combined probability distribution, is obtained when there is no correlation between the transmission-loss fluctuations in the various channels. This entails adequate antenna separations in space diversity, adequate frequency separation in frequency diversity and adequate time intervals between message repetitions in time diversity.

In order that the fading be uncorrelated with frequency diversity it is necessary that the frequency separation be greater than the correlation bandwidth $B_c = 1/2\Delta$ where Δ is as previously defined. With time diversity the interval τ between repetitions must be such that $R(\tau) \ll 1$, where $R(\tau)$ is given by (9.12). This will be the case if $\gamma_0^2 \tau^2 / 2 \geq 3$, where γ_0 is in radians/second. With the fading bandwidth γ_0 taken in c/s, the interval τ in seconds becomes $\tau \geq 1/4\gamma_0$. To insure uncorrelated fading with space diversity, the receiving or transmitting antenna must be moved along the great circle for a distance $\lambda/(\alpha + \beta)\theta$, where λ is the wavelength and the other quantities are defined in Fig. 9.8.

With any one of the above methods different combining or decision procedures can be used at the receivers, as discussed in considerable detail by Brennan [12] who also gives relations for the probability distributions of the combined signal-to-noise ratios for the three methods considered below.

9.18 Maximal Ratio Combining

A basic requirement for optimum performance with any combining method is in-phase addition of the signals received over the various channels. This requirement cannot be fully realized since complete adjustment cannot be made with any practicable receiver to unavoidable and unpredictable phase fluctuations in the received signals.

Let $V_j(t)$ represent the signal voltage in channel j, $\alpha_j(t)$ the momentary amplification factor and N_j the average noise power in channel j. With in-phase addition of the signals in m channels, the momentary ratio of average signal power to average noise power is

$$\rho_m(t) = \frac{[\sum_{j=1}^{m} V_j(t)\alpha_j(t)]^2}{\sum_{j=1}^{m} N_j \alpha_j^2(t)}. \tag{9.64}$$

The optimum $\alpha_j(t)$ resulting in the maximum $\rho_m(t)$ is obtained by the same method as employed in section 4.9 for matched filter reception in time-invariant channels. Relation (9.64) corresponds to (4.100) in combination with (4.101), with the integral signs replaced by summation signs. The optimum value of $\alpha_j(t)$ is given by [12]

$$\alpha_j(t) = c \frac{V_j(t)}{N_j}, \tag{9.65}$$

which corresponds to (4.102), with c an arbitrary constant.

Insertion of (9.65) in (9.64) yields the following expression for the maximum value of $\rho_m(t)$

$$\rho_m^0(t) = \sum_{j=1}^{m} \frac{V_j^2(t)}{N_i} = \sum_{j=1}^{m} \rho_j(t). \tag{9.66}$$

In accordance with (9.65), optimum performance is obtained when the voltage amplification is made proportional to the received voltage and inversely proportional to the average noise power. In the usual case of equal average noise powers, the optimum voltage amplification is proportional to the received voltage, and maximal-ratio combining corresponds to square-law addition.

The foregoing relations apply for equal average transmission losses in the channels. When the average losses vary among the channels, optimum performance entails different average transmitted powers. The optimum division of transmitter power among the channels can then be obtained by an optimization procedure similar to that used in section 4.4 to determine the optimum transmitting filter characteristic.

Of basic importance in both analog and digital transmission is the probability distribution of the time-variant optimum signal-to-noise ratio $\rho_m^0(t)$. It is obtained by considering the probability densities of the individual channels and determining the combined probability density, as outlined below.

The probability density of the envelope r of the carrier in any channel is obtained by differentiation of the probability distribution (9.5). With $r_1/\bar{r} = z$, the probability density of the voltage becomes

$$p(z) = 2ze^{-z^2} \tag{9.67}$$

With $(r_1/\bar{r})^2 = \rho_1/\bar{\rho} = x$, the probability density of the signal-to-noise power ratio fluctuation becomes

$$p(x) = e^{-x} \tag{9.68}$$

Maximal Ratio Combining

The probability that the signal-to-noise power ratio exceeds the mean value $\bar{\rho}$ by a factor k is thus

$$P(\rho_1 \geq k\bar{\rho}) = \int_k^\infty e^{-x}\, dx = e^{-k}. \tag{9.69}$$

The probability density of a sum of random variables such as $\rho_m(t)$ given by (9.66) is obtained with the aid of the characteristic function of $p(x)$ by the method described in Appendix 2. The characteristic function obtained with (9.68) in (19) of Appendix 2 is $C_1 = (1 - iy)^{-1}$. Using (21), the following relation is obtained for the combined probability density function:

$$p_m(x) = \frac{1}{2\pi} \int_{-\infty}^\infty \frac{e^{-iyx}\, dy}{(1-iy)^m}, \tag{9.70}$$

applying for $x > 0$. For $x < 0$, $p_m(x) = 0$.

Evaluation of the integral by the residue theorem yields

$$p_m(x) = \frac{x^{m-1}}{(m-1)!} e^{-x} \tag{9.71}$$

The probability that $\rho_m < k\bar{\rho}$ is thus with $\bar{\rho} = \bar{\rho}_1$

$$P(\rho_m < k\bar{\rho}_1) = \int_0^k \frac{x^{m-1}}{(m-1)!} e^{-x}\, dx \tag{9.72}$$

$$= 1 - \sum_{j=1}^m \frac{k^j}{(j)!} e^{-k}. \tag{9.73}$$

The latter relation yields

$$P(\rho_1 < k\bar{\rho}_1) = 1 - e^{-k}, \tag{9.74}$$

$$P(\rho_2 < k\bar{\rho}_1) = 1 - (1 + k)e^{-k}, \tag{9.75}$$

$$P(\rho_3 < k\bar{\rho}_1) = 1 - \left(1 + k + \frac{k^2}{2!}\right)e^{-k}. \tag{9.76}$$

By way of example, for $k = 0.5$ the last three relations yield 0.39, 0.09 and 0.02, respectively.

If the combined transmitter power is the same as for a single channel, rather than m times greater as above, the factor k is replaced by km. In this case, the foregoing three numerical values for $k = 0.5$ are replaced by 0.39, 0.26, and 0.09. Hence even without increase in total transmitter power, some advantage is derived from the nonsimultaneous fading of the channels.

9.19 Equal-Gain Combining

Maximal-gain combining is difficult to implement and is largely of theoretical interest, for the reason that almost as good a performance can be realized with the simpler method of equal-gain combining. The latter method entails common gain control of all receivers.

With equal-gain combining, the resultant signal-to-noise ratio is given by the following relation in place of (9.66)

$$\rho_m(t) = \frac{[\sum_{j=1}^{m} v_j(t)]^2}{mN}, \qquad (9.77)$$

where N is the average noise power per channel in the demodulator output, and $v_j(t)$ the output voltage in channel j.

With equal average signal powers per channel, the average combined signal-to-noise ratio becomes

$$\bar{\rho}_m = \frac{1}{mN}\left(\sum_{i=1}^{m} \overline{v_1^2} + \sum_{i \neq j} \bar{v}_i \bar{v}_j\right)$$

$$= \frac{ms_1 + m(m-1)(\bar{v}_1)^2}{mN}$$

$$= \bar{\rho}_1 + (m-1)\bar{\rho}_1 c, \qquad (9.78)$$

where $\bar{v}_1$ is the mean amplitude of the voltage in one channel, $s_1 = \overline{v_1^2}$ is the average message power in the output and

$$c = \frac{(\bar{v}_1)^2}{\overline{v_1^2}} = \frac{(\bar{v}_1)^2}{s_1}. \qquad (9.79)$$

The constant c depends on the probability distribution of the voltage fluctuation. With a Rayleigh distribution this factor is obtained from relations (3.57) and (3.59) as $c = \pi/4$ so that (9.78) can be written

$$\bar{\rho}_m = \bar{\rho}_1\left(\left(1 - \frac{\pi}{4}\right) + m\frac{\pi}{4}\right). \qquad (9.80)$$

The mean output signal-to-noise ratio per channel is

$$\frac{\bar{\rho}_m}{m} = \frac{\pi}{4}\bar{\rho}_1 + \frac{1}{m}\left(1 - \frac{\pi}{4}\right)\bar{\rho}_1. \qquad (9.81)$$

The probability distribution of a sum of random variables with a Rayleigh probability density (9.67) cannot be obtained in terms of simple functions. An approximate solution can be obtained in terms of the mean ratio $\bar{\rho}_m/m$ given by (9.81) by noting that for $m \to \infty$ the mean ratio differs from that with maximal ratio combining by a factor $\pi/4$. Hence the following approximate relation applies

$$P\left(\rho_m < \frac{\bar{\rho}_m}{m}\right) \simeq 1 - \sum_{i=1}^{m} \frac{k^i}{i!} e^{-k}, \qquad (9.82)$$

where the right-hand side is the same as (9.73) for maximal ratio combining. The above relation yields the correct result for $m = 1$ since then $\bar{\rho}_m/m = \bar{\rho}_1$. It also applies for $m \to \infty$. For $m = 2$ the above relation yields

$$P(\rho_2 \leq 0.9\bar{\rho}_1) = 1 - (1 + k)e^{-k}. \qquad (9.83)$$

For any value of k, the value of ρ_2 for equal-gain combining should thus be less than for maximal ratio combining by a factor 0.9, corresponding to 0.5 dB reduction in the output signal-to-noise ratio. This conforms well with results of numerical integration given by Brennan [12].

9.20 Pre-detection versus Post-detection Combining

The previous probability distributions of the combined output signal-to-noise ratios for maximal ratio and equal-gain combining apply only with linear coherent demodulators. With nonlinear demodulators a less favorable probability distribution is obtained, both with pre-detection combining of the carrier signals prior to common demodulation and with post-detection combining of the baseband output signals of the demodulators. The adverse modification of the probability distribution is more pronounced with post-detection than pre-detection combining. The difference can be significant in analog FM systems with a large frequency deviation ratio where nonlinearity in the demodulator is pronounced for input signal-to-noise ratio below threshold. Hence a significant improvement in performance can be realized in such systems with pre-detection combining over the simpler method of post-detection combining that does not require in-phase addition of the carrier signals.

The foregoing adverse threshold effect is less important for small frequency-deviation ratios as used in digital and in some analog FM systems. However, the probability distribution is adversely modified even in the absence of a threshold with noncoherent demodulators, such as frequency discriminators, dual filter envelope detectors and differential phase product demodulators as used in digital systems. With such demodulators some improvement in performance can be realized with pre-detection over post-detection combining. However, because of the moderate improvement and the complexity of pre-detection combiners, they would hardly be considered in conjunction with the foregoing type of demodulators, which are used principally to avoid the problems of synchronous demodulation in fading channels. For these reasons only post-detection combining is considered in Chapter 10 in determining the error probabilities in digital systems with diversity transmission.

9.21 Selection Diversity

Selection diversity is simpler to implement than maximal ratio and equal gain combining, but is less effective for systems with linear modulators. However, when the demodulators have a pronounced nonlinearity, the performance with selection diversity may surpass that with post-detection combining, for the reason that the adverse modification of signal-to-noise ratios discussed in the preceding section is largely avoided. Thus selection diversity may be preferable to equal gain post-detection combining for FM systems with a large frequency deviation ratio [12]. With selection diversity, transients occur during switchovers. These do not ordinarily cause excessive transmission impairments in analog systems, but may contribute significantly to the error rate in digital systems.

Let the probability that the signal-to-noise ratio in a particular channel is $\rho_i < \bar{\rho}_1$, where $\bar{\rho}_1$ is the mean signal-to-noise ratio of one channel. The probability that the signal-to-noise ratios of all channels are simultaneously less than $x\bar{\rho}_1$ is then

$$P(\rho_m < x\bar{\rho}_1) = [P(\rho_1 < x\bar{\rho}_1)]^m. \tag{9.84}$$

For Rayleigh probability distribution

$$P(\rho_m \leq x\bar{\rho}_1) = (1 - e^{-x})^m. \tag{9.85}$$

The probability density function is accordingly

$$p_m(x) = m(1 - e^{-x})^{m-1} e^{-x}. \tag{9.86}$$

The average signal-to-noise ratio is

$$\bar{\rho}_m = \bar{\rho}_1 \int_0^\infty x p_m(x)\, dx = \bar{\rho}_1 \int_0^\infty mx(1 - e^{-x})^{m-1} e^{-x}\, dx$$

$$= \bar{\rho}_1 \left(1 + \frac{1}{2} + \frac{1}{3} + \cdots + \frac{1}{m}\right). \tag{9.87}$$

For $x = k \ll 1$, relation (9.85) yields for $m = 2$

$$P(\rho_2 < k\bar{\rho}_1) \simeq k^2. \tag{9.88}$$

This compares with the following approximation for maximal ratio combining and $k \ll 1$

$$P(\rho_2 \leq k\bar{\rho}_1) = \tfrac{1}{2}k^2. \tag{9.89}$$

9.22 Intermodulation Distortion with Diversity Transmission

The ratio of average message power to intermodulation distortion, taken over an adequately long period, is independent of transmitter power. It depends on the frequency selectivity of the fading, the message-modulating power, and the frequency-deviation ratio. Hence for a fixed deviation ratio and message power, corresponding to a fixed rms frequency deviation, diversity transmission should have no effect on the ratio of average message power to average intermodulation distortion power. However, the probability distribution of intermodulation distortion will be modified, so that the probability of strong intermodulation distortion is reduced in exchange for an increase in the probability of weak distortion, as in the case of additive Gaussian noise. As shown elsewhere [11] the probability distribution of intermodulation distortion depends on the rms frequency deviation. The shape of the probability distribution curves is such that for small frequency deviations median intermodulation distortion is less than the average, while the converse applies for large frequency deviations. Consequently an increase in median intermodulation distortion would be expected for small frequency deviations, and a decrease at large deviations. This conforms with experimental results for dual diversity, which show a small increase, about 2 dB, for small deviations and a decrease of some 4 dB for large frequency deviations [9].

The probability distribution curves of intermodulation distortion as related to the frequency deviation [11] indicates that for large deviations the probability of strong intermodulation distortion cannot be reduced by diversity transmission to the same degree as indicated by the various previous relations applying for Gaussian noise.

10
Digital Transmission over Random Multipath Channels

10.0 General

The preceding chapter dealt with certain statistical properties of variations with time and with frequency in the transmittance of random multipath channels. Certain experimental data reviewed there indicated that random multipath transmission is a legitimate approximation for troposcatter channels, and is assumed here in determination of error probability in digital transmission.

With any digital modulation method this error probability in general depends on (a) the signal-to-noise ratio at the receiver, averaged over the time interval under consideration; (b) the change in carrier phase or frequency during a pulse interval; and (c) pulse distortion, owing to frequency-selective fading. When the probability of error on each of these accounts is known, it is possible by a simple graphical method to estimate the combined error probability. This method is illustrated here and charts are presented of error probabilities in digital transmission by binary PM and FM as related to various basic parameters of tropospheric scatter links.

In Chapter 9 various methods of diversity transmission were considered from the standpoint of improving performance of analog systems, where output signal-to-noise ratios is a basic criterion. The same methods will be considered here as regards their error-reduction performance in binary transmission.

With signal-to-noise ratios that are practicable for troposcatter links, only binary carrier modulation methods would come into consideration, such as binary FSK with dual filter envelope detection or frequency-discrimination detection or binary PSK with differential phase product demodulation (Chapter 5). In virtually all probabilistic work on error probability with these methods, rectangular pulses are assumed with matched filter correlation integration over a signal interval T. As discussed in Section 1.13, this method entails infinite channel bandwidth, and offers no advantage over the simpler method of instantaneous sampling assumed here and ordinarily used in digital systems. With both methods the error probability owing to additive Gaussian noise is the same for equal signal-to-noise ratios and optimum design. However, this does not apply in the presence of distortion owing to frequency-selective fading; and instantaneous sampling of bandlimited pulses in this case affords improved performance. As in previous chapters, pulses with raised cosine spectra will be assumed here in some numerical evaluation, because of various advantages cited in Section 1.12.

In determination of error probabilities in digital transmission, it is necessary to consider variations in the average path loss over a convenient period, such as an hour, relative to the average over a much longer period, say a month. These slow fluctuations in loss are closely approximated by the log-normal law; that is, the loss in dB follows the normal law. In addition, consideration must be given to rapid fluctuations in loss relative to the above hourly averages. These are closely approximated by the Rayleigh law, and are ordinarily more important than slow fluctuations, particularly in digital transmission, in that they are more pronounced and cannot be fully compensated for by automatic gain control. Nearly all theoretical analyses of error probabilities are based on a Rayleigh distribution together with various other simplifying assumptions.

The simplest assumption is flat or nonselective Rayleigh fading over the channel band, in conjunction with a sufficiently slow fading rate such that changes over a few pulse intervals can be disregarded. On this premise Turin [1] and Pierce [2,3] have determined the probability of error, owing to additive Gaussian noise for binary FSK with dual filter coherent and noncoherent detection.

The error probability with two phase and four-phase modulation with differential phase detection has been determined by Voelcker [4] on the premise of flat Rayleigh fading at such a rate that the change in phase over a pulse interval must be considered. Moreover, he considers the probability of both single and double digital errors, with both single- and dual-diversity transmission.

Voelcker's analysis is applicable to transmission at a sufficiently slow rate such that amplitude and phase distortion can be ignored over the relatively narrow band of the pulse spectra. However, it does not apply to high-speed digital transmission that requires sufficiently wide pulse spectra such that the amplitude and phase distortion indicated in Fig. 9.1 must be considered. For this case the duration of pulses will be so short that the phase changes considered by Voelcker can be disregarded. Instead, it now becomes necessary to take into account pulse distortion and resultant intersymbol interference caused by the erratic variations with frequency in the amplitude and phase characteristics illustrated in Fig. 9.1. The error probability for this case has been determined by Bello and Nelin [5] for binary FSK with dual filter envelope detection and binary PSK with differential phase detection. Their numerical results apply for rectangular pulses and a Gaussian autocorrelation function of the frequency selectivity. For pulses with a raised cosine spectrum and instantaneous sampling an approximate evaluation has been made of error probability owing to frequency-selective fading [6]. In this evaluation, the quadratic component of phase nonlinearity was assumed to be the principal source of pulse distortion, and an autocorrelation function as determined in Chapter 9 was used. A more exact determination has been made by Bailey and Lindenlaub [7], for the same autocorrelation function and raised cosine pulse spectra, by appropriate modification of the analytical method devised by Bello and Nelin.

The rather extensive literature on error probability in digital transmission over fading channels includes several comprehensive expositions [8,9,10]. These contain further references to many papers dealing with other situations than considered here of less importance as regards troposcatter systems.

10.1 Errors from Gaussian Noise in Binary FM and PM

It will be assumed that the fading is nonselective and occurs at a sufficiently slow rate so that the received signal amplitude and phase can be considered constant over a pulse interval T.

Let r be the carrier amplitude and $P_e^0(r)$ the error probability of errors owing to random noise in transmission over a stable channel with signal amplitude r. With a time-varying channel transmittance, let the probability density of various amplitudes be $p(r)$. The error probability in transmission over a channel with time-varying transmittance is then

$$P_e = \int_0^\infty P_e^0(r)\, p(r)\, dr. \qquad (10.1)$$

With Rayleigh fading, the probability density $p(r)$ is the derivative of (9.5) with respect to r_1. With r in place of r_1 the probability density is

$$p(r) = \left(\frac{2r}{\bar{r}^2}\right) \exp\left(\frac{-r^2}{\bar{r}^2}\right) \qquad (10.2)$$

$$= \left(\frac{r}{S}\right) \exp\left(\frac{-r^2}{2S}\right) \qquad (10.3)$$

where $S = \bar{r}^2/2$ is the average signal power.

In binary PM, marks and spaces are transmitted by phase reversals. With ideal coherent or synchronous detection the error probability in transmission over a stable channel is then in accordance with (5.87).

$$P_e^0(r) = \tfrac{1}{2} \operatorname{erfc}(\rho_i^{1/2}), \qquad (10.4)$$

where

$$\rho_i = \frac{(r^2/2)}{N_i} \qquad (10.5)$$

$r^2/2$ is the average carrier power at the input to the receiving filter, and N_i is the average noise power in a band $B_i = 2B = 1/T$, as indicated in Fig. 5.10.

The error probability with Rayleigh fading as obtained from (10.1) is, in this case

$$P_e = \frac{1}{2}\left(1 - \left(\frac{\bar{\rho}}{\bar{\rho}+1}\right)^{1/2}\right) \approx \frac{1}{4\bar{\rho}} \qquad (10.6)$$

where $\bar{\rho} = S/N_i$ = ratio of average received signal power with Rayleigh fading to average noise power as previously defined.

With binary PM and differential phase detection the error probability in transmission over a stable channel is in accordance with (5.92)

$$P_e^0(r) = \tfrac{1}{2} e^{-\rho_i}, \qquad (10.7)$$

and (10.1) yields

$$P_e = \frac{1}{2(\bar{\rho}+1)} \simeq \frac{1}{2\bar{\rho}}. \qquad (10.8)$$

For binary FM with dual filter synchronous detection, (5.96) applies and

$$P_e^0(r) = \tfrac{1}{2} \operatorname{erfc}\left(\frac{\rho_i}{2}\right)^{1/2} \qquad (10.9)$$

Comparison of (10.9) and (10.4) shows that in this case (10.6) is replaced by

$$P_e = \frac{1}{2}\left(1 - \left(\frac{\bar{\rho}/2}{1+\bar{\rho}/2}\right)^{1/2}\right) \simeq \frac{1}{2}\frac{1}{\bar{\rho}+2}. \qquad (10.10)$$

For binary FM with dual filter envelope detection (5.100) applies, so that

$$P_e^0(r) = \tfrac{1}{2} e^{-\rho_i/2}. \qquad (10.11)$$

Comparison with (10.7) shows that (10.8) is replaced by

$$P_e = \frac{1}{\bar{\rho}+2}. \qquad (10.12)$$

As discussed in Section 5.15, frequency-discriminator detection is preferable to dual filter detection in binary FM because of a two-fold reduction in channel bandwidth. This is particularly important in high-speed digital transmission where frequency-selective fading is a primary source of errors, as discussed in Section 10.6. Binary FM, or frequency shift keying, with frequency-discriminator detection requires the same bandwidth as binary PM, but about 3 dB increase in signal-to-noise ratio over binary PM with synchronous detection. Hence in this case

$$P_e \simeq \frac{1}{2\bar{\rho}}. \qquad (10.13)$$

10.2 Binary AM versus Binary PM and FM

Because of the great range of fluctuation in transmission loss with Rayleigh fading, it is essential to provide automatic gain control at the receiver to prevent overloading and resultant adverse effects. Such gain control is activated by circuitry that integrates the received wave over a number of signal intervals T. With FM and PM, only a few pulse intervals are required, for the reason that the received carrier wave is nearly independent of the pulse patterns. It is thus possible to provide effective gain control against rapid variations in the received carrier wave that occurs over a few signal intervals. Moreover, with FM and PM the distinction between marks and spaces is made by positive and negative deviations from zero threshold level in the detection process. This permits the use of limiters at the input to the detectors, to prevent the adverse effect of rapid fluctuations in the amplitude of the received carrier wave owing to fading. These advantages in applications to fading channels are not shared by AM, for reasons outlined next.

In binary AM or on-off carrier transmission, the received wave may be absent over a large number of consecutive signal intervals T. Hence automatic gain control must be activated by circuitry that integrates the received pulse train over a very large number of signal intervals T; otherwise gain would be increased during long spaces, regardless of the fading condition. For this reason, automatic gain control is inherently slow, in relation to the duration of a signal interval. It may thus be ineffective as applied to transmission at slow rates. With transmission at high rates, however, such that variations in the received wave owing to fading are inappreciable even over a large number of signal intervals, it may be possible to implement effective gain control.

At low transmission rates, such that fading is virtually constant over the band of the pulse spectrum, intersymbol interference can be made inappreciable. In this case it is possible to employ limiting prior to detection, and this method may then be more effective than automatic gain control, or could be used in conjunction with it. The limiter would slice the received wave at an appropriately selected level L. In the choice of the optimum slicing level it is necessary to consider the probability of errors during a mark owing to fading such that the received wave is less than L. In accordance with (9.5) this probability is with equal probability of marks and spaces

$$P_e(r \leq L) = \frac{1}{2}\left[1 - \exp\left(\frac{-L^2}{\bar{r}^2}\right)\right]. \tag{10.14}$$

A second consideration in the choice of L is the probability of errors owing to noise during a space, which is increased as L is reduced. In accordance with (5.78) the probability of error on this account is

$$P_e(n \geq L) = \tfrac{1}{2} \exp\left(\frac{-L^2}{2N_i}\right), \tag{10.15}$$

where the substitution $\rho_i = L^2/N_i$ has been made, n is the instantaneous noise amplitude and N_i the average noise power.

The combined error probability is

$$P_e = \frac{1}{2}\left[1 - \exp\left(\frac{-\mu}{2}\right) + \exp\left(\frac{-\bar{\rho}\mu}{2}\right)\right], \tag{10.16}$$

where

$$\mu = L^2/S; \qquad \bar{\rho} = S/N_i. \tag{10.17}$$

The optimum L or μ is obtained from the condition $dP_e/d\mu = 0$. This yields the following relation for the optimum value μ_0

$$\exp\left(\frac{-\mu_0}{2}\right) = \bar{\rho} \exp\left(\frac{-\bar{\rho}\mu_0}{2}\right), \tag{10.18}$$

or

$$\mu_0 = \frac{2 \ln \bar{\rho}}{\bar{\rho} - 1} = \frac{4.606 \log_{10} \bar{\rho}}{\bar{\rho} - 1}. \tag{10.19}$$

In practicable systems $\bar{\rho} \gg 1$, in the order of 100 or more, $\mu_0 \ll 1$. With (10.19) in (10.16), the following approximation is obtained for the minimum error probability

$$P_{e,\min} \approx \frac{1}{2}\left(\frac{\ln \bar{\rho}}{\bar{\rho} - 1} + \exp(-\ln \bar{\rho})\right). \tag{10.20}$$

The foregoing error probability is significantly greater than with binary PM or FM. The error probability is thus greater than for binary FM with dual filter coherent detection by a factor of at least $\ln \bar\rho$. For $\bar\rho = 1000$ (30 dB) this factor is about $\ln \bar\rho \approx 7$. Hence about $10 \log_{10} 7 \approx 8.5$ dB greater average signal power would be required than with binary FM.

This assumes no intersymbol interference. Owing to even small intersymbol interference, the use of a limiter as postulated above may be precluded in actual systems. For example, let L be 10 per cent of the rms signal amplitude $\bar r$, and let intersymbol interference be 5 percent of L when the received signal is just equal to L. When the received signal is increased by a factor 20, intersymbol interference would be increased correspondingly and would be equal to L. Hence errors would occur even in the absence of noise. This is the inherent reason why limiting is generally ineffective as applied to binary AM.

10.3 Combined Rayleigh and Slow Log-Normal Fading

In the previous determination of error probabilities, rapid Rayleigh fading was assumed, with a fixed mean signal-to-noise ratio $\bar\rho$ over the interval under consideration. It will now be assumed that in this interval there is a slow log-normal variation in path loss and thus in signal-to-noise ratio at the receiver, in conjunction with rapid Rayleigh fading.

Let P_e be the error probability with Rayleigh fading as previously related to the mean signal-to-noise ratio $\bar\rho = \bar s^2/\bar n^2$, where $\bar s$ is the rms signal amplitude and $\bar n$ the rms noise amplitude. If $p(\bar s)$ is the probability density of the rms amplitudes with slow fading, the probability of error in an interval during which the rms amplitude exceeds $\bar s_1$ is

$$P_e = \int_{\bar s_1}^{\infty} P_e(\bar s)\, p(\bar s)\, d\bar s. \tag{10.21}$$

For $\bar\rho \gg 1$, the expression for $P_e(\bar s)$ is of the general form

$$P_e(\bar s) \approx c/\bar\rho = \frac{c}{\bar s^2/\bar n^2}. \tag{10.22}$$

For binary PM with differential phase detection and for binary PM with coherent dual filter detection, $c = \tfrac{1}{2}$.

The probability density $p(\bar s)$ is given by (72) of Appendix 2, or in the present notation

$$p(\bar s) = \frac{1}{\sqrt{2\pi}} \frac{1}{\sigma \bar s} \exp\left[-(\ln \bar s/\bar s_0)^2/2\sigma^2\right] \tag{10.23}$$

where $\bar s_0$ is the median rms amplitude and σ is the standard deviation of the fluctuation in $\bar s$.

With (10.22) and (10.23) in (10.21)

$$P_e = c \frac{1}{\sqrt{2\pi}} \frac{1}{\sigma} \int_{\bar{s}_1}^{\infty} \frac{1}{\bar{s}^2/\bar{n}^2} \frac{1}{\bar{s}} \exp\left[-(\ln \bar{s}/\bar{s}_0)^2/2\sigma^2\right] d\bar{s},$$

$$= \frac{c}{2} \frac{1}{\sqrt{2\pi}} \int_{\rho_1}^{\infty} \frac{1}{\rho^2} \exp\left[-(\tfrac{1}{2}\ln \rho/\rho_0)^2/2\sigma^2\right] d\rho, \qquad (10.24)$$

where $\rho_0 = \bar{s}_0^2/\bar{n}^2$ and $\rho_1 = \bar{s}_1^2/\bar{n}^2$.

Solution of (10.24) yields the relation

$$P_e = P_e \cdot \eta(\sigma, k), \qquad (10.25)$$

where

$$k = \rho_1/\rho_0,$$

and

$$\eta(\sigma, k) = \tfrac{1}{2} \exp(2\sigma^2) \operatorname{erfc}\left\{\frac{1}{\sqrt{8}\sigma}[4\sigma^2 + \ln k]\right\}. \qquad (10.26)$$

For $\rho_1 = 0$, $\ln k = -\infty$ and $\operatorname{erfc}(-\infty) = 2$. Hence for this case

$$\eta = \exp(2\sigma^2). \qquad (10.27)$$

This is the factor by which the error probability taken over a long interval is greater than without a log-normal variation in signal-to-noise ratio and only rapid Rayleigh fading.

Instead of modifying the error probability by (10.27), an alternative method is to use an equivalent mean signal-to-noise ratio $\bar{\rho}_e$ that is smaller than $\bar{\rho}$ by the factor $\exp(-2\sigma^2)$. Thus

$$\bar{\rho}_e = \bar{\rho} \exp(-2\sigma^2). \qquad (10.28)$$

When $\bar{\rho}_e$, $\bar{\rho}$ and σ are all expressed in dB, expression (10.28) can alternatively be written

$$\bar{\rho}_{e,\mathrm{dB}} = \bar{\rho}_{\mathrm{dB}} - \sigma_{\mathrm{dB}}^2/8.69. \qquad (10.29)$$

For example, with a representative value $\sigma_{\mathrm{dB}} = 8$ dB, the last term in (10.29) is 7.4 dB. Thus the charts in the later Figs. 10.3 and 10.4 apply when $\bar{\rho}$ is taken 7.4 dB less than the median signal-to-noise ratios with log-normal fading.

10.4 Errors in PM from Phase Variations over Signal Interval

In transmission at low rates, the bandwidth of the pulse spectra will be narrow, so that fading can be regarded as constant over the spectrum band. Errors from selective fading, can then be disregarded. On the other hand, the duration of a signal interval T may then be sufficiently long so that consideration must be given to random fluctuations in the amplitude, phase, and frequency of the carrier between one signal interval and the next. Errors may occur, owing to such fluctuations even in the absence of noise. The probability of errors on this account is evaluated here for PM systems.

The probability density of the carrier phase is $1/2\pi$, such that any phase may be encountered unless the carrier phase wander is limited in the demodulation process. In a digital phase modulation system where appreciable phase wander may be expected, the preferable demodulation method is differential phase detection as assumed here. With this method the phase wander will be limited to that encountered over a signal interval T.

From (9.33) it is possible to determine the probability of an error for a given maximum tolerable phase change θ over an interval T. For $k \gg 1$ the following relation applies with $\varphi_1' = \theta/T$

$$P(|\varphi'| \geq |\varphi_1'|) = \frac{1}{2k^2} = \frac{T^2}{2\theta^2}(\varphi_m')^2. \tag{10.30}$$

With a Gaussian fading power spectrum (9.13), relation (9.36) applies and

$$P(|\varphi'| \geq (\varphi_1')) = (\gamma_0^2 T^2/2\theta^2). \tag{10.31}$$

With two-phase modulation $\theta = \pm(\pi/2)$, while with four-phase modulation $\theta = \pm(\pi/4)$. Hence the probability of error with these methods as obtained from (10.31) is, for two-phase modulation

$$P_e \approx (2/\pi^2)\gamma_0^2 T^2 \approx 0.2\,\gamma_0^2 T^2, \tag{10.32}$$

and for four-phase modulation

$$P_e \approx (8/\pi^2)\gamma_0^2 T^2 \approx 0.82\,\gamma_0^2 T^2. \tag{10.33}$$

These expressions apply provided that the signal duration is sufficiently short so that the change in phase is small and can be considered linear over the interval. More accurate expressions that do not involve this assumption have been derived by Voelcker [4] for the error probability. Thus, with two-phase modulation the error probability is actually

$$P_e = \tfrac{1}{2}[1 - k(T)], \tag{10.34}$$

and with four-phase modulation

$$P_e = \frac{1}{2} - \frac{2}{\pi} k(T)\,[2 - k^2(T)]^{-\frac{1}{2}} \tan^{-1}\frac{k(T)}{[2 - k^2(T)]^{\frac{1}{2}}}, \tag{10.35}$$

where $k(T) = k(\tau)$ for $\tau = T$, that is, the autocorrelation function for each quadrature component as defined by (9.10).

Errors from Phase Variations

For a Gaussian fading spectrum, $k(T)$ as obtained from (9.12) is

$$k(T) = \exp\left(\frac{-\gamma_0^2 T^2}{2}\right). \tag{10.36}$$

For $\gamma_0 T \ll 1$:

$$k(T) \approx 1 - \frac{\gamma_0^2 T^2}{2}. \tag{10.37}$$

With the latter approximation in (10.34) and (10.35), the error probability with two-phase modulation becomes

$$P_e \approx 0.25 \, \gamma_0^2 T^2, \tag{10.38}$$

and with four-phase modulation

$$P_e = \left(\frac{1}{2} + \frac{1}{\pi}\right) \gamma_0^2 T^2 \approx 0.82 \, \gamma_0^2 T^2 \tag{10.39}$$

which are to be compared with (10.32) and (10.33), respectively. The somewhat greater inaccuracy with two-phase than with four-phase modulation comes about since the phase change $\pm(\pi/2)$ cannot be considered small as required for (10.31) to apply.

In the above relations, T is the interval between phase changes, which is related to the bandwidth of the baseband pulse spectrum. With idealized spectra of the type shown in Fig. 5.10 the interval is

$$T = 1/2B \quad \text{(two-phase)} \tag{10.40}$$

$$= 1/4B \quad \text{(four-phase)}, \tag{10.41}$$

where B is the equivalent mean bandwidth.

In terms of the bandwidth B the error probabilities (10.32) and (10.33) are thus the same for two-phase and four-phase modulation and given by

$$P_e \simeq 0.05(\gamma_0/B)^2 \tag{10.42}$$

$$\simeq 2(\gamma_0/\Omega)^2 \tag{10.43}$$

$$= 1.25(\bar{\gamma}/\Omega)^2, \tag{10.44}$$

where γ_0 is in radians per second, $\Omega = 2\pi B$ and $\bar{\gamma}$ is the equivalent fading bandwidth (9.14). Relation (10.42) applies for any number of phases. For this reason the capacity of a noiseless channel could be increased indefinitely by increasing the number of phases. There will, however, be certain limitations in this respect, owing to intersymbol interference, as in stable channels.

The foregoing error probability is shown by curve 1 in Fig. 10.1 as a function of the ratio $\Omega/\bar{\gamma}$.

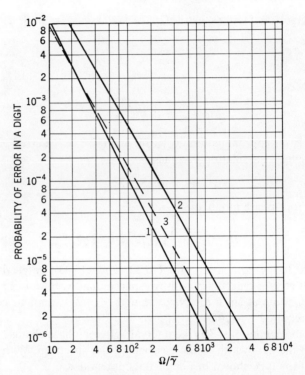

Figure 10.1 Error probabilities in binary PM and FM owing to change in carrier phase or frequency over a pulse interval T. 1. Binary PM with differential phase detection. 2. Binary FM with frequency discriminator detection. 3. Binary FM with dual filter detection.

10.5 Errors in FM from Frequency Variations over Signal Interval

It will be assumed that the carrier frequency excursion is limited with the aid of a signal-tracking oscillator, or that a demodulation process is used in binary FM in which the change from mark to space is based on comparison of the frequencies in adjacent signal intervals of duration T. If the separation between mark and space frequencies is $2\Omega_{01}$, an error will occur if the frequency is changed by $+\Omega_{01}$ for a space and by $-\Omega_{01}$ for a mark.

From (9.46) it is possible to determine the probability of errors, owing to frequency changes $\pm \Omega_{01}$ over a signal interval of duration T. The maximum permissible value of φ'' is determined from

$$\varphi''_{\max}T = \pm \Omega_{01}, \tag{10.45}$$

where the positive sign applies for a space and the negative sign for a mark.

With an ideal pulse spectrum the pulse interval is given by $T = \pi/\Omega$, that (10.45) can be written

$$\varphi''_{max} = \pm \Omega_{o1}\Omega/\pi. \tag{10.46}$$

The error probability is, in this case,

$$P_e = \tfrac{1}{2}P(|\varphi''| \geq |\varphi''_{max}|), \tag{10.47}$$

where the factor $\tfrac{1}{2}$ occurs when the probability function is defined in terms of the absolute values as (9.46).

The parameter k in this case becomes

$$\begin{aligned} k_{max} &= \varphi''_{max}/\gamma_0^2 \\ &= \Omega_{o1}\Omega/\pi\gamma_0^2. \end{aligned} \tag{10.48}$$

With frequency discriminator detection, $\Omega_{o1} = \Omega$, and

$$k_{max} = \Omega^2/\pi\gamma_0^2. \tag{10.49}$$

With the latter value of k in (9.46)

$$P_e \simeq \left(\frac{\gamma_0}{\Omega}\right)^2\left[1 + \ln\left(1 + \frac{1}{2\pi}\frac{\Omega^2}{\gamma_0^2}\right)\right] \tag{10.50}$$

$$\simeq 0.64\left(\frac{\bar{\gamma}}{\Omega}\right)^2\left[1 + \ln\left(1 + \frac{1}{4}\frac{\Omega^2}{\bar{\gamma}^2}\right)\right], \tag{10.51}$$

where $\bar{\gamma}$ is the equivalent fading bandwidth (9.14).

The foregoing error probability is shown by curve 2 in Fig. 10.1 as a function of $\Omega/\bar{\gamma}$.

With dual filter envelope detection and a raised cosine pulse spectrum, an error will occur with $\Omega_{o1} = 2\Omega$ and (10.51) is modified into

$$P_e \simeq 0.16\left(\frac{\bar{\gamma}}{\Omega}\right)^2\left[1 + \ln\left(1 + \frac{\Omega^2}{\bar{\gamma}^2}\right)\right]. \tag{10.52}$$

As indicated in Fig. 10.1, the error probability is then nearly the same as for binary PM with differential phase detection.

10.6 Errors from Frequency-Selective Fading

With frequency-selective fading the amplitude and phase characteristics can be represented by the power series (9.56) and (9.57). In a first approximation, coarse-structure amplitude deviations can be represented by a linear component $a_1 u$. As shown in Section 6.10, this component does not give rise to distortion in analog FM. In Section 7.9 it was demonstrated that the resultant intersymbol interference is negligible at sampling instants.

The first component of phase distortion that gives rise to transmission impairments is the quadratic component $b_2 u^2$. In Section 9.16, it was shown that intermodulation distortion in FM, owing to this quadratic component, conforms well with measurements on troposcatter links, employing narrow beam antennas for path lengths up to 200 miles. Hence it appears a legitimate first approximation to determine pulse distortion and resultant errors in digital transmission on the premise of quadratic phase distortion, and this approximation is used here.

In Chapter 7 evaluation was made of maximum transmission impairments, owing to quadratic phase distortion, that is, linear delay distortion. The results for a raised cosine pulse spectrum and various modulation methods are given in Fig. 7.10, as a function of the ratio $\lambda = d/T = d\hat{B}$.

It will be noted that the noise margin is reduced to zero for certain values λ_0 of λ. These values apply for certain combinations of baseband pulses in about four pulse positions. The probability of this and other pulse patterns must be considered in evaluating error probability.

As λ is increased slightly above the value λ_0 just mentioned, intersymbol interference increases rapidly. Thus errors will occur for a value λ_e of λ only slightly greater than λ_0, for certain combinations of two pulses, occurring at times $-T$ and $+T$ relative to the sampling instant $t = O$. There are four possible combinations of these two pulses. For one of these (say 1,1), an error will occur if $\lambda \geq \lambda_e$. For another (say $-1,-1$), an error will occur if $\lambda \leq -\lambda_e$. For the other combinations $(-1,1)$ and $(1,-1)$, intersymbol interference will cancel so that the probability of error is zero. The probability of error is thus

$$Pe = \tfrac{1}{2}(\tfrac{1}{4} + \tfrac{1}{4})P(|\lambda| \geq |\lambda_e|)$$

$$= \tfrac{1}{4}P(|\lambda| \geq |\lambda_e|), \qquad (10.53)$$

where $P(|\lambda| \geq |\lambda_e|)$ is the probability that the absolute value of λ is greater than λ_e.

For a given value $\lambda_e = d_e \hat{B}$ the corresponding slope $\ddot{\varphi}$ of the linear delay distortion is

$$\ddot{\varphi}_e = \frac{d_e}{2\pi \hat{B}}$$

$$= \frac{\lambda_e}{2\pi \hat{B}^2}. \qquad (10.54)$$

Errors from Frequency-Selective Fading

The following relation applies

$$P(|\lambda| \geq |\lambda_e|) = P(|\ddot{\varphi}| \geq |\ddot{\varphi}_e|). \tag{10.55}$$

The probability distribution represented by the right-hand side of (10.55) is given by (9.47), so that the following relation applies:

$$\ddot{\varphi}_e = k_e \left(\frac{\dot{\sigma}}{\sigma}\right)^2, \tag{10.56}$$

where k_e is the value of k for which an error occurs. In view of (10.54) and (10.56)

$$k_e = \frac{\lambda_e(\sigma/\dot{\sigma})^2}{2\pi\hat{B}^2}. \tag{10.57}$$

The error probability as obtained from relation (9.48) is

$$P_e = \tfrac{1}{4} P(|\ddot{\varphi}| \geq |\ddot{\varphi}_e|)$$

$$= \frac{1}{2\pi k_e} \left[1 + \ln\left(\frac{k_e}{2} + 1\right)\right]. \tag{10.58}$$

In the particular case of a flat power spectrum of the frequency-selective fading relation (9.50) applies and

$$P_e = \frac{\Delta^2 \hat{B}^2}{3\lambda_e} \left[1 + \ln\left(1 + \frac{3\lambda_e}{4\pi \Delta^2 \hat{B}^2}\right)\right]. \tag{10.59}$$

From Fig. 7.10, it will be noted that for binary AM and FM, and for binary PM with differential phase detection, $\lambda_0 \approx 1.8$. For these cases it appears a legitimate approximation to take $\lambda_e = 2$. Figure 7.12 indicates that this is also an appropriate choice for pure binary PM with frequency discriminator detection. On this premise the error probabilities are as shown in Fig. 10.2 as a function of $\Delta \hat{B} = \Delta/T$.

A more exact determination of error probability, owing to frequency-selective fading, can be made by a more elaborate analytical procedure devised by Bello and Nelin and applied by them to binary FSK with dual-filter envelope detection and binary FSK with differential phase detection. Their numerical results apply for rectangular pulses and a Gaussian frequency correlation function. With certain modifications, their method has been used by Bailey and Lindenlaub [7] to determine error probability with binary PSK and pulses with a raised cosine spectrum, both for a Gaussian frequency correlation function and a correlation function (9.21) applying for narrow beam troposcatter antennas.

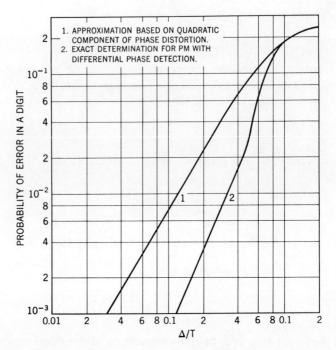

Figure 10.2 Error probability in absence of noise owing to pulse distortion from frequency-selective fading, for binary PM with differential phase detection.

In the foregoing papers, intersymbol interferences were considered for all combinations of four consecutive pulses as in the present determination of the factor λ_e. This yields a slightly lower error probability than would be obtained for an infinite sequence of pulses, and thus a lower bound.

In Fig. 10.2 the foregoing lower bound is compared with approximation (10.59). The two curves converge for large Δ/T, but diverge for small Δ/T and small error probabilities. In substance, the comparison indicates that median phase distortion can be approximated by the quadratic term in (9.57), while this is not true for more pronounced phase distortion of smaller probability. This conforms with expectations based on the measured probability distribution of intermodulation distortion in FM analog systems, as compared with predictions based on the quadratic term (Section 9.16).

With binary FSK and frequency discriminator detection approximation (10.59) is virtually the same as for binary PSK with differential phase detection. Hence, the curves in Fig. 10.2 would in a first approximation apply for both methods.

10.7 Combined Error Probabilities

As a first approximation, the combined error probability is given by

$$P_e = P_e^{(1)} + P_e^{(2)} + P_e^{(3)}, \tag{10.60}$$

$P_e^{(1)}$ = probability of errors, owing to random noise with nonselective Rayleigh fading,

$P_e^{(2)}$ = probability of errors in the absence of noise, owing to random variations in carrier phase or frequency,

$P_e^{(3)}$ = probability of errors in the absence of noise, owing to intersymbol interference caused by frequency-selective fading.

As will be evident from the preceding discussion, and from charts that follow, $P_e^{(2)}$ can be disregarded when $P_e^{(3)}$ must be considered, and, conversely, for error probabilities $P_e^{(1)}$ in the range of practical interest. Hence in actual applications (10.61) will take one of the following forms

$$P_e \approx P_e^{(1)} + P_e^{(2)}, \tag{10.61}$$

$$P_e \approx P_e^{(1)} + P_e^{(3)}. \tag{10.62}$$

In addition, there are intermediate cases in which $P_e \approx P_e^{(1)}$.

The validity of approximation (9.61) is evidenced by the following exact relation derived by Voelcker [4] for binary PSK with differential phase detection:

$$P_e = [\bar{\rho}/(\bar{\rho} + 1)]P_e^{(1)} + P_e^{(2)}. \tag{10.63}$$

Since $\bar{\rho}$ would ordinarily exceed 10, it follows that in this case (10.61) is a legitimate approximation.

That (10.62) is a valid approximation is indicated by numerical results given in reference [7] for the combined error probability (10.62), for rectangular pulses.

In Fig. 10.3, are shown the foregoing three error probabilities as a function of the transmission rate $\hat{B} = 1/T$ for binary PM with differential phase detection and a raised cosine pulse spectrum. The error probability $P_e^{(1)}$ is obtained from (10.8) and is shown for a number of values of the mean signal-to-noise ratio $\bar{\rho}$ with Rayleigh fading. The error probability $P_e^{(2)}$ depends on the equivalent fading bandwidth $\bar{\gamma}$ and is obtained from (10.45), with $\Omega = \pi\hat{B} = \pi/T$. The error probability $P_e^{(3)}$ is obtained from curve 2 in Fig. 10.2.

By way of illustration, the combined error probability obtained from (10.60) is shown by the dashed curve for a particular case. An unusually high value of $\bar{\gamma}$ was assumed to facilitate illustration.

With binary FM and frequency discriminator detection, the minimum required bandwidth for a given pulse transmission rate is the same as for binary PM, and half as great as that required with dual filter detection.

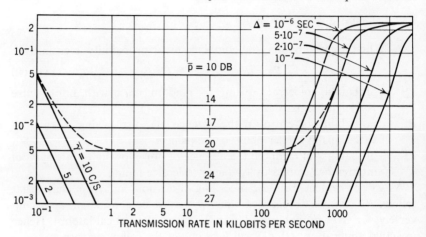

Figure 10.3 Combined error probability in binary PM with differential phase detection. Dashed curve shows combined error probability with $\bar{\rho} = 20\,\text{dB}, \bar{\gamma} = 10\text{c/s}$, and $\Delta = 5\cdot10^{-7}\text{sec}$.

In Fig. 10.4 are shown the error probabilities $P_e^{(1)}$, $P_e^{(2)}$, and $P_e^{(3)}$ for binary FM as a function of the transmission rate. The curves apply for a raised cosine pulse spectrum, and the same basic parameters as shown in Fig. 10.3 for binary PM. The error probability for the particular set of parameters previously assumed is shown by the dashed curve.

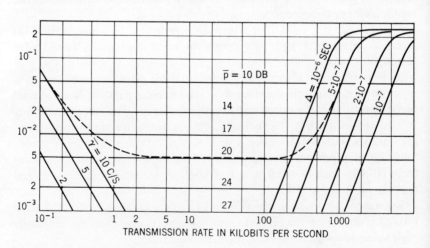

Figure 10.4 Combined error probability in binary FM with frequency discriminator detection. Dashed curve shows combined error probability with $\bar{\rho} = 20\text{dB}$, $\bar{\gamma} = 10\text{c/s}$, $\Delta = 5\cdot10^{-7}\text{sec}$.

Diversity Transmission 407

The curves in Figs. 10.3 and 10.4 differ from similar ones in reference [6] in that the latter are based on an error probability $P_e^{(3)}$ shown by curve 1 of Fig. 10.2.

The curves in Fig. 10.3 apply for binary PM by the method dealt with in Chapter 5, in which envelope fluctuations are accepted in exchange for a bandwidth no greater than required by binary AM. With pure binary PM without envelope fluctuations, the factor λ_e discussed in Section 10.6 would remain substantially unchanged, for reasons considered further in Section 7.15. Hence the curves in Fig. 10.3 would also apply for pure binary PM. Similarly, the curves in Fig. 10.4 should afford a satisfactory approximation for binary FM. In transmission over troposcatter links designed for analog transmission by PM or FM, binary pulse trains would be applied at the baseband inputs, with detection at the baseband outputs. The curves in Figs. 10.3 and 10.4 should thus apply for this condition with adequate engineering accuracy.

10.8 Diversity Transmission for Error Reduction

As discussed in Section 9.17, the objective in simultaneous or selective diversity operation is to reduce the liability to deep signal fades and resultant momentary transmission impairments. In digital transmission, error probability is a basic performance criterion and will be determined here for the three diversity methods considered in Chapter 9. With two of these methods, maximal ratio and equal-gain combining, post-detection rather than pre-detection combining would be used in digital systems, for reasons mentioned in Section 9.20. With this method, the baseband pulse trains at the outputs of the individual channel demodulators would be combined into a single pulse train prior to sampling.

The third method, selection diversity, is simpler to implement than the first two methods, but is less efficient, particularly with a large number of diversity channels. Moreover, the liability to errors during switchovers may render this method impracticable for digital transmission. If switch-over errors are disregarded, relations for error probability with Rayleigh fading are more readily determined with this than with the other two methods, and yield a lower bound on performance with diversity transmission.

With maximal ratio pre-detection combining, and with selection diversity, the error probability is obtained in the same manner as for a single channel, by replacing in (10.1) the probability density $p(r)$ with that for maximal ratio combining, as derived in Section 9.18, or for selection diversity as derived in Section 9.21. With post-detection combining, it is necessary to follow a different procedure for other than synchronous demodulators. The probability density for the signal-to-noise ratio at the output of a single demodulator must first be determined, and, in turn, the probability density for the sum of the m demodulator outputs. With the aid of the latter probability density, which differs from that with predetection combining, the error probability for the combined baseband pulse train can be determined. This error probability has been determined by Pierce [2] for binary FSK with dual-filter envelope detection and by Voelcker [4] for binary PM with differential phase product demodulation.

For reasons outlined in Section 9.17, it is customary in dealing with diversity transmission to assume that the transmitter power per channel is the same as without diversity, so that part of the diversity improvement resides in increased total transmitter power. This assumption is also made in the various relations that follow, which are confined to the various binary methods considered in Section 10.1 and in other literature on diversity transmission.

A more general solution has been given by Lindsey [9] for incoherent diversity reception of correlated binary waveforms when the received signal contains a fixed component together with a slowly varying component with a Rayleigh probability distribution. This reference also contains a comprehensive bibliography.

10.9 Error Probabilities with Optimum Diversity Methods

Let the input to the various channels be synchronously demodulated, in which case the amplitude probability distribution of the noise is the same at the demodulator outputs as at the inputs. Under this condition the error probability with parallel combination of the m baseband outputs is given by

$$P_{e,m} = \int_0^\infty P_e^0(\rho) p_m(\rho) \, d\rho. \qquad (10.64)$$

Here $P_e^0(\rho)$ is the error probability for a single time-invariant channel with synchronous demodulation as a function of the signal-to-noise ratio ρ at the receiver input. The probability density $p_m(\rho)$ of the combined signal-to-noise ratio at the demodulator output depends on the amplification of the individual channels. The optimum condition is realized with maximal ratio combining, in which the power amplification of the various input signals is made proportional to the input signal power of each channel, as shown in Section 9.18 for analog transmission. Under this condition, the following relation is obtained for the probability density $p_m(\rho)$.

$$p_m(\rho) = \frac{1}{\bar{\rho}}\left(\frac{\rho}{\bar{\rho}}\right)^{m-1} \frac{1}{(m-1)!} e^{-\rho/\bar{\rho}}. \tag{10.65}$$

The latter relation is obtained from (9.71) with $x = \rho/\bar{\rho}$ and $dx = d\rho/\bar{\rho}$.

With binary PM and synchronous demodulation, the error probability $P_e^0(\rho)$ is given by (10.4), with ρ in place of ρ_i for convenience in the relations that follow. Thus

$$P_e^{(0)}(\rho) = \tfrac{1}{2}\,\text{erfc}\,(\rho)^{\frac{1}{2}}. \tag{10.66}$$

Hence in this case insertion of (10.65) and (10.66) in (10.64) gives the relation

$$P_{e,m} = \frac{1}{2\bar{\rho}} \int_0^\infty \text{erfc}\,(\rho^{\frac{1}{2}}) \left(\frac{\rho}{\bar{\rho}}\right)^{m-1} \frac{1}{(m-1)!} e^{-\rho/\bar{\rho}}\, d\rho \tag{10.67}$$

$$= \frac{1}{2\pi^{\frac{1}{2}}(m-1)!} \int_\rho^\infty \frac{e^{-y}}{y^{\frac{1}{2}}} \int_0^\infty \left(\frac{\rho}{\bar{\rho}}\right)^{m-1} e^{-\rho/\bar{\rho}}\, d\rho. \tag{10.68}$$

Evaluation of this integral yields the following relations. For $m = 1$:

$$P_{e,1} = \frac{1}{2}\left(1 - \left(\frac{\bar{\rho}}{1+\bar{\rho}}\right)^{\frac{1}{2}}\right) \tag{10.69}$$

$$\simeq \frac{1}{4\bar{\rho}} \quad \text{for} \quad \bar{\rho} \gg 1. \tag{10.70}$$

For $m = 2$:

$$P_{e,2} = \frac{1}{2}\left(1 - \left(\frac{\bar{\rho}}{1+\bar{\rho}}\right)^{\frac{1}{2}} - \frac{1}{2\bar{\rho}}\left(\frac{\bar{\rho}}{1+\bar{\rho}}\right)^{\frac{3}{2}}\right) \tag{10.71}$$

$$\simeq \frac{3}{16\bar{\rho}^2} = 3P_{e,1}^2 \quad \text{for} \quad \bar{\rho} \gg 1. \tag{10.72}$$

Integral (10.68) can be evaluated for any number m and the evaluation results in a continuation of the series given by (10.71) for $m = 2$. Since only values of $\bar{\rho} \gg 1$ are of interest, an asymptotic expansion of the integrand of (10.68) can be used to yield the following relations for small values of $P_{e,1}$:

$$P_{e,m} \approx \frac{(2m-1)!}{m!(m-1)!} P_{e,1}^m, \tag{10.73}$$

$$P_{e,2} \approx 3P_{e,1}^2, \tag{10.74}$$

$$P_{e,3} \approx 10P_{e,1}^3, \tag{10.75}$$

$$P_{e,4} \approx 35P_{e,1}^4, \tag{10.76}$$

where $P_{e,1}$ is given by (10.6).

The approximate relation (10.73) also applies for other than synchronous demodulation of the individual channels, with optimum post-detection combining of the several demodulated baseband pulse trains prior to sampling. Thus (10.73) applies for binary FM with dual-filter envelope detection [2] when $P_{e,1}$ is given by (10.12) and for binary PM with differential phase product demodulation [4], when $P_{e,1}$ is given by (10.8). With these methods the error probability is derived by the different procedure outlined in the previous section, for post-detection combining with other than linear synchronous demodulators.

With equal average noise powers in all channels, maximal ratio combining is obtained with square law weighing in the combination of the demodulated binary pulse trains, except for binary PM with differential phase product demodulation, where linear addition is employed [4]. In this case the multiplication of the received carrier wave by a delayed wave in each product demodulator yields a square-law output in each demodulator. Hence equal amplifier gains with linear addition of demodulator outputs corresponds to maximal ratio post-detection combining.

The error probability with differential phase product demodulation and diversity transmission has been determined by Voelcker, [4] considering both errors ($P_e^{(1)}$) from Gaussian noise with Rayleigh fading and errors ($P_e^{(2)}$) from time variations in the phase of the channel transmittances. In addition, Voelcker has also determined the error probability with dual diversity and four-phase modulation, considering errors ($P_e^{(2)}$) from time variations in the phase only. For the latter case, Voelcker's more exact expression, when reduced to small error probabilities, gives a factor $4\pi(3 + \pi)/(2 + \pi)^2 \simeq 3.13$ in place of 3 in (10.74). Thus in all of the foregoing cases (10.74) affords a close approximation for dual diversity transmission over independently fading channels, even when the errors are not caused by additive Gaussian noise.

As discussed in Section 9.17, the foregoing relations assume that the total transmitter power is increased in proportion to the number of diversity paths and that a single receiver is used for each path. For a given total transmitter power there will be a certain optimum number of paths. Alternately, for a specified error probability $P_{e,m}$ it is possible to obtain a minimum combined transmitter power for a certain optimum number of paths m.

Pierce [3] and Harris [11] have shown that the minimum total transmitter power is obtained for any specified error probability when m is so chosen that in each diversity channel $\bar{\rho} \approx 3$, or about 5 dB, for binary FM with dual filter envelope detection. The number of diversity paths required to realize the minimum total transmitter power is rather large, and the power reduction that can be realized with more than four paths is

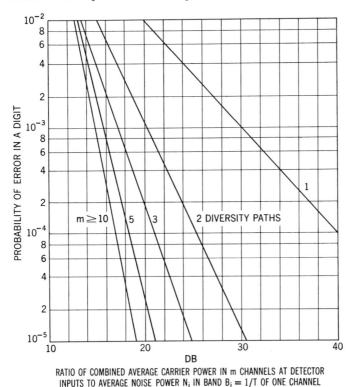

Figure 10.5 Error probabilities with diversity transmission for binary FM with dual filter envelope detection, for independently fading diversity paths.

fairly small. For example, Pierce [3] shows that for an error probability $P_{e,m} = 10^{-4}$, the minimum average transmitter power is realized with $m = 16$. The total signal-to-noise ratio is in this case 16.7 dB, corresponding to a signal-to-noise ratio per channel of 4.7 dB ($\bar{\rho} = 2.95$). With $m = 1$ the average signal-to-noise ratio is 40 dB, with $m = 4$ it is 19.4 dB, so that the reduction in transmitter power is 20.6 dB. An additional reduction of only 2.7 dB is realized when the number of diversity paths is increased from $m = 4$ to $m = 16$. The error probability for various numbers of diversity paths as given by Pierce [3] for binary FM with dual filter envelope detection is shown in Fig. 10.5 as a function of $m\bar{\rho}$.

10.10 Error Probabilities with Equal-Gain Diversity

The minimum error probabilities given in the preceding section apply for individual receiver power gains that are proportional to the momentary input signal-to-noise power ratios, except for PM with differential phase detection, which requires equal receiver gains, for reasons mentioned previously. With the other binary methods, the error probability with equal receiver gains is difficult to determine, for the reason that the probability density $p_m(\rho)$ cannot be expressed by a simple function, such as (10.65) for maximal ratio combining. However, as discussed in Section 9.19 equal gain operation is only slightly less efficient than maximal ratio combining. The penalty in combined output signal-to-noise ratio is about 0.5 dB for dual diversity, increasing to about 1 dB for a large number of diversity channels. Thus, the error probability is greater than obtained from (10.73) by a factor 1.1 for $m = 2$, increasing to $4/\pi \simeq 1.25$ for large values of m.

10.11 Error Probabilities with Selection Diversity

As noted in Section 10.8, selection diversity may be impracticable for digital transmission, owing to errors during switchovers. The relations that follow apply to other than switching errors, and give an upper bound on error probability with diversity transmission.

The probability density of the signal-to-noise ratio with selection diversity is obtained from (9.86) with $x = \rho/\bar{\rho}$ and is

$$p_m(\rho) = \frac{m}{\bar{\rho}} (1 - e^{-\rho/\bar{\rho}})^{m-1} e^{-\rho/\bar{\rho}}. \qquad (10.77)$$

The error probability with any of the binary methods considered previously is obtained with (10.77) in (10.64). By way of illustration, with binary PM and differential phase detection P_e^0 is given by (5.92) with $\rho_i = \rho$, or by

$$P_e^0 = \tfrac{1}{2} e^{-\rho}. \qquad (10.78)$$

In this case (10.64) becomes

$$P_{e,m} = \frac{m}{2\bar{\rho}} \int_0^\infty e^{-\rho(1+1/\bar{\rho})} (1 - e^{-\rho/\bar{\rho}})^{m-1} \, d\rho. \qquad (10.79)$$

For $m = 1$:

$$P_{e,1} = \frac{1}{2\bar{\rho}} \int_0^\infty e^{-\rho(1-1/\bar{\rho})} \, d\rho$$

$$= \frac{1}{2(1 + \bar{\rho})}. \qquad (10.80)$$

For $m = 2$:

$$P_{e,2} = \frac{1}{\bar{\rho}} \int_0^\infty (e^{-\rho(1+1/\bar{\rho})} - e^{-\rho(1+2/\bar{\rho})}) \, d\rho$$

$$= \frac{1}{\bar{\rho}} \left(\frac{1}{1+\bar{\rho}} - \frac{1}{2+\bar{\rho}} \right). \tag{10.81}$$

For $\bar{\rho} \gg 1$ or $P_{e,1} \ll 1$:

$$P_{e,2} \simeq 4P_{e,1}^2. \tag{10.82}$$

By solution of (10.79) the following relations are obtained for $\bar{\rho} \gg 1$ or $P_{e,1} \ll 1$:

$$P_{e,m} \approx 2^{m-1} m! P_{e,1}^m, \tag{10.83}$$

$$P_{e,2} \approx 4P_{e,1}^2, \tag{10.84}$$

$$P_{e,3} \approx 24 P_{e,1}^3, \tag{10.85}$$

$$P_{e,4} \approx 192 P_{e,1}^4. \tag{10.86}$$

These relations apply for all methods considered previously, and were published by Pierce for frequency-shift keying with dual-filter detection [2]. The error probabilities $P_{e,1}$ with these various methods are given in Section 10.1.

For equal error probability, the average signal power with selection diversity must be greater than with optimum diversity by a factor equal to the mth root of the ratio of the factors in (10.83) and (10.73). The power must thus be increased by 0.62, 1.27, and 1.85 dB for $m = 2$, 3, and 4, respectively.

10.12 Diversity Improvements with Selective Fading

The relations in the preceding sections apply for errors owing to additive Gaussian noise, with nonselective or flat Rayleigh fading. The mechanism responsible for error reduction in the foregoing cases also applies to transmission over channels with frequency-selective fading such that the errors are caused principally by intersymbol interference. With independently fading transmission paths there will be no correlation between intersymbol interference in the various channels, even though the signals are the same. A similar situation is encountered when the errors are caused by phase changes over a signal interval, as considered in Sections 10.4 and 10.5. As mentioned in Section 10.9, relations given by Voelcker for dual-diversity transmission with binary PM and differential phase detection shows that the same relation (10.74) applies for errors, owing to phase changes, as for errors caused by additive Gaussian noise. In general, the error probability with diversity transmission would be expected to be of the same form $K(m) P_{e,1}^m$ as (10.73) except that $K(m)$ may be somewhat different for errors caused by frequency-selective fading.

The error probability $P_{e,m}^{(3)}$ has been determined in reference [7] for binary PSK with differential phase detection and a raised cosine pulse spectrum. The results are shown in Fig. 10.6 and indicate that for $m = 2$ $K(m) \simeq 4$ as compared with 3 in (10.74). For $m = 4$ the error probability is nearly the same as obtained from (10.76). It is thus a legitimate approximation to use (10.73) also for errors due to frequency-selective fading. The inaccuracy in this approximation would be quite small for the combined error probability.

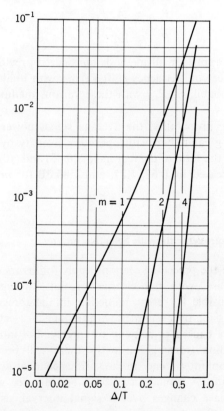

Figure 10.6 Irreducible error probability $P_{e,m}^{(3)}$ without diversity, with dual and quadruple diversity, for pulses with raised cosine spectrum.

10.13 Error Reduction by Error-Correcting Codes

As shown in Chapter 5, optimum error-correcting codes yield a somewhat lower error probability than direct binary transmission without coding, or than repeated binary transmission over channels with additive Gaussian noise. The situation is different in the case of time-variant channels as considered here, where repeated transmission takes the form of diversity operation in the time, frequency, or space domain. In the case of Rayleigh fading, it turns out that optimum diversity operation yields a much lower error probability than do error correcting codes used without diversity transmission, as shown in this section.

Let a binary sequence consist of n information digits and e error correcting digits capable of correcting errors in d digits. Let further $p = P_e^0(\rho)$ be the probability of an error in a digit with direct transmission over a time invariant channel, as related to the signal-to-noise ratio ρ. For a given information rate the channel bandwidth must then be increased by a factor $k = (n + e)/n$, and the signal power must be increased by the same factor k in order that the signal-to-noise ratio ρ remains the same as with direct transmission of n digits.

In accordance with (5.119) the probability of an error in a character of $n + e$ digits and n information digits can then be written

$$\epsilon_k^{(n)} = \sum_{u=d+1}^{n+e} \binom{n+e}{u} p^u (1-p)^{n+e-u}$$

$$= E(n+e, d+1, p), \tag{10.87}$$

where E is the cumulative binomial distribution for which comprehensive tables are available.*

Let $p_1(\rho)$ be the probability density of the channel transmittance variation without diversity transmission, as obtained from (10.65) with $m = 1$, for the particular case of Rayleigh fading. The probability of an error in a character of n information digits is then for very slow fading

$$P_{e,k}^{(n)} = \int_0^\infty p_1(\rho) \, E(n+e, d+1, p) \, d\rho. \tag{10.88}$$

With Rayleigh fading $p_1(\rho) = \dfrac{1}{\bar\rho} e^{-\rho/\bar\rho}$ and (10.88) can be written:

$$P_{e,k}^{(n)} = C_k^{(n)}(\bar\rho) \frac{1}{\bar\rho}, \tag{10.89}$$

---

* Staff of Harvard Computation Laboratory: "Table of the Cumulative Binomial Distribution," Annals 35, Harvard University Press, Cambridge.

where

$$C_k^{(n)}(\bar{\rho}) = \int_0^\infty e^{-\rho/\bar{\rho}} E(n + e, d + 1, p) \, dp. \qquad (10.90)$$

For binary PM with differential phase detection $p = \tfrac{1}{2} e^{-\rho}$ and numerical integration yields the values given in the table below for (15,5), (23,12) and (73,45) error correcting codes (see Chapter 5 for references).

Factor $C_k^{(n)}(\bar{\rho})$ for Some Error-Correcting Codes

n	e	d	k	$\bar{\rho} = 2(3\text{ dB})$	$\bar{\rho} = 10(10\text{ dB})$	$\rho = 100(20\text{ dB})$
5	10	3	3	0.64	0.71	0.77
12	11	3	23/12	0.85	1.00	1.20
45	28	4	4/45	1.4	2.10	2.40

The foregoing factors apply for a k-fold increase in signal power and bandwidth compared to direct transmission of the n information digits, such that $\bar{\rho}$ is the same in all cases. With the foregoing k-fold increase in signal power with direct transmission, such that the transmitter power is the same, the probability of an error in n digits becomes

$$P_{e,1}^{(n)} \simeq \frac{n}{k} P_{e,1}^{(1)} = \frac{n}{k} P_{e,1}, \qquad (10.91)$$

where $P_{e,1}$ is the probability of an error in a digit, as given by (10.8) for binary PM with differential phase detection. For this case (10.91) can be written

$$P_{e,1}^{(n)} = D_k^{(n)}(\bar{\rho}) \frac{1}{\bar{\rho}}, \qquad (10.92)$$

where

$$D_k^{(n)}(\bar{\rho}) = \frac{n}{2k} \frac{\bar{\rho}}{1 + \rho} = \frac{n^2}{2(n + e)} \frac{\bar{\rho}}{1 + \bar{\rho}}. \qquad (10.93)$$

This factor is given in the table below for n and k corresponding to the foregoing error-correction codes.

Factor $D_k^{(n)}(\bar{\rho})$ for Direct Transmission and Same Signal Power as with Error-Correcting Codes

n	k	$\bar{\rho} = 2(3\text{ dB})$	$\bar{\rho} = 10(10\text{ dB})$	$\bar{\rho} = 100(20\text{ dB})$
5	3	0.56	0.76	0.84
12	23/12	2.10	2.80	3.10
45	73/45	9.20	12.60	14.00

Comparison of $D_k^{(n)}$ and $C_k^{(n)}$ shows that the performance improves with increasing block length of the codes, and thus increasing complexity of implementation. While the error reduction is insignificant with the (15,5) code, the (23,12) code affords about a 2.6-fold reduction in errors and the (73,45) code about a 6-fold reduction.

In the foregoing comparisons, binary PM with differential phase detection was assumed. With synchronous detection $p = \frac{1}{2}\,\text{erfc}\,\rho$ in (10.90) while for binary FM with dual filter envelope detection $p = \frac{1}{2}e^{-\rho/2}$. This will cause relatively minor modifications in the above results for binary PM.

With the aid of previous relations for the error probability with diversity transmission, it is readily demonstrated that this is a far more effective method of error reduction than error correcting codes, for equal total transmitter power and channel bandwidth. By way of example, for optimum post-detection dual diversity (10.74) applies, and for binary differential phase detection yields the relation $P_{e,2} = (3/4)\cdot(1 + \bar{\rho})^{-2}$. For $\bar{\rho} = 10$ this error probability is 0.0063 as compared with 0.10 obtained from (10.89) for a closely comparable error correcting code with $k = 23/12$.

The above results apply for flat fading at such a slow rate that the carrier amplitude is essentially constant over the interval of the $n + e$ digits of a code block. Under this condition errors owing to the random phase and frequency fluctuations in the channel transmittance that accompany fading can be neglected in comparison with errors caused by random noise, as assumed in the above comparisons. The other extreme would be fading at such a fast rate that there is no correlation between the carrier amplitudes of successive digits. Under this condition the random phase fluctuations in the channel transmittance would give rise to an excessive error rate even in the absence of noise, as discussed in Section 10.4, and error correcting codes would be of no avail. However, even with slow fading it is possible to insure that there is no correlation between the carrier amplitudes of successive digits, provided diversity transmission is used in conjunction with error correcting codes, to be considered next.

10.14 Combination of Diversity Transmission and Error Correction

Improved performance can be realized if error-correcting codes are used in appropriate manner in conjunction with diversity transmission. Let the $n + e$ digits considered previously be transmitted over $n + e$ independently fading channels established by adequate frequency, time or space separation. In this case an independent Rayleigh amplitude distribution will exist among the $n + e$ digits of a character. Hence the probability of error in a character is obtained from (10.87) except that now p is the probability $P_{e,1}$ of an error in a digit with Rayleigh fading, as previously determined for various binary modulation and detection methods.

By way of illustration, for a (15,5) error-correcting code the probability of error in a character with n information digits and $e = 10$ error-correction digits becomes, in accordance with (10.87):

$$\epsilon_k^{(n)} = \epsilon_3^{(5)} = E(15,4,p) \tag{10.94}$$

For $p < 10^{-2}$

$$\epsilon_3^{(5)} \simeq 1365p^4 = 1365 P_{e,1}^4 \tag{10.95}$$

This compares with the following relation for the error probability with optimum triple diversity obtained from (10.75) for a sequence of $n = 5$ digits

$$P_{e,3}^{(5)} = n10 P_{e,1}^3 = 50 P_{e,1}^3 \tag{10.96}$$

For $P_{e,1} = 10^{-2}$

$$P_{e,3}^{(5)} = 5 \cdot 10^{-5} \quad \text{and} \quad \epsilon_3^{(5)} \simeq 1.36 \cdot 10^{-5}$$

Thus improved performance is obtained for a given total bandwidth and average signal power when error correction is combined with diversity transmission over $n + e = 15$ channels. This improvement is realized in exchange for the greater complexity of error-correcting codes combined with the greater complexity of 15 in place of 3 diversity paths.

The above analysis assumes that fading is flat over the band of each of the $n + e$ diversity channels. Moreover, that fading occurs at such a slow rate that phase or frequency changes between digits in each diversity channel can be disregarded. To this end it is necessary that the bandwidth Ω of each diversity channel be much larger than the fading bandwidth $\bar{\gamma}$ as indicated in Fig. 10.1. Otherwise the additional errors, owing to phase or frequency changes, must be taken into account.

The performance of certain error-correcting or failure-correcting codes, when used in the foregoing manner in conjunction with diversity transmission, has been determined in a paper by White [12.]

10.15 Multiband Transmission

The curves in Figs. 10.3 and 10.4 suggest that for a given total transmitter power and channel bandwidth, the error probability can be reduced by transmitting at a slower rate over each of a number of narrower channels in parallel. An approximate optimum bandwidth for each channel would be such that $P_e^{(2)} + P_e^{(3)}$ is minimized. This can be accomplished with separate transmitters and receivers for each channel, such that mutual interference between channels is avoided. Hence the adverse effects of selective fading can be overcome with the aid of more complicated terminal equipment, without the need for increased signal power or channel bandwidth.

An alternative method that is simpler in implementation is to transmit the combined digital wave from the parallel channels by frequency or phase modulation of a common carrier, as ordinarily used for transmission of voice channels in frequency division multiplex. This method entails some mutual interference between channels, owing to intermodulation distortion, as well as greater channel bandwidth and carrier power than with direct digital carrier modulation, as will be discussed presently.

With the foregoing method, the spectrum of the modulated carrier wave will have greater bandwidth than with direct digital carrier modulation. To avoid excessive transmission distortion of the combined wave, the bandwidth between transmitter and receiver must be at least twice that with digital carrier modulation. Hence at least 3 dB greater average carrier power is required in order that the noise threshold level of the common channel be comparable with that of direct digital carrier modulation.

With such multiband transmission, intersymbol interference owing to selective fading is avoided, in exchange for mutual interference between the various channels owing to intermodulation distortion caused by frequency-selective fading. For a modulating wave with the properties of random noise, which is approximated with a large number of binary channels in frequency division multiplex, the results of measurements are given in Section 9.16. The results indicate that under this condition intermodulation distortion will cause less transmission impairment than does intersymbol interference in direct digital transmission. Hence multiband transmission by common carrier modulation permits a reduction in error probability in exchange for at least a two-fold increase in bandwidth and carrier power. However, this reduction in error probability may be less than can be realized with direct digital carrier modulation in conjunction with a two-fold increase in bandwidth and signal power with dual diversity.

Error probabilities in binary multiband transmission by frequency modulation of a common carrier are dealt with by Barrow [13] on the premise of slow flat fading over the combined band, so that only errors caused by noise need be considered, and intermodulation distortion can be disregarded.

10.16 Experimental Results

In Section 9.16, comparisons were made of intermodulation distortion observed in analog transmission by FM over four troposcatter links and, on the other hand, first-order predictions based on the present idealized random multipath model. These comparisons indicated satisfactory conformance with measurements for links 185 and 194 miles in length.

Comparisons have also been made of measurements with predicted error probabilities for three methods of digital transmission over a 375-mile troposcatter link [14]. One of the digital systems employed binary FM with transmission at a rate of 376 kilobits/sec. The value of Δ determined from (9.54) was $\Delta = 0.43$ microsecond, so that $\Delta/T = 0.29$. Measurements were made of error probability with dual diversity for various values of $\bar{\rho}$ in the range between 10 and 20 dB. The irreducible error probability $P_{e,2}^{(3)}$ obtained by extrapolation was in the range between $3 \cdot 10^{-3}$ and 10^{-2}, corresponding to values of $P_{e,1}^{(3)}$ between $3 \cdot 10^{-2}$ and $6 \cdot 10^{-2}$. These values conform fairly well with the approximate curve 1 in Fig. 10.2 for $\Delta/T = 0.29$, but are appreciably greater than the presumably more accurate curve 2. This could be fortuitous and further experimental data would be required for verification.

With the other two digital methods used in the tests the transmission rate was much lower so that errors from frequency selective fading were negligible.

10.17 High-Frequency Ionospheric Transmission

In HF radio systems the signal is ordinarily received over two or more main ionospheric paths, with such differences in delay that a single transmitted pulse is received as two or more distinct pulses. Each such main path can be resolved into a large number of minor paths with much smaller differences in delay. The latter differences give rise to frequency-selective fading in each main path and resultant distortion in each of the received pulses.

The transmission delay over each main path is nearly constant over appreciable periods, and fading occurs at a relatively slow rate. Hence it is feasible to employ automatic equalization in order to substantially eliminate pulses over unwanted main paths. After such equalization, the transmission rate is limited by frequency-selective fading and resultant distortion of the pulses received over the selected main path, similar to the manner discussed for troposcatter links. However, in HF ionospheric transmission, the multipath differential delay Δ is ordinarily much greater than in troposcatter systems, so that the maximum transmission rate is much smaller.

The expected performance of HF ionospheric pulse systems with adaptive equalization is dealt with in a paper by DiToro that contains a comprehensive bibliography [15].

11
Intersystem Interference

11.0 General

Transmission impairments considered in previous chapters reside in various kinds of imperfections in the channel, and could in principle be overcome by an increase in signal power in conjunction with reduction in characteristic distortion of various kinds. This is not generally the case with intersystem interference, which is caused by overlaps of the signal spectra of two systems of the same or different kinds, owing to electromagnetic coupling of the transmission paths. Evaluation and control of such interference is a problem of first importance in extensive communication systems employing various transmission media, such as open-wire lines, balanced cable pairs, coaxial cable pairs, radio relays and communication satellites. As an aid in the determination of such interference, certain basic considerations and analytical methods are outlined in this chapter. Further detailed considerations and theoretical relations for engineering applications are presented in literature dealing with particular situations, which also contain additional bibliography [1–13].

In addition to interference between communication systems, it is necessary to consider the possibility of disruptive voltages owing to interference from power systems and lightning discharges. Analytical methods for the evaluation of such disturbances together with protective measures are dealt with elsewhere [14].

One of the more important interference problems is encountered when several systems of the same kind occupy the same transmission band on different metallic cable pairs in close proximity, such as open wire pairs on the same pole line or insulated pairs of various kinds in the same cable. Electromagnetic coupling between transmission paths in these cases comes about because of small unbalances remaining after transpositions or twisting of pairs or because of imperfect shielding of coaxial units. Such coupling gives rise to "far-end" interference between transmission paths in the same direction and "near-end" interference or crosstalk

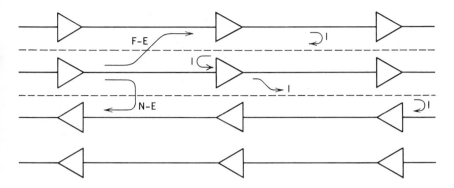

Figure 11.1 Coupling between transmission paths in the same and opposite directions. *F-E:* far-end coupling path. *N-E:* near-end coupling path. *I:* indirect or interaction coupling path.

between paths in opposite direction, as indicated in Fig. 11.1. In addition, it is possible to have coupling between the outputs and inputs of repeaters via "tertiary" circuits, which gives rise to "interaction" interference or crosstalk.

Near-end interference can be avoided by using different frequency bands for opposite directions of transmission. To avoid interaction interference it is necessary, in addition, to transpose or "frogg" the frequency bands between one repeater section and the next, as indicated in Fig. 11.2.

In microwave radio relay systems it is not feasible to employ the same two frequency bands for several two-way systems along the same route, even for transmission in the same direction, since adequate interference suppression cannot be secured with reasonable antenna separations and directivity. Hence different frequency bands are used for systems of the same kind. In this case interference is encountered, owing to the coupling

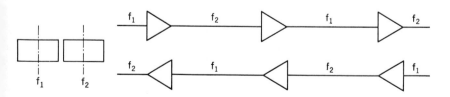

Figure 11.2 Frequency transpositions to avoid near-end and interaction interference.

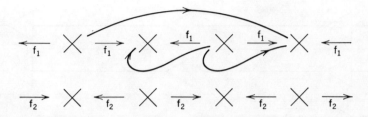

Figure 11.3 Transmission and interference paths between antennas of radio relay systems.

of different repeater sections of the same system. As indicated in Fig. 11.3, two coupling paths between opposite directions of transmission are encountered, owing to the incomplete suppression of backward antenna radiation, while a third path results from overreach of antenna beams in the same direction of transmission.

Other situations arise, involving systems of different kinds rather than essentially the same kind. By way of example, a PCM system may be provided on some pairs in a cable with an analog AM or SSB system in a lower transmission band on other pairs. Coupling between pairs may then

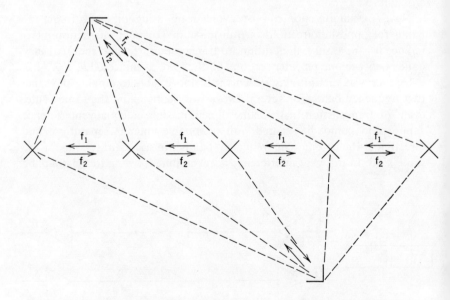

Figure 11.4 Coupling paths between single link and multilink radio relay systems. ———— Main antenna lobe paths between repeaters of each system. – – – – – Mutual antenna sidelobe paths of repeaters of both systems.

give rise to excessive interference in the usually more susceptible AM systems, unless the pulse spectrum of the PCM system is modified to reduce low-frequency components falling into the band of the AM system.

In the above examples, more or less continual intersystem interference would be experienced. Other situations arise where two systems of the same or different kinds cause occasional interference. For example, a satellite communication link may employ FM with a large deviation ratio in nearly the same frequency band as used by multilink FM radio relays. There may then be occasional mutual interference, owing to overlapping antenna patterns, as indicated in Fig. 11.4.

The situations outlined above and considered in this chapter involve interference between extensive communication systems of a kind that can usually be anticipated and taken into account in systems design and operation. In addition, interference of a more spurious nature may be encountered, particularly in radio systems, but is not considered here.

11.1 Basic Considerations

Intersystem interference and resultant transmission impairments depend on several factors. Of first importance is the power transfer or coupling factor $K(\omega)$, between the output of the transmitter of the interfering system and the input to the receiving filter of the disturbed system. Other primary quantities are the signal power spectra of the disturbed system, 1, and the interfering system, 2.

Ordinarily two carrier systems are involved, and it is convenient to choose the carrier frequency ω_1 of the disturbed systems as a reference. The following designations indicated in Fig. 11.5, will be used:

$$K_{21}(\omega) = K_{21}(\omega_1 + u) = C_{21}(u) = C(u),$$

$$W_{11}(\omega) = W_{11}(\omega_1 + u) = W_1(u),$$

$$W_{22}(\omega) = W_{22}(\omega_1 + u) = W_2(u - \omega_{12}),$$

$$W_{11}^{(i)}(\omega) = W_{11}^{(i)}(\omega_1 + u) = W_1^{(i)}(u),$$

$$W_{22}^{(i)}(\omega) = W_{22}^{(i)}(\omega_1 + u) = W_2^{(i)}(u - \omega_{12}).$$

W_{11} and W_{22} designate the power spectra at the output of the transmitters, $W_{11}^{(i)}$ and $W_{22}^{(i)}$ those at the input to the receiving filters.

The power spectrum of interference at the input to the receiving filter of system, 1, is thus

$$W_{21}^{(i)}(u) = K_{21}(\omega) W_{22}(\omega) = C(u) W_2(u - \omega_{12}). \tag{11.1}$$

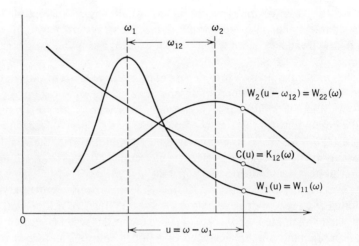

Figure 11.5 Designations of signal power spectra $W_1(u)$ of disturbed system and $W_2(u-\omega_{12})$ of interfering system, and of power coupling factor $C(u)$.

At the baseband output of system, 1, the power spectrum is of the general form

$$W_1^{(0)}(\omega) = W_{11}^{(0)}(\omega) + W_{21}^{(0)}(\omega). \tag{11.2}$$

$W_{11}^{(0)}(\omega)$ is related to $W_{11}^{(i)}(\omega) = W_1^{(i)}(u)$ in a manner that depends on the modulation method of system 1 and its implementation. With linear modulation methods $W_{21}^{(0)}(\omega)$ depends on $W_{21}^{(i)}(u)$ alone, but with nonlinear methods (such as FM) it also depends on $W_1^{(i)}(u)$, owing to intermodulation effects.

At the baseband output, the ratio of average interference power to average message power at the frequency ω is

$$\eta_0(\omega) = \frac{W_{21}^{(0)}(\omega)}{W_{11}^{(0)}(\omega)}. \tag{11.3}$$

This ratio is of interest when the modulating wave consists of a number of narrow-band message waves in frequency-division multiplex, as in multichannel voice transmission.

In the case of broad-band analog messages, such as TV pictures, the ratio of average interference power to average message power in a band Ω_0 is of interest and is given by

$$\mu_0 = \int_0^{\Omega_0} W_{21}^{(0)}(\omega)\,d\omega / \int_0^{\Omega_0} W_{11}^{(0)}(\omega)\,d\omega. \tag{11.4}$$

Basic Considerations

With m interfering systems of the same kind the factor $C(u)$ in (11.1) is taken as the sum of the power transfer factors of the individual systems, that is,

$$C(u) = \sum_{i=2}^{m} C_{i,1}(u) = m\underline{C}(u), \qquad (11.5)$$

where $\underline{C}(u)$ is the average power-coupling factor.

When the interfering and disturbed systems are of the same kind, the power spectrum $W_{21}^{(0)}(\omega)$ will in general contain a component that is linearly related to the message-power spectrum of the interfering system. This component represents coherent interference. In the case of speech transmission this gives rise to intelligible crosstalk, which is more disturbing than random noise of the same average power. As the number of interfering systems is increased, the combined interference will become similar to random noise and thus incoherent. In speech transmission, the number of interfering systems required to this end may be quite large, because of the great variation in speech volume among talkers discussed in Chapter 2. The permissible maximum ratios $\eta_0(\omega)$ and μ_0 will depend on the nature of the interference, since this may have a significant effect on the transmission impairments.

In the case of digital systems, the foregoing average interference ratios $\eta_0(\omega)$ and μ_0 do not generally suffice for determination of the effect of interference. It is also necessary to establish the probability distribution of the instantaneous amplitudes in order to determine error probability. When the interference originates from a large number of disturbers, the probability distribution can ordinarily be assumed to be Gaussian, so that the error probability can be determined in the same manner as for additive Gaussian noise. In some cases, however, this may entail such a large number of disturbers that it renders the foregoing rule inapplicable in actual situations. The other extreme would be represented by interference from a single system, in which case the probability distribution of interference would be nearly the same as that of the interfering signal, and in most cases could be determined. For intermediate cases, determination of the probability distribution of interference and thus of error probability in digital transmissions is a vexing analytical or computational problem. Determination of the probability distributions of interference is beyond the scope of this presentation.

In accordance with the foregoing general formulation, evaluation of mutual intersystem interference entails determination of three basic quantities: (1) the coupling factor, (2) the signal power spectra in both systems, and (3) the interference ratios η_0 and μ_0 in analog systems or error probability in digital systems.

Determination of the coupling factors is a problem in itself beyond the scope of this presentation. It may involve analytical and experimental determination of electromagnetic shielding, as considered elsewhere [12]. The coupling of balanced pairs and coaxial units are discussed further in Sections 11.9 and 11.10. The coupling factor of radio systems depends on the antenna radiation patterns and on the orientation of the systems. For example, in Fig. 11.4 the multilink radio relay may be exposed at several repeater inputs and it is necessary to determine the coupling factor for each repeater.

Determination of the signal power spectra at the transmitter output and receiver input may present a problem with some modulation methods. For a given method, the signal power required at the receiver depends on the density and the nature of additive random noise, together with various kinds of characteristic distortion, as discussed in previous chapters. The corresponding transmitter power depends on path loss, which in radio transmission may exhibit pronounced fluctuations, owing to fading.

The signal power spectrum at the transmitter output is determined in part by the transmitting filter, which also has the important function of limiting the power of spectral components outside the prescribed band so as to avoid excessive interference in other systems. At the receiver, the signal and interference spectra are controlled in part by the receiving filter. Hence appropriate filter design is an important factor in the control of intersystem interference [13]

Finally, the allowable interference ratios may depend significantly on the nature of the interference and the probability of occurrence. A case in point would be mutual interference between certain kinds of satellite communications and radio relay systems. Another example is crosstalk between carrier-system voice circuits, owing to coupling between cable pairs. Here allowance must be made not only for the greater disturbing effect of intelligible interference, but also for the pronounced variations in speech power among individuals (Section 2.2) and in the coupling of cable pairs (Section 11.9).

The exposition that follows is concerned principally with a more specific formulation of the ratios $\eta_0(\omega)$ and μ_0, defined by (11.3) and (11.4), for the more important modulation methods and interference situations encountered in extensive communication networks.

11.2 Interference between Single-Sideband Systems

The more efficient method of analog transmission over metallic pairs is by single-sideband operation. In Fig. 11.6, let $W_1(u)$ and $W_2(u - \omega_{12})$ be

Interference between SSB Systems

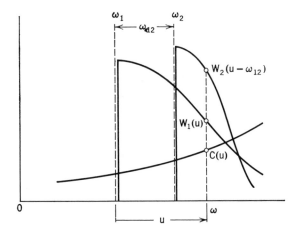

Figure 11.6 Interference between single sideband systems. Interference power in system 1 in narrow band du at $\omega = u + \omega_1$ at receiver input is $W_{12}^{(i)}(u) = C(u)\, W_2(u - \omega_{12})\, du$.

the sideband power spectra of a disturbed and an interfering carrier system with original carriers at ω_1 and ω_2. If $C(u)$ designates the power coupling factor between the two transmission paths, the power spectrum of interference in system 1 at the sideband frequency u is

$$C(u) W_2(u - \omega_{12}).$$

Let the baseband receiving filter of system 1 have a power transmittance $r_1(\omega)$. The ratio of average interferences power to average message power in a narrow band at the baseband frequency $\omega = u$ is then

$$\eta_0(u) = \frac{C(u)\, W_2(u - \omega_{12}) r_1(u)}{W_1^{(i)}(u) r_1(u)}$$

$$= \frac{C(u)\, W_2(u - \omega_{12})}{W_1^{(i)}(u)}. \qquad (11.6)$$

With two like systems having the same carrier frequencies and message power spectra

$$\eta_0(u) = \frac{C(u)\, W_1(u)}{W_1^{(i)}(u)}$$

$$= C(u)\, G_1(u) \qquad (11.7)$$

where $C_1(u)$ is the power amplification required at the receiver to compensate for attenuation between transmitter and receiver in system 1. Ordinarily this is the same as the power amplification provided by the amplifier at the receiver.

In the case of broad-band analog messages, the ratio of interference power to message power in a band Ω_0 is of interest, and is given by

$$\mu_0 = \frac{\int_0^{\Omega_0} C(u)\ W_2(u - \omega_{12}) r_1(u)\ du}{\int_0^{\Omega_0} W_1^{(i)}(u) r_1(u)\ du}. \tag{11.8}$$

For two like systems of equal average power

$$\mu_0 = \frac{\int_0^{\Omega_0} C(u)\ G_1(u) W_m(u)\ du}{\int_0^{\Omega_0} W_m(u)\ du}, \tag{11.9}$$

where $W_m = r_1(u) W_1^{(i)}(u)$ is the message power spectrum at $\omega = u$.

Interference in single-sideband transmission would be coherent or intelligible in a channel occupying a band over which the coupling $C(u)$ is essentially constant. This applies even with a small difference in carrier frequencies, since this only results in a corresponding frequency shift in the interference spectrum, but not in pronounced distortion. (The situation is somewhat different in double-sideband AM systems, discussed in the next section.)

The coupling factors $K_{21}(\omega) = C(u)$ in the above relations are discussed in further detail in Section 11.9 for unshielded metallic pairs in close proximity, as in conventional multipair cable. The near-end coupling, that is, between oppositely directed transmission paths, is in this case much greater than far-end coupling between transmission paths in the same direction. Hence near-end coupling places a limitation on the maximum permissible number of systems in one cable. This limitation is removed by using separate cables for opposite directions of transmission, in which case far-end coupling between pairs in the same cable dictates the permissible number of systems. Such far-end coupling can be reduced by the use of balancing coils at repeaters [4], and the effect of the resultant interference can also be reduced with the aid of compandors [5].

Coupling between coaxial units is discussed in further detail in Section 11.10. It turns out that in this case far-end rather than near-end coupling places a limitation in the permissible number of systems or on maximum system length. A significant advantage of coaxial pairs is that opposite directions of transmission are feasible in one cable.

11.3 Interference between AM Systems

For reasons set forth in Section 2.16, amplitude modulation is used almost exclusively for radio broadcasting, but has been employed to advantage for some carrier communication system intended principally for routes 25 to 100 miles in length [5]. To insure greater flexibility in this application it is desirable to provide opposite directions of transmission in the same cable. To this end the frequency transposition as indicated in Fig. 11.2 has been used between one repeater section and the next, to eliminate near-end and interaction crosstalk. With this arrangement, far-end interference imposes a principal limitation on the permissible number of systems in one cable. In place of balancing coils, as used in single-sideband systems, syllabic compandors have been used in each voice channel to reduce the effect of far-end crosstalk, as well as additive random noise.

With amplitude modulation the combined received wave is given by

$$V_1(t) = A_1[1 + m_1 V_1(t)] \cos \omega_1 t + cA_2[1 + m_2 V_2(t)] \cos [\omega_1 t + \theta(t)].$$

(11.10)

The envelope of this wave is for $cA_2 \ll A_1$

$$\overline{V}(t) \simeq A_1[1 + m_1 V(t)] + cA_2[1 + m_2 V_2(t)] \cos \theta(t). \qquad (11.11)$$

In the particular case of a frequency difference ω_{12} between carriers, $\theta(t) = \omega_{12} t$ and the interference will contain a component $cA_2 \cos \omega_{12}(t)$ together with a component $cA_2 m_2 V_2(t) \cos \omega_{12} t$. With speech transmission, the modulation index must be small, because of the high peak factor of speech waves. Hence the interference from the steady-state component will predominate. However, if the frequency difference ω_{12} is fixed, this component can in principle be eliminated with the aid of a bandpass filter that is sufficiently narrow so that the resultant distortion of the signal spectrum can be disregarded.

When the two carrier frequencies are nominally the same, $\theta(t)$ will in general vary slowly and the resultant low-frequency component can be eliminated in the output. The intelligible interference will then be of the form $cA_2 m_2 V_2(t) \cos \theta(t)$, and will vary slowly with time, with insignificant distortion from the factor $\cos \theta(t)$. With some difference in carrier frequencies, distortion will be greater than in single-sideband transmission and will tend to render the interference unintelligible.

With amplitude modulation the sideband power spectra at the transmitter are as indicated in Fig. 11.7 and can be written

$$W_1(u) = \frac{1}{2}\left(\frac{A_1^2}{2}\right) \delta(u) + \tilde{W}_1(u), \qquad (11.12)$$

$$W_2(u - \omega_{12}) = \frac{1}{2}\left(\frac{A_2^2}{2}\right) \delta(u - \omega_{12}) + \tilde{W}_2(u - \omega_{12}), \qquad (11.13)$$

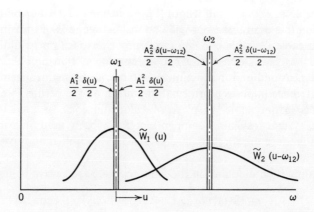

Figure 11.7 Signal power spectra in double sideband AM systems at transmitter outputs.

where $\delta(u)$ is the delta function. Integration of (11.12) between $u = -\omega_1$ and $u = \infty$ yields

$$\int_{-\omega_1}^{\infty} W_1(u) = \frac{A_1^2}{2} \delta(0) + \int_{-\omega_1}^{\infty} \widetilde{W}_1(u) \, du$$

$$= \frac{A_1^2}{2} + \tilde{P}_1, \tag{11.14}$$

where $A_1^2/2$ is the average carrier power and $\tilde{P}_1$ the combined average power of the sidebands.

The spikes in the signal power spectra (11.12) and (11.13) are translated into spikes in the spectrum of the demodulated wave at the baseband frequencies $u = O$ and $u = \omega_{12}$ as represented by

$$\frac{A_1^2}{2} \delta(u) \, r_1(O) + \frac{A_2^2}{2} C(u) \, \delta(u - \omega_{12}) \, r_1(\omega_{12}) \tag{11.15}$$

where $r_1(u)$ is the power transmittance of the receiving filter at $\omega = u$. In the time domain the first component represents a direct current and the second component a sine wave of frequency ω_{12}.

The continuous one-sided power spectrum of the baseband output at $\omega = u$ is given by

$$W_i^{(0)}(u) = 4\widetilde{W}_1^{(i)}(u) r_1(u) + C(u) \, \widetilde{W}_2(u - \omega_{12}) r_1(u)$$
$$+ C(-u) \, \widetilde{W}_2(-u - \omega_{12}) r_1(u). \tag{11.16}$$

Interference between FM Systems

The factor 4 rather than 2 comes about because the two sideband spectra are completely correlated and combine by direct amplitude addition in the demodulator output. The ratio of average interference power to average message power in a narrow band at $\omega = u$ is thus

$$\eta_0(u) = \tfrac{1}{4} C(u) \frac{\tilde{W}_2(u - \omega_{12})}{W_1^{(i)}(u)} + \tfrac{1}{4} C(-u) \frac{\tilde{W}_2(u + \omega_{12})}{W_1^{(i)}(u)}. \quad (11.17)$$

For equal carrier frequencies and power spectra, the interference ratio at $\omega = u$ becomes

$$\eta_0(u) = \tfrac{1}{4}[C(u)\, G_1(u) + C(-u)\, G_1(-u)], \quad (11.18)$$

where $G_1(u)$ is defined as in connection with (11.7). In (11.18) interference from each of the two sideband spectra is less by a factor $\tfrac{1}{2}$ than given by (11.7) for single-sideband transmission. This factor $\tfrac{1}{2}$ represents the average value of $\cos^2 \theta$ considering all possible phase relations between the two carriers.

Taking the continuous parts of the signal power spectra, the ratio of average interference power to average message power in a band Ω_0 at the baseband output becomes

$$\mu_0 = \frac{\int_{-\Omega_0}^{\Omega_0} C(u)\, \tilde{W}_2(u - \omega_{12}) r_1(u)\, du}{4 \int_0^{\Omega_0} \tilde{W}_1^{(i)}(u) r_1(u)\, du}. \quad (11.19)$$

With equal carrier frequencies and power spectra

$$\mu_0 = \frac{\int_0^{\Omega_0} \tfrac{1}{4}[C(u)\, G_1(u) + C(-u)\, G_1(-u)]\, W_m(u)\, du}{\int_0^{\Omega_0} W_m(u)\, du} \quad (11.20)$$

where $W_m(u) = \tilde{W}_1^{(i)}(u) r_1(u)$ is the message power spectrum.

11.4 Interference between FM Systems

Frequency modulation is ordinarily employed in microwave transmission, as used in multilink radio relays, in single-link troposcatter and satellite communication systems. Ordinarily several systems may be provided along the same routes in different transmission bands. With several systems of the same or different kinds along two different routes, for example as indicated in Fig. 11.4, the carrier frequencies of the systems along the two different routes will not in general be the same. If the carriers were unmodulated, the worst interference would arise if the difference in carrier frequencies were equal to or less than that of the maximum baseband frequency of the disturbed system, but greater than the lower cutoff frequency. Interference in this case will be greater than for the worst condition with fully modulated carriers, which is encountered when the carrier frequencies coincide. Determination of interference with unmodulated carriers is fairly simple and not considered here. Moreover, if interference with unmodulated carriers should be excessive it can be reduced by appropriate measures to insure nearly full carrier modulation at all times. Hence the following discussion is concerned with relations applying for modulated carriers.

In microwave systems the channel band is sufficiently narrow in relation to the carrier frequency, so that the coupling can be regarded as constant over the channel band. This simplifies the relations that follow, which can readily be generalized to a varying $C(u)$ in a manner indicated by previous relations for AM. This modification might be required in dealing with mutual interference between FM or PM systems in the same band on different metallic pairs.

In general the two systems will have different carrier frequencies and modulating waves with different power spectra and different frequency pre-emphasis. Thus one might be an FM system with a small frequency deviation ratio and the other a PM system with a large deviation ratio. When the power spectra of the modulated carriers are known, the power spectrum of the interference in the baseband output can be obtained by a convolution of the power spectra of the disturbed and the interfering carrier waves. This applies provided the disturbed and interfering signals are uncorrelated and provided the interfering signal is small in relation to the disturbed signal. On this premise the convolution procedure applies for interference into an FM or PM system with any pre-emphasis, regardless of the origin of the interfering power spectrum, and regardless of the statistical properties of the message or modulating waves in the disturbed and interfering systems. However, with FM or PM determination of the power spectra of the modulated carriers is difficult for modulating or message waves with continuous power spectra, except when the modulating waves have the statistical properties of a Gaussian random variable. In the latter case, the power spectra can be obtained from general relations (6.146) through (6.148), in which $R_\varphi(\tau)$ is given by (6.110).

For the particular case of two systems with the same carrier frequency, explicit relations are readily obtained by appropriate modifications in the relations for echo distortion.

Let $\varphi_1(t)$ and $\varphi_2(t)$ be the modulating waves of the disturbed and the interfering carriers. In the absence of correlation between $\varphi_1(t)$ and $\varphi_2(t)$, relation (6.95) is modified into

$$v(t) = \varphi_2(t) - \varphi_1(t), \qquad (11.21)$$

and (6.96) becomes

$$R_v(t) = R_{\varphi_1}(t) + R_{\varphi_2}(t). \qquad (11.22)$$

Interference between FM Systems

The power spectrum $W_i^{(0)}(u)$ of the interference in the demodulated message wave is obtained from (6.102), applying for equal carrier amplitudes, by taking

$$r^2 = r_{21}^2 = C \frac{A_2^2}{A_1^2}, \tag{11.23}$$

where A_1 and A_2 are the carrier amplitudes. Relation (6.102) represents the integral of the product of two autocorrelation functions, since $\exp R_v(t) = \exp R_{\varphi_1}(t) \cdot \exp R_{\varphi_2}(t)$. In accordance with relation (18) of Appendix 3, (6.102) can therefore alternately be written

$$W_i^{(0)}(u) = 4r^2 \int_{-\omega_1}^{\infty} W_1(\lambda) W_2(u + \lambda) \, d\lambda. \tag{11.24}$$

Here $W_i^{(0)}(u)$ is the single-sided power spectrum of the interference at the baseband output at $\omega = u$ while $W_1(u)$ and $W_2(u)$ are the double-sided power spectra of the modulated carriers.

When the carrier frequencies differ by $\omega_2 - \omega_1 = \omega_{12}$, relation (11.24) is modified into

$$W_i^{(0)}(u) = 4r^2 \int_{-\omega_1}^{\infty} W_1(\lambda) W_2(u - \omega_{12} + \lambda) \, d\lambda. \tag{11.25}$$

In general the power spectra contain a spike at each carrier frequency and a continuous component related to the modulating waves as indicated in Fig. 11.8. The normalized power spectra with $A_1 = A_2 = 1$ are

$$W_1(u) = \tfrac{1}{4} Q_1 \delta_1(u) + \tilde{W}_1(u), \tag{11.26}$$

$$W_2(u) = \tfrac{1}{4} Q_2 \delta_2(u - \omega_{12}) + \tilde{W}_2(u - \omega_{12}) \tag{11.27}$$

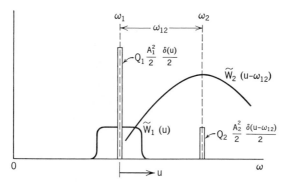

Figure 11.8 Signal power spectra in FM or PM systems at transmitter output. Factors Q_1 and Q_2 depend on frequency deviation ratios.

where Q_1 and Q_2 depend on the degree of carrier modulation and the subscripts for the delta function are for identification in equations below.

With (11.26) and (11.27) in (11.25) the following relation is obtained:

$$W_i^{(0)}(u) = \frac{r^2}{4} Q_1 Q_2 [\delta_1(O) \delta_2(u - \omega_{12}) + \delta_1(-u + \omega_{12}) \delta_2(O)]$$
$$+ 2r^2 Q_1 \delta_1(O) W_2(u - \omega_{12}) + 2r^2 Q_2 \delta_2(O) W_1(u - \omega_{12})$$
$$+ 4r^2 \int_{-\infty}^{\infty} \tilde{W}_1(\lambda) \tilde{W}_2(u - \omega_{12} + \lambda) \, d\lambda \quad (11.28)$$

or

$$W_i^{(0)}(u) = \frac{r^2}{2} Q_1 Q_2 \delta(u - \omega_{12}) + 2r^2 Q_1 \tilde{W}_2(u - \omega_{12}) + 2r^2 Q_2 \tilde{W}_1(u - \omega_{12})$$
$$+ 4r^2 \int_{-\infty}^{\infty} \tilde{W}_1(\lambda) \tilde{W}_2(u - \omega_{12} + \lambda) \, d\lambda. \quad (11.29)$$

As indicated in Fig. 11.9, the first term is a spike at $u = \omega_{12}$, that in the time domain corresponds to a sine wave of frequency ω_{12}. The second

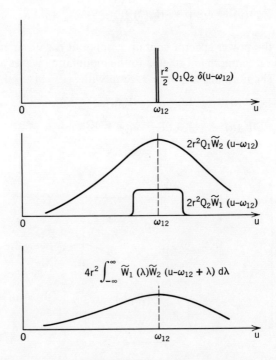

Figure 11.9 Four components of interference power spectrum at baseband output in FM or PM.

term represents the reduced continuous power spectrum of the interfering signal centered on ω_{12}. The third term represents the reduced continuous power spectrum of the disturbed carrier wave displaced in frequency by ω_{12} and thus centered on ω_{12}. The fourth term gives the interference power spectrum resulting from the convolution of the continuous power spectra of the two carriers.

The ratio of average interference power to average message power in a narrow band at $\omega = u$ in the disturbed system becomes

$$\eta_0(\omega) = \frac{W_i^{(0)}(\omega)}{W_{\varphi_1}(\omega)}, \qquad (11.30)$$

where $W_{\varphi_1}(\omega)$ is given by (6.109) or

$$W_{\varphi_1}(\omega) = \Delta_1^2 \frac{s_1(\omega) t_1(\omega)/\omega^2}{\int_0^{\Omega_0} s_1(\omega) t_1(\omega)\, d\omega} \qquad (11.31)$$

in which $s_1(\omega)$ is the message power spectrum and $t_1(\omega)$ the power transmittance of the transmitting filter.

The ratio of average interference power to average message power in the band Ω_0 is given by (6.122) with $W_i^{(0)}(\omega)$ in place of $W_\epsilon(\omega)$ or

$$\mu_0 = \frac{1}{\Delta_1^2} \left(\int_0^{\Omega_0} \frac{\omega^2 W_i^{(0)}(\omega)\, d\omega}{t_1(\omega)} \right) \frac{\int_0^{\Omega_0} s_1(\omega) t_1(\omega)\, d\omega}{\int_0^{\Omega_0} s_1(\omega)\, d\omega}. \qquad (11.32)$$

The above relations apply for any amplitude probability distribution of the message waves. However, explicit expressions for the power spectra of the modulated carriers are difficult to obtain for modulating waves with continuous power spectra. Published results for this case have been confined to modulating waves having the statistical properties of Gaussian random variables, as considered next.

11.5 Interference between FM Systems with Gaussian Modulation

With a Gaussian modulating wave $\varphi(t)$ the power spectrum of the modulated carrier is obtained from (6.148) for a carrier of unit amplitude. In accordance with the latter relation the expressions for Q and $\tilde{W}(u)$ appearing in (11.29) are

$$Q = e^{-R_\varphi(0)} \qquad (11.33)$$

$$\tilde{W}(u) = \frac{1}{2\pi} e^{-R_\varphi(0)} \int_0^\infty [\exp R_\varphi(\tau) - 1] \cos u\tau\, d\tau. \qquad (11.34)$$

The function $R_\varphi(\tau)$ for any frequency pre-emphasis is obtained from (6.110). For pure FM (6.112) applies and for pure PM (6.113).

Some special cases of (11.29) will now be considered. Maximum interference with fully modulated carriers is obtained when the carrier frequencies virtually coincide, and this condition is assumed here. In this case $\omega_{12} = O$ and (11.29) in conjunction with (11.33) and (11.34) yields

$$W_i^{(O)}(u) = \frac{r^2}{2} \exp - [R_{\varphi_1}(O) + R_{\varphi_2}(O)] + 2r^2 \tilde{W}_1(u) e^{-R_{\varphi_2}(O)}$$

$$+ 2r^2 \tilde{W}_2(u) e^{-R_{\varphi_1}(O)} + 4r^2 \int_{-\infty}^{\infty} \tilde{W}_1(\lambda) \tilde{W}_2(u + \lambda) \, d\lambda. \quad (11.35)$$

This relation is equivalent to (6.102) for echo distortion with $R_v(\tau) = R_{\varphi_1}(\tau) + R_{\varphi_2}(\tau)$ in accordance with (11.22). Following the same procedure as in relations (6.104) through (6.107), it is possible to write (11.35) in the alternate form

$$W_i^{(O)}(u) = \frac{r^2}{2} \exp - [R_{\varphi_1}(O) + R_{\varphi_2}(O)]$$

$$+ \frac{r^2}{2} [W_{\varphi_1}(u) + W_{\varphi_2}(u)] \exp - [R_{\varphi_1}(O) + R_{\varphi_2}(O)] + r^2 F(u)$$

(11.36)

where $F(u)$ is given by (6.107) with $R_v(\tau)$ defined in accordance with (11.22).

The power spectrum represented by the second term in (11.36) is of the same shape as the sum of the message power spectra in the disturbed and the interfering systems. These represent coherent interference, provided the phase difference $\theta(t)$ between the carriers varies at a slow rate compared to that of the fluctuations in the message power spectra.

The ratio of average coherent interference power to average message power in a narrow band at $\omega = u$ becomes

$$\eta_c(\omega) = \frac{r^2}{2} e^{-R_v(O)} \frac{W_{\varphi_2}(\omega)}{W_{\varphi_1}(\omega)}. \quad (11.37)$$

The ratio of average coherent power to average message power in a band Ω_0 becomes

$$\mu_c = \frac{r^2}{2} e^{-R_v(O)} \frac{\int_O^{\Omega_0} W_{\varphi_2}(u) \, du}{\int_O^{\Omega_0} W_{\varphi_1}(u) \, du}. \quad (11.38)$$

For adequately small rms deviation ratios $\underline{D}$, $R_v(O) \to O$. Furthermore, the continuous parts of the power spectra become so small that the convolution of these spectra as represented by the last term in (11.35) vanishes. The above interference ratios (11.37) and (11.38) are in this case the same as the corresponding ratios (11.17) and (11.19) for AM, with $\omega_{12} = O$.

The more general case will now be considered, in which the rms deviation ratios are sufficiently large so that intermodulation effects of the power spectra must be taken into account. In this case there is also a component of incoherent interference as represented by the last term in (11.36). The power spectrum of this component is obtained from (6.106) and (6.107) for equal power spectra of the disturbed and interfering signals. The ratio of average incoherent interference power to average message power at the frequency ω in the baseband output is obtained from (6.108). In the expression for $F(\omega)$, $R_v(\tau) = 2R_\varphi(\tau)$ for equal power spectra. Under this condition the curves in Fig. 6.10 and 6.11 apply for the ratio $\eta(\omega)$ of average incoherent interference power to average message power in a narrow band at $\omega = a\Omega_0$.

These curves apply for equal carrier frequencies, equal modulating waves and equal rms deviation ratios. Relations will now be given in which the modulating waves and the rms deviation ratios may differ.

As a first case let $\underline{D}_1 \ll 1$ and $\underline{D}_2 > 1$. The power spectrum $\tilde{W}_1(u)$ in this case tends to vanish, together with $\exp - R_{\varphi_2}(O)$. Hence (11.35) becomes in a first approximation, for $\underline{D}_1 \ll 1$, $\underline{D}_2 > 1$:

$$W_i^{(O)}(u) \simeq 2r^2 \tilde{W}_2(u) e^{-R_{\varphi_1}(O)} \simeq 2r^2 \tilde{W}_2(u). \tag{11.39}$$

Similarly, when $\underline{D}_1 > 1$ and $\underline{D}_2 \ll 1$

$$W_i^{(O)}(u) \simeq 2r^2 \tilde{W}_1(u) e^{-R_{\varphi_2}(O)} \simeq 2r^2 \tilde{W}_1(u). \tag{11.40}$$

With $\underline{D}_1 > 1$ and $\underline{D}_2 > 1$, $\exp - R_{\varphi_1}(O) \to O$ and $\exp - R_{\varphi_2}(O) \to O$. Hence (11.35) becomes in a first approximation

$$W_i^{(O)}(u) \simeq 4r^2 \int_{-\infty}^{\infty} \tilde{W}_1(u) \tilde{W}_2(u + \lambda) \, d\lambda. \tag{11.41}$$

In the latter case the double-sided power spectra as obtained with the aid of (6.149) are for both FM and PM.

$$\tilde{W}_1(u) = \frac{1}{(8\pi)^{1/2}} \frac{1}{\Delta_1} e^{-u^2/2\Delta_1^2}, \tag{11.42}$$

$$\tilde{W}_2(u) = \left(\frac{1}{8\pi}\right)^{1/2} \frac{1}{\Delta_2} e^{-u^2/2\Delta_2^2}. \tag{11.43}$$

When the above approximations are introduced in (11.30) and (11.32) the following relations are obtained for pure FM and pure PM in the disturbed system, regardless of whether FM or PM is used in the interfering system.

$\underline{D}_1 > 1$ *and* $\underline{D}_2 \ll 1$
Disturbed system pure FM:

$$\eta_0(\omega) = \frac{r^2 a^2}{(2\pi)^{1/2} \underline{D}_1^3} e^{-a^2/2\underline{D}_1^2}. \tag{11.44}$$

Disturbed system pure PM:

$$\eta_0(\omega) = \frac{r^2}{(2\pi)^{1/2} 3 \underline{D}_1^3} e^{-a^2/2\underline{D}_1^2}. \tag{11.45}$$

Disturbed system pure FM or PM:

$$\mu_0 = \frac{r^2}{(2\pi)^{1/2} 3 \underline{D}_1^2}, \tag{11.46}$$

where $a = \omega/\Omega_0$, $\underline{D}_1 = \underline{\Delta}_1/\Omega_0$, $\underline{D}_2 = \underline{\Delta}_2/\Omega_0$. Note that Ω_0 of the disturbed system is used in specifying $\underline{D}_2$.

$\underline{D}_1 \ll 1$ and $\underline{D}_2 > 1$:
Disturbed system pure FM:

$$\eta_0(\omega) \simeq \frac{r^2 a^2}{(2\pi)^{1/2} \underline{D}_1^2 \underline{D}_2} e^{-a^2/2\underline{D}_2^2}. \tag{11.47}$$

Disturbed system pure PM:

$$\eta_0(\omega) = \frac{r^2}{(2\pi)^{1/2} 3 \underline{D}_1^2 \underline{D}_2} e^{-a^2/2\underline{D}_2^2}. \tag{11.48}$$

Disturbed system pure FM or pure PM:

$$\mu_0 = \frac{r^2}{(2\pi)^{1/2} 3 \underline{D}_1^2 \underline{D}_2}. \tag{11.49}$$

$\underline{D}_1 > 1$ and $\underline{D}_2 > 1$
Disturbed system pure FM:

$$\eta_0(\omega) = \frac{r^2 a^2}{2\pi^{1/2} \underline{D}_1^2 \underline{D}_e} e^{-a^2/4\underline{D}_e^2} \tag{11.50}$$

Disturbed system pure PM:

$$\eta_0(\omega) = \frac{r^2}{2\pi^{1/2} 3 \underline{D}_1^2 \underline{D}_e} e^{-a^2/4\underline{D}_e^2} \tag{11.51}$$

Disturbed system pure FM:

$$\mu_0 = \frac{r^2}{(4\pi)^{1/2} 3 \underline{D}_1^2 \underline{D}_e} \tag{11.52}$$

Disturbed system pure PM:

$$\mu_0 = \frac{r^2}{4 \cdot 3 \underline{D}_1^2 \underline{D}_e} \tag{11.53}$$

where

$$\underline{D}_e^2 = \tfrac{1}{2}(\underline{D}_1^2 + \underline{D}_2^2). \tag{11.54}$$

It is interesting to note that relations (11.50) and (11.51) give the same results as relations (11.47) and (11.48) if in (11.54) $\underline{D}_1 \ll \underline{D}_2$; and gives the same results as (11.44) and (11.45) if in (11.54) $\underline{D}_1 \gg \underline{D}_2$. Thus, provided $\underline{D}_e > 1$, relations (11.50) through (11.54) yield approximate results for all cases. For $\underline{D}_e \ll 1$, relations (11.37) and (11.38) apply.

In the foregoing relations $r^2 = CA_2^2/A_1^2$ is the ratio of the interfering signal power $CA_2^2/2$ to the desired signal power $A_1^2/2$ at the input to the receiving filter of the disturbed systems. When the permissible ratios μ_0 and η_0 are specified, together with $\underline{D}_1$ and $\underline{D}_2$, it is possible to determine the required ratios r^2. For the important case of mutual interference of satellite communication and surface radio relay systems, the procedure is exemplified in further detail in Ref. 10.

11.6 Interference between Baseband Pulse Systems in Multipair Cable

In analog pulse transmission the average interference power is of principal interest, while in digital systems the probability distribution is also required for determination of error probability. The average interference power can be determined by the same methods as outlined previously for conventional analog modulation methods. To this end it is necessary to determine the power spectra $W_1(u)$ and $W_2(u - \omega_{12})$ of the disturbed and the interfering systems. This will be done here for baseband transmission with pulses of representative shape.

Baseband transmission entails the use of metallic transmission paths, such as balanced pairs or coaxial units. Interference between systems on different coaxial units diminishes with increasing frequency because of the increased shielding afforded by the outer coaxial conductor. Hence the transmission capacity of multiunit coaxial cable is not limited by interference. By contrast, the coupling between unshielded balanced pairs increases with frequency and mutual interference may place a limitation on the transmission capacity of multipair cable, whether used for analog or digital transmission. Explicit relations will be given here for far-end interference in transmission over balanced pairs for representative pulse shapes, and of the transmission capacity as limited by such interference.

Let pulses with a spectrum $S(i\omega)$ be transmitted at intervals T. As discussed in Section 4.7, the power spectrum will in general contain a discrete component of average power P_1 and a continuous component of average power P_0. The continuous part of the power spectrum is related to $S(i\omega)$ by

$$s(\omega) = \underline{a}^2 |S(i\omega)|^2, \tag{11.55}$$

where $\underline{a} = \underline{A}^2/\pi T$ and $\underline{A}$ is the rms peak amplitude of the pulses. In the case of a bipolar binary pulse train $\underline{A} = 1$.

The ratio μ_0 given by (11.9) becomes, with $\omega_1 = O$ as reference frequency, so that $C(u) = K(\omega)$. Thus

$$\mu_0 = \frac{\int_O^{\Omega_0} K(\omega) \, G_1(\omega) s(\omega) \, d\omega}{\int_O^{\Omega_0} s(\omega) \, d\omega}. \tag{11.56}$$

Here $s(\omega)$ corresponds to the message power spectrum in analog baseband transmission. $G_1(\omega)$ is defined as in connection with (11.7).

With a raised cosine spectrum of the received baseband pulses, the following relation applies for the amplitude of the spectrum

$$|S(i\omega)| = T \cos^2 \frac{\pi \omega}{2\Omega_0} \quad \text{for} \quad O < \omega < \Omega_0 \tag{11.57}$$

$$= O \qquad \omega > \Omega_0.$$

The pulse interval in this case is $T = 2\pi/\Omega_0$ and the integral of $|S(i\omega)|$ between O and Ω_0 is 1.

With (11.57) in (11.55) and (11.56)

$$\mu_0 = \frac{\int_O^{\Omega_0} K(\omega) \, G_1(\omega) \cos^4 (\pi \omega/2\Omega_0) \, d\omega}{\int_O^{\Omega_0} \cos^4 (\pi \omega/2\Omega_0) \, d\omega}. \tag{11.58}$$

In the case of interference between systems on balanced cable pairs, the factor $K(\omega) G_1(\omega)$ is of the following form

$$K(\omega) \, G_1(\omega) = k\omega^2. \tag{11.59}$$

The constant k is related to capacitive and inductive unbalances between cable pairs, as discussed further in Section 11.9.

With (11.59) in (11.58)

$$\mu_0 = k \frac{4\Omega_0^2}{\pi^2} \frac{\int_O^{\pi/2} x^2 \cos^4 x \, dx}{\int_O^{\pi/2} \cos^4 x \, dx}$$

$$= k \frac{\Omega_0^2}{6\pi^2} (2\pi^2 - 15) \simeq 0.078 k \Omega_0^2. \tag{11.60}$$

With m interfering systems the factor K is of the form

$$k = \sum_{i=1}^{m} K_i = m\underline{k}, \tag{11.61}$$

where $\underline{k}$ is the average value.

The above relations apply for analog as well as digital pulse transmission with any number of pulse amplitudes. The permissible maximum value of

μ_0 in digital transmission depends on the number of pulse amplitudes and on the probability distribution of the interference. With a large number n of interfering systems, a Gaussian probability distribution is approached, so that the interference power can be combined directly with that of additive Gaussian noise in determining error probability.

As discussed in Chapter 5, dicode transmission may be employed to reduce the adverse effect of low-frequency cutoff in baseband transmission. With this method, an on-off binary pulse train is transmitted, together with a pulse train of opposite polarity delayed by a pulse interval T. For equal probabilities of marks and spaces in each pulse train, the power spectrum of the combined pulse train is obtained by multiplying the power spectrum of the component pulse trains by $|1 - e^{-i\omega T}|^2 = 2(1 - \cos \omega T)$, where $T = 2\pi/\Omega_0$. Relation (11.60) is then modified into

$$\mu_0 = k4 \frac{\Omega_0^2}{\pi^2} \frac{\int_0^{\pi/2} x^2 \cos^4 x(1 - \cos 4x)\, dx}{\int_0^{\pi/2} \cos^4 x(1 - \cos 4x)\, dx}$$

$$= k\Omega_0^2 \frac{5}{15\pi^2}\left(\pi^2 - \frac{2}{3} - \frac{219}{32}\right)$$

$$\simeq .083k\Omega_0^2 = .083m\underline{k}\Omega_0^2 \qquad (11.62)$$

Comparison with (11.60) shows that μ_0 is increased by only 5 percent. However, with dicode transmission, the tolerable μ_0 for a given error probability is only half that with random bipolar transmission (Section 5.8).

From (11.60) and (11.61) can be determined the maximum permissible bandwidth, given μ_0, m, and $\underline{k}$. Thus

$$\Omega_0^2 = \frac{\mu_0}{0.078m\underline{k}}. \qquad (11.63)$$

For a specified number of pulse amplitudes r, the permissible value of μ_0 can be determined from (4.114) with μ_0^{-1} in place of ρ. Thus

$$\mu_0 = \frac{3}{\kappa^2(r^2 - 1)}. \qquad (11.64)$$

The corresponding channel capacity is

$$C = \frac{\Omega_0}{4\pi} \log_2 r^2 \qquad (11.65)$$

where $\Omega_0/4\pi = B_0/2 = B$, the mean bandwidth.

In the case of Gaussian noise the factor κ would be taken as about $\kappa = 4$, corresponding to an rms noise amplitude $\frac{1}{4}$ the maximum before an error occurs. It will be assumed here that $\kappa = 40$, that is, that crosstalk interference is 20 dB below the average power of random noise. It will further be assumed that there are $m = 100$ interfering systems and that $\underline{k} = 10^{-20}$ per radian per second. This corresponds to about 64 dB far-end crosstalk loss in 1 mile at 1 megacycle. The channel capacities per pair given in Table 1 are then obtained for various numbers of pulse amplitudes.

Table 1 *Channel Capacities of Multipair Cable as Limited by Far-End Interference for* $m = 100$, $\underline{k} = 10^{-20}$

r	μ_0	Ω_0	C bits/sec
2	$6.3 \cdot 10^{-4}$	$9 \cdot 10^7$	$14.1 \cdot 10^6$
4	$1.3 \cdot 10^{-4}$	$4 \cdot 10^7$	$12.8 \cdot 10^6$
8	$3 \cdot 10^{-5}$	$1.1 \cdot 10^7$	$5.3 \cdot 10^6$
64	$4.6 \cdot 10^{-7}$	$2.4 \cdot 10^6$	$2.3 \cdot 10^6$

This comparison applies for a single repeater section and shows that binary transmission affords the maximum channel capacity as limited by far-end interference. In addition binary transmission facilitates the use of regenerative repeaters, to prevent the cumulation of far-end interference along a repeater chain. However, from the standpoint of maximum repeater spacing in the presence of additive random noise, the preferable number of pulse amplitudes would be $r = 4$, as discussed previously in Section 4.13.

In determining the allowable value of μ_0, not only the mean coupling factor $\underline{k}$ must be considered, but also the probability distribution. As discussed in Section 11.9 it is possible for the coupling between individual pairs to exceed the above average $\underline{k}$ by a very large amount. In systems with a large number of regenerative repeaters in tandem, it is thus possible for interference to be excessive in a particular repeater section, though well below the tolerable values in other sections. Hence the allowable value of μ_0 must be kept rather small, as above. With non-regenerative repeaters, the average value $\underline{k}$ is used to obtain the cumulative interference in a long repeater chain. Far-end and near-end interference is discussed in further detail elsewhere, for a particular binary baseband system employing the equivalent of dicode transmission [6].

11.7 Interference between Digital PM Systems

As discussed in Chapter 5, pulses can be transmitted by amplitude, phase, or frequency modulation of carrier waves. In dealing with inter-

Interference between Digital PM Systems

ference in carrier pulse transmission, it is necessary to distinguish between transmission over facilities primarily intended for analog transmission and over facilities designed exclusively for digital transmission. By way of example various methods of carrier modulation and detection are used for digital transmission over voice channels. In this case far-end and near-end interference will be encountered between systems on different cable pairs. Such interference can, however, be disregarded in digital transmission for the reason that the tolerable interference with any practicable number of digital amplitudes is significantly greater than for analog transmission. This also applies when broad-band microwave relays designed for analog PM or FM are used for high-speed digital transmission with any practicable number of pulse amplitudes at the baseband input.

The present discussion will be concerned with interference between systems designed for digital transmission by carrier phase modulation. This is the more efficient and practicable method as applied to high-speed digital transmission over microwave channels as afforded by waveguides, radio relays, troposcatter links or communication satellites. In any one of these applications it may be desirable to provide a number of systems over adjacent frequency bands, as indicated in Fig. 11.10, with separate repeaters for each channel. The same arrangement may be used for multichannel transmission with a common amplifier for all channels. There may then be interchannel interference from other sources than considered here, such as attenuation and phase distortion over the transmission band or amplifier nonlinearity, as dealt with in previous chapters.

As discussed in Chapter 5, the more efficient practicable method of carrier pulse transmission is by phase modulation in conjunction with differential phase detection. This method requires the same channel bandwidth as double-sideband AM, but the phase modulation is accompanied by amplitude modulation. This may give rise to excessive unwanted phase

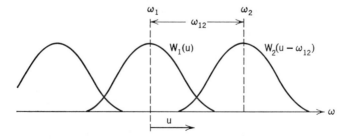

Figure 11.10 Spectrum overlaps with digital PM systems at carrier frequency intervals ω_{12}.

modulation in applications to channels having highly nonlinear repeaters with AM-PM conversion, particularly with a large number of repeaters in tandem. An alternate method that would be preferable in the latter case is pure PM in conjunction with frequency discriminator detection. This method entails an increase in the sideband spectrum and some unwanted amplitude and phase modulation will arise from channel bandwidth limitations. However, transmission impairments on the latter account will be less than with the first method. The first method will be considered here, since the sideband spectrum and intersystem interference can be determined more readily with this method.

With the arrangement in Fig. 11.10, there will in general be interference in a given system from two adjacent systems or channels. Evaluation of the ratio μ_0 will be exemplified here for an assumed raised cosine spectrum of the pulses at the detector input, that is, at the output of the receiving filter. With optimum division of spectrum-shaping between transmitting and receiving filter, the pulse spectrum at the input to the receiving filter will then have a cosine shape given by

$$S_1(u) = A_1 \frac{T}{2} \cos \frac{\pi u}{2\Omega_0}, \tag{11.66}$$

$$S_2(u - \omega_{12}) = A_2 \frac{T}{2} \cos \frac{\pi(u - \omega_{12})}{2\Omega_0} \tag{11.67}$$

where $T = 2\pi/\Omega_0$. The first relation applies for $-\Omega_0 < u < \Omega_0$ and the last relation in the frequency band $-\Omega_0 < u - \omega_{12} < \Omega_0$. Outside these bands the spectrum vanishes.

With the same bandwidth restrictions the amplitude characteristic of the receiving filter of system 1 is

$$R_1(u) = \cos \frac{\pi u}{2\Omega_0}. \tag{11.68}$$

With $C(u) = C$ relation (11.19) takes the form

$$\mu_0 = \frac{C \int_{-\Omega_0}^{\Omega_0} S_2^2(u - \omega_{12}) R_1^2(u) \, du}{4 \int_0^{\Omega_0} S_1^2(u) R_1^2(u) \, du} \tag{11.69}$$

$$= \frac{CA_2^2}{A_1^2} \frac{1}{3} \left(\left(1 - \frac{\omega_{12}}{2\Omega_0}\right)\left(1 + \frac{1}{2} \cos \frac{\pi \omega_{12}}{\Omega_0}\right) + \frac{3}{\pi} \sin \frac{\pi \omega_{12}}{\Omega_0} \right). \tag{11.70}$$

This ratio is shown in Fig. 11.11 as a function of $\omega_{12}/\Omega_0 = \omega_{12}T/2\pi = f_{12}T$.

With a Gaussian pulse spectrum and optimum division of channel shaping between transmitting and receiving filters, (11.66) through (11.70)

Interference between Digital PM Systems

are modified into

$$S_1(u) = A_1 T \left(\frac{1}{2}\left(\frac{c}{\pi}\right)^{1/2}\right)^{1/2} e^{-bu^2/2\Omega_0^2} \tag{11.71}$$

$$S_2(u - \omega_{12}) = A_2 T \left(\frac{1}{2}\left(\frac{c}{\pi}\right)^{1/2}\right)^{1/2} e^{-b(u-\omega_{12})^2/2\Omega_0^2} \tag{11.72}$$

$$R_1(u) = e^{-bu^2/2\Omega_0^2} \tag{11.73}$$

$$\mu_0 = C \frac{A_2^2}{A_1^2} \frac{1}{2\sqrt{2}} e^{-b\omega_{12}^2/2\Omega_0^2}. \tag{11.74}$$

The constant b depends on intersymbol interference allowed between adjacent pulses at sampling instants. With 1 percent intersymbol interference it follows from (1.85) with $\Omega_0 = 2\Omega$ that $b = 4 \times 0.54 = 2.16$. On this premise the ratio μ_0 is as shown in Fig. 11.11.

The maximum ratio μ_0 depends on the number of digital levels or corresponding phase positions and also on the allowed interference power. Let

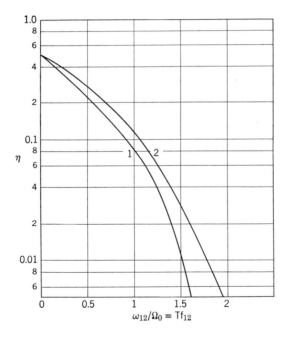

Figure 11.11 Interference in digital PM with overlapping pulse spectra as in Fig. 11.10, for equal division of spectrum shaping between transmitting and receiving filters. 1. Raised cosine pulse spectrum at output of receiving filter. 2. Gaussian pulse spectrum with 1 % intersymbol interference from each adjacent pulse.

the allowed average interference power be $\frac{1}{4}$ that of the average additive noise powers, so that the combined interference from two adjacent channels is 3 dB below the average noise power. For an error probability 10^{-6} owing to additive random noise, the values of μ_0 given in the table below are then obtained with the aid of Table 2 of Chapter 5. In the same table are given the corresponding ratios ω_{12}/Ω_0 obtained from Fig. 11.11 for raised cosine and Gaussian pulse spectra.

Table 2 *Ratios μ_0 and ω_{12}/Ω_0 for Average Interference Power 6 dB below Additive Random Noise Power*

$r =$	2	4	8	Phases
$\mu_0 =$ 0.023		0.011	0.0033	
$\omega_{12}/\Omega_0 \simeq$ 1.4		1.5	1.65	(Raised cosine)
$\omega_{12}/\Omega_0 =$ 1.6		1.75	2.10	(Gaussian)

The above results apply for digital PM systems with synchronous or differential phase detection. These require the same bandwidth as doublesideband AM systems, but the phase modulation is accompanied by envelope fluctuations. With pure PM and frequency-discriminator detection a greater bandwidth will be required, as discussed in Section 7.14, particularly for large values of r. The required ratios ω_{12}/Ω_0 will then be somewhat greater than given in Table 2.

In the above example 1 percent intersymbol interference from a single pulse was assumed, taken in relation to the peak pulse amplitude. With r amplitudes, or phase positions, peak intersymbol interference from two adjacent pulses is then $2(r-1)$ percent, when taken in relation to the maximum tolerable distortion from all sources. Thus, with $r = 8$, 14 percent peak distortion is obtained, which is not an excessive value. It is, of course, possible to reduce intersymbol interference by widening the pulse spectrum in exchange for increased intersystem interference, and to determine an optimum channel spacing. An analysis of this kind has been made for various filter characteristics in connection with binary pulse transmission with envelope detection over wave guides [11].

11.8 Interference into Analog PM from Digital PM Systems

The relations given in Section 11.4 apply for interference into an analog FM or PM system, for known power spectra of the disturbed and interfering systems. The relations are valid regardless of the origin of the interfering power spectrum, and are thus applicable to interference into an analog FM or PM system from a digital system of any kind. More specific relations are given here, for interference from a digital PM system into an analog FM or PM system when the modulating waves of the latter have the properties of Gaussian random variables. In the latter case the more special relations in Section 11.5 apply, when $W_2(u - \omega_{12})$ represents the power spectrum of the interfering digital PM system.

When the interfering system employs digital PM with differential phase detection, the carrier is suppressed and the sideband power spectrum is the same as for AM. For purposes of numerical evaluation, it is convenient and sufficiently accurate to assume that the sideband spectrum is of the same shape as (11.71) for digital transmission. With equal division of channel-shaping between transmitting and receiving filters, the amplitude spectrum of the envelope of the transmitted signal becomes

$$S_2(u - \omega_{12}) = A_2 T_2 \left(\frac{1}{2}\left(\frac{b}{\pi}\right)^{1/2}\right)^{1/2} e^{-b(u-\omega_{12})^2/2\Omega_2^2} \qquad (11.75)$$

where $T_2 = 2\pi/\Omega_2$ is the interval between pulses and the constant b depends on intersymbol interference allowed at sampling points. The peak amplitude of the envelope of the received carrier wave at the output of the receiving filter prior to demodulation is

$$\frac{1}{\pi}\int_{-\infty}^{\infty} S_2^2(u - \omega_{12})\, du = A_2. \qquad (11.76)$$

The power spectrum of the transmitted wave becomes

$$W_2(u - \omega_{12}) = \frac{1}{\pi T_2} S_2^2(u - \omega_{12})$$

$$= \frac{A_2^2}{2} \frac{T_2}{2\pi} \cdot \left(\frac{b}{\pi}\right)^{1/2} e^{-b(u-\omega_{12})^2/\Omega_2^2}. \qquad (11.77)$$

The corresponding average signal power is

$$\int_{-\infty}^{\infty} W_2(u - \omega_{12})\, du = \frac{A_2^2}{2}. \qquad (11.78)$$

The envelope power spectrum is $2W_2(u - \omega_{12})$ and the average envelope power is A_2^2. Comparison of (11.77) with (11.43) shows that the relations in Section 11.5 for a Gaussian power spectrum of the interfering signal apply provided that the following substitutions are made:

$$\underline{\Delta}_2 = \frac{\Omega_2^2}{2b} \qquad (11.79)$$

$$\underline{D}_2 = \frac{\underline{\Delta}_2}{\Omega_0} = \left(\frac{\Omega_2}{\Omega_0}\right)\left(\frac{1}{2b}\right)^{1/2}. \qquad (11.80)$$

With 1 percent intersymbol interference at sampling instants from each adjacent pulse, $b = 2.16$ and the equivalent value of $\underline{D}_2$ to be used in relations (11.47) through (11.54) becomes

$$\underline{D}_2 \simeq \frac{0.5\Omega_2}{\Omega_0}. \tag{11.81}$$

With a Gaussian power spectrum as assumed above, the equivalent mean input bandwidth of the receiving filter is

$$\Omega_i = 2\pi B_i = \left(\frac{\pi}{b}\right)^{1/2} \Omega_2$$

$$\simeq 1.2\Omega_2. \tag{11.82}$$

The required ratio ρ_i of average signal power to average noise in the band Ω_i at the input to the receiving filter depends on the number of phases and on the demodulation method. For a probability $\epsilon_p = 10^{-6}$ of an error in a digit, the above ratio ρ_i is obtained from relations given in Chapter 5 and is about as follows

Table 3 *Ratio ρ_i for Error Probability 10^{-6}*

Number of Phase Positions:	2	4	8
Synchronous Detection:	11	22	77
Differential Phase Detection:	14	48	150
The ratios in dB are $10 \log_{10} \rho_i$			

With pure PM and frequency-discriminator detection, relation (11.80) applies in a first approximation, but the required ratio ρ_i is about 22, rather than 11 with synchronous detection. With multiphase PM the sideband power spectrum is approximated by (11.77) provided the maximum phase excursion is limited to $\hat{\theta} < \pm\pi/2$, as discussed in Sections 7.13 and 7.14. Hence relation (11.80) also applies in this case, but the required ratio ρ_i will then be significantly greater than for synchronous or differential phase detection.

By way of numerical illustration, let the interfering system be a binary 7 digit PCM system with the same bandwidth Ω_0 of the message wave as the disturbed system. In this case $\Omega_2 = 14\Omega_0$ and the equivalent $\underline{D}_2$ obtained from (11.81) is $\underline{D} = 7$.

For comparison, consider an interfering analog PM system. To insure operation above threshold, the required ratio ρ_i is in this case about 10. In order that the ratio of peak message power to average noise power in the baseband output be the same as for $2^7 = 128$ quantizing levels in the PCM system, that is, $3 \cdot 128^2 \simeq 50{,}000$, it is necessary that $\hat{D}_2 \simeq 12$. The required bandwidth Ω_i is in this case $\Omega_i \simeq 2(1 + \hat{D}_2)\Omega_0 \simeq 26\Omega_0$, as compared with $14\Omega_0$ for the PCM systems. In order to insure operation above threshold it is necessary to have a ratio $\rho_i \simeq 10$, which is obtained provided the average signal power at the receiving filter input is greater than for the PCM system by a factor 1.3, or about 1 dB.

With a message wave having a Gaussian amplitude probability distribution, the rms deviation ratio in analog FM or PM would be $\underline{D}_2 \simeq \hat{D}_2/4 \simeq 3$, as compared with an equivalent ratio $\underline{D}_2 = 7$ for the PCM system. Relations (11.47) through (11.49) show that in this case interference into an analog FM or PM system with $\underline{D}_1 \ll 1$ would be greater with an interfering analog FM or PM system than with an interfering PCM system, by a factor $(7/3) \times 1.3 \simeq 3$ or about 5 dB. For $\underline{D}_1 > 1$, relations (11.50) through (11.54) show that interference from an analog system would be greater by about 4.2 dB.

The foregoing comparisons apply for a message wave with a Gaussian amplitude probability distribution. The principal reason for the greater interference from the analog PM system in this case is the relatively small rms deviation ratio $\underline{D}_2 \simeq \hat{D}_2/4$. This results in a Gaussian signal power spectrum with narrower bandwidth than for binary PM, with higher spectral density near the carrier for the same average signal power. With somewhat greater rms deviation ratios, interference from an analog PM system would be reduced, but would be greater than from a binary PM system.

11.9 Coupling Factors of Twisted Pairs

To reduce mutual coupling, pairs on open-wire lines and in cable are usually transposed or twisted, with the twist length differing among pairs. In spite of these measures, some residual coupling is encountered, because of the finite twist lengths and small residual mutual admittances $\Delta Y(x)$ and impedance $\Delta Z(x)$. As indicated in Fig. 11.12, the resultant far-end and near-end currents are given by*

$$\Delta i_f(x) = i\omega\left(\frac{K}{2}\Delta C(x) - \frac{1}{2K}\Delta L(x)\right)I(O)e^{-\Gamma l} \qquad (11.84)$$

$$\Delta i_n(x) = i\omega\left(\frac{K}{2}\Delta C(x) + \frac{1}{2K}\Delta L(x)\right)I(O)e^{-2\Gamma x} \qquad (11.85)$$

* The mutual admittance $\Delta Y(x)$ is $\frac{1}{4}$ the direct admittance unbalance used in Ref. 1.

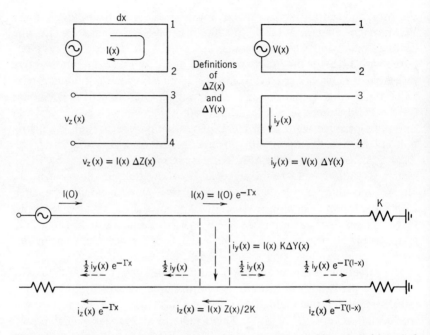

Figure 11.12 Far-end and near-end interference between cable pairs of length l owing to mutual admittances $\Delta Y(x)$ and mutual impedances $\Delta Z(x)$. Far-end interference current: $\Delta i_f = [I(O)/2][K\,\Delta Y - \Delta Z/K]\,e^{-\Gamma l}$. Near-end interference current: $\Delta i_n = -[I(O)/2][K\,\Delta Y + \Delta Z/K]\,e^{-2\Gamma x}$

where $K = (ZY)^{1/2}$ is the characteristic impedance of the pairs and it is assumed that the pairs are terminated in their characteristic impedances. It is further assumed that resistance and conductance unbalances can be neglected, so that $\Delta Y = i\omega \Delta C$ and $\Delta Z = i\omega \Delta L$.

For two particular pairs, the far-end and near-end coupling factors obtained by integration over the length l of the pairs become

$$K_f = \frac{i_f}{I(O)} = i\omega e^{-\Gamma l} \int_0^l \left(\frac{K}{2}\Delta C(x) - \frac{1}{2K}\Delta L(x)\right) dx, \tag{11.86}$$

$$K_n = \frac{i_n}{I(O)} = i\omega \int_0^l \left(\frac{K}{2}\Delta C(x) + \frac{1}{2K}\Delta L(x)\right) e^{-2\Gamma x}\, dx \tag{11.87}$$

where the unit length propagation constant Γ is given by

$$\Gamma = [(R + i\omega L)(G + i\omega C)]^{1/2}$$
$$= \alpha + i\beta. \tag{11.88}$$

Coupling Factors of Twisted Pairs

At sufficiently high frequencies so that coupling need be considered, the resistance R will vary as $(i\omega)^{\frac{1}{2}}$ and the attenuation as

$$\alpha = \alpha_0(\omega)^{\frac{1}{2}}. \qquad (11.89)$$

In (11.86) and (11.87), $\Delta C(x)$ and $\Delta L(x)$ will vary at random with x, with virtually no correlation between unbalances a short distance apart (say 1 ft.). On this premise it can be established that the following relations apply for the far-end and near-end power coupling factors of two particular pairs of length l

$$K_f(\omega) = k_f \omega^2 l e^{-2\alpha l} \qquad (11.90)$$

$$K_n(\omega) = k_n \omega^{3/2} l. \qquad (11.91)$$

The corresponding "equal-level" far-end power coupling factor expressed in dB is

$$X_f(\omega) = 10 \log_{10} K_f(\omega) e^{2\alpha l}$$
$$= \boxed{10 \log_{10} (k_f \omega^2 l)}. \qquad (11.92)$$

The near-end power coupling factor in dB is

$$X_n(\omega) = 10 \log_{10} K_n(\omega)$$
$$= \boxed{10 \log_{10} k_n \omega^{3/2} l}. \qquad (11.93)$$

The corresponding coupling losses are $-X_f(\omega)$ and $-X_n(\omega)$.

The foregoing variations with frequency conform well with results of measurements on various kinds of cable. These measurements also indicate a large variation in coupling between different pairs, and that the probability distribution of the coupling factors follows closely the log-normal law discussed in Chapter 2 and in Appendix 2.

The probability that X exceeds a certain specified value X_1 is thus

$$P(X \geq X_1) = \frac{1}{2}\left(1 - \operatorname{erf} \frac{X_1 - X_0}{\sqrt{2}\sigma_x}\right) \qquad (11.94)$$

when X_0 is the mean value of X_f or X_n, as the case may be, and σ_x is the standard deviation in dB.

From measurements of the mean value X_0 and the standard deviation σ_x it is possible to determine the corresponding average coupling factor $\overline{K}(\omega)$ or $\overline{X}(\omega)$ to be used in calculations of interference. Thus, with the aid of relation (2.18)

$$\overline{X} = X_0 + 0.115\sigma_x^2. \qquad (11.95)$$

By way of numerical example, the following values apply at a frequency of 3 mc, $(\omega = 6 \cdot \pi \cdot 10^6)$ for crosstalk coupling within a 50 pair unit of a

22 gauge paper insulated cable with a capacitance of 0.084 mf/mile (1 mile = 1.61 km).

Table 4 *Coupling Factors in dB at 3 Mc for a Length of One Mile*

	X_0	σ_x	$\underline{X}$
Far-End (Eq. Lev.)	−52	9	−43
Near-End	−68	10	−56

The values of $\underline{K}_f e^{2\alpha l}$ and $\underline{K}_n$ corresponding to the above values of $\underline{X}$ are obtained from (11.92) and (11.93).

For a cable of the foregoing type, the attenuation at a frequency of 3 Mc is about 50 dB/mile. Thus, with a repeater spacing of one mile, average near-end interference power would be only 6 dB below the received signal power for a single disturber and a single repeater section, while average far-end interference would be 43 dB below the signal power.

In cable carrier systems, frequencies much lower than 3 Mc would be used, and separate cables would be employed for opposite directions of transmission to avoid excessive near-end crosstalk. For example, if the frequency were 100 kc, the average far-end coupling $\underline{X}$ would be less than given in Table 4 by $20 \log_{10} 30 \simeq 30$ dB. In this case the repeater spacing with 50 dB loss would be about 5.5 miles. With $n = 50$ repeater sections and $m = 20$ disturbers, the average far-end crosstalk interference would be 43 dB below the signal power. This assumes the same average signal power for systems on different pairs.

In the case of speech transmission, the average power will vary markedly among individuals, as discussed in Section 2.2. For this reason, and because of the large variation in coupling between pairs, there will be pronounced departures from the above average interference among systems on different pairs. Moreover, for a given system, large fluctuations in interference will be encountered, owing to the pronounced variation in speech power among individuals. These variations must be recognized in specifying the permissible average interference power determined on the premise of average coupling for all pairs and the average signal power of all talkers.

11.10 Coupling Factors of Coaxial Lines

Coupling between coaxial units comes about because of the finite conductivity of the outer return conductor of each coaxial. Let $k = k(\omega)$

be the unit-length current or voltage transfer factor of two coaxial lines terminated in their characteristic impedance K. Relations (11.86) and (11.87) are then replaced by

$$\frac{i_f}{I(O)} = k(\omega)le^{-\Gamma l} \tag{11.96}$$

$$\frac{i_n}{I(O)} = \int_0^l k(\omega)e^{-2\Gamma x} = k(\omega)l\frac{1 - e^{-2\Gamma l}}{2\Gamma l}$$

$$= k(\omega)le^{-\Gamma l}\frac{\sinh \Gamma l}{\Gamma l}. \tag{11.97}$$

The ratio of near-end to far-end coupling factors is accordingly

$$\frac{K_n(\omega)}{K_f(\omega)} = \left(\frac{\sinh \Gamma l}{\Gamma l}\right)^2 \tag{11.98}$$

$$= \frac{1}{2}\frac{\cosh 2\alpha l + \sin^2 \beta l - \cos^2 \beta l}{(\alpha^2 + \beta^2)l^2}, \tag{11.99}$$

where the last relation is obtained with $\Gamma = \alpha + i\beta$.

The factor $k(\omega)$ depends on the construction of the outer coaxial conductor or shield, on the number of coaxial units and their configuration. Exact analytical determination is difficult except for the simple case of two adequately separated coaxials [3]. Hence $k(\omega)$ is ordinarily determined by measurements on short samples of the configuration under consideration. However, with any configuration, $k(\omega)$ diminishes exponentially with increasing frequency, so that interference at low frequencies is of principal importance. To obtain adequate shielding at low frequencies, a combination of copper and permalloy tape has been used on coaxial systems where the lowest message frequency is about 30 c/s. By using an adequately high lower frequency it is possible to avoid excessive interference without the need for a substantial return conductor, in exchange for a somewhat greater top frequency and some reduction in repeater spacing.

It turns out that at low frequencies ratio (11.99) is substantially less than 1 for the reason that $\alpha \ll i\beta$. Thus far-end crosstalk interference prevails over near-end interference, in a single-repeater section as assumed above.

With n repeaters in tandem, far-end interference will increase by direct amplitude addition, provided the coaxials and repeaters are identical so that there is no phase difference between the transmittances of parallel coaxial lines. This is a valid approximation in the more important lower frequency range. The average far-end interference power thus increases as n^2, whereas average near-end interference power is proportional to n.

Thus with n repeaters in tandem and m disturbers, the combined power transfer factors become

$$K_f(\omega)e^{2\alpha l} = k^2(\omega)l^2 n^2 m, \tag{11.100}$$

$$K_n(\omega)e^{2\alpha l} = k^2(\omega)\left|\frac{\sinh \Gamma l}{\Gamma l}\right|^2 l^2 nm. \tag{11.101}$$

Since far-end interference will prevail over near-end interference even for a single repeater section, $n = 1$, it is evident that near-end interference becomes progressively less important the greater the number of repeater sections. Hence it is feasible to provide opposite directions of transmission over coaxial units in the same cable. Moreover, since the coupling factor diminishes with increasing frequency, coaxial units are particularly suited to broadband transmission in that intersystem interference imposes no limitation on maximum bandwidth.

Appendix 1
Basic Transform Concepts and Relations

In transmission and modulation theory extensive use is made of integral transforms in one form or another. This technique is embodied in Fourier and Laplace transforms as employed in network and transmission theory in translating from time functions to their spectra or conversely. In statistical analysis the same technique is encountered in the translation from probability density functions to their spectra or characteristic functions and also in the conversion from power spectra to correlation functions, and vice versa. Moreover, this method is used in statistical analysis of distortion owing to nonlinear devices, where in essence the spectrum of the transfer function of the device or communication channel is determined by the transform method and combined with the spectrum of the probability density function.

Various important transforms are reviewed here, together with certain basic propositions derived from them that are often encountered in transmission and modulation theory. (Bibliography on page 467.)

1. Fourier Integral Transforms

Fourier integral transforms are used in various analytical disciplines, as in boundary value problems in mathematical physics, in electric circuit theory or in probability theory. Let $g(x)$ be a function of a variable x, the nature of which will depend on the particular application. In boundary-value problems, x may denote a spatial coordinate, or a spatial derivative of such a coordinate. In electric circuit theory, x may denote time or a derivative with respect to time, while in probability theory, x may denote the amplitude of some quantity or its derivative with respect to amplitude.

Subject to certain restrictions that are necessary to insure the convergence of integrals, which are satisfied in applications to physical problems, the function $g(x)$ can be represented by the following Fourier integral transform of the function $f(y)$

$$g(x) = \frac{1}{2\pi} \int_{-\infty}^{\infty} f(y) e^{ixy} \, dy \tag{1}$$

where

$$f(y) = \int_{-\infty}^{\infty} g(x) e^{-ixy} \, dx \tag{2}$$

The first integral can be multiplied by an arbitrary constant c if the second integral is multiplied by $1/c$, and in some cases c is chosen as $c = (2\pi)^{1/2}$ to give the transforms a symmetrical appearance.

In the foregoing relations both $g(x)$ and $f(y)$ could be complex functions, provided that appropriate restrictions are imposed to make the integrals converge or to insure proper definition of their meaning at points of discontinuity in $g(x)$ or $f(y)$. However, in applications to ordinary problems $g(x)$ is a real function, in which case $f(y)$ in general is a complex function. The complex function $f(y)$ is usually referred to as the spectrum of the real function, a designation that originates from application to optics.

The complex function $f(y)$ can be represented as

$$f(y) = U(y) - iV(y) \tag{3}$$

$$= a(y) e^{-i\beta(y)} \tag{4}$$

$$= e^{\alpha(y) - i\beta(y)}, \tag{5}$$

where $U(y)$, $a(y)$, and $\alpha(y)$ must be even functions of y while $V(y)$ and $\beta(y)$ must be odd functions of y in order that $g(x)$ be a real function.

Relations (1) and (2) can now be written

$$g(x) = \frac{1}{\pi} \int_0^{\infty} U(y) \cos xy \, dy + \frac{1}{\pi} \int_0^{\infty} V(y) \sin xy \, dy \tag{6}$$

$$= \frac{1}{\pi} \int_0^{\infty} a(y) \cos[xy - \beta(y)] \, dy \tag{7}$$

where

$$U(y) = a(y) \cos \beta(y) = \int_0^{\infty} [g(x) + g(-x)] \cos xy \, dx, \tag{8}$$

$$V(y) = a(y) \sin \beta(y) = \int_0^{\infty} [g(x) - g(-x)] \sin xy \, dx. \tag{9}$$

If $g(x)$ is an even function $g(x) - g(-x)$ vanishes so that $V(y)$ and $\beta(y) = O$. For an odd function $g(x)$, $U(y)$ vanishes and $\beta = \pi/2$. In general $g(x)$ can be regarded as the sum of an even and odd function. In some applications physical considerations require that $g(-x)$ be zero, as discussed in the next section.

2. Laplace Integral Transforms

In application of Fourier transforms to electric circuits or other dynamical systems, the variable $x = t$ represents the time and $y = \omega = 2\pi f$ represents the radian frequency of a cisoidal oscillation $e^{i\omega t}$. In this case $g(t)$ represents a force and the time origin can be so chosen that $g(t) = 0$ for $t < 0$. This condition must then also prevail in the Fourier integral representation of $g(t)$. From (6) it follows that this will be the case provided for $t \geq 0$.

$$\int_0^\infty U(\omega) \cos \omega t \, d\omega = \int_0^\infty V(\omega) \sin \omega t \, d\omega, \qquad (10)$$

and (6) can then be written

$$g(t) = \frac{2}{\pi} \int_0^\infty U(\omega) \cos \omega t \, d\omega \qquad (11)$$

$$= \frac{2}{\pi} \int_0^\infty V(\omega) \sin \omega t \, d\omega, \qquad (12)$$

where now (8) and (9) become

$$U(\omega) = 2 \int_0^\infty g(t) \cos \omega t \, dt, \qquad (13)$$

$$V(\omega) = 2 \int_0^\infty g(t) \sin \omega t \, dt. \qquad (14)$$

With the restriction imposed above, that $g(t) = g(x)$ is of such nature that an origin for x or t can be so chosen that $g(t) = 0$ for $t < 0$, the lower limit in (2) can be replaced by 0. With the further introduction of the new variable $p = i\omega$ the integrals take the appearance of Laplace transforms. Ordinarily such transforms are written in the more general form

$$g(t) = \frac{1}{2\pi i} \int_{c-i\infty}^{c+i\infty} f(p) e^{pt} \, dp, \qquad (15)$$

$$f(p) = \int_0^\infty g(t) e^{-pt} \, dt, \qquad (16)$$

with proper choice of c to insure convergence.

The above condition $g(x) = g(t) = 0$ for $x < 0$, with appropriate choice of origin, cannot be introduced for all functions $g(x)$. For example with $g(x) = \sin x/x$ it is not possible to choose a finite negative origin for the ordinate x.

3. Hilbert Integral Transforms

The two integrals in (10) are known as conjugate integrals. From (10) it is possible to derive explicit relations between $U(\omega)$ and $V(\omega)$ that hold for Laplace transforms and must be imposed in application of Fourier transforms to electric circuits. The following relations apply and are known as Hilbert transforms

$$U(\omega) = -\frac{1}{\pi}\int_{-\infty}^{\infty} \frac{V(z)}{\omega - z} dz, \quad (17)$$

$$V(\omega) = \frac{1}{\pi}\int_{-\infty}^{\infty} \frac{U(z)}{\omega - z} dz, \quad (18)$$

$$\alpha(\omega) = -\frac{1}{\omega}\int_{-\infty}^{\infty} \frac{\beta(z)}{\omega - z} dz, \quad (19)$$

$$\beta(\omega) = \frac{1}{\omega}\int_{-\infty}^{\infty} \frac{\alpha(z)}{\omega - z} dz, \quad (20)$$

where $\alpha(\omega) = \ln a(\omega)$ and $\ln = \log_e$ and where the principal values of the integrals are to be used. That is, the path of integration is indented around poles on the real axis [6]. Thus results of the form $\ln(-z)$ are to be taken as $\ln |-z|$ rather than as $\ln |-z| + i\pi$. The simplest examples of Hilbert transforms are $\cos \omega t$ and $\sin \omega t$. Other examples are the real and imaginary components of network impedances and transfer characteristics, often designated in-phase and quadrature components.

The amplitude $a(\omega)$ is not modified if the function $f(\omega)$ is multiplied by a factor

$$e^{-i\beta_0(\omega)} = \frac{1 - if_0(\omega)}{1 + if_0(\omega)}, \quad (21)$$

where $f_0(\omega)$ is a real function. For this reason, the phase characteristic cannot be uniquely determined from (20) but may contain a component $\beta_0(\omega)$. The phase function obtained from (20) is known as the minimum phase function. An additional positive component $\beta_0(\omega)$ can be present when $a(\omega)$ represents the transfer characteristic of an electrical or mechanical transducer that may serve as a communication channel. Such an additional component can be provided with the aid of so called lattice structures, which have a transfer characteristic as represented by the right-hand side of (21) and are often used for purposes of phase correction to insure increased phase linearity.

4. Convolution Integral Transforms

In various applications of linear analysis where the superposition theorem applies, so-called convolution integrals of the following form are encountered:

$$g_{1,2}(x) = \int_{-\infty}^{\infty} g_1(z) g_2(x - z) \, dz \tag{22}$$

$$= \int_{-\infty}^{\infty} g_1(x - z) g_2(z) \, dz. \tag{23}$$

For example, $g_1(x)$ and $g_2(x)$ may be two independent probability density functions, in which case $g_{1,2}(x)$ is the combined probability density function. Or, $g_1(x)$ may represent the current at a distance x in a conductor, owing to unit electric force applied over an infinitesimal interval; in which case $g_{1,2}(x)$ represents the current in response to an electric force varying with distance as $g_2(x)$. In applications to boundary value problems $g_1(x - z)$ is referred to as Green's function of the problem and often denoted $K(x - z)$.

In application to electric circuit analysis $g_2(x) = g_2(t)$ can be regarded as a wave, which can be assumed to be made up of rectangular pulses of infinitesimal duration, $dx = dt$. The function $g_1(x) = g_1(t)$ would then represent the response at the time t, owing to such a rectangular pulse, while $g_{1,2}(t)$ would be the response owing to a wave $g_2(t)$, as indicated in Fig. A1.1. In this case it is necessary to invoke the same restrictions as

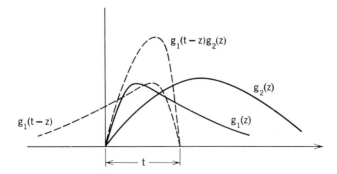

Figure A1-1 Function $g_1(z)$, its convolution $g_1(t - z)$ and the product $g_1(t - z)g_2(z)$. Area under the product curve represents the convolution integral $g_{1,2}(t)$.

discussed for Laplace transforms. That is, $g_1(t - z) = O$ for $t - z < O$. Hence the upper limit in (23) can be replaced by $z = t$. Moreover, if $g_2(t) = O$ for $t < O$, the lower limit in (23) can be replaced by O. Thus the convolution integral takes the following form, often referred to as Duhamel's superposition integral:

$$g_{1,2}(t) = \int_0^t g_1(t - z)g_2(z)\,dz. \tag{24}$$

The upper and lower limits in the latter expression can, however, be formally replaced by ∞ and $-\infty$ since the above restrictions automatically insure that there is no contribution to the integral for $z > t$ and $z < O$. The convolution integral represents the area under the curve $g_1(t - z)g_2(z)$ shown in Fig. A1.1.

A basic theorem on Fourier integrals states that

$$\int_{-\infty}^{\infty} g_1(z)g_2(x - z)\,dz = \frac{1}{2\pi}\int_{-\infty}^{\infty} f_1(y)f_2(y)e^{ixy}\,dy, \tag{25}$$

or conversely

$$\frac{1}{2\pi}\int_{-\infty}^{\infty} f_1(z)f_2(y - z)\,dz = \int_{-\infty}^{\infty} g_1(x)g_2(x)e^{-ixy}\,dx, \tag{26}$$

where

$$f_1(y) = \int_{-\infty}^{\infty} g_1(x)e^{-ixy}\,dx, \tag{27}$$

$$f_2(y) = \int_{-\infty}^{\infty} g_2(x)e^{-ixy}\,dx. \tag{28}$$

In accordance with (25), (22) can alternately be written

$$g_{1,2}(x) = \frac{1}{2\pi}\int_{-\infty}^{\infty} f_1(y)f_2(y)e^{ixy}\,dy. \tag{29}$$

In certain applications $g_{1,2}(x)$ will be involved in a second convolution integral

$$g_{1,2,3}(x) = \int_{-\infty}^{\infty} g_{1,2}(z)g_3(x - z)\,dz, \tag{30}$$

and this, in turn, in a third integral

$$g_{1,2,3,4}(x) = \int_{-\infty}^{\infty} g_{1,2,3}(z)g_4(x - z)\,dz. \tag{31}$$

The solution in the general case is given by

$$g_{1,\ldots,n}(x) = \frac{1}{2\pi}\int_{-\infty}^{\infty} f_{1,\ldots,n}(y)e^{ixy}\,dy, \tag{32}$$

where
$$f_{1,\ldots,n}(y) = f_1(y)f_2(y),\ldots,f_n(y), \tag{33}$$
and
$$f_m(y) = \int_{-\infty}^{\infty} g_m(x)e^{-ixy}\,dx, \tag{34}$$

where the lower limits can be replaced by O when $g_m(x) = O$ for $x < O$ so that the Fourier transform can be replaced by a Laplace transform.

An important advantage of Fourier transforms, as applied to network analysis and synthesis, is that the various successive convolution integrals (29), (30), (31), etc., can be replaced by a single integral (32), involving the spectral function defined by (33) and (34). This is particularly the case as applied to communication systems network synthesis where it is necessary to attain a certain overall response characteristic as represented by $f_{1,\ldots,n}(y)$ by appropriate combinations of individual response characteristics f_m of network components. Aside from the foregoing convenience from the standpoint of analysis, the latter steady-state response characteristics can, in general, be determined more readily and accurately by calculation or measurement than by the transient response characteristics $g_n(t)$ of the network components.

5. Fourier Integral Energy Relations

Relation (25) with $x = O$, or (26) with $y = O$, is known as Parseval's theorem, that is,
$$\int_{-\infty}^{\infty} g_1(x)g_2(x)\,dx = \frac{1}{2\pi}\int_{-\infty}^{\infty} f_1(z)f_2(-z)\,dz. \tag{35}$$
In the special case where $g_1 = g_2 = g$ so that $f_1 = f_2 = f$, the following relation applies with $z = y$
$$\int_{-\infty}^{\infty} g^2(x)\,dx = \frac{1}{2\pi}\int_{-\infty}^{\infty} f(y)f(-y)\,dy. \tag{36}$$
In view of (3), $f(y)f(-y) = U^2 + V^2 = |f(y)|^2$. Hence
$$\int_{-\infty}^{\infty} g^2(x)\,dx = \frac{1}{2\pi}\int_{-\infty}^{\infty} |f(y)|^2\,dy. \tag{37}$$
This is the Fourier integral energy theorem due to Rayleigh.

In applications to electric circuits, $g(x) = g(t) = O$ for $t < O$. The lower limit on the left-hand side of (37) can then be replaced by O and (37) with $y = \omega$ becomes
$$\int_{O}^{\infty} g^2(t)\,dt = \frac{1}{\pi}\int_{O}^{\infty} |f(\omega)|^2\,d\omega. \tag{38}$$

6. Resolution of Functions into Components

With the aid of Fourier and Laplace transforms, a function $g(x)$ can be resolved into a continuum of elementary cyclic components e^{ixy} of amplitudes $f(y)$. This has certain advantages on applications to communication systems synthesis as noted before. From the standpoint of analysis of communication systems performance in response to various signals, which are inherently transient in nature, a preferable resolution is ordinarily into elementary transient constituents or elements. In general, however, the functions to be resolved into elementary components may represent other quantities than variations in signal amplitude with time; for example, the variation with distance of the electric force along a circuit.

Returning to the convolution integral (22), let $g_{1,2}(x)$ represent a function which is assumed known in the interval between $x = -\infty$ and $x = \infty$. In accordance with (25) the function can be represented by a Fourier integral as

$$g_{1,2}(x) = \frac{1}{2\pi} \int_{-\infty}^{\infty} f_1(y) f_2(y) e^{ixy} \, dy. \tag{39}$$

If it is assumed that the Fourier transform of $g_{1,2}(x)$ vanishes for $|y| > y_1$, (39) can be written

$$g_{1,2}(x) = \frac{1}{2\pi} \int_{-y_1}^{y_1} f_1(y) f_2(y) e^{ixy} \, dy. \tag{40}$$

If it is further assumed that

$$\begin{aligned} f_1(y) &= 1 & 0 < |y| < y_1 \\ &= 0 & |y| > y_1, \end{aligned} \tag{41}$$

the corresponding function $g_1(x)$ is

$$g_1(x) = \frac{1}{2\pi} \int_{-y_1}^{y_1} e^{ixy} \, dy$$

$$= \frac{y_1}{\pi} g_0(u), \tag{42}$$

where $u = xy_1$ and

$$g_0(u) = \frac{\sin u}{u}. \tag{43}$$

When $g_1(x)$ is given by (43), the convolution integral (23) becomes

$$g_{1,2}(x) = \int_{-\infty}^{\infty} g_1(x - y) g_2(y) \, dy$$

$$= \frac{1}{2\pi} \int_{-y_1}^{y_1} f_2(y) e^{ixy} \, dy. \tag{44}$$

Basic Transform Concepts and Relations

The latter relation shows that if the spectrum of the function $g(x) = g_2(x)$ is confined to the range $0 < |y| < y_1$, the function can be represented by an integral in terms of the elementary function $g_1(x)$ as

$$g(x) = \int_{-\infty}^{\infty} g_1(x - z) g(z) \, dz \tag{45}$$

where $g(z) = 0$ for $|y| > y_1$ and

$$g_1(x - z) = \frac{y_1}{\pi} \frac{\sin y_1(x - z)}{y_1(x - z)}. \tag{46}$$

In applications to signals $xy_1 = \omega_1 t = 2\pi f_1 t$, where f_1 is the maximum frequency component of the signal. The function $g_1(x - z)$ has the property that $g_1(0) = y_1/\pi$ and that

$$\int_{-\infty}^{\infty} g_1(x - z) \, dz = 1. \tag{47}$$

For $y_1 \to \infty$, the function $g_1(x - z)$ can be identified with the unit impulse function, or the Dirac delta function $\delta(x - z)$, which has the further property that

$$\delta(x - z) = \frac{y_1}{\pi} = \infty \quad \text{for} \quad x = z \tag{48}$$

$$= 0 \quad \text{for} \quad x \neq z$$

Since the unit impulse function was obtained by taking the limit $y_1 \to \infty$, it follows that the spectrum of the unit impulse function is infinite and can be represented by

$$f_\delta(y) = 1 \tag{49}$$

For $y_1 \to \infty$, integral (45) is modified into

$$g(x) = \int_{-\infty}^{\infty} \delta(x - z) g(z) \, dz, \tag{50}$$

which is often referrred to as the sifting integral.

7. Series Representation of Band-Limited Functions

In the foregoing representation the function was shown as an integral in terms of an elementary function. With the same restrictions on the function $g(x)$, as before, it can also be represented by an infinite series as

$$g(x) = \sum_{n=-\infty}^{\infty} g\left(\frac{n\pi}{y_1}\right) g_0(xy_1 - \pi n), \tag{51}$$

where $g_0(u)$ is defined by (43). In (51), $g(n\pi/y_1)$ represents the coefficients in a Fourier series analysis of the function $g(x)$. In applications to signals, the variable x can be identified with time $x = t_0$ chosen far from the origin of the signal and the variable y with $\omega = 2\pi f$, so that $y_1 = \omega_1 = 2\pi f_1$. In this case (51) takes the form

$$g(t_0) = \sum_{n=-\infty}^{\infty} g\left(\frac{n}{2f_1}\right) \frac{\sin \pi(2f_1 t_0 - n)}{\pi(2f_1 t_0 - n)}, \qquad (52)$$

which is the sampling theorem. In accordance with this theorem, a function of time with a Fourier transform or spectrum of bandwidth f_1 can be uniquely represented by signal samples $g(t_0)$ taken at uniform sampling intervals $t_1 = 1/2f_1$, known as the Nyquist interval. These samples represent the coefficients that would be obtained in a Fourier series analysis of the function and hence suffice for reconstruction of the function when the maximum frequency is f_1.

In accordance with (52) the signal at a time t_0 from a sampling instant depends on the signal values at other sampling instants, except when $t_0 = O$. In the latter case (52) becomes

$$g(O) = \sum_{n=-\infty}^{\infty} g\left(\frac{n}{2f_1}\right) \frac{\sin 2\pi n}{2\pi n} \qquad (53)$$

$$= g(O) \quad \text{for} \quad n = O$$

$$= O \quad \text{for} \quad n \neq O,$$

where the latter relations apply since $\sin 2\pi n/2\pi n = O$ for $n \neq O$ and 1 for $n = O$. Thus if the signal is assumed to be made up of elementary components as in (53) there is no mutual effect between signal elements, at sampling instants $t_0 = O$ or no intersymbol interference.

Since the elementary signal components have infinite duration, extending from $t_0 = -\infty$ to $t_0 = \infty$, all bandlimited signals have infinite duration. This is a condition that cannot be encountered with actual signals produced by physical devices. Hence (53) only represents a convenient approximation in signal analysis. The same approximation is ordinarily used in dealing with signal transmission over bandlimited channels, when $g_0(u)$ as given by (43) with $u = \omega_1 t_0$ is the impulse characteristic of an idealized bandlimited channel with a sharp cutoff at the frequency ω_1. It should be noted, however, that this impulse characteristic is an abstraction that entails infinite transmission delay, as discussed in Section 1.5.

General Bibliography

1. G. A. Campbell and R. M. Foster, "Fourier Integrals for Practical Application," *BSTJ*, September, 1931. D. Van Nostrand Company, 1948.
2. E. C. Titchmarch, *Introduction to the Theory of Fourier Integrals*, Oxford University Press, 1937.
3. R. V. Churchill, *Modern Operational Mathematics in Engineering*, McGraw-Hill Book Company, 1944.
4. E. J. Scott, *Transform Calculus*, Harper and Row, 1955.
5. W. R. LePage, *Complex Variables and the Laplace Transform for Engineers*, McGraw-Hill Book Company, 1961.
6. E. T. Whittaker and G. N. Watson, *A Course of Modern Analysis*, Cambridge University Press, 1935 (p. 117).

Appendix 2
Basic Statistical Concepts and Relations

In dealing with signals and unwanted disturbances or distortion in communication channels, it would be possible to consider a number of specific waveforms of the signals and the interference. Because of the large number of possible waveforms, this would be a cumbersome and not very informative procedure. A preferable method is to introduce statistical analysis of performance, which, in effect, considers all possible waveforms of signals and disturbances of a particular kind, with appropriate weighting. This is accomplished by introducing certain basic parameters that specify the properties of the signals and disturbances, and certain probability functions involving these parameters, as reviewed here. A more comprehensive exposition will be found in any of several books.*

1. Random Variables and Probability Functions

Let x designate a particular value of the variable under consideration. A basic concept is the probability density function $p(x)$, which designates the small probability that the variable is in the range between x and $x + dx$, as indicated in Fig. A2.1. A second basic concept is the probability distribution function defined by

$$P(x \leq X) = \int_{-\infty}^{X} p(x)\, dx, \qquad (1)$$

which gives the probability that the variable is less than a specified value X, as indicated in Fig. A2.1.

---

* For example: (1) W. B. Davenport, Jr., and W. L. Root, *An Introduction to Random Signals and Noise*, McGraw-Hill Book Company, 1958. (2) A. Papoulis, *Probability, Random Variables and Stochastic Processes*, McGraw-Hill Book Company, 1965.

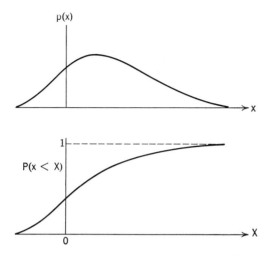

Figure A2-1 Probability density function $p(x)$ and probability distribution function $P(x < X)$.

The complementary probability distribution function is

$$P(x \geq X) = 1 - P(x \leq X) = \int_X^\infty p(x)\, dx. \qquad (2)$$

With two variables, x and y, the probability density function $p(x,y)$ designates the joint probability that x is between x and dx while y is between y and dy. The variable x might be the amplitude of a signal at a given instant and y that at a different instant; and in general, there would be some correlation between these amplitudes.

The joint probability that $x \leq X$ and $y \leq Y$ is

$$P(x \leq X, y \leq Y) = \int_{-\infty}^X \int_{-\infty}^X p(x,y)\, dx\, dy. \qquad (3)$$

Allowing each limit to become infinite separately, the individual probability distributions are obtained as

$$P(x \leq X) = \int_{-\infty}^\infty \int_{-\infty}^X p(x,y)\, dx\, dy \qquad (4)$$

$$P(y \leq Y) = \int_{-\infty}^Y \int_{-\infty}^\infty p(x,y)\, dx\, dy \qquad (5)$$

The distribution functions with respect to one of several variables are referred to as marginal distribution functions. When both sides of the latter equations are differentiated, the individual or marginal probability densities are obtained as

$$p(X) = \int_{-\infty}^{\infty} p(X,y)\,dy \tag{6}$$

$$p(Y) = \int_{-\infty}^{\infty} p(x,Y)\,dx. \tag{7}$$

Related to the foregoing joint and marginal probability densities are the conditional probability densities defined by

$$p_x(y) = \frac{p(x,y)}{p(x)} \quad \text{and} \quad p_y(x) = \frac{p(x,y)}{p(y)}. \tag{8}$$

These give the probability density for y when x is given and for x when y is given.

The corresponding probability distributions are

$$\begin{aligned} P(x \le X, Y) &= \frac{1}{p(X)} \int_{-\infty}^{X} p(x,Y)\,dx, \\ P(X, y \le Y) &= \frac{1}{p(Y)} \int_{-\infty}^{Y} p(X,y)\,dy. \end{aligned} \tag{9}$$

2. Averages and Moments of Random Variables

From the probability functions it is possible to derive various average values that characterize the random variable, such as the mean value and the standard deviation from the mean. Various kinds of averages can be identified with the moments of the probability density functions.

The nth moment of the probability density function is defined as the statistical average of the nth power of x. It is variously designated in literature by "av x^n," by "$\langle x^n \rangle$" and by the mathematical expectation $E(x^n)$. In accordance with this definition.

$$m_n = E(x^n) = \int_{-\infty}^{\infty} x^n p(x)\,dx, \tag{10}$$

$m_0 = 1 = P(x < \infty)$,

$m_1 = E(x) =$ mean value of the variable,

$m_2 = E(x^2) =$ mean squared value of the variable.

In most applications it is desirable to determine moments with respect to the mean value m_1. These moments about the mean, or central moments, are

$$\mu_n = E[(x - m_1)^n] = \int_{-\infty}^{\infty} (x - m_1)^n p(x)\,dx. \tag{11}$$

The more important moments are

$$\mu_1 = 0,$$
$$\mu_2 = m_2 - 2m_1^2 + m_1^2 = m_2 - m_1^2$$
$$= E(x^2) - [E(x)]^2 = \sigma_x^2.$$

The second central moment is commonly denoted by σ^2 and referred to as the variance or dispersion, while σ is known as the standard deviation or root-mean square (rms) value of the variable.

The moments of the bivariate joint probability distribution include the moments of the individual distributions, which are special cases of the cross-moment of order $n + k$ defined by

$$m_{nk} = E(x^n, y^k) = \int_{-\infty}^{\infty} \int_{-\infty}^{\infty} x^n y^k p(x,y) \, dx \, dy. \tag{12}$$

It will be noted that $m_{00} = 1$ and that the mean of the individual distributions are

$$x_O = m_{10} = E(x) = \int_{-\infty}^{\infty} \int_{-\infty}^{\infty} x p(x,y) \, dx \, dy$$
$$= \int_{-\infty}^{\infty} x p(x) \, dx \tag{13}$$

$$y_O = m_{01} = E(y) = \int_{-\infty}^{\infty} y p(y) \, dy \tag{14}$$

The joint central moments are

$$\mu_{nk} = E[(x - x_0)^n (y - y_0)^k]$$
$$= \int_{-\infty}^{\infty} (x - x_0)^n (y - y_0)^k p(x,y) \, dx \, dy. \tag{15}$$

The more important joint moments are

$$\mu_{20} = \sigma_x^2; \quad \mu_{02} = \sigma_y^2.$$

The joint moment μ_{11} is called the covariance of the random variables x and y and the normalized covariance or correlation coefficient is defined by

$$\rho = \frac{\mu_{11}}{\sigma_x \sigma_y} = \frac{\mu_{11}}{\sqrt{\mu_{20}\mu_{20}}}. \tag{16}$$

Various general expressions can be postulated for the probability density functions, with an appropriate number of arbitrary constants. When such an expression for $p(x)$ is inserted in (10), one equation for the choice of constants is obtained from the relation $m_0 = 1$. By determining m_1 and

m_2 two additional equations are obtained from which the arbitrary constants can be expressed in terms of m_1 and m_2. For a bivariate probability density the same general procedure is used, but becomes more complicated.

3. Sums of Random Variables

A random variable $x(t)$ is in general composed of a sum of random variables, each with different probability density functions. A problem often encountered is to determine the probability density for a sum of random variables when the individual probability functions are known. To this end it is often convenient to employ Fourier transforms of the individual probability functions, or so called *characteristic* functions.

Let $p_1(x)$ and $p_2(x)$ be the probability density functions for two independent random variables. For the sum of these variables there is then a probability density $p_{1,2}(x)$ that the sum is between z and $z + dz$, given by

$$p_{1,2}(x) = \int_{-\infty}^{\infty} p_1(z) p_2(x - z)\, dz. \tag{17}$$

This relation follows when it is considered that to obtain a given value of x for a specified value z of one of the variables, the other variable must equal $x - z$. The total probability is obtained by integrating with respect to z as in (17).

From the relations given in Appendix 1, it follows that the foregoing integral can be solved by the convolution integral transform method. The solution thus obtained is

$$p_{1,2}(x) = \frac{1}{2\pi} \int_{-\infty}^{\infty} C_1(y) C_2(y) e^{iyx}\, dy, \tag{18}$$

where

$$C_1(y) = \int_{-\infty}^{\infty} p_1(x) e^{ixy}\, dx, \tag{19}$$

$$C_2(y) = \int_{-\infty}^{\infty} p_2(x) e^{ixy}\, dx. \tag{20}$$

The latter functions are known as the characteristic functions of the probability density functions, and represent the spectrum of these functions with respect to a dummy variable y.

For any number n of independent variables, it follows from (32) and (33) in Appendix 1 that the probability density function for their sums is given by

$$p_{1,\ldots,n}(x) = \frac{1}{2\pi} \int_{-\infty}^{\infty} C_{1,\ldots,n}(y)\, e^{ixy}\, dy, \tag{21}$$

Basic Statistical Concepts and Relations

where

$$C_{1,\ldots,n}(y) = C_1(y)\, C_2(y)\, C_3(y),\ldots, C_n(y), \tag{22}$$

and

$$C_m(y) = \int_{-\infty}^{\infty} p_m(x)\, e^{ixy}\, dx. \tag{23}$$

In (23) the function $u = e^{ixy}$ represents a nonlinear transformation of a random function. $C_m(y)$ is the average value of $u = e^{ixy}$ for a probability density $p_m(x)$ of x (see (49).) Similarly the combined characteristic function is, by Fourier transformation,

$$C_{1,\ldots,n}(y) = E(e^{ixy}) = \int_{-\infty}^{\infty} p_{1,\ldots,n}(x)\, e^{iyx}\, dx. \tag{24}$$

After the combined characteristic function $C_{1,\ldots,n}(y)$ has been determined, the probability that $-X < x < X$ is obtained from the relation

$$\begin{aligned}P(-X < x < X) &= \int_{-X}^{X} p_{1,\ldots,n}(x)\, dx \\ &= \frac{1}{2\pi}\int_{-X}^{X}\int_{-\infty}^{\infty} C_{1,\ldots,n}(y)\, e^{ixy}\, dy\, dx \\ &= \frac{2}{\pi}\int_{0}^{\infty} C_{1,\ldots,n}(y)\, \frac{\sin yX}{y}\, dy.\end{aligned} \tag{25}$$

For a bivariate probability density function $p(x,y)$ the corresponding joint characteristic functions are

$$\begin{aligned}C(u,v) &= E(e^{iux+ivy}) \\ &= \int_{-\infty}^{\infty}\int_{-\infty}^{\infty} p(x,y)\, e^{iux} e^{ivy}\, dx\, dy\end{aligned} \tag{26}$$

The inverse Fourier transformation gives

$$p(u,v) = \left(\frac{1}{2\pi}\right)^2 \int_{-\infty}^{\infty}\int_{-\infty}^{\infty} C(u,v)\, e^{-iux} e^{-ivy}\, du\, dv. \tag{27}$$

The moments of the resultant probability distribution can be obtained with the aid of expression (10) with $p_{1,\ldots,n}$ in place of $p(x)$. An alternate and simpler method can be used when the characteristic function is known as above and can be expanded in a Taylor's series. The characteristic function for a probability density $p(z)$ is

$$C(y) = \int_{-\infty}^{\infty} e^{ixy} p(x)\, dx, \tag{28}$$

and

$$\left(\frac{d^n C(y)}{d^n y}\right)_{y=0} = \int_{-\infty}^{\infty} (ix)^n p(x)\, dx = i^n m_n \tag{29}$$

where m_n is the nth moment as given by (10).

The Taylor's series expansion of $C(y)$ is

$$C(y) = \sum_{n=0}^{\infty} \left[\frac{d^n C(y)}{dy^m}\right]_{y=0} \frac{y^n}{n!}$$

$$= \sum_{n=0}^{\infty} \frac{m_n (iy)^n}{n!} \tag{30}$$

$$= m_0 + m_1 iy + m_2 \frac{i^2}{2!} y^2 + m_3 \frac{i^3}{3!} y^3 + \cdots,$$

$$= a_0 + a_1 y + a_2 y^2 + a_3 y^3 + \cdots. \tag{31}$$

Thus, from the coefficients in the power series expansion of $C(y)$ the moments can be determined from the relation

$$m_n = \frac{a_n n!}{i^n}. \tag{32}$$

The characteristic function is closely related to the moment generating function used in statistical analysis. The latter function is defined by $M_x(y) = E(e^{xy})$ as compared to $C_x(y) = E(e^{ixy})$. Expression (32) is then modified to $m_n = a_n n!$

The second moment equals the variance, that is, $m_2 = \sigma^2$. The ratio m_4/m_2^2 gives an indication of the peakedness of the distribution. For a Gaussian probability density the ratio is $m_4/m_2^2 = 3$ as shown later.

In a similar manner, the joint moments can be obtained from the bivariate characteristic function given by (26). In this case

$$C(u,v) = \sum_{j=0}^{\infty} \sum_{k=0}^{\infty} \frac{i^{j+k} u^j v^k}{j!\, k!} m_{jk}. \tag{33}$$

Hence m_{jk} is given by

$$m_{jk} = a_{jk} \frac{j!\, k!}{i^{(j+k)}}, \tag{34}$$

where a_{jk} is the coefficient for $u^j v^k$ in the power series expression of $C(u,v)$.

An important proposition in statistics relating to the sum of random variables is the central limit theorem. It states that as the number of statistically independent variables is increased without limit, a Gaussian probability distribution is approached for the sum, regardless of the probability distributions of the various random variables, except under certain conditions as when a few of the variables predominate in amplitude. For this reason, the Gaussian probability function is of special importance and is discussed further in Section 6.

4. Time Correlation Functions and Power Spectra

Let $x_i(t)$, $i = 1, 2, 3, \ldots, n$ be a number of random variables such as the noise in a number of similar circuits. If a large number of such random variables are sampled at a particular instant t_0, then the resultant mean value, $[x_1(t_0) + x_2(t_0) + \cdots, x_n(t_0)]/n$ for $n \to \infty$ is referred to as the statistical or ensemble average of the random process designated by $x(t)$. If the same mean and variance from the mean is obtained in repeated trials at different times t_0, the process is said to be stationary in the wide sense.*

Alternately, the average of each function could be determined by averaging with respect to time as

$$av\, x_i(t) = \lim_{T \to \infty} \frac{1}{2T} \int_{-T}^{T} x_i(t)\, dt \tag{35}$$

If the average thus obtained is the same for all variables, that is, $i = 1, 2, \ldots, n$, the process is again said to be stationary in the wide sense. When, in addition, this average is the same as the ensemble average, the process is called ergotic. This assumption is ordinarily made in statistical signal analysis and permits the interchange of statistical and ensemble averages with the time average for a single "representative" function $x(t)$ from the ensemble.

Let $x(t)$ and $y(t)$ be such representative variables of time extending between $t = -\infty$ and $t = \infty$. The time average of the product $x(t + \tau)y(t)$ is defined as the time cross correlation function of the two random variables, that is,

$$R_{xy}(\tau) = \lim_{T \to \infty} \frac{1}{2T} \int_{-T}^{T} x(t + \tau) y(t)\, dt. \tag{36}$$

With the restrictions imposed on $x(t)$ and $y(t)$ for ergotic random processes, the same result is obtained by taking the product $x(t - \tau)y(t)$ or

---

* Stationarity in the strict sense entails equality not only of the first and second moments, but also of all higher order moments of the function $x(t)$. For Gaussian random variables wide sense also entails strict sense stationarity.

$x(t)y(t \pm \tau)$. Moreover, the cross correlation functions can then be obtained from (12) as

$$m_{11} = R_{xy}(\tau) = \int_{-\infty}^{\infty} \int_{-\infty}^{\infty} xyp(x,y,\tau)\,dx\,dy, \qquad (37)$$

where the time interval τ now appears in the joint probability density functions. It will be noted that for $x = y$ and $\tau = O$, $R_{xx}(O) = \sigma_x^2$.

With $x = y$ in (36) the autocorrelation function of the random variable x is obtained. From (25) of Appendix 1 it follows, with $g_1(z) = x(t)$ and $g_2(x - z) = y(t - \tau)$, that (36) can then be evaluated by the following Fourier transform as $T \to \infty$.

$$\begin{aligned} R_{xx}(\tau) &= \lim_{T \to \infty} \frac{1}{4\pi T} \int_{-T}^{T} f_x^2(\omega)\,e^{i\omega\tau}\,d\omega \\ &= \int_{-\infty}^{\infty} S_x(\omega)\,e^{i\omega\tau}\,d\omega \end{aligned} \qquad (38)$$

where $R_{xx}(\tau)$ is the autocorrelation function and $S_x(\omega)$ is the power spectral density of the random variable x defined by

$$S_x(\omega) = \lim_{T \to \infty} \frac{f_x^2(\omega)}{4\pi T}. \qquad (39)$$

The power spectral density of a random function is ordinarily referred to as the power spectrum, and is defined as the mathematical expectation or average of the spectral densities of the individual functions. The power spectrum $S(\omega)$ is sometimes defined for both positive and negative frequencies, and at other times for positive frequencies only, as

$$W(\omega) = S(-\omega) + S(\omega) = 2S(\omega). \qquad (40)$$

Expression (38) can alternately be written

$$\begin{aligned} R(\tau) &= 2\int_{O}^{\infty} S(\omega)\cos\omega\tau\,d\omega \\ &= \int_{O}^{\infty} W(\omega)\cos\omega\tau\,d\omega, \end{aligned} \qquad (41)$$

conversely the power spectrum is given by

$$S(\omega) = \frac{1}{\pi}\int_{O}^{\infty} R(\tau)\cos\omega\tau\,d\tau,$$

or $\qquad (42)$

$$W(\omega) = \frac{2}{\pi}\int_{O}^{\infty} R(\tau)\cos\omega\tau\,d\tau.$$

It may be noted that $R(O) = \sigma^2$, the variance of the random variable.

Basic Statistical Concepts and Relations

The correlation function for the sum of two random variables x and y is given by

$$R_{x+y} = R_{xx}(\tau) + R_{yy}(\tau) + R_{xy}(\tau) + R_{yx}(\tau) \tag{43}$$

The corresponding power spectrum is

$$S_{x+y} = S_{xx}(\omega) + S_{yy}(\omega) + S_{xy}(\omega) + S_{yx}(\omega) \tag{44}$$

Here $R_{xx}(\tau)$ and $R_{yy}(\tau)$ are obtained from (41) while $S_{xx}(\omega)$ and $S_{yy}(\omega)$ are obtained from (42). The cross correlation functions are defined in terms of the cross-power spectra, and conversely, by

$$R_{xy}(\tau) = \int_{-\infty}^{\infty} S_{xy}(\omega)e^{i\omega\tau}\,d\omega,$$

$$R_{yx}(\tau) = \int_{-\infty}^{\infty} S_{yx}(\omega)e^{i\omega\tau}\,d\omega, \tag{45}$$

$$S_{xy}(\omega) = \frac{1}{2\pi}\int_{-\infty}^{\infty} R_{xy}(\tau)e^{-i\omega\tau}\,d\tau,$$

$$S_{yx}(\omega) = \frac{1}{2\pi}\int_{-\infty}^{\infty} R_{yx}(\tau)e^{-i\omega\tau}\,d\tau. \tag{46}$$

In the foregoing relations, the cross correlation functions and power spectra vanish when x and y are uncorrelated random variables.

The functions $R_{xx}(\tau)$ and $R_{yy}(\tau)$ are even functions of τ, and the corresponding power spectra are hence real functions of ω. However, $R_{xy}(\tau)$ and $R_{yx}(\tau)$ are not necessarily even functions of τ. For this reason, relations (46) cannot be written in the same form as (42) with lower limit O, and the cross-power spectra $S_{xy}(\omega)$ and $S_{yx}(\omega)$ are in general complex quantities. The following relations apply: $R_{xy}(\tau) = R_{yx}^*(-\tau)$ and $S_{xy}(\omega) = S_{yx}^*(\omega)$, when the asterisks denote complex conjugate quantities.

5. Functions of Random Variables

In the previous section, certain basic parameters of random variables were discussed, such as the probability distribution, its various moments and correlation functions. An important problem in application to communication systems is to determine the corresponding quantities for a function of one or two random variables. This function may represent the transfer characteristic of a nonlinear communication channel or a modulation process applied to the input signal.

Let $u = u(x)$ be a real function of the random variable x, for example the instantaneous output amplitude of a nonlinear device in response to an input amplitude x. The inverse relation obtained by solution of the above relation will be designated $x = h(u)$. For example, if $u = x^{1/2}$, then $x = h(u) = u^2$.

The probability of a value between u and $u + du$ will be designated $p_u(u) \, du$ and that of a value between x and $x + dx$, $p_x(x) \, dx$. It follows that $p_u(u) \, du = p_x(x) \, dx$. With $dx/du = dh(u)/du = h'(u)$ the following equality applies for monotonic function $h(u)$:*

$$p_u(u) \, du = p_x[h(u)]h'(u) \, du. \tag{47}$$

The probability of a value less than u_1 becomes

$$P_u(u \leq u_1) = \int_{-\infty}^{u_1} p_x[h(u)]h'(u) \, du$$

$$= P_x[h(u_1)] \tag{48}$$

For example, let $P_x(x \leq x_1) = 1 - e^{-ax_1^{1/2}}$ and $u = x^{1/2}$. Then $h(u) = u^2$ and $h(u_1) = u_1^2 = x_1$. Hence

$$P_u(u < u_1) = P_x(x_1 < u_1^2) = 1 - e^{-au_1}.$$

The nth moment of the modified probability distribution becomes

$$m_n(u) = E(u^n) = \int_{-\infty}^{\infty} u^n(x)p(x) \, dx. \tag{49}$$

The central moments are

$$\mu_n(u) = E(u - m_1(u)^n) = \int_{-\infty}^{\infty} [u(x) - m_1(u)]^n p(x) \, dx. \tag{50}$$

Thus

$$\mu_2(u) = E(u^2) - [E(u)]^2 = \sigma_u^2.$$

With two random variables the joint moments become

$$m_{nk}(u) = E(u^n u^k) = \int_{-\infty}^{\infty} \int_{-\infty}^{\infty} u^n(x)u^k(y)p(x,y) \, dx \, dy. \tag{51}$$

The modified autocorrelation function is

$$R_u(\tau) = \int_{-\infty}^{\infty} \int_{-\infty}^{\infty} u(x)u(y)p(x,y) \, dx \, dy. \tag{52}$$

The modified power spectrum is given by

$$S_u(\omega) = \frac{1}{\pi} \int_0^{\infty} R_u(\tau) \cos \omega\tau \, d\tau. \tag{53}$$

* For a more general discussion, see reference [2].

Basic Statistical Concepts and Relations

6. Gaussian Probability Functions

The most important probability functions in statistical analysis as applied to communication systems is the Gaussian single and bivariate distributions. The probability density function for a single variable is in this case given by

$$p(x) = \frac{1}{\sigma\sqrt{2\pi}} e^{-(x-x_0)/2\sigma^2}, \tag{54}$$

and is illustrated in Fig. A2.2. The parameters have been normalized to satisfy the following conditions:

$$P(x \leq \infty) = \int_{-\infty}^{\infty} p(x)\, dx = 1, \tag{55}$$

$$m_1 = \int_{-\infty}^{\infty} x p(x)\, dx = x_0, \tag{56}$$

$$m_2 = \int_{-\infty}^{\infty} x^2 p(x)\, dx = x_0^2 + \sigma^2, \tag{57}$$

$$\mu_2 = m_2 - m_1^2 = \sigma^2. \tag{58}$$

The probability distribution function is

$$P(x \leq X) = \int_{-\infty}^{X} p(x)\, dx$$
$$= \frac{1}{2}\left(1 + \operatorname{erf} \frac{X - x_0}{\sigma\sqrt{2}}\right), \tag{59}$$

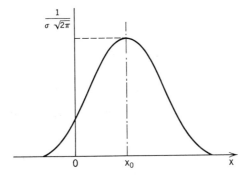

Figure A2-2 Gaussian probability density function $p(x) = 1/\sigma\sqrt{2\pi}\, e^{-(x-x_0)^2/2\sigma^2}$.

where the error function is defined as

$$\operatorname{erf} z = \frac{2}{\sqrt{\pi}} \int_0^z e^{-\lambda^2} d\lambda, \qquad (60)$$

The corresponding characteristic function is

$$C(y) = \int_{-\infty}^{\infty} p(x) e^{ixy} du$$

$$= e^{iyx_0} e^{-\sigma^2 y^2/2}$$

$$= e^{iyx_0}\left(1 - \frac{\sigma^2 y^2}{2} + \frac{\sigma^4 y^4}{4\cdot 2!} - \cdots\right). \qquad (61)$$

The characteristic function is thus also a Gaussian function for $x_0 = O$. The central moments obtained from the foregoing series by application of (32) are

$$\mu_2 = \sigma^2; \qquad \mu_3 = 0; \qquad \mu_4 = 3\sigma^4$$

It may be noted that in this case

$$\frac{\mu_4}{\mu_2^2} = 3.$$

The bivariate Gaussian probability density is given by

$$p(x,y) = \frac{1}{2\pi A} \exp -\frac{\mu_{02}(x-x_0)^2 + \mu_{20}(y-y_0)^2 - 2\mu_{11}(x-x_0)(y-y_0)}{2A^2}, \qquad (62)$$

where

$$A^2 = \mu_{02}\mu_{20} - \mu_{11}^2. \qquad (63)$$

Here

$$\mu_{20} = \sigma_x^2 = R_x(O), \qquad \mu_{02} = \sigma_y^2 = R_y(O),$$

and

$$\mu_{11} = R_{xy}(\tau).$$

In the special case where $\sigma_x = \sigma_y$ and $\mu_{11} = R_{xx}(\tau) = R_{yy}(\tau) = R(\tau)$ the following expression applies with $x_0 = O$ and $y_0 = O$.

$$p(x,y) = \frac{[R^2(O) - R^2(\tau)]^{1/2}}{2\pi} \exp -\frac{R(O)(x^2 + y^2) - 2R(\tau)xy}{2[R^2(O) - R^2(\tau)]}, \qquad (64)$$

The corresponding bivariate characteristic function obtained from (26) is

$$C(u,v) = \exp\left(-\frac{R(O)}{2}(u^2 + v^2) - R(\tau)uv\right). \qquad (65)$$

Basic Statistical Concepts and Relations

7. Rayleigh Probability Functions

In the foregoing expressions the joint probability density $p(x,y)$ was expressed in rectangular coordinates. With polar coordinates, such that $x = r \cos \theta$ and $y = r \sin \theta$, the joint probability density $p(r,\theta)$ can be determined by change of variables. Of particular interest in many applications is the radial probability density $p(r)$ when $p(x)$ and $p(y)$ are independent. In this case change of variables by the usual methods employing the so-called Jacobian gives

$$\int_{-\infty}^{\infty}\int_{-\infty}^{\infty} p(x)p(y)\,dx\,dy = \int_{r=0}^{\infty}\left[\int_{\theta=0}^{2\pi} p(r\cos\theta)p(r\sin\theta)\,d\theta\right]r\,dr$$

$$= \int_{0}^{\infty} p_0(r)\,dr, \tag{66}$$

where

$$p_0(r) = r\int_{0}^{2\pi} p(r\cos\theta)p(r\sin\theta)\,d\theta \tag{67}$$

With Gaussian probability densities, the following relations are obtained

$$p(r\cos\theta)p(r\sin\theta) = \frac{1}{2\pi\sigma^2}\exp{-\frac{r^2(\cos^2\theta + \sin^2\theta)}{2\sigma^2}}$$

$$= \frac{1}{2\pi\sigma^2} e^{-r^2/2\sigma^2}, \tag{68}$$

and (67) gives

$$p_0(r) = \frac{r}{\sigma^2} e^{-r^2/2\sigma^2}. \tag{69}$$

This is known as the Rayleigh probability density. The corresponding probability that r exceeds R becomes

$$P(r \geq R) = \int_{R}^{\infty} p_0(r)\,dr = e^{-R^2/2\sigma^2}. \tag{70}$$

8. The Log-Normal Probability Function

In statistical analyses as applied to communication systems, a log-normal probability function is often encountered. It is obtained from (54) with the transformation

$$x = \ln y; \quad y \geq 0, \tag{71}$$

in which case with $\sigma = \sigma_x$

$$p(y) = \frac{1}{\sqrt{2\pi}\sigma y} e^{-(\ln y - x_0)^2/2\sigma^2}, \qquad (72)$$

and

$$P(y < y_1) = \int_0^{y_1} p(y)\,dy = \int_{-\infty}^{x_1} p(x)\,dx \qquad (73)$$

$$= \frac{1}{2}\left(1 + \operatorname{erf}\frac{x_1 - x_0}{\sqrt{2}\sigma}\right) \qquad (74)$$

$$= \frac{1}{2}\left(1 + \operatorname{erf}\frac{\ln(y_1/y_0)}{\sqrt{2}\sigma}\right), \qquad (75)$$

where $x_0 = \ln y_0$ is the mean value of x. The last integral in (73) follows in view of (54), while (74) follows from (59).

The moments of $p(y)$ are

$$m_n = \int_0^\infty y^n p(y)\,dy = \frac{e^{nx_0}}{\sqrt{2\pi}}\int_{-\infty}^\infty e^{n\sigma u} e^{-u^2/2}\,du = e^{nx_0} e^{n^2\sigma^2/2}. \qquad (76)$$

The second integral is obtained with the transformation $(\ln y - x_0)/\sigma = u$.

The mean value of y is thus

$$m_1 = \bar{y} = e^{x_0 + \sigma^2/2}, \qquad (77)$$

or

$$\ln \bar{y} = x_0 + \sigma^2/2, \qquad (78)$$

The second moment of $p(y)$ is

$$\mu_2 = \sigma_y^2 = m_2 - m_1^2$$

$$= e^{2x_0} e^{+\sigma^2}(e^{\sigma^2} - 1). \qquad (79)$$

Appendix 3
Nonlinear Transformations of Power Spectra

In Section 5 of Appendix 2 a general formulation was given for the autocorrelation function and corresponding power spectrum after a nonlinear transformation $z(x)$ of a random variable $x(t)$. Actual determination of these quantities is in general a formidable problem, but explicit solutions have been obtained for the important case of a Gaussian random variable $x(t)$ and transformations $z(x)$ that approximate those performed by certain nonlinear devices encountered in communication channels. Aside from being nonlinear, actual devices will have other important properties, such as limited bandwidth together with attenuation and phase distortion, which will be disregarded here. The nonlinear transformations considered here correspond to those of an idealized device of infinite bandwidth, that is, zero transmission delay or memory.

For Gaussian random variables the joint probability density function is given by (64) of Appendix 2. Explicit relations for the corresponding autocorrelation function $R_z(\tau)$ can in this case be obtained directly by solution of (52), and this "direct method" has been used for power law transformations $z(x) = x^\nu$. However, a more powerful procedure is the "transform method" initiated by Bennett and Rice and expounded in a number of tutorial publications.* This method is reviewed here, together with results obtained for a number of special functions $z(x)$ of importance in connection with nonlinear distortion in communication systems. The same results can be obtained by an alternate method introduced by Price [1,2] but not considered here.

---

[1] R. Price, "A Useful Theorem for Non-Linear Devices Having Gaussian Inputs," *IRE Trans. on Information Theory*, Vol. I 1–4, 1958.

[2] E. L. McMahon: "An Extension of Price's Theorem," *IEEE Trans. on Information Theory*, Vol. I 1–10, 1964.

* For example, W. B. Davenport and W. L. Root, *Random Signals and Noise*, McGraw-Hill Book Company, 1958.

1. Basic Relations

Let $z(x)$ be a function that can be represented by a Fourier integral as

$$z(x) = \frac{1}{2\pi} \int_{-\infty}^{\infty} f(i\lambda) e^{i\lambda x} \, d\lambda, \tag{1}$$

in which case

$$f(i\lambda) = \int_{-\infty}^{\infty} z(x) e^{-i\lambda x} \, dx. \tag{2}$$

Relation (52) of Appendix 2 can then be written

$$R_z(\tau) = \left(\frac{1}{2\pi}\right)^2 \int_{-\infty}^{\infty} f(iv) \, dv \int_{-\infty}^{\infty} f(iu) \, du \int_{-\infty}^{\infty} \int_{-\infty}^{\infty} p(x,y) e^{iux + ivy} \, dx \, dy \tag{3}$$

$$= \left(\frac{1}{2\pi}\right)^2 \int_{-\infty}^{\infty} f(iv) \, dv \int_{-\infty}^{\infty} f(iu) \, du \, C(u,v), \tag{4}$$

where $C(u,v)$ is the joint characteristic function given by (26) of Appendix 2.

For Gaussian random variables $x(t)$ and $y(t)$ relation (65) for $C(u,v)$ applies, and in this case solutions of (4) assume the following general forms for even functions, that is, $z(-x) = z(x)$:

$$R_z(\tau) = \sum_{k=0,2,4,\ldots}^{\infty} \frac{h_{0k}^2 R^k(\tau)}{k!}. \tag{5}$$

For odd functions, that is, $z(-x) = -z(x)$:

$$R_z(\tau) = \sum_{k=1,3,5,\ldots}^{\infty} \frac{h_{0k}^2 R^k(\tau)}{k!}, \tag{6}$$

where $R(\tau) = R_x(\tau) = R_y(\tau)$ and $R(O) = \sigma^2$.

The coefficients h_{0k} are given by

$$h_{0k} = \frac{1}{2\pi} \int_{-\infty}^{\infty} f(i\lambda)(i\lambda)^k e^{-\sigma^2\lambda^2/2} \, d\lambda. \tag{7}$$

The double-sided power spectrum becomes

$$S_z(\omega) = \frac{1}{2\pi} \int_{-\infty}^{\infty} R_z(\tau) e^{-i\omega\tau} \, d\tau$$

$$= \frac{1}{2\pi} \int_{-\infty}^{\infty} R_z(\tau) \cos \omega\tau \, d\tau. \tag{8}$$

The one-sided power spectrum is $W_z(\omega) = 2S_z(\omega)$. With $\omega = 2\pi f$, $S_z(f) = 2\pi S_z(\omega)$ and $W_z(f) = 2\pi W_z(\omega)$.

When $x(t)$ is a random variable with zero mean value, $R(\infty) = 0$. Hence $R_z(\infty)$ is obtained by taking the term in (5) for $k = 0$ and becomes

$$R_z(\infty) = h_{00}^2. \tag{9}$$

The corresponding component of the power spectrum $S_z(\omega)$ represents its average value and is given by

$$\bar{S}_z(\omega) = \frac{1}{2\pi} \int_{-\infty}^{\infty} h_{00} \cos \omega \tau \, d\tau$$

$$= \delta(\omega) h_{00}. \tag{10}$$

The variable component of the power spectrum is obtained by subtracting (9) from (8) and is

$$\tilde{S}_z(\omega) = \frac{1}{2\pi} \int_{-\infty}^{\infty} [R_z(\tau) - R_z(\infty)] \cos \omega \tau \, d\tau. \tag{11}$$

2. Power Law Transformations

Let ν be an integer and

$$z(x) = x^\nu. \tag{12}$$

In this case (2) and (7) yield

$$f(i\lambda) = \frac{\nu!}{(i\lambda)^{\nu+1}}, \tag{13}$$

$$h_{0k}^2 = \frac{[R(0)]^{\nu-k}}{2^{\nu-k} k!} \left[\frac{\nu!}{\left(\frac{\nu-k}{2}\right)!} \right]^2. \tag{14}$$

For even values of ν, (5) becomes

$$R_z(\tau) = R_\nu(\tau) = \sum_{k=0,2,4,\ldots}^{\infty} \frac{[R(0)]^{\nu-k} R^k(\tau)}{2^{\nu-k} k!} \left[\frac{\nu!}{\left(\frac{\nu-k}{2}\right)!} \right]^2. \tag{15}$$

For odd values of ν, (15) applies with $k = 1, 3, 5, \ldots$.

For various values of ν the following autocorrelation functions are obtained:

ν	$R_\nu(\tau)$
1	$R(\tau)$
2	$R^2(0) + 2R^2(\tau)$
3	$3[3R(\tau)R^2(0) + 2R^3(\tau)]$
4	$3[3R^4(0) + 24R^2(\tau)R^2(0) + 8R^4(\tau)]$
5	$15[15R(\tau)R^4(0) + 40R^3(\tau)R^2(0) + 8R^5(\tau)]$

For $\nu = 1$

$$R_1(\tau) = R(\tau) = \int_{-\infty}^{\infty} S_1(\omega)e^{i\omega\tau}\,d\tau. \tag{16}$$

For $\nu = 2$, (10) and (11) yield for the power spectra

$$\bar{S}_2(\omega) = \delta(\omega)R_1^2(O), \tag{17}$$

$$\tilde{S}_2(\omega) = \frac{2}{2\pi}\int_{-\infty}^{\infty} R_1^2(\tau)e^{-i\omega\tau}\,d\tau$$

$$= \frac{2}{2\pi}\int_{-\infty}^{\infty} S_1(\omega_1)\,d\omega_1 \int_{-\infty}^{\infty} R_1(\tau)e^{-i(\omega-\omega_1)\tau}\,d\tau$$

$$= 2\int_{-\infty}^{\infty} S_1(\omega_1)\,S_1(\omega - \omega_1)\,d\omega_1. \tag{18}$$

The corresponding relation for the one-sided power spectrum is

$$\tilde{W}_2(\omega) = \int_{-\infty}^{\infty} W_1(\omega_1)\,W_1(\omega - \omega_1)\,d\omega_1. \tag{19}$$

Relations (18) and (19) also apply with ω and ω_1 replaced by f and f_1. For $\nu = 3$, $\bar{S}_3(\omega) = O$ and

$$\tilde{S}_3(\omega) = 9R_1^2(O)\,S_1(\omega) + \frac{6}{2\pi}\int_{-\infty}^{\infty} S_1(\omega_1)\,d\omega_1 \int_{-\infty}^{\infty} R_1^2(\tau)e^{-i(\omega-\omega_1)\tau}\,d\tau$$

$$= 9R_1^2(O)\,S_1(\omega) + 3\int_{-\infty}^{\infty} S_1(\omega_1)\,S_2(\omega - \omega_1)\,d\omega_1. \tag{20}$$

The corresponding relation for the one-sided power spectrum is

$$\tilde{W}_3(\omega) = 9R_1^2(O)W_1(\omega) + \frac{3}{2}\int_{-\infty}^{\infty} W_1(\omega_1)\,W_2(\omega - \omega_1)\,d\omega_1. \tag{21}$$

Relations (20) and (21) also apply when ω and ω_1 are replaced by f and f_1.

3. Cosine and Sine Transformations

Let

$$z = \cos x, \tag{22}$$

in which case

$$f(i\lambda) = \frac{1}{2}\int_{-\infty}^{\infty} (e^{ix(1-\lambda)} + e^{-ix(1+\lambda)})\,d\lambda$$

$$= \frac{1}{2}[\delta(1) + \delta(-1)]. \tag{23}$$

Hence

$$h_{0k} = \frac{1}{4\pi} \int_{-\infty}^{\infty} [\delta(1) + \delta(-1)](i\lambda)^k e^{-\sigma^2 \lambda^2/2} \, d\lambda$$

$$= \frac{i^k}{2} [1^k e^{-\sigma^2/2} + (-1)^k e^{-\sigma^2/2}]. \tag{24}$$

For $k = 1, 3, 5, \ldots$, $h_{0k} = O$ and for $k = 0, 2, 4, \ldots$

$$h_{0k} = i^k e^{-\sigma^2/2}; \qquad h_{0k}^2 = e^{-\sigma^2}. \tag{25}$$

Relation (5) yields:

$$R_z(\tau) = \sum_{k=0,2,4,\ldots}^{\infty} \frac{e^{-R(O)}}{k!} R^k(\tau)$$

$$= e^{-R(O)} \cosh R(\tau), \tag{26}$$

$$R_z(O) = e^{-R(O)}. \tag{27}$$

A similar derivation for

$$z = \sin x \tag{28}$$

gives

$$R_z(\tau) = \sum_{k=1,3,5,\ldots} \frac{e^{-R(O)}}{k!} R^k(\tau)$$

$$= e^{-R(O)} \sinh R(\tau) \tag{29}$$

$$R(\infty) = O. \tag{30}$$

The corresponding power spectra are obtained from (10) and (11). Let

$$z(t) = \cos[\omega_0 t + x(t)]$$
$$= \cos x(t) \cos \omega_0 t - \sin x(t) \sin \omega_0 t. \tag{31}$$

The corresponding autocorrelation function obtained by taking the time average of the produce $z(t)z(t + \tau)$ then becomes

$$R_z(\tau) = \tfrac{1}{2}[R_{\cos}(\tau) + R_{\sin}(\tau)]$$
$$= \tfrac{1}{2} e^{-R(O)} e^{R(\tau)}. \tag{32}$$

4. Ideal Limiter Transformations

An ideal linear limiter is represented by

$$z(x) = x \quad \text{for} \quad -c < x < c \tag{33a}$$
$$= c \quad \text{for} \quad -c > x > c \tag{33b}$$

This function can be represented by (1) with

$$f(i\lambda) = 2i \frac{\sin c\lambda}{\lambda^2}. \tag{34}$$

Solution of (7) yields the relation

$$h_{0k} = \frac{2}{\sqrt{\pi}} e^{-a^2} \frac{H_{k-2}(a)}{2^{(k-1)/2}}, \tag{35}$$

where $k = 1, 3, 5, \ldots$, $a^2 = c^2/2\sigma^2 = c^2/2R(O)$ and $H_m(a)$ is a Hermite function defined by

$$H_m(a) = (-1)^m e^{a^2} \frac{d^m}{da^m} e^{-a^2}, \tag{36}$$

$$H_1(a) = \frac{\sqrt{\pi}}{2} e^{a^2} \operatorname{erf}(a). \tag{37}$$

With (35) in (6)

$$R_z(\tau) = R(\tau)[\operatorname{erf}(a)]^2 + \frac{4}{\pi} e^{-2a^2} \sum_{k=3,5,7,\ldots}^{\infty} \frac{R^k(\tau) H_{k-2}^2(a)}{2^{k-1} k!}. \tag{38}$$

The function $e^{-a^2/2} H_m(a)$ has been tabulated in Reference 14 of Chapter 8.

Appendix 4
Determination of Optimum Filter Characteristics

1. General

In the optimization problems considered here, the product, ratio, or differences of integrals are involved, which contain the variable function $X(\omega)$ in the integrands. The optimum variation of X as a function of the variable of integration ω is to be determined.

Let Y be the product of two such integrals

$$Y = J_1[X(\omega)]J_2[X(\omega)], \tag{1}$$

where

$$J_1 = \int_{-\infty}^{\infty} H_1(\omega)\, H_2[X(\omega)]\, d\omega, \tag{2}$$

$$J_2 = \int_{-\infty}^{\infty} H_3(\omega)\, H_4[X(\omega)]\, d\omega. \tag{3}$$

The functions H under consideration have even symmetry about ω, so that the lower integration limits can be replaced by O.

With a small variation $\delta X(\omega)$, the resultant variation in Y is

$$\delta Y = J_2\, \delta J_1 + J_1\, \delta J_2 \tag{4}$$

$$= J_2 \int_{-\infty}^{\infty} H_1(\omega)\, H_2'[X(\omega)] \delta X(\omega)\, d\omega$$

$$+ J_1 \int_{-\infty}^{\infty} H_3(\omega)\, H_4'[X(\omega)] \delta X(\omega)\, d\omega$$

$$= \int_{-\infty}^{\infty} \{J_2 H_1(\omega)\, H_2'[X(\omega)]\, d\omega + J_1 H_3(\omega)\, H_4'[X(\omega)]\} \delta X(\omega)\, d\omega, \tag{5}$$

where H_2' and H_4' are derivatives with respect to X.

The minimum or maximum value of Y is obtained with $\delta Y = O$, which is the case when

$$H_1(\omega) H'_2[X(\omega)] + aH_3(\omega) H'_4[X(\omega)] = O, \qquad (6)$$

where the constant a is

$$a = \frac{J_1}{J_2}. \qquad (7)$$

When the ratio

$$Z = \frac{J_1^m[X(\omega)]}{J_2^n[X(\omega)]}, \qquad (8)$$

is to be minimized or maximized, (4) is replaced by

$$\begin{aligned}\delta Z &= J_2^n(J_1 + \delta J_1)^m - J_1^m(J_2 + \delta J_2)^n \\ &= J_2^n J_1^m + mJ_1^{m-1}\delta J_1 - J_1^m J_2^n - nJ_2^{n-1}\delta J_2 \\ &= mJ_2^n J_1^{m-1}\delta J_1 - nJ_2^{n-1}\delta J_1^m \delta J_2.\end{aligned} \qquad (9)$$

It follows that condition (6) again applies, except

$$a = -\frac{nJ_1^m J_2^{n-1}}{mJ_1^{m-1} J_2^n}.$$

2. Optimum Receiving Filter

The signal-to-noise in this case is given by (4.33) or

$$\rho_0 = \frac{PP_0}{J_1[X(\omega)]J_2[X(\omega)]}, \qquad (10)$$

where

$$J_1 = \int_0^{\Omega_0} \frac{s(\omega)\,d\omega}{l(\omega)r(\omega)}, \qquad (11)$$

$$J_2 = \int_0^{\Omega_0} n_i(\omega)r(\omega). \qquad (12)$$

Comparison with (2) and (3) shows that in this case

$$H_1(\omega) = \frac{s(\omega)}{l(\omega)}; \qquad H_2(\omega) = \frac{1}{r(\omega)} \qquad (13)$$

$$H_3(\omega) = n_i(\omega); \qquad H_4(\omega) = r(\omega). \qquad (14)$$

Relation (6) becomes

$$-\frac{s(\omega)}{l(\omega)}\frac{1}{r^2(\omega)} + cn_i(\omega) = O, \qquad (15)$$

or
$$r(\omega) = c\left(\frac{s(\omega)}{l(\omega)n_i(\omega)}\right)^{1/2}, \qquad (16)$$

where $c = (1/a)^{1/2}$. It will be noted that ρ_0 as given by (10) is independent of c. That the above solution gives the minimum value of $J_1 J_2$, that is, the maximum ρ_0 can be established by noting that it will always be possible to have a greater product $J_1 J_2$, except in the special case where $r(\omega) = $ constant.

3. Matched Filter Detection

In this case the signal-to-noise ratio is given by (4.100) or:
$$\rho(t) = Z = \frac{J_1^2}{J_2} \frac{1}{2\pi^2} \qquad (17)$$

where
$$J_1 = \int_{-\infty}^{\infty} \bar{S}_i(\omega)\, \bar{R}(\omega) e^{i[\omega t - \varphi(\omega) - \psi(\omega)]}\, d\omega \qquad (18)$$

$$J_2 = \int_{-\infty}^{\infty} n_i(\omega)\, \bar{R}^2(\omega)\, d\omega. \qquad (19)$$

Hence
$$H_1(\omega) = \bar{S}_i(\omega) e^{-i[\omega t - \varphi(\omega) - \psi(\omega)]}; \qquad H_2(\omega) = \bar{R}(\omega) \qquad (20)$$
$$H_3(\omega) = n_i(\omega); \qquad H_4(\omega) = \bar{R}^2(\omega). \qquad (21)$$

Relation (6) gives
$$H_1(\omega) + 2an_i(\omega)\, \bar{R}(\omega) = 0, \qquad (22)$$
$$a = -J_1/2J_2, \qquad (23)$$
or
$$\bar{R}(\omega) = c \cdot \frac{\bar{S}_i(\omega)}{n_i(\omega)} e^{-i[\omega t - \varphi(\omega) - \psi(\omega)]}, \qquad (24)$$

where $c = \frac{1}{2}a$. The ratio ρ given by (17) is independent of c.

For $\bar{R}(\omega)$ to be real it is necessary that $\psi(\omega) = \omega\tau - \varphi(\omega)$, and that $t = \tau$.

4. Optimum Smoothing Filter

Let $\bar{R}(\omega)$ be the amplitude characteristic of a receiving filter that gives distortion-free transmission in the absence of noise. The signal power spectrum at the input to the smoothing filter in Fig. 4.1 is then $s(\omega) = s_i(\omega)r(\omega)$ and the noise power spectrum is $n(\omega) = n_i(\omega)r(\omega)$.

The ratio of total interference power to average undistorted signal power in the output is given by (4.26) or

$$\mu_0 = \frac{N'}{P'} = \frac{J_1}{J_2}, \tag{25}$$

where

$$N' = \int_0^\infty [\bar{R}_0^2[s(\omega) + n(\omega)] - 2\bar{R}_0 s(\omega) + s(\omega)] \, d\omega, \tag{26}$$

$$P' = \int_0^\infty [2\bar{R}_0(\omega) - 1]s(\omega) \, d\omega. \tag{27}$$

In this case comparison with (2) and (3) shows that

$$H_1(\omega) = 1,$$

$$H_2(\omega) = \bar{R}_0^2[s(\omega) + n(\omega)] - 2\bar{R}_0(\omega)s(\omega) + s(\omega), \tag{29}$$

$$H_3(\omega) = s(\omega), \tag{30}$$

$$H_4(\omega) = 2\bar{R}_0(\omega) - 1. \tag{31}$$

Relation (6) gives

$$2\bar{R}_0(\omega)[s(\omega) + n(\omega)] - 2s(\omega) - 2\mu_0 s(\omega) = O \tag{32}$$

$$\bar{R}_0(\omega) = \frac{s(\omega)}{s(\omega) + n(\omega)} (1 + \mu_0) \tag{33}$$

The minimum ratio μ_0 obtained with (33) in (25) becomes

$$\mu_0 = \frac{\int_0^\infty \frac{s(\omega)}{s(\omega) + n(\omega)} [n(\omega) + \mu_0^2 s(\omega)] \, d\omega}{\int_0^\infty \frac{s(\omega)}{s(\omega) + n(\omega)} [(1 + 2\mu_0)s(\omega) - n(\omega)] \, d\omega}. \tag{34}$$

5. Channel Capacity with Multiband Transmission

The channel capacity with multiband transmission is given by (4.125), or

$$C = \frac{1}{2\pi} \int_0^{\Omega_i} \log_2 \left(1 + a \frac{s_i(\omega)}{n_i(\omega)}\right) d\omega. \tag{35}$$

The average transmitter power is

$$P_0 = \int_0^{\Omega_i} s_0(\omega) \, d\omega = \int_0^{\Omega_i} \frac{s_i(\omega)}{l(\omega)} \, d\omega. \tag{36}$$

Optimization of Filter Characteristics

The ratio to be maximized is

$$Z = \frac{C}{P_0} = \frac{J_1}{J_2}. \tag{37}$$

In this case, comparison with (2) and (3) gives

$$H_1(\omega) = 1; \quad H_2(\omega) = \log_2\left(1 + a\frac{s_i(\omega)}{n_i(\omega)}\right) \tag{38}$$

$$H_3(\omega) = 1; \quad H_4(\omega) = s_0(\omega) = \frac{s_i(\omega)}{l(\omega)}. \tag{39}$$

Relation (6) in this case gives

$$\frac{\frac{a}{n_i(\omega)}}{1 + a\frac{s_i(\omega)}{n_i(\omega)}} - \frac{c}{l(\omega)} = 0, \tag{40}$$

or

$$s_0(\omega) + \frac{n_i(\omega)}{al(\omega)} = \lambda, \tag{41}$$

where

$$\lambda = \frac{1}{c} = \frac{J_2}{J_1} = \text{constant}.$$

Integration of (41) with respect to ω yields the following relation for determination of λ

$$P_0 + \int_0^{\Omega_i} \frac{n_i(\omega)}{al(\omega)} d\omega = \lambda\Omega_i. \tag{42}$$

When λ as obtained from (42) is inserted in (41), relation (4.129) is obtained.

References

1. Basic Transmission Concepts and Relations

[1] H. Nyquist, "Certain Topics in Telegraph Transmission Theory," *AIEE Trans.*, 47, p. 617, April, 1928.
[2] C. E. Shannon, "A Mathematical Theory of Communication," *B.S.T.J.*, 27, p. 379, Oct., 1948.
[3] D. Gabor, "Theory of Communication," *JIEE*, 93, Part III, p. 429, 1946.
[4] H. W. Bode, *Network Analysis and Negative Feedback Amplifier Design*, D. Van Nostrand Company, Inc., 1945.
[5] Y. W. Lee, "Synthesis of Electric Networks by Means of Fourier Transforms of Laguerre's Functions," *Journal of Math. and Phys.*, 11, p. 83, June, 1932.
[6] E. D. Sunde, "Theoretical Fundamentals of Pulse Transmission," *B.S.T.J.*, 33, p. 987, May and July, 1954.
[7] ———, "Ideal Binary Pulse Transmission by AM and FM," *B.S.T.J.*, 38, p. 1357, Nov., 1959.
[8] W. F. McGee, "Bandlimiting Distortionless FSK Filters," *IEEE Trans. on Comm. Technology*, COM-15, p. 721, Oct., 1967.
[9] B. R. Saltzberg and L. Kurz, "Design of Bandlimited Signals for Binary Communications Using Simple Correlation Detection," *B.S.T.J.*, 44, p. 235, Feb., 1965.
[10] D. Slepian and H. O. Pollack, "Prolate Spheroidal Wave Functions, Fourier Analysis and Uncertainty," *B.S.T.J.*, 40, p. 43, Jan., 1961.
[11] H. J. Landau and H. O. Pollack, "Prolate Spheroidal Wave Functions, Fourier Analysis and Uncertainty," *B.S.T.J.*, 40, p. 65, Jan., 1961.
[12] D. M. DiToro and K. Steiglitz, "Application of the Maximum Principle to the Design of Minimum Bandwidth Pulses," *IEEE Trans. on Comm. Technology*, COM-13, p. 433, Dec., 1965.

2. Basic Modulation Concepts and Objectives

[1] M. Schwartz, W. R. Bennett, and S. Stein, *Communication Systems and Techniques*, McGraw-Hill Book Company, 1966.
[2] C. E. Shannon, "A Mathematical Theory of Communication," *B.S.T.J.*, 27, p. 379, Nov., 1948.

[3] *Proc. IEEE*, 55, March, 1967. Special Issue on "Redundancy Reduction Techniques and Applications."
[4] H. B. Holbrook and J. T. Dixon, "Load Rating Theory of Multichannel Amplifiers," *B.S.T.J.*, 18, p. 624, Oct., 1939.
[5] R. F. Purton, "A Survey of Telephone Speech-Signal Statistics and their Significance in the Choice of a PCM Companding Law," *Proc. IEEE*, 109, Part B, p. 60, Jan., 1962.
[6] E. E. David, M. R. Schroeder, B. F. Logan and A. J. Prestigiacomo, "Voice Exited Vocoders for Practical Bandwidth Compression," *IRE Trans. on Information Theory*, IT-8, p. 101, Sept., 1962.
[7] C. W. Carter, A. C. Dickieson, and D. Mitchell, "Application of Compandors to Telephone Circuits," *Trans. AIEE*, 65, p. 1079, 1946.
[8] C. B. Feldman and W. R. Bennett, "Bandwidth and Transmission Performance," *B.S.T.J.*, 28, p. 490, July, 1949.
[9] B. Smith, "Instantaneous Companding of Quantized Signals," *B.S.T.J.*, 36, p. 635, May, 1957.
[10] J. B. O'Neal, "Delta Modulation Quantizing Noise—Analytical and Computer Simulation Results for Gaussian and Television Input Signals," *B.S.T.J.*, 45, p. 117, Jan., 1966.
[11] E. Bedrosian, "The Analytic Signal Representation of Modulated Wave Forms," Proc. *IRE*, 50, p. 2071, Oct., 1962.
[12] H. B. Voelcker, "Toward a Unified Theory of Modulation,"—Part I, "Phase Envelope Relationships," *Proc. IEEE*, 54, p. 340, March, 1966.—Part II, "Zero Manipulation," 54, p. 735, May, 1966.
[13] A. J. Giger and J. G. Chaffee, "Telstar Experiment—The FM Demodulator with Negative Feedback," *B.S.T.J.*, 42, p. 1109, July, 1963, Part I.
[14] M. J. Cyr and A. Thuswalder, "Multichannel Load Calculation using the Monte Carlo Method," *IEEE Trans. on Comm. Technology*, COM-14, p. 177, April, 1966.
[15] W. R. Bennett, "Distribution of the Sum of Randomly Phased Components," *Quarterly of Applied Math.*, 5, p. 385, Jan., 1948.
[16] ———, "Cross-Modulation Requirements on Multichannel Amplifiers Below Overload," *B.S.T.J.*, 19, p. 587, Oct., 1940.
[17] O. E. DeLange, "The Timing of High-Speed Regenerative Repeaters," *B.S.T.J.*, 37, p. 1455, Nov., 1958.
[18] W. R. Bennett, "Statistics of Regenerative Repeaters," *B.S.T.J.*, 37, p. 1501, Nov., 1958.
[19] C. J. Byrne, B. J. Karafin, and D. B. Robinson, "Systematic Jitter in a Chain of Regenerators," *B.S.T.J.*, 42, p. 2679, Nov., 1963.
[20] R. S. Caruthers, "The Type N1 Carrier System—Objectives and Transmission Features," *B.S.T.J.*, 30, p. 1, Jan., 1951.
[21] C. H. Elmendorf, R. D. Ehrbar, R. H. Klie, and A. J. Grossman, "The L3 Coaxial System—System Design," *B.S.T.J.*, 32, p. 781, July, 1953.
[22] M. R. Aaron, "PCM Transmission in the Exchange Plant," *B.S.T.J.*, 41, p. 99, Jan., 1962.
[23] E. D. Sunde, "Theoretical Fundamentals of Pulse Transmission," *B.S.T.J.*, 33, p. 987, May and July, 1954.
[24] B. M. Oliver, J. R. Pierce, and C. E. Shannon, "The Philosophy of PCM," *Proc. IRE*, 36, p. 1234, Nov., 1948.
[25] J. M. Wozencraft and I. M. Jacobs, *Principles of Communication Engineering*, John Wiley & Sons, Inc., 1965.

[26] S. O. Rice, "Communication in the Presence of Noise," *B.S.T.J.*, 29, p. 60, Jan., 1950.
[27] W. W. Peterson, *Error Correcting Codes*, John Wiley & Sons, Inc., 1961.

3. Statistical Properties of Gaussian Random Variables

[1] S. O. Rice, "Mathematical Analysis of Random Noise," *B.S.T.J.*, 23, p. 282, July, 1944; and 24, Jan., 1945.
[2] ———, "Properties of Sine Wave Plus Random Noise," *B.S.T.J.*, 27, p. 109, Jan., 1948.
[3] W. R. Bennett, "Methods of Solving Noise Problems," *Proc. IRE*, 44, p. 609, May, 1956.
[4] ———, "Response of Linear Rectifier to Signals and Random Noise," *Acoust. Soc. Am. Jour.*, Jan., 1944; and *B.S.T.J.*, 23, p. 97, Jan., 1944.
[5] Robert Price, "A Note on Envelope and Phase Modulated Components of Narrow-Band Gaussian Noise," *IRE Trans. on Information Theory*, IT-1, p. 9, Sept., 1955.
[6] A. Levine and R. B. McGee, "Cumulative Distribution Functions of Sinoid Plus Gaussian Noise," *IRE Trans. on Information Theory*, IT-5, p. 90, June, 1959.
[7] J. Salz and S. Stein, "Distribution of Instantaneous Frequency for Signals Plus Noise," *IEEE Trans. on Information Theory*, IT-10, p. 272, Oct., 1964.
[8] S. O. Rice, "Noise in F. M. Receivers," *Time Series Analysis* (M. Rosenblatt, Editor), John Wiley & Sons, Inc., 1963.
[9] W. B. Davenport, Jr. and W. L. Root, *Random Signals and Noise*, McGraw-Hill Book Company, 1956.
[10] J. H. Laning and R. H. Battin, *Random Processes in Automatic Control*, McGraw-Hill Book Company, 1956.
[11] A. Papoulis, *Probability, Random Variables and Stochastic Processes*, McGraw-Hill Book Company, 1965.

4. Transmission Systems Optimization Against Gaussian Noise

[1] J. R. Pierce, "Physical Sources of Noise," *Proc. IRE*, 44, p. 601, May, 1956.
[2] W. R. Bennett, *Electrical Noise*, McGraw-Hill Book Company, 1960.
[3] D. A. Bell, *Electrical Noise*, D. Van Nostrand Company, London, 1959.
[4] B. M. Oliver, "Thermal and Quantum Noise," *Proc. IEEE*, 53, p. 436, May, 1965.
[5] C. E. Shannon, "Communication in the Presence of Noise," *Proc. IRE*, 37, p. 10, Jan., 1949.
[6] N. Wiener, *Extrapolation, Interpolation and Smoothing of Stationary Time Series*, John Wiley & Sons, Inc., 1949.
[7] H. W. Bode and C. E. Shannon, "A Simplified Derivation of Linear Least-Square Smoothing and Prediction Theory," *Proc. IRE*, 38, p. 417, April, 1950.
[8] J. W. Smith, "The Joint Optimization of Transmitted Signal and Receiving Filter for Data Transmission," *B.S.T.J.*, 44, p. 2363, Dec., 1965.
[9] T. Berger and D. W. Tufts, "Optimum Pulse Amplitude Modulation—Part I, Transmitter-Receiver Design and Bounds from Information Theory," *IEEE Trans. on Information Theory*, IT-13, p. 196, April, 1967.

[10] M. R. Aaron and D. W. Tufts, "Intersymbol Interference and Error Probability," *IEEE Trans. on Information Theory*, IT-12, p. 26, Jan., 1966.
[11] G. L. Turin, "An Introduction to Matched Filter," *IRE Trans. on Information Theory*, IT-6, p. 311, June, 1960.
[12] B. M. Oliver, J. R. Pierce, and C. E. Shannon, "The Philosophy of PCM," *Proc. IRE*, 36, p. 1324, Nov., 1948.
[13] J. R. Pierce, "Information Rate of a Coaxial Cable with Various Modulation Systems," *B.S.T.J.*, 45, p. 1197, Oct., 1966.
[14] R. W. Chang and S. L. Freeny, "Hybrid Digital Transmission Systems—Part 1: Joint Optimization of Analog and Digital Repeaters," *B.S.T.J.*, 47, p. 1663, Oct., 1968.
[15] S. L. Freeny and R. W. Chang, "Hybrid Digital Transmission Systems—Part 2: Information Rate of Hybrid Coaxial Cable Systems," *B.S.T.J.*, 47, p. 1687, Oct., 1968.

5. Digital Transmission and Modulation Methods

[1] W. R. Bennett and J. R. Davey, *Data Transmission*, McGraw-Hill Book Company, 1965.
[2] M. R. Aaron, "PCM Transmission in the Exchange Plant," *B.S.T.J.*, 41, p. 99, Jan., 1962.
[3] W. R. Bennett and S. O. Rice, "Spectral Density and Autocorrelation Functions with Binary Frequency Shift Keying," *B.S.T.J.*, 42, p. 2355, Sept., 1963.
[4] R. R. Anderson and J. Salz, "Spectra for a Class of Asynchronous FM Waves," *B.S.T.J.*, 44, p. 2149, July–August, 1965.
[5] E. D. Sunde, "Self-Timing Regenerative Repeaters," *B.S.T.J.*, 36, p. 891, July, 1957.
[6] W. R. Bennett, "Statistics of Regenerative Digital Transmission," *B.S.T.J.*, 37, p. 1501, Nov., 1958.
[7] H. E. Rowe, "Timing in a Long Chain of Regenerative Binary Repeaters," *B.S.T.J.*, 37, p. 1543, Nov., 1958.
[8] C. J. Byrne, B. J. Karafin, and D. B. Robinson, "Systematic Jitter in a Chain of Regenerators," *B.S.T.J.*, 42, p. 2679, Nov., 1963.
[9] J. M. Sipress, "A New Class of Ternary Pulse Transmission Plans for Digital Transmission Lines," *IEEE Trans. on Comm. Technology*, COM-13, p. 366, Sept., 1965.
[10] R. L. Didday and W. C. Lindsey, "Subcarrier Tracking Methods and Communication Systems Design," *IEEE Trans. on Comm. Technology*, COM-16, p. 541, August, 1968.
[11] A. Lender, "Correlative Digital Transmission Techniques," *IEEE Trans. on Comm. Systems*, CS-12, p. 128, Dec., 1964.
[12] R. D. Howson, "An Analysis of the Capabilities of Polybinary Data Transmission," *IEEE Trans. on Comm. Technology*, COM-13, p. 312, Sept., 1965.
[13] W. R. Bennett, "Methods of Solving Noise Problems," *Proc. IRE*, 44, p. 609, May, 1956.
[14] J. G. Lawton, "Comparison of Binary Data Transmission Systems," *Proc. of the Second National Conference on Military Electronics*, 1958.
[15] C. R. Cahn, "Combined Digital Amplitude and Phase Modulation," *IRE Trans. on Comm. Systems*, CS-8, p. 150, Sept., 1960.

[16] S. Reiger, "Error Probability of Binary Data Transmission Systems in the Presence of Noise," *IRE Convention Record*, Part 8, p. 72, 1953.
[17] W. R. Bennett and J. Salz, "Binary Data Transmission over a Real Channel," *B.S.T.J.*, 42, p. 2387, Sept., 1963.
[18] A. J. Viterby, "On Coded Phase-Coherent Communications," *IRE Trans. on Space Electronics and Telemetry*, SET-7, p. 3, March, 1961.
[19] E. N. Gilbert, "A Comparison of Signaling Alphabets," *B.S.T.J.*, 31, p. 504, May, 1952.
[20] R. W. Hamming, "Error Detecting and Correcting Codes," *B.S.T.J.*, 29, p. 147, April, 1950.
[21] R. C. Bose and D. K. Ray-Chauduri, "On a Class of Error Correcting Binary Group Codes," *Information and Control*, 3, p. 68, March, 1960.
[22] W. W. Peterson, "Encoding and Error Correcting Procedures for Bose-Chauduri Codes," *IRE Trans. on Information Theory*, IT-6, p. 459, Sept., 1960.
[23] M. E. Mitchell, "Performance of Error Correcting Codes," *IRE Trans. on Comm. Systems*, CS-10, p. 72, March, 1962.
[24] Martin Nesenbergs, "Comparison of 3 out of 7 ARQ with Bose-Chauduri-Hocqenghem Coding Systems," *IEEE Trans. on Comm. Systems*, CS-11, p. 202, June, 1963.
[25] R. J. Benice and A. H. Frey, "An Analysis of Retransmission Systems," and "Comparison of Error Control Techniques," *IEEE Trans. on Comm. Technology*, COM-12, pp. 135, 146, Dec., 1964.
[26] M. Horstein, "Efficient Communication Through Burst-Error Channels by Means of Error Detection," *IEEE Trans. on Comm. Technology*, COM-14, p. 117, April, 1966.
[27] R. W. Chang and R. A. Gibby, "A Theoretical Study of Performance of an Orthogonal Multiplexing Data Transmission Scheme," *IEEE Trans. on Comm. Technology*, COM-16, p. 529, August, 1968.
[28] M. L. Doelz, E. T. Heald and D. L. Martin, "Binary Data Transmission Techniques for Linear Systems," *Proc. IRE*, 45, p. 656, May, 1957.
[29] A. A. Alexander, D. W. Nast and R. M. Cryb, "Capabilities of the Telephone Network for Data Transmission," *B.S.T.J.*, 39, p. 431, May, 1960.
[30] P. Mertz, "Model of Impulsive Noise for Data Transmission," *IRE Trans. on Comm. Systems*, CS-9, p. 130, June, 1961.
[31] J. H. Halton and A. D. Spaulding, "Error Rates in Differentially Coherent Phase Systems in Non-Gaussian Noise," *IEEE Trans. on Comm. Technology*, COM-14, p. 594, Oct., 1966.

6. Attenuation and Phase Distortion in Analog Systems

[1] P. Mertz and K. W. Pfleger, "Irregularities in Broadband Wire Transmission Circuits," *B.S.T.J.*, 16, p. 541, Oct., 1937.
[2] H. Kaden, "Advances in Statistics of Impedance Irregularities in Television Cable," *Archiv der Elek. Übertragung*, 8, p. 523, 1954.
[3] A. Rosen, "The Unit Treatment of Impedance Irregularities and its Application to Long Lines," *Proc. IEE*, 101, Part III, p. 394, 1954.
[4] A. Kaufman, "Bibliography on Nonuniform Transmission Lines," *IRE Trans. on Ant. and Propagation*, AP-3, p. 218, Oct., 1955.

[5] H. E. Rowe and W. D. Waters, "Transmission in Multimode Wave Guide with Random Imperfections," *B.S.T.J.*, 41, p. 1031, May, 1962.
[6] S. O. Rice, "Distortion Produced by a Noise Modulated Signal by Nonlinear Attenuation and Phase Shift," *B.S.T.J.*, 36, p. 879, July, 1957.
[7] W. R. Bennett, H. E. Curtis, and S. O. Rice, "Interchannel Interference in FM and PM Systems under Noise Loading Conditions," *B.S.T.J.*, 34, p. 601, May, 1955.
[8] R. G. Medhurst and G. F. Small, "Distortion in Frequency Modulation Systems Due to Small Sinusoidal Variations of Transmission Characteristics," *Proc. IRE*, 44, p. 1608, Nov., 1956.
[9] E. Bedrosian and S. O. Rice, "Distortion and Crosstalk in Linearly Filtered Angle-Modulated Signals," *Proc. IEEE*, 56, p. 2, Jan., 1968.
[10] E. D. Sunde, "Intermodulation Distortion in Analog FM Troposcatter Systems," *B.S.T.J.*, 43, Part II. p. 399, Jan., 1964,
[11] C. D. Beach and J. M. Trecher, "A Method for Predicting Interchannel Modulations Due to Multipath Propagation in FM and PM Tropospheric Radio Systems," *B.S.T.J.*, 42, p. 1, Jan., 1963.
[12] N. Abrahamson, "Bandwidth and Spectra of Phase-and-Frequency Modulated Waves," *IEEE Trans. on Comm. Systems*, CS-11, p. 407, Dec., 1963.
[13] H. E. Rowe, *Signals and Noise in Communication Systems*, D. Van Nostrand Company, 1965.
[14] C. C. Ferris, "Spectral Characteristics of FDM-FM Signals," *IEEE Trans. on Comm. Technology*, COM-16, p. 233, April, 1968.
[15] T. G. Cross, "Intermodulation Noise in FM Systems Due to Transmission Deviations and AM/PM Conversion," *B.S.T.J.*, 45, p. 1749, Dec., 1966.
[16] M. L. Liou, "Noise in an FM System Due to an Imperfect Linear Transducer," *B.S.T.J.*, 45, p. 1537, Nov., 1966.
[17] G. J. Garrison, "Intermodulation Distortion in Frequency-Division-Multiplex FM Systems—A Tutorial Summary," *IEEE Trans. on Comm. Technology*, COM-16, p. 289, April, 1968.

7. Attenuation and Phase Distortion in Digital Systems

[1] E. D. Sunde, "Pulse Transmission by AM, FM, and PM in the Presence of Phase Distortion," *B.S.T.J.*, 40, p. 359, March, 1961.
[2] R. A. Gibby, "An Evaluation of AM Data System Performance by Computer Simulation," *B.S.T.J.*, 39, p. 657, May, 1960.
[3] M. A. Rappaport, "Digital Computer Simulation of a Four-Phase Data Transmission System," *B.S.T.J.*, 43, p. 927, May, 1964.
[4] R. W. Lucky and H. R. Rudin, "An Automatic Equalizer for General-Purpose Communication Channels," *B.S.T.J.*, 46, p. 2179, Nov., 1967. (Contains comprehensive bibliography on adaptive equalization.)
[5] B. R. Saltzberg, "Error Probabilities for Binary Signal Perturbed by Intersymbol Interference and Gaussian Noise," *IEEE Trans. on Comm. Systems*, CS-12, p. 117, March, 1964.
[6] M. L. Doelz, E. T. Heald, and D. L. Martin, "Binary Data Transmission Techniques for Linear Systems," *Proc. IRE*, 45, p. 656, May, 1957.
[7] B. R. Saltzberg and M. K. Simon: "Data Transmission Error Probabilities in the Presence of Low-Frequency Removal and Noise," *B.S.T.J.*, 48, p. 255 Jan. 1969.

8. Intermodulation Distortion in Nonlinear Channels

[1] W. R. Bennett, "Cross-Modulation Requirements on Multi-Channel Amplifiers below Overload," *B.S.T.J.*, 19, p. 587, Oct., 1940.
[2] R. A. Brockbank and C. A. A. Wass, "Nonlinear Distortion in Transmission Systems," *JIEE*, 92, Part III, p. 45, March, 1945.
[3] O. T. M. Smith, "Statistical Spectral Output of Power Law Nonlinearity," and "Spectral Output of Piecewise Linear Nonlinearity," *AIEE Trans. on Communication and Electronics*, 78, p. 535, Nov., 1959.
[4] W. R. Bennett and S. O. Rice, "Note on Methods of Computing Modulation Products," *Phil. Magazine*, Series 7, 18, p. 422, Sept., 1934.
[5] S. O. Rice, "Mathematical Analysis of Random Noise," *B.S.T.J.*, 23, p. 282, July, 1944; and 24, Jan., 1945.
[6] W. B. Davenport and W. L. Root, *Random Signals and Noise*, McGraw-Hill Book Company, 1956.
[7] J. H. Laning and R. H. Battin, *Random Processes in Automatic Control*, McGraw-Hill Book Company, 1956.
[8] C. R. Cahn: "Crosstalk Due to Finite Limiting of Frequency Multiplexed Signals," *Proc. IRE*, 48, p. 53, Jan., 1960.
[9] R. G. Medhurst, J. H. Roberts, and W. R. Walsh, "Distortion of SSB Transmission Due to AM-PM Conversion," *IEEE Trans. on Comm. Systems*, CS-12, p. 166, June 1964.
[10] E. D. Sunde, "Intermodulation Distortion in Multicarrier FM Systems," *IEEE Int. Convention Record*, Part II, p.130, 1965.
[11] W. R. Bennett, "Distribution of the Sum of Randomly Phased Components," *Quarterly of Mathematics*, 5, p. 385, Jan., 1948.
[12] ———, "The Biased Ideal Rectifier," *B.S.T.J.*, 26, p. 139, Jan., 1947.
[13] W. Magnus and F. Oberhettinger, *Special Functions of Mathematical Physics*, Chelsea Publishing Company, 1949.
[14] J. B. Russel, "A Table of Hermite Functions," *Journal of Mathematics and Physics*, 12, p. 91, 1933.
[15] P. D. Shaft, "Limiting of Several Signals and its Effect on Communication Systems Performance," *IEEE Trans. on Comm. Technology*, COM-13, p. 504, Dec., 1965.
[16] T. P. Laico, H. L. McDowell, and R. C. Moster, "A Medium Power Traveling Wave Tube for 6000 mc. Radio Relay," *B.S.T.J.*, 35, p. 1285, Nov., 1956.
[17] M. G. Bodmer, T. P. Laico, E. G. Olsen, and A. T. Ross, "The Satellite Traveling Wave Tube," *B.S.T.J.*, 42, Part III, p. 1703, July, 1963.
[18] R. J. Westcott, "Investigation of Multiple FM/FDM Carriers Through a Satellite TWT Operating Near to Saturation," *Proc. IEE*, 114, p. 726, June, 1967.
[19] R. C. Chapman and J. B. Millard, "Intelligible Crosstalk between Frequency Modulated Carriers Through AM-PM Conversion," *IEEE Trans. on Comm. Systems*, CS-12, p. 160, June, 1964.

9. Random Multipath and Troposcatter Transmittances

[1] L. A. Zadeh, "Frequency Analysis of Variable Networks," *Proc. IRE*, 38, p. 291, March, 1950.

References

[2] P. A. Bello, "Characterization of Randomly Time-Variant Linear Channels," *IEEE Trans. on Comm. Systems*, CS-11, p. 360, Dec., 1963.

[3] S. O. Rice, "Radio Field Strengths Statistical Fluctuations beyond the Horizon," *Proc. IRE*, 41, p. 379, Feb., 1953.

[4] S. O. Rice, "Distribution of the Duration of Fades in Radio Transmission," *B.S.T.J.*, 37, p. 581, May, 1958.

[5] K. Bullington, "Radio Propagation Fundamentals," *B.S.T.J.*, 36, p. 593, May, 1957.

[6] A. B. Crawford, D. C. Hogg, and W. H. Kummer, "Studies in Tropospheric Propagation beyond the Horizon," *B.S.T.J.*, 38, p. 1607, Sept., 1959.

[7] N. R. Ortwein, R. U. F. Hopkins, and J. E. Pohl, "Properties of Tropospheric Scatter Fields," *Proc. IRE*, 49, p. 788, April, 1961.

[8] W. S. Patrick and M. J. Wiggins, "Experimental Studies of the Correlation Bandwidth of the Troposcatter Medium at Five Gigacycles," *IEEE Trans. on Aerosp. and Nav. Electronics*, ANE-10, p. 133, June, 1963.

[9] C. E. Clutts, R. N. Kennedy and J. M. Trecker, "Results of Bandwidth Tests on the 185-Mile Florida–Cuba Scatter Radio System," *IRE Trans. on Comm. Systems*, CS-9, p. 434, Dec., 1961.

[10] C. D. Beach and J. M. Trecher, "A Method for Predicting Interchannel Modulation Due to Multipath Propagation in FM and PM Tropospheric Radio Systems," *B.S.T.J.*, 42, p. 1, Jan., 1963.

[11] E. D. Sunde, "Intermodulation Distortion in Analog FM Troposcatter Systems," *B.S.T.J.*, 43, Part II, p. 399, Jan., 1964.

[12] D. C. Brennan, "Linear Diversity Combining Techniques," *Proc. IRE*, 47, p. 1075, June, 1959.

10. Digital Transmission over Random Multipath Channels

[1] G. L. Turin, "Error Probabilities for Binary Symmetric Ideal Reception through Nonselective Slow Fading and Noise," *Proc. IRE*, 46, p. 1603, Sept., 1958.

[2] J. N. Pierce, "Theoretical Diversity Improvement in Frequency Shift Keying," *Proc. IRE*, 46, p. 903, May, 1958.

[3] J. N. Pierce, "Theoretical Limitations on Frequency and Time Diversity for Fading Binary Transmission," *IRE Trans. on Comm. Systems*, CS-9, p. 186, June, 1961.

[4] H. B. Voelcker, "Phase–Shift Keying in Fading Channels," *Proc. IEEE*, 107, Part B, p. 31, Jan., 1960.

[5] P. A. Bello and D. B. Nelin, "The Effect of Frequency-Selective Fading on Binary Error Probabilities of Incoherent and Differentially Coherent Matched Filter Receivers," *IEEE Trans. on Comm. Systems*, CS-11, p. 170, June, 1963; and CS-12, p. 230, Dec., 1964. (Corrections.)

[6] E. D. Sunde, "Digital Troposcatter Transmission and Modulation Theory," *B.S.T.J.*, 43, p. 143, Jan., 1964, Part I.

[7] C. C. Bailey and J. C. Lindenlaub, "Further Results Concerning the Effect of Frequency-Selective Fading on Differentially Coherent Matched Filter Receivers," *IEEE Trans. on Comm. Systems*, COM-16, p. 749, Oct., 1968.

[8] B. B. Barrow, *Error Probabilities for Data Transmission over Fading Channels*, Van Gorcum & Co., Assen, Netherlands.

[9] W. C. Lindsey, "Error Probabilities for Incoherent Diversity Reception," *IEEE Trans. on Information Theory*, IT-11, p. 491, Oct., 1965.
[10] M. Schwartz, W. R. Bennett, and S. Stein, "Communication Systems and Techniques," McGraw-Hill Book Company, 1966.
[11] D. P. Harris, "Techniques for Incoherent Scatter Communication," *IRE Trans. on Comm. Systems*, CS-10, p. 154, June, 1962.
[12] H. E. White, "Failure Correction Decoding," *IEEE Trans. on Comm. Technology*, COM-15, p. 23, Feb., 1967; Addendum COM-15, p. 726, Oct., 1967.
[13] B. B. Barrow, "Error Probabilities for Telegraph Signals Transmitted on a Fading FM Carrier," *Proc. IRE*, 48, p. 1613, Sept., 1960.
[14] "Use of Sunde Tropochannel Model to Predict Error Probabilities of AMRT Tests," Final Report RADC-TR-65-427 by The Staff of Signal Processing Section-Communication Research Branch; Rome Air Development Center, Air Force Systems Command, Griffiss Air Force Base, New York, Dec., 1965.
[15] M. J. DiToro, "Communication in Time-Frequency Spread Media Using Adaptive Equalization," *Proc. IEEE*, 56, p. 1653, Oct., 1958.

11. Intersystem Interference

[1] G. A. Campbell, "Crosstalk Formulas for Non-Loaded Circuits," *B.S.T.J.*, 14, p. 559, Oct., 1935.
[2] A. G. Chapman, "Open-Wire Crosstalk," *B.S.T.J.*, 13, p. 19, April, 1934.
[3] S. A. Schelkunoff and T. M. Odarenko, "Crosstalk Between Coaxial Transmission Lines," *B.S.T.J.*, 16, p. 144, April, 1937.
[4] M. A. Weaver, R. S. Tucker, and P. S. Darnell, "Crosstalk and Noise Features of Cable Carrier Telephone," *B.S.T.J.*, 17, p. 137, Jan., 1938.
[5] R. S. Caruthers, "The Type N-1 Carrier System–Objectives and Transmission Features," *B.S.T.J.*, 30, p. 1, Jan., 1951.
[6] H. Cravis and T. V. Crater, "Engineering of T1 Carrier System Repeatered Lines," *B.S.T.J.*, 42, p. 431, March, 1963.
[7] H. E. Curtis, "Radio Frequency Interference Considerations in TD2 Radio Relay System," *B.S.T.J.*, 39, p. 369, March, 1960.
[8] R. G. Medhurst, "RF Spectra and Interfering Carrier Distortion in FM Trunk Radio Systems with Low Modulation Ratios," *IRE Trans. on Comm. Systems*, CS-9, p. 107, June, 1961.
[9] G. J. Garrison, "An Extended Analysis of RF Interference in FDM-FM Radio Relay Systems," *IEEE Trans. on Communication Technology*, COM-15, p. 705, Oct. 1967.
[10] H. E. Curtis, "Interference between Satellite Communication Systems and Common Carrier Surface Systems," *B.S.T.J.*, 41, p. 921, May, 1962.
[11] E. A. Marcatili, "Time and Frequency Crosstalk in Pulse Modulation Systems," *B.S.T.J.*, 40, p. 951, May, 1961.
[12] *IEEE Trans. on Electromagnetic Compatibility*, EMC-10, March, 1968; Special Issue on Shielding.
[13] *IEEE Trans. on Electromagnetic Compatibility*, EMC-10, June 1968; Special Issue on Filters.
[14] E. D. Sunde, *Earth Conduction Effects in Transmission Systems*, D. Van Nostrand Company, 1949. Reprint by Dover Publications, Inc., New York, 1968.

Index

Acoustic speech spectrum, 66
Adaptive equalization, 268, 420
Addition laws, see Combining laws
Additive noise, see Gaussian random variables
AM analog systems, applications, 97
  attenuation and phase deviations, 240
  equalization, see Filter optimization
  demodulation methods, 84, 97
  filter optimization, 143
  fine-structure distortion, 240
  intermodulation distortion, 308
  intersystem interference, 428, 430
  random noise, 143
AM digital systems, 176
  attenuation and phase deviations, 266, 272, 274
  carrier extraction, 189
  channel capacity, 86
  detection methods, 84, 85, 86, 177, 201
  error probabilities (random noise), 53, 200
  filter optimization, 154
  fine-structure deviations, 240
  intersystem interference, 441
  noise, 200
  nonlinear distortion, 356
  pulse shapes, 53, 154
  timing jitter, 187
  timing wave extraction, 184
Amplification nonlinearity, see Nonlinear amplification
Amplitude distributions, see Probability distributions
Amplitude-frequency characteristics, 16, 22, 226
  amplitude-phase relations, 23, 28
  in-phase and quadrature components, 19
  see also Idealized channels; Spectra
Amplitude modulation (AM), 79
AM-PM combined, 179, 180
AM-PM conversion, 310
  conversion factors, 340
  intelligible crosstalk, 354
  intermodulation distortion, 262, 342, 446
Analog modulation methods, AM, 79
  FM, 80
  PM, 80
  single sideband AM, 79
  single sideband FM, 81
Analog pulse transmission, 17, 148
Analytic (complex) signals, 6, 21
Angle modulation, 86
Antenna beam angles (troposcatter), 375
Attenuation, metallic pairs, 165
Attenuation and phase distortion, effects, 224, 266
  see also AM analog systems; AM digital systems; Baseband analog systems; Baseband digital systems; FM and PM analog systems; FM digital systems; and PM digital systems
Attenuation deviations, 225, 226
  coarse structure, 226, 228
  cosine and sine deviations, 28, 232
  fine structure, 225, 228
  Fourier series deviations, 236
  linear deviations, 243, 282
  power series deviations, 237, 265
Attenuation-phase relations, 23
  band-pass channels, 26
  high-pass channels, 26
  low-pass channels, 23
Autocorrelation functions, 62, 476
  Gaussian random variables, 109
  nonlinear transducers, 317, 483
  time-variant channels, 364, 365, 374
  see also Power spectra
Average power, 61, 72
Averages (statistics), 470

Band-limited functions, 9
Bandpass channel filters, 5
  amplitude-phase relations, 23, 25
  impulse characteristic, 17
  in-phase and quadrature components, 18
  signal transmission, 17
  translation of reference frequency, 17
Bandpass transmittance deviation effects, see AM analog systems; AM digital systems; FM and PM analog systems; FM digital systems; and PM digital systems
Bandpass transmittance deviations, coarse structure, 228
  cosine and sine, 234
  fine structure, 228, 291
  linear amplitude, 243

linear delay, 244
power series, 237
quadratic delay, 283
Bandwidth requirements, analog AM, 79
   analog FM and PM, 81, 260
   analog single sideband AM, 80
   Carson rule for FM, 5, 81, 262
   digital AM, 200
   digital baseband, 194
   digital FM, 206, 208
   digital PM, 203
   pure digital FM and PM, 299
Baseband analog systems, attenuation and phase distortion, 224, 238
   intermodulation distortion, 308, 322, 325
   intersystem interference, 454
   noise impairments, 142
   optimum filters, 142
   see also Baseband channel transmittance deviations
Baseband channel transmittance deviations, coarse structure, 228
   cosine amplitude deviations, 28, 227, 232
   fine structure, 238
   Fourier series deviations, 227, 236
   low-frequency cutoff, 301, 305
   power series deviations, 237
   quadratic delay deviation, 228, 233
   sinusoidal phase deviations, 228, 233
Baseband channel transmittances, amplitude-phase relations, 22, 28
   impulse characteristic, 16
Baseband digital systems, attenuation and phase distortion, 266, 285, 291
   channel capacity, 102, 194
   intersystem interference, 441
   low-frequency cutoff, 301, 305
   noise impairments, 193
   nonlinear distortion, 356
   optimum filters, 151, 194
   optimum pulse shapes, 53
Baseband signal transmission, 16, 77
Basic statistical concepts, 468; see also Statistical concepts
Basic transmission parameters, 99, 101, 134, 225, 311, 358, 390
Binary pulse modulation advantages, 75, 167, 173; see also Digital baseband transmission
Biorthogonal and orthogonal codes, 213
Bipolar transmission, constrained binary bipolar, 196
   conventional multilevel bipolar, 195
Bit (binary digit), 62, 75
Bivariate probability functions, 469, 471
Bose-Chaudhuri codes, 217

Carrier analog modulation, 78; see also AM analog systems; FM and PM analog systems; and Vestigial sideband digital transmission
Carrier digital modulation, see AM digital systems; FM digital systems; PM digital systems; and Vestigial sideband digital transmission
Carrier extraction, 189
Carrier pulse transmission characteristics, basic formulation, 18
   double sideband AM and PM, 48
   double sideband FM (FSK), 48
   flat spectrum, 33
   Gaussian spectrum, 36
   partial raised cosine spectrum, 37
   raised cosine spectrum, 41
Carrier signal waves, 78, 82
Carrier wave demodulation, 84, 178
Carson bandwidth rule (FM), 5, 81, 262
CCIR and CCIT, 100
Central limit theorem, 474
Channel bandwidth requirements, see Bandwidth requirements
Channel capacity, instantaneous detection, AM, 199
   baseband, 159, 194
   FM, 206
   PM, 203
   quadrature carrier, 202
   vestigial sideband, 209
   statistical detection, 103, 163
Channel characterization, amplitude-frequency characteristics, 22, 225
   attenuation deviations, 225, 237
   coarse and fine-structure deviations, 228
   nonlinearity, 311
   phase deviations, 225, 237
   random multipath, 358
   troposcatter, 373
Channel nonlinearity, see Intermodulation distortion; Nonlinear amplification
Characteristic distortion, 83, 224; see also Attenuation and phase distortion; Intermodulation distortion; Nonlinear amplification; and Random multipath channels
Characteristic functions, 113, 472
Characterization of message waves, 61, 101
Choice of modulation method, 97
Circuit elements, 2
Clamping circuits, 305
Coarse-structure transmittance deviations, 225, 228, 265
Coaxial lines, 1
   crosstalk coupling, 454
   digital transmission capacity, 164
   transmission loss (attenuation), 165
Coding in digital transmission, 75
   biorthogonal and orthogonal codes, 213
   Bose-Chaudhuri codes, 217
   error-correcting codes, 216
   Hamming codes, 216
   Shannon's statistical encoding, 103
   see also Digital baseband transmission; Digital signal modulation
Coherent (homodyne, synchronous) detection, 84, 177
Coherent interference, 427, 430, 438

Index

Combined digital AM and PM, 179
Combining laws, distortion from various sources, 99, 133, 352
  distortion in repeater chains, 96
  errors in repeater chains, 96
  transmittances in tandem, 22
Combining methods, diversity transmission, *see* Diversity combining
Common antenna volume, 372
Compandors, instantaneous, 5, 70, 95
  bandwidth requirements, 71
  phase linearity requirements, 71
  reduction of quantizing noise, 95
  time division advantage, 95
  syllabic, 67
    bandwidth requirements, 68
    crosstalk reduction, 68, 70
    noise reduction, 69
Complex (analytic) signals, 6, 21
Composing functions, *see* Elementary composing functions
Communication gaps, 4
Communication networks, 1, 58
Communication satellites, 1, 98, 244, 348, 353
Communication systems, 1, 4
  basic diagrams, 58, 60
  engineering objectives, 2, 59
  information sources, 58
  performance criteria, 99, 101, 134
  see also Transmission impairments
Constrained binary bipolar transmission, 183, 189, 196, 305
Conversion, AM to PM, 310, 340, 342, 354
Convolution integrals, 14, 18
Correlation coefficient, 471
Correlation detection (integration), 54, 215
Cosine amplitude deviations, 28
Cosine attenuation deviations, 28
Coupling factors, analytical determination, 428, 451, 454
  balanced pairs, 451
  coaxial pairs, 454
  far-end, 422
  interaction, 423
  near-end, 422
Covariance, 471
Criteria of performance, *see* Performance criteria
Cross correlation, 475
Crosstalk interference, balanced pairs, 451
  coaxial pairs, 454
  far-end, 423, 451
  incoherent, 427, 431
  intelligible (coherent), 354, 427
  interaction (indirect), 423
  near-end, 423, 451
  reduction, by compandors, 70
    by frogging, 423, 431
  see also Intersystem interference
Cube-law amplification distortion, Gaussian narrow-band signals, 328
  Gaussian wide-band signals, 325

multiple sine wave signals, 322
Cube-law phase distortion, analog FM and PM systems, 285, 287
  digital AM systems, 287
  digital FM systems, 287
  digital PM systems, 286
Cumulation in repeater chains, distortion and noise, 96
  errors in digital systems, 96

Data transmission, 1, 174; *see also* Digital transmission
dBm, 65
Delay distortion, linear, 244, 287, 368
  quadratic, 283
Delta (Dirac) function, 465
Delta (differential) modulation, 76
Demodulation methods, *see* Detection methods
Density function (probability), 468
Detection methods, coherent, homodyne, synchronous, 84, 203
  correlation, maximum likelihood, 215
  differential phase, 178, 180, 204
  dual filter, 206
  envelope, 86, 201
  error detection and correction, 216
  frequency discriminator, 86, 208
  product demodulation, 84
Diagram of systems, 58, 60
Dicode transmission, 183, 305
Differential delay (troposcatter), 374
Differential (delta), PCM, PDM, PPM, 76
Differential phase distortion (color TV), 309
Differential phase product demodulation, 178, 180, 204
Digital baseband transmission, advantages of binary modulation, 75, 167, 204
  constrained binary bipolar, 183, 189, 196, 305
  conventional multilevel bipolar, 195
  dicode, 183, 305
  duobinary, 197
  unipolar (on-off), 200, 201
Digital carrier detection methods, 177
  AM, envelope detection, 201
  synchronous detection, 177
  FM, dual filter detection, 206
  frequency discriminator detection, 208
  PM, differential phase detection, 178, 180
  synchronous detection, 177
Digital carrier modulation methods, 174, 176, 179, 180, 206, 208
Digital signal modulation, PCM, pulse code modulation, 76
  PDM, pulse duration modulation, 76
  PPM, pulse position modulation, 76
Digital systems transmission considerations, 173; *see also* AM digital systems; Baseband digital systems; FM digital systems; and PM digital systems
Digital transmission capacities, baseband systems, 102, 196, 444

carrier systems, 201
sub-binary systems, 214
troposcatter systems, 390, 406
see also AM digital systems; Baseband digital systems; FM digital systems; and PM digital systems
Dirac delta function, 465
Distortion, subjective effects, 224, 308
Distortion addition, see Combining laws
Distortion sources, 2, 56, 83, 93, 95, 99, 101, 133, 172, 224, 260; see also AM-PM conversion; Attenuation deviations; Intermodulation distortion; Intersystem interference; Nonlinear amplification; and Phase variations
Distribution function (probability), 468
Diversity combining, see Equal gain combining; Maximal ratio combining; and Selection diversity
Double sideband modulation, 79, 97
Duhamel's integral, 462
Duobinary transmission, 197

Echo distortion, analog AM, 234
  analog baseband, 228, 234
  analog FM and PM, 246
  subjective effects, 224
Echoes, 232, 234
Electrical transducers, 4
Elementary composing functions, cisoidal function, 13, 464
  delta function, 465
  $\sin x/x$ function, 15, 464
  unit impulse, 15
  unit step, 15
Energy relation (Fourier transforms), 463
Envelope detection, analog transmission, 85
  digital transmission, 201
Envelope variations, see Gaussian random variables; Sine wave plus Gaussian noise
Equal gain combining, error reduction, 411
  intermodulation distortion, 389
  signal-to-noise ratios, 386
Equalization, adaptive, 269, 420
  of transmission lines, noise absent, 139
  noise present, 140
  see also Filter optimization
Ergoticity, 475
Error function, erf, 112, 479
Error function complement, erfc, 112
Error noise in PCM, 87, 101
Error probabilities in fading channels, flat fading and random noise, binary FSK, dual filter detection, 393
  binary FSK, frequency discriminator detection, 394
  binary PSK, differential phase detection, 393
  binary PSK, synchronous detection, 393
  frequency-selective fading, no noise, 401
  binary FSK, 404
  binary PSK, 404
  time varying fading, no noise, binary FSK, 400
  binary PSK, 397
  total error probability, binary FSK, 405
  binary PSK, 405
Error probabilities in time invariant channels, analytical procedures, 159, 193
  baseband, constrained binary bipolar, 196
  duobinary, 198
  random bipolar, 53, 160, 194
  carrier AM, 53, 200
  carrier FM, dual filter detection, 206
  frequency discriminator detection, 208
  carrier PM, coherent detection, 203
  differential phase detection, 204
  carrier quadrature modulation, 202
  carrier vestigial sideband modulation, 202
  error-correcting codes, 216
  feedback error correction, 218
  sub-binary transmission PPM, 212
  orthogonal codes, 213, 215
  reduced rate, 210
Error probabilities in troposcatter system, 405; see also Error probabilities in fading channels

Fading-multipath channels, basic considerations, 358
  fading rate (bandwidth), 374
  flat fading, 391
  frequency-selective fading, 359, 390
  log-normal fading, 373, 391
  Rayleigh fading, 362, 391
  see also Random multipath channels
Far-end crosstalk, see Crosstalk interference
Feedback in FM detection, 90
Filter optimization, 72, 489
  analog AM, 143
  analog baseband, 142
  analog FM and PM, 144
  analog pulse systems, baseband, 148
  carrier, 152
  digital AM, 154
  digital baseband, 159, 161
  digital FM (FSK), 208
  digital PM (PSK), 203
Filter types, matched filter, 134, 157, 491
  objectives, 5, 72, 133, 428
  receiving filter, 134, 490
  smoothing filter, 134, 140, 491
  transmitting filter, 134
Fine-structure deviations, 225, 228
Fine-structure transmittance distortion, 225
  analog AM and SSB systems, 240
  analog baseband systems, 238
  analog FM and PM systems, 251, 254
  digital AM systems, 293
  digital baseband systems, 291
  digital FM systems, 251
  digital PM systems, 251
FM and PM analog systems, advantages, 82, 97
  attenuation and phase deviations, 242

Index

bandwidth requirements, 81, 93, 260
echo distortion, 246
envelope fluctuations, 243, 244, 263
fine-structure distortion, 251, 254
filter optimization, 144
intermodulation distortion, 242
intersystem interference, 433, 437
linear attenuation distortion, 243
linear delay distortion, 244
random noise, 144
signal power spectra, 257
threshold performance, 89
troposcatter transmission, 377; *see also* Troposcatter transmission systems
FM digital systems, attenuation and phase deviations, 266
bandwidth requirements (transmission capacity), 81, 206, 208
envelope fluctuations–random noise, 176, 208
error probabilities, 206, 209
filter optimization, 208, 209
fine-structure deviations, 252, 254
gain control, 206, 384, 386, 394
intersystem interference, 444
linear attenuation distortion, 282
linear delay distortion, 291
quadratic delay distortion, 286
troposcatter transmission; *see* Troposcatter transmission
Fourier integral energy relation, 83, 463
Fourier integrals (transforms), 13, 457
Fourier series transmittance deviations, 227, 236
Four-phase modulation, 180, 203, 204, 399
Framing digit, 195
Frequency deviation ratio, 89
Frequency discriminator detection, 80, 86
analog FM and PM systems, 86
digital FM systems, 208
threshold extension (feedback), 87
Frequency division multiplexing, 91, 220
Frequency frogging, 97, 423
Frequency modulation (FM), 80
Frequency shift pulse transmittance, 48
Frequency variations, sine wave plus Gaussian noise, power spectrum, 126
probability distribution, 121
variance, 123
Functions of random variables, 477

Gain control, 384, 386, 394
Gain deviations, *see* Attenuation deviations
Gating in digital transmission, 53, 184
Gaussian autocorrelation function, 365, 366, 374
Gaussian impulse characteristic, 41
Gaussian noise, in analog systems, AM, 143
baseband, 142
FM and PM, 144

in digital systems, *see* Error probabilities
sources of, 133
see also Gaussian random variables
Gaussian power spectrum, 365
Gaussian pulse spectrum, 41
Gaussian random variables (noise), 107
amplitude distribution, 111
autocorrelation functions, 109
bivariate characteristic function, 480
characteristic function, 480
envelope derivative distribution, 129
envelope distribution, 114
phase derivative distribution, 121
phase distribution, 119
phase second derivative distribution, 129
power spectra, 109
variance, of envelope fluctuations, 116
of frequency fluctuations, 123
of phase fluctuations, 119
Generating functions (statistics), 474
Green's function, 461

Hamming codes, 216
Hard limiting, 340
Hermite functions, 337, 488
HF ionospheric transmission, 420
High-pass filters, 27
Hilbert transforms, 6, 18, 23, 78, 460
Homodyne detection, *see* Coherent detection

Idealized channels, 29, 33, 37, 225
Idealized impulse transmittance, basic requirements, 16
Gaussian spectrum, 41
partial raised cosine spectrum, 35
raised cosine spectrum, 37
rectangular spectrum, 29
Idealized transmission-frequency characteristic, 13, 225
Impedance irregularities, 225
Imperfections in transmittance, 56, 225
Impulse characteristic (transmittance), 12, 16
Impulse equivalent transmittance, bandpass channels, 44
baseband channels, 43
Impulse noise, 72, 169
Incoherent interference, 427, 431
Information content, 62
Information rate, 75
Information signal power, 62
Information sources, 58
In-phase component of transmittance, 18, 20, 40
Instantaneous compandors, *see* Compandors
Instantaneous sampling, 52
Interference, *see* Intersystem interference
Intermodulation distortion, 93, 308, 377
in AM systems, basic formulation, 314
conversion, AM to PM, 301, 392
cubic amplification nonlinearity, 322, 325, 328
linear limiter nonlinearity, 332, 337, 339
power series nonlinearity, 330
quadratic amplification nonlinearity, 320

variation with number of carriers, 348
effects, 93, 308
  crosstalk, 93, 308
  differential phase distortion (color TV), 309
  distortion, in broadband transmission, 242, 308
  distortion in time division systems, 93
  intelligible interference, 354
  in FM and PM systems, 242
    AM-PM conversion, 262
    attenuation and phase deviations, 242
    bandwidth limitation, 5, 260
    echoes, 246
    fine-structure deviations, 251
    linear amplitude deviations, 243
    linear delay deviations, 244, 377
    power series deviations, 265
  mechanism, 320, 322
  in multicarrier FM systems, amplification nonlinearity, 335, 336
    AM-PM conversion, 340, 342
    combined intermodulation, 346, 348
    combined output distortion, 353
    intelligible interference, 354
    second-order products, 320, 323
  in single sideband AM, *see* Intermodulation distortion in AM
  third-order products, 323, 349
Intersymbol interference, 95, 102
Intersystem coupling paths, far-end, 423
  near-end, 423
  interaction, 423
Intersystem coupling relations, balanced pairs, 451
  coaxial pairs, 454
Intersystem interference, 100, 422
  AM double sideband systems, 430
  AM single sideband systems, 428
  digital baseband systems, 441
  digital PM systems, 444
  digital to analog PM systems, 448
  FM and PM systems, 433, 437
  intelligible interference, 354, 427, 430, 438
  unintelligible interference, 427

Ladder structures, 23
Laplace transforms, 6, 459
Lattice structures, 23
Limiter characterization, 312
Limiter distortion, multiple sine waves, 332
  narrow-band Gaussian signals, 339
  wide-band Gaussian signals, 337
Linear amplitude deviation effects, digital systems, 282
  FM and PM systems, 243
Linear delay deviation effects, analog FM and PM, 244, 377
  digital AM, 287, 289, 291, 292
  digital FM, 292
  digital PM, 289, 292
  digital quadrature AM, 291, 292
  digital vestigial sideband AM, 291
  troposcatter analog FM and PM, 377
  troposcatter digital FM and PM, 402
Linear delay distortion, pulse shapes, 288
Linear limiter, *see* Limiter
Linear phase channel transmittances, 25
Linear rectifier, 312
Log-normal distribution, 64, 373, 453, 481
Log-normal fading, 373
Low-frequency cutoff, 172, 301
  compensation methods, 305
  pulse distortion, 301
  reduction by dicode transmission, 305
Low-pass channels, 16; *see also* Baseband channel

McLaurin's theorem, 311
Margin against noise, digital systems, 269
Marginal probability functions, 470
Matched filters, receivers, 4, 134, 157, 491
Mathematical channels, filters, 5
Mathematical communication theory, 3
Mathematical models, 4
Maximal ratio combining, 383, 408
Maximum likelihood detection, 215
Message wave characterization, 61, 101
Message wave distortion, *see* Distortion sources
Message wave modification, filters, 72
  instantaneous compandors, 70
  syllabic compandors, 67
Metallic circuits, attenuation, 165
  crosstalk coupling, balanced pairs, 451
  coaxial pairs, 454
  digital baseband transmission, 164, 441
  optimum filters, 161
Microwave radio relays, 1
  choice of modulation method, 216
  interference considerations, 454, 411; *see also* Intersystem interference
Microwave transmitters, *see* Traveling wave tubes
Minimum phase shift transducers, 22
Models, 4
Modulation methods, 97; *see* Analog modulation methods; Choice of modulation method; and Digital modulation methods
Moment generating function, 474
Moments, statistics, 470
Multiband transmission, random multipath channels, 418
  time invariant channels, 134, 161
Multicarrier FM systems, intelligible crosstalk, 354
  intermodulation distortion, 351; *see also* Intermodulation distortion, in multicarrier FM systems
  noise plus intermodulation distortion, 353
Multipath channels, *see* Random multipath channels; Troposcatter transmission
Multiplexing, frequency division, 91, 220
  special methods, 220
  time division, 94, 221
Multiplicative noise, *see* Attenuation and

Index

phase distortion; Intermodulation distortion; and Nonlinear amplification
Narrow-band channels, 46
Narrow-band Gaussian noise, *see* Gaussian random variables
Noise, in analog systems, *see* Gaussian noise, in analog systems
  in digital systems, *see* Error probabilities
  Gaussian, 107; *see also* Gaussian random variables
  general, 72
Noise figure, temperature, 138
Noise loading, 101, 309
Noise margin, digital systems, 269
Noise waves, 83
Noisy sine wave, *see* Sine wave plus Gaussian noise
Nonlinear amplification, analytical representations, 311
  cube law, 322
  power series, 311
  linear limiter, 312
  square law, 320
Nonlinear amplification effects, 93, 101, 308; *see also* Intermodulation distortion, in AM systems; Intermodulation distortion, effects
Nonlinear phase characteristic, cube law, 283
  Fourier series, 236
  power series, 237
  sinusoidal, 228
  square law, 287
Nonlinear phase effects, *see* Intermodulation distortion, in FM and PM systems; Phase distortion, effects in digital systems
Nonlinear transformations, of power spectra, 483; *see also* Power spectrum nonlinear transformations
  of probability functions, 71
Nyquist criteria, first and second, 39
  third, 55
Nyquist interval, 15
Number of third order intermodulation products, 324

On-off digital transmission, 48
Optimum digital transmission methods, 52
Optimum filters, *see* Filter optimization
Optimum pulse shapes, 53
Orthogonal binary codes, 213
Orthogonal carrier modulation, 202
Orthogonal signal shapes, 213, 221
Oscillator, phase controlled, 193
Output signal power, 136

Parameters of systems, 134, 374
Parity check codes, 219
Parseval's theorem, 463
PCM, *see* Pulse code modulation systems
Peak frequency deviation, 92

Peak power, in frequency division systems, 92
  in time division systems, 95
Performance criteria, analog systems, 99, 138
  digital systems, 101, 138
Performance evaluations, 99
Phase distortion, analog systems, 224; *see also* Intermodulation distortion, in FM and PM systems
  effects in digital systems, 266
    AM, envelope detection, 287, 292
    AM, synchronous detection, 285, 289
    baseband transmission, 285
    binary FM, frequency discrimination detection, 287, 292
    binary PM, differential phase detection, 286, 287, 292
    binary PM, synchronous detection, 285, 286, 292
    four-phase, differential phase detection, 286, 292
    four-phase, synchronous detection, 285, 286, 292
    quadrature AM, synchronous detection, 285, 291
    vestigial sideband AM, synchronous detection, 285, 286, 291
Phase equalization, 5
Phase intercept, 30
Phase-locked oscillators, 193
Phase modulation (PM), 80
Phase variations with frequency, attenuation-phase relations, 23, 28
  coarse-structure deviations, 228
  cube-law deviation, 283
  fine-structure deviations, 225
  Fourier series deviations, 227, 236
  power series deviations, 237
  quadratic deviation, 244, 287
  sinusoidal deviation, 28
Phase variations, sine wave plus Gaussian noise, *see* Sine wave plus Gaussian noise
Physical realizability, 6
Picture transmission, 6, 10
Pilot carrier, 86
PM analog systems, *see* FM and PM analog systems
PM digital systems, 203, 204
  attenuation and phase deviations, 266
  bandwidth requirements, 202
  channel capacity, 203
  error probabilities, 203, 205
  fine-structure deviations, 251, 254
  filter optimization, 204
  gain control, 203
  intersystem interference, 444
  linear amplitude distortion, 282
  linear delay distortion, 292
  nonlinear distortion, 356
  quadratic delay distortion, 285, 286
  troposcatter transmission, *see* Troposcatter transmission
Polar modulation, combined AM-PM, 179
Post-detection combining, 287

Power, average, 61, 72
Power addition, 7, 99
Power series nonlinearities, amplification characteristic, 311
  attenuation and phase characteristics, 237
Power spectra, of signal waves, 62, 181
  analog AM, 79
  analog FM and PM, 257
  digital AM, 184
  digital baseband, 181
  digital binary PM, 182
  digital FM, 184
  digital pulse trains, 181
  sine wave plus Gaussian noise, envelope variations, 126
  frequency variations, 127
  phase variations, 126
Power spectrum, nonlinear transformations, 483
  basic formulation, 317, 483
  cosine and sine transformations, 258, 486
  linear limiter transformations, 337, 487
  power law transformations, 330, 485
  of speech, 66
Pre-detection combining, 383, 387
Pre-emphasis, *see* Filter optimization
Probability density functions, 468
Probability distribution functions, 486
Probability distributions, Gaussian noise, *see* Gaussian random variables
  sine wave plus Gaussian noise, *see* Sine wave plus Gaussian noise
Product demodulation, 84
  coherent (synchronous), 178
  differential phase, 178, 180
Products, intermodulation, 178, 180
  second order, 320, 323
  third order, 323
Pulse code modulation (PCM) systems, basic considerations, 76, 93, 98
  channel capacities, 103
  coding method, 75
  error noise, 87
  quantizing noise, 73
  threshold signal-to-noise ratio, 87; *see also* AM digital systems, Baseband digital systems; FM digital systems; and PM digital systems
Pulse distortion from transmittance deviations, cosine and sine deviations, 232, 234
  fine-structure deviations, 240, 291
  linear amplitude deviations, 282
  linear delay deviations, 287
  low-frequency cutoff, 301
  quadratic delay deviations, 283
Pulse duration modulation (PDM), 76
Pulse position modulation (PPM), 76
Pulse spectra, *see* Spectra
Pulse trains, baseband transmission, 77, 174
  carrier transmission, 175

Pulse transmission characteristics, bandpass channels, AM and PM, 17
  bandpass channels, binary FM (FSK), 48
  baseband channels, 16, 43; *see also* Digital baseband transmission; Pulse distortion

Quadratic delay distortion, *see* Cube-law phase distortion
Quadratic phase distortion, *see* Linear delay deviation effects
Quadrature (orthogonal) carrier modulation, 209
  attenuation and phase distortion, 274
  error probabilities, 202
Quadrature component of transmittance, 18, 20, 40
Quantization of signals, 73
Quantizing levels, 73
Quantizing noise, 73
Quantum noise, 133

Radar echoes, 4
Radio relay systems, *see* Relay systems
Raised cosine pulses, 53
Raised cosine spectrum, 37
Random Gaussian variables, *see* Gaussian random variables
Random multipath channels, 358
  autocorrelation function of variations, with respect to frequency, 365
  with respect to time, 364
  envelope variations, with frequency, 367
  with time, 367
  fading, rapid (Rayleigh), 362
  slow (log-normal), 373
  fading rate, 374
  phase derivative, with respect to frequency, 369
  with respect to time, 369
  phase second derivative, with respect to frequency, 369
  with respect to time, 370
  see also Diversity combining; Troposcatter transmission
Random variables, *see* Statistical concepts
Rayleigh probability distribution, 116, 362, 363, 481
Receiving filters, 72; *see also* Filter optimization
Redundancy, 63
Redundancy reduction, 63
Reference carrier extraction, 189
Regenerative repeaters, 74
  applications, 98
  digital baseband system application, 167
  error cumulation, 196
  timing wave deviations, 187
  timing wave extraction, 184
Relay systems, microwave, 1, 98
  choice of modulation method, 97
  intermodulation distortion, *see* Intermodulation distortion in FM and PM systems

Index

intersystem interference, 424, 441
Repeaters, 1, 58, 96
Resolution of signals, 13
Resonator parameters, 187
Response functions, see Elementary composing functions

Sampling and quantization, 73
Sampling theorem, 16, 466
Satellite systems, 1, 98, 224, 348, 353, 445
Second-order modulation products, 320, 323
Selection diversity combining, 388, 412
Selective fading, 353
Shot noise, 133
Sideband spectrum, 79
Sifting integral, 465
Signal components, 12
Signal distortion, 73; see also Transmission impairments
Signal quantization, 73
Signal resolution into components, 13
Signal sampling, 13, 16
Signal-to-noise ratios, in analog systems, AM, 143
  baseband, 142
  FM and PM, 146
  definitions, 137
  digital systems, 137
  AM, 153, 200, 202
  baseband, 152, 195
  FM, 206
  PM, 203, 205, 208
  with diversity combining, equal gain combining, 386
  maximal ratio combining, 383
  selection diversity, 388
Signal wave characterization, 77, 78, 82
Sine and cosine attenuation deviations, 28
Sine and cosine phase deviations, 28
Sine wave plus Gaussian noise, 108
  amplitude variations, 112
  autocorrelation functions, 109
  envelope variations, 114
  frequency variations, 121
  mean envelope variation, 117
  mean frequency variation, 123
  phase variations, 119
  power spectra, 109
    of frequency variations, 127
    of phase variations, 126
  variance, of envelope fluctuations, 117
    of frequency variations, 123
    of phase variations, 119
Single sideband AM, 79, 97
Single sideband FM, 81
Slicing levels, 269
Slope overload noise, 76
Smear-desmear technique, 222
Smoothing filters, 140
Solar batteries, 106
Space telemetry, 4, 106
Spectra, of idealized baseband pulses, 34
  Gaussian spectrum, 41

raised cosine spectrum, 38
rectangular spectrum, 16
$\sin x/x$ spectrum, 29
of idealized carrier pulses, 45
  raised cosine spectrum, 40
  rectangular spectrum, 46
  $\sin x/x$ spectrum, 31
Spectral density, see Power spectra
Spectrum definition, 458
Spectrum modification, see Filter types; Power spectrum, nonlinear transformations
Speech properties, 63
  acoustic spectrum, 66
  information rate, 66
  instantaneous amplitude distribution, 65
  power (volume) distribution, 64
  speech components, phonemes, 67
  voice coders (vocoders), 63, 67
Square law nonlinearities, amplification characteristic, 320
  delay characteristic, 283
  phase characteristic, 244, 287
Stationarity (statistics), 475
Statistical concepts, autocorrelation functions, 476
  averages, 470
  bivariates, 469, 471
  central limit theorem, 474
  characteristic functions, 472
  covariance, 471
  correlation functions, 475
  cross correlation function, 475
  density functions, 468
  dispersion (variance), 471
  distribution functions, 468
  ergoticity, 475
  error function, 112, 479
  expectation, 470
  functions of random variables, 477
  Gaussian probability functions, 475
  log-normal distribution function, 64, 373, 481
  marginal probabilities, 470
  moment generating function, 474
  moments, 470
  power spectra, 476
  Rayleigh probability distribution, 116, 481
  standard deviation, 471
  stationarity, 475
  time correlation, 475
  variance (dispersion), 471
Statistical detection, 3, 4, 103
Sub-binary transmission, biorthogonal and orthogonal codes, 213
  error correcting codes, 216
  feedback error correction, 218
  pulse position modulation, 211
  reduced rate or repetition, 210
Subcarrier modulation, 91
Subjective effects of interference, 224
Submarine cable, 1
Suboptimum digital transmission, 54
Sums of random variables, 472
Syllabic compandors, 67
Symmetrical sideband transmission, 79

Synchronization of channels, 185
Synchronous (coherent, homodyne) demodulation, 84, 203
Systems engineering, 2, 59; see also Communication systems

Taylor's series, 474
Telecommunication, 1
Telemetry, 1, 106
Telephoto, 1, 100
Television, 1, 100, 308
Temperature of noise, 138
Thermal noise, 11, 133
Third order intermodulation products, 323, 349
Threshold signal-to-noise ratios, FM, 89
  with feedback, 90
  PCM, 87
Time division multiplexing, 94, 221
Time origin, choice of, 14, 30, 459
Time variant channels, see Random multipath channels; Troposcatter transmission
Timing wave deviations (jitter), 57, 101, 187
Timing wave extraction, 184
Transform method for nonlinear channels, 310, 318, 483
Transforms, Fourier, 13, 457
  Hilbert, 6, 18, 23, 460
  Laplace, 459
Translation of reference frequency, 17
Transmission capacity, see Channel capacity
Transmission delay, 83
Transmission-frequency characteristics, 12, 22; see also Attenuation deviations; Attenuation-phase relations; Phase deviations; and Random multipath channels
Transmission impairments, sources of, 1, 56, 93, 95, 99, 101; see also AM-PM conversion; Attenuation and phase deviations; Gaussian noise; Intermodulation distortion; Intersymbol interference; Intersystem interference; Nonlinear amplification; Random multipath channels; Timing wave deviations; and Troposcatter transmission
Transmission loss, metallic pairs, 165
Transmission loss fluctuations, see Fading-multipath channels

Transmission parameters, see Basic transmission parameters
Transmission quality, 1, 7
Transmittance deviations, 225; see also Transmission impairments
Traveling wave tube characteristics, amplification, 334
  AM-PM conversion, 340, 356
Troposcatter transmission, 1, 98, 372, 390
  antenna beam angles, 375
  common antenna volume, 372
  differential time delay, 374
  digital transmission capacities, 406
  error probabilities, binary FM and PM, 406
  frequency-selective fading, 401
  random noise, 392
  time-selective fading, 400
  intermodulation distortion in analog FM systems, 377
  multiband transmission, 418
  see also Diversity combining; Random multipath channels

Unit impulse (delta) function, 12, 465
Unit impulse response characteristic, 12, 15; see also Idealized impulse transmittance; Impulse equivalent transmittance
Unit step response characteristic, 15
Upper bound on channel capacity, 103, 163

Variables, see Statistical concepts
Variance (dispersion), 61
Variances, sine wave plus Gaussian noise, envelope fluctuations, 117
  frequency fluctuations, 123
  phase fluctuations, 119
Vestigal sideband digital transmission, 48, 209
Voice coders (vocoders), 63, 67
Volume (power) of speech, 64
Volume units (VU), 65

Wave characterization, 61
  baseband waves, 77
  carrier waves, 78
  message waves, 61
White (flat) noise, 3, 11, 133
Wideband signals, 325

Zero crossings of pulse trains, 184, 186